Methods in Enzymology

Volume 373
LIPOSOMES
Part C

METHODS IN ENZYMOLOGY

EDITORS-IN-CHIEF

John N. Abelson   Melvin I. Simon

DIVISION OF BIOLOGY
CALIFORNIA INSTITUTE OF TECHNOLOGY
PASADENA, CALIFORNIA

FOUNDING EDITORS

Sidney P. Colowick and Nathan O. Kaplan

Methods in Enzymology

Volume 373

Liposomes

Part C

EDITED BY

Nejat Düzgüneş

DEPARTMENT OF MICROBIOLOGY
UNIVERSITY OF THE PACIFIC SCHOOL OF DENTISTRY
SAN FRANCISCO, CALIFORNIA

ELSEVIER
ACADEMIC
PRESS

AMSTERDAM • BOSTON • HEIDELBERG • LONDON
NEW YORK • OXFORD • PARIS • SAN DIEGO
SAN FRANCISCO • SINGAPORE • SYDNEY • TOKYO

Academic Press is an imprint of Elsevier

Academic Press
An Elsevier Imprint.
525 B Street, Suite 1900, San Diego, California 92101-4495, USA
http://www.academicpress.com

Academic Press
84 Theobald's Road, London WC1X 8RR, UK
http://www.academicpress.com

International Standard Book Number: 0-12-182276-1

PRINTED IN THE UNITED STATES OF AMERICA
03   04   05   06   07   08   9   8   7   6   5   4   3   2   1

Table of Contents

Section I. Liposomes in Immunology

v

Section II. Liposomes in Diagnostics

Section III. Liposomes in Gene Delivery and Gene Therapy

Contributors to Volume 373

Article numbers are in parentheses and following the names of contributors.
Affiliations listed are current.

SALVADOR F. ALIÑO (26), *Departamento de Famacologia, Facultad de Medicina, Universidad de Valencia, Avda Blasco Ibanez 15, 46010 Valencia, Spain*

CARL R. ALVING (2, 3, 10), *Department of Membrane Biochemistry, Walter Reed Army Institute of Research, Washington, D.C. 20307*

M. A. ARANGOA (22), *Department of Pharmacology and Pharmaceutical Technology, School of Pharmacy, University of Navarra, 31080 Pamplona, Spain*

UDO BAKOWSKY (18), *Department of Pharmaceutical Technology and Biopharmacy, University of Saarbruecken, Germany*

RICHARD R. BANKERT (33), *Department of Microbiology, SUNY at Buffalo, 138 Farber Hall, 3435 Main Street, Buffalo, New York 14214*

LAJOS BARANYI (10), *Department of Membrane Biochemistry, Walter Reed Army Institute of Research, Washington, D.C. 20307*

MARTA BENET (26), *Departamento de Famacologia, Facultad de Medicina, Universidad de Valencia, Avda Blasco Ibanez 15, 46010 Valencia, Spain*

MICHAEL BODO (10), *Department of Membrane Biochemistry, Walter Reed Army Institute of Research, Washington, D.C. 20307*

OTTO C. BOERMAN (15), *Department of Nuclear Medicine (565), University Medical Center Nijmegen, P.O. Box 9101, 6500 HB Nijmegen, The Netherlands*

ELENA BOGDANENKO (28), *V N Orekhovich Institute of Biomedical Chemistry, Russian Academy of Medical Sciences, 10, Pogodinska ya Street, 119832 Moscow, Russia*

JEFF W.M. BULTE (12), *Department of Radiology, The Johns Hopkins University School of Medicine, Baltimore, Maryland 21205*

LAURA BUNGENER (5), *Department of Medical Microbiology, Molecular Virology Section, University of Groningen, 9713 AV Groningen, The Netherlands*

ROLF BUNGER (10), *Department of Membrane Biochemistry, Walter Reed Army Institute of Research, Washington, D.C. 20307*

GERARDO BYK (23), *Laboratory of Peptidomimetics and Genetic Chemistry, Bar Ilan University, Department of Chemistry, 52900 Ramat Gan, Israel*

JIN-SOO CHANG (9), *Morgan Biotechnology Research Institute, 341 Pojung-Ri, Koonsung-Myon, Youngin City, Kyonggi-Do 449-910, South Korea*

MYEONG-JUN CHOI (9), *Charmzone Co.,Ltd., Bioimaterial Research Center, 301 Hankang Building, 184-11 Kwangjang-dong, Kwangjin-ju, Seoul, Korea*

JAIME CRESPO (26), *Departamento de Famacologia, Facultad de Medicina, Universidad de Valencia, Avda Blasco Ibanez 15, 46010 Valencia, Spain*

TOOS DAEMEN (5), *Department of Medical Microbiology, Molecular Virology Section, University of Groningen, 9713 AV Groningen, The Netherlands*

SUMEET DAGAR (13), *Departments of Pharmaceutics and Pharmacodynamics, University of Illinois at Chicago, 833 Wood Street, Chicago, Illinois 60612*

FRANCISCO DASÍ (26), *Departamento de Famacologia, Facultad de Medicina, Universidad de Valencia, Avda Blasco Ibanez 15, 46010 Valencia, Spain*

ROBERT J. DEBS (34), *Geraldine Brush Cancer Research Institute, 2330 Clay Street, San Francisco, California 94115*

MARCEL DE CUYPER (12), *Interdisciplinary Research Center, Katholieke Universiteit Leuven, Campus Kortrijk, B-8500 Kortrijk, Belgium*

C. TROS DE ILARDUYA (22), *Department of Pharmacology and Pharmaceutical Technology, School of Pharmacy, University of Navarra, 31080 Pamplona, Spain*

NEJAT DÜZGÜNEŞ (19, 22, 24, 28), *Department of Microbiology, University of the Pacific School of Dentistry, 2155 Webster Street, San Francisco, California 94115*

NEJAT K. EĞILMEZ (33), *Department of Microbiology, SUNY at Buffalo, 138 Farber Hall, 3435 Main Street, Buffalo, New York 14214*

ABDELATIF ELOUAHABI (20), *Center for Structural Biology and Bioinformatics, CP 206/2, Campus Plaine-ULB, Blv du Triomphe, 1050 Brussels, Belgium*

HENRIQUE FANECA (19), *Department of Biochemistry, Faculty of Sciences and Technology, University of Coimbra, 3000 Coimbra, Portugal*

SYLVIA FONG (34), *Geraldine Brush Cancer Research Institute, 2330 Clay Street, San Francisco, California 94115*

BENOÎT FRISCH (4), *Laboratoire de Chimie Bioorganique, UMR 7514 CNRS-ULP, Faculte de Pharmacie, 74 Route du Rhin, Illkirch 67400, France*

STEPHEN J. FROST (16), *Department of Clinical Biochemistry, The Princess Royal Hospital, Lewes Rd. Haywards Heath, West Suxxes RH16 3LU, England*

M. TERESA GIRÃO DA CRUZ (24), *Department of Biochemistry, Faculty of Sciences and Technology, University of Coimbra, 3000 Coimbra, Portugal*

LAURENT GIRAUDO (7), *Centre d'Immunologie de Marseille-Luminy, Campus de Luminy, Case 906, 13288 Marsielle Cedex 09, France*

MITSURU HASHIDA (25), *Graduate School of Pharmaceutical Sciences, Kyoto University, Sakyo-ku, Kyoto 606-850, Japan*

KAZUYA HIRAOKA (30), *Division of Gene Therapy Science, Graduate School of Medicine, Osaka University, 2-2 Yamada-oka, Suita City, Osaka 565-0871, Japan*

DICK HOEKSTRA (18), *Department of Membrane Cell Biology, University of Groningen, Antonius Deusinglaan 1, 9713 AV Groningen, The Netherlands*

LEAF HUANG (21), *Center for Pharmacogenetics, School of Pharmacy, University of Pittsburgh, 633 Salk Hall, Pittsburgh, Pennsylvania 15213*

ANKE HUCKREIDE (5), *Department of Medical Microbiology, Molecular Virology Section, University of Groningen, 9713 AV Groningen, The Netherlands*

YASUFUMI KANEDA (30), *Division of Gene Therapy Science, Graduate School of Medicine, Osaka University, 2-2 Yamada-oka, Suita City, Osaka 565-0871, Japan*

SHIGERU KAWAKAMI (25), *Faculty of Pharmaceutical Sciences, Nagasaki University, Magaski 852-8521, Japan*

CHONG-KOOK KIM (17), *College of Pharmacy, Seoul National University, San 56-1, Shinlim-Doug, Kwanak-Gu, Seoul, South Korea*

KENJI KONO (27), *Department of Applied Materials Science, Graduate School of Engineering, Osaka Prefecture University, 1-1, Gakuencho, Sakai, Osaka 599-8531, Japan*

KRYSTYNA KONOPKA (31), *Department of Microbiology, University of the Pacific School of Dentistry, 2155 Webster Street, San Francisco, California 94115*

LAKSHMI KRISHNAN (11), *Institute for Biological Sciences, National Research Council of Canada, 100 Sussex Drive, Ottawa, Ontario K1A 0R6, Canada*

PETER E. JENSEN (8), *Department of Pathology and Laboratory Medicine, Emory University School of Medicine, Atlanta, Georgia 30322*

LAWRENCE B. LACHMAN (6), *Department of Bioimmunotherapy, Box 422, The University of Texas MD Anderson Cancer Center, 1515 Holcombe Blvd., Houston, Texas, 77030*

OLIVIER LAMBERT (29), *Institut Curie, Section de Recherche, UMR-CNRS 168 et LRC-CEA 8, 11 rue Pierre et Marie Curie, 75231 Paris, France*

PETER LAVERMAN (15), *Department of Nuclear Medicine (565), University Medical Center Nijmegen, P.O. Box 9101, 6500 HB Nijmegen, The Netherlands*

PAUL J. LEE (14), *Vitreoretinal Surgical Fellow, Tulane University Medical Center, 1430 Tulane Avenue, New Orleans, Louisiana 70112*

LEE LESERMAN (7), *Centre d'Immunologie de Marseille-Luminy, Campus de Luminy, Case 906, 13288 Marsielle Cedex 09, France*

SONG LI (21), *Center for Pharmacogenetics, School of Pharmacy, University of Pittsburgh, 633 Salk Hall, Pittsburgh, Pennsylvania 15213*

SOO-JEONG LIM (17), *College of Pharmacy, Seoul National University, San 56-1, Shinlim-Doug, Kwanak-Gu, Seoul, South Korea*

YONG LIU (34), *Geraldine Brush Cancer Research Institute, 2330 Clay Street, San Francisco, California 94115*

PATRICK MACHY (7), *Centre d'Immunologie de Marseille-Luminy, Campus de Luminy, Case 906, 13288 Marsielle Cedex 09, France*

MIGUEL MANO (19), *Department of Biochemistry, Faculty of Sciences and Technology, University of Coimbra, 3000 Coimbra, Portugal*

GARY R. MATYAS (3), *Department of Membrane Biochemistry, Walter Reed Army Institute of Research, Washington, D.C. 20307*

NATHALIE MIGNET (23), *UMR 7001, Laboratoire de Chimie Bioorganiquet et de Biotechnologie Moleculaire et Cellulaire, Ecole National Superieure de Chimie de Paris, 13 Quai Jules Guesde, BP 14, 94403 Vitry sur Siene, France*

JANOS MILOSEVITS (10), *Department of Membrane Biochemistry, Walter Reed Army Institute of Research, Washington, D.C. 20307*

ALEXEY MOSKOVTSEV (28), *V N Orekhovich Institute of Biomedical Chemistry, Russian Academy of Medical Sciences, 10, Pogodinska ya Street, 119832 Moscow, Russia*

JEAN M. MUDERHWA (3), *Department of Membrane Biochemistry, Walter Reed Army Institute of Research, Washington, D.C. 20307*

MAKIYA NSHIKAWA (25), *Graduate School of Pharmaceutical Sciences, Kyoto University, Sakyo-ku, Kyoto 606-850, Japan*

VOLKER OBERLE (18), *Department of Membrane Cell Biology, University of Groningen, Antonius Deusinglaan 1, 9713 AV Groningen, The Netherlands*

HAYAT ÖNKYÜKSEL (13), *Departments of Pharmaceutics and Pharmacodynamics, University of Illinois at Chicago, 833 Wood Street, Chicago, Illinois 60612*

BÜLENT ÖZPOLAT (6), *Department of Bioimmunotherapy, Box 422, The University of Texas MD Anderson Cancer Center, 1515 Holcombe Blvd., Houston, Texas, 77030*

WILLIAM M. PARDRIDGE (32), *University of California-Los Angeles, Warren Hall, 13-164, 900 Veteran Avenue, Los Angeles, California 90024*

GIRISHCHANDRA B. PATEL (11), *Institute for Biological Sciences, National Research Council of Canada, 100 Sussex Drive, Ottawa, Ontario K1A 0R6, Canada*

VÉRONIQUE PECTOR (20), *Center for Structural Biology and Bioinformatics, CP 206/2, Campus Plaine-ULB, Blv du Triomphe, 1050 Brussels, Belgium*

MARIA C. PEDROSO DE LIMA (19, 24), *Department of Biochemistry, Faculty of Sciences and Technology, University of Coimbra, 3000 Coimbra, Portugal*

NUNO PENACHO (19), *Department of Biochemistry, Faculty of Sciences and Technology, University of Coimbra, 300 Coimbra, Portugal*

GHOLAM A. PEYMAN (14), *Ophthalmology Department (SL 69), Tulane University Medical Center, 1430 Tulane Avenue, New Orleans, Louisiana 70112*

PEDRO PIRES (24), *Center for Neuroscience and Cell Biology, University of Coimbra, 3000 Coimbra, Portugal*

OLGA PODOBED (28), *V N Orekhovich Institute of Biomedical Chemistry, Russian Academy of Medical Sciences, 10, Pogodinska ya Street, 119832 Moscow, Russia*

MANGALA RAO (2), *Department of Membrane Biochemistry, Walter Reed Army Institute of Research, Washington, D.C. 20307*

JEAN-LOUIS RIGAUD (29), *Institut Curie, Section de Recherche, UMR-CNRS 168 et LRC-CEA 8, 11 rue Pierre et Marie Curie, 75231 Paris, France*

AUDREY ROTH (4), *Laboratoire de Chimie Bioorganique, UMR 7514 CNRS-ULP, Faculte de Pharmacie, 74 Route du Rhin, Illkirch 67400, France*

STEPHEN W. ROTHWELL (2), *Department of Membrane Biochemistry, Walter Reed Army Institute of Research, Washington, D.C. 20307*

ISRAEL RUBINSTEIN (13), *Departments of Pharmaceutics and Pharmacodynamics, University of Illinois at Chicago, 833 Wood Street, Chicago, Illinois 60612*

JEAN-MARIE RUYSSCHAERT (20), *Center for Structural Biology and Bioinformatics, CP 206/2, Campus Plaine-ULB, Blv du Triomphe, 1050 Brussels, Belgium*

SANDOR SAVAY (10), *Department of Membrane Biochemistry, Walter Reed Army Institute of Research, Washington, D.C. 20307*

DANIEL SCHERMAN (23), *UMR 7001, Laboratoire de Chimie Bioorganiquet et de Biotechnologie Moleculaire et Cellulaire, Ecole National Superieure de Chimie de Paris, 13 Quai Jules Guesde, BP 14, 94403 Vitry sur Siene, France*

FRANCIS SCHUBER (4), *Laboratoire de Chimie Bioorganique, UMR 7514 CNRS-ULP, Faculte de Pharmacie, 74 Route du Rhin, Illkirch 67400, France*

KARINE SERRE (7), *Centre d'Immunologie de Marseille-Luminy, Campus de Luminy, Case 906, 13288 Marsielle Cedex 09, France*

SÉRGIO SIMÕES (19, 24), *Department of Biochemistry, Faculty of Sciences and Technology, University of Coimbra, 3000 Coimbra, Portugal*

G. DENNIS SPROTT (11), *Institute for Biological Sciences, National Research Council of Canada, 100 Sussex Drive, Ottawa, Ontario K1A 0R6, Canada*

GERT STORM (15), *Department of Pharmaceutics, Utrecht Institute for Pharmaceutical Sciences (UIPS), Utrecht University, The Netherlands*

JANOS SZEBENI (10), *Department of Membrane Biochemistry, Walter Reed Army Institute of Research, Washington, D.C. 20307*

TORU TAKAGISHI (27), *Department of Applied Materials Science, Graduate School of Engineering, Osaka Prefecture University, 1-1, Gakuencho, Sakai, Osaka 599-8531, Japan*

MARC THIRY (20), *Laboratory of Cell and Tissue Biology, University of Liege, Rue de Pitteurs, Liege, Belgium*

MICHEL VANDENBRANDEN (20), *Center for Structural Biology and Bioinformatics, CP 206/2, Campus Plaine-ULB, Blv du Triomphe, 1050 Brussels, Belgium*

ESTHER VAN KESTEREN-HENDRIKX (1), *Department of Cell and Immunology, Faculty of Medicine, Vrije Universiteit Medical Center, Van de Boechhorststraat 7, 1081 BT Amsterdam, The Netherlands*

NICO VAN ROOIJEN (1), *Department of Cell and Immunology, Faculty of Medicine, Vrije Universiteit Medical Center, Van de Boechhorststraat 7, 1081 BT Amsterdam, The Netherlands*

LARRY E. WESTERMAN (8), *VGS/DVRD/NCID, Centers for Disease Control and Prevention, Atlanta, Georgia 30322*

BARBARA WETZER (23), *UMR 7001, Laboratoire de Chimie Bioorganiquet et de Biotechnologie Moleculaire et Cellulaire, Ecole National Superieure de Chimie de Paris, 13 Quai Jules Guesde, BP 14, 94403 Vitry sur Siene, France*

JAN WILSCHUT (5), *Department of Medical Microbiology, Molecular Virology Section, University of Groningen, 9713 AV Groningen, The Netherlands*

SEIJI YAMAMOTO (30), *Division of Gene Therapy Science, Graduate School of Medicine, Osaka University, 2-2 Yamada-oka, Suita City, Osaka 565-0871, Japan*

FUMIYOSHI YAMASHITA (25), *Graduate School of Pharmaceutical Sciences, Kyoto University, Sakyo-ku, Kyoto 606-850, Japan*

JING-SHI ZHANG (21), *Center for Pharmacogenetics, School of Pharmacy, University of Pittsburgh, 633 Salk Hall, Pittsburgh, Pennsylvania 15213*

RENAT ZHDANOV (28), *V N Orekhovich Institute of Biomedical Chemistry, Russian Academy of Medical Sciences, 10, Pogodinska ya Street, 119832 Moscow, Russia*

Preface

The origins of liposome research can be traced to the contributions by Alec Bangham and colleagues in the mid 1960s. The description of lecithin dispersions as containing "spherulites composed of concentric lamellae" (A.D. Bangham and R.W. Horne, J. Mol. Biol. 8, 660, 1964) was followed by the observation that "the diffusion of univalent cations and anions out of spontaneously formed liquid crystals of lecithin is remarkably similar to the diffusion of such ions across biological membranes (A.D. Bangham, M.M. Standish and J.C. Watkins, J. Mol. Biol. 13, 238, 1965). Following early studies on the biophysical characterization of multilamellar and unilamellar liposomes, investigators began to utilize liposomes as a well-defined model to understand the structure and function of biological membranes. It was also recognized by pioneers including Gregory Gregoriadis and Demetrios Papahadjopoulos that liposomes could be used as drug delivery vehicles. It is gratifying that their efforts and the work of those inspired by them have lead to the development of liposomal formulations of doxorubicin, daunorubicin and amphotericin B now utilized in the clinic. Other medical applications of liposomes include their use as vaccine adjuvants and gene delivery vehicles, which are being explored in the laboratory as well as in clinical trials. The field has progressed enormously in the 38 years since 1965.

This volume includes applications of liposomes in immunology, diagnostics, and gene delivery and gene therapy. I hope that these chapters will facilitate the work of graduate students, post-doctoral fellows, and established scientists entering liposome research. Other volumes in this series cover additional subdisciplines in liposomology.

The areas represented in this volume are by no means exhaustive. I have tried to identify the experts in each area of liposome research, particularly those who have contributed to the field over some time. It is unfortunate that I was unable to convince some prominent investigators to contribute to the volume. Some invited contributors were not able to prepare their chapters, despite generous extensions of time. In some cases I may have inadvertently overlooked some experts in a particular area, and to these individuals I extend my apologies. Their primary contributions to the field will, nevertheless, not go unnoticed, in the citations in these volumes and in the hearts and minds of the many investigators in liposome research.

I would like to express my gratitude to all the colleagues who graciously contributed to these volumes. I would like to thank Shirley Light of Academic Press for her encouragement for this project, and Noelle Gracy of Elsevier Science for her help at the later stages of the project. I am especially thankful to my wife Diana Flasher for her understanding, support and love during the seemingly never-ending editing process, and my children Avery and Maxine for their unique curiosity, creativity, cheer, and love. Finally, I wish to dedicate this volume to two other members of my family who have been influential in my life, with their love and support since my childhood days, my aunt Sevim Uygurer and my brother Dr. Arda Düzgüneş

NEJAT DÜZGÜNEŞ

METHODS IN ENZYMOLOGY

VOLUME LIV. Biomembranes (Part E: Biological Oxidations)
Edited by SIDNEY FLEISCHER AND LESTER PACKER

VOLUME LV. Biomembranes (Part F: Bioenergetics)
Edited by SIDNEY FLEISCHER AND LESTER PACKER

VOLUME LVI. Biomembranes (Part G: Bioenergetics)
Edited by SIDNEY FLEISCHER AND LESTER PACKER

VOLUME LVII. Bioluminescence and Chemiluminescence
Edited by MARLENE A. DELUCA

VOLUME LVIII. Cell Culture
Edited by WILLIAM B. JAKOBY AND IRA PASTAN

VOLUME LIX. Nucleic Acids and Protein Synthesis (Part G)
Edited by KIVIE MOLDAVE AND LAWRENCE GROSSMAN

VOLUME LX. Nucleic Acids and Protein Synthesis (Part H)
Edited by KIVIE MOLDAVE AND LAWRENCE GROSSMAN

VOLUME 61. Enzyme Structure (Part H)
Edited by C. H. W. HIRS AND SERGE N. TIMASHEFF

VOLUME 62. Vitamins and Coenzymes (Part D)
Edited by DONALD B. MCCORMICK AND LEMUEL D. WRIGHT

VOLUME 63. Enzyme Kinetics and Mechanism (Part A: Initial Rate and Inhibitor Methods)
Edited by DANIEL L. PURICH

VOLUME 64. Enzyme Kinetics and Mechanism (Part B: Isotopic Probes and Complex Enzyme Systems)
Edited by DANIEL L. PURICH

VOLUME 65. Nucleic Acids (Part I)
Edited by LAWRENCE GROSSMAN AND KIVIE MOLDAVE

VOLUME 66. Vitamins and Coenzymes (Part E)
Edited by DONALD B. MCCORMICK AND LEMUEL D. WRIGHT

VOLUME 67. Vitamins and Coenzymes (Part F)
Edited by DONALD B. MCCORMICK AND LEMUEL D. WRIGHT

VOLUME 68. Recombinant DNA
Edited by RAY WU

VOLUME 69. Photosynthesis and Nitrogen Fixation (Part C)
Edited by ANTHONY SAN PIETRO

VOLUME 70. Immunochemical Techniques (Part A)
Edited by HELEN VAN VUNAKIS AND JOHN J. LANGONE

VOLUME 71. Lipids (Part C)
Edited by JOHN M. LOWENSTEIN

Section I

Liposomes in Immunology

[1] "*In Vivo*" Depletion of Macrophages by Liposome-Mediated "Suicide"

By NICO VAN ROOIJEN and ESTHER VAN KESTEREN-HENDRIKX

Introduction

Macrophages are multifunctional cells. They play a key role in natural and acquired host defense reactions, in homeostasis, and in the regulation of numerous biological processes. The main tools they use to achieve these goals are phagocytosis followed by intracellular digestion, and production and release of soluble mediators such as cytokines, chemokines, and nitric oxide (NO). Macrophages can be found as resident cells in all organs of the body, and they can be recruited to sites of inflammation. Their immediate precursors are monocytes, which are released in the blood circulation from the bone marrow. After some time, monocytes leave the circulation, cross the barrier formed by the walls of blood vessels, and enter into one of the organs, where their final differentiation into mature macrophages will take place.

Depletion of macrophages followed by functional studies in such macrophage-depleted animals forms a generally accepted approach to establish their role in any particular biomedical phenomenon. Early methods for depletion of macrophages were based on the administration of silica, carrageenan, or by various other treatments. However, incompleteness of depletion, and even stimulation of macrophages, as well as unwanted effects on nonphagocytic cells, were obvious disadvantages.[1]

For that reason, we have developed a more sophisticated approach, based on the liposome-mediated intracellular delivery of the bisphosphonate clodronate.[2,3] In this approach, liposomes are used as a Trojan horse to get the small clodronate molecules into the macrophage. Once ingested by macrophages, the phospholipid bilayers of liposomes are disrupted under the influence of lysosomal phospholipases. The strongly hydrophilic clodronate molecules intracellularly released in this way do not escape from the cell, because they will not easily cross its cell membranes. As a result, the intracellular clodronate concentration increases as more liposomes are ingested and digested. At a certain clodronate concentration, irreversible damage causes the macrophage to be killed by apoptosis.[4,5] Clodronate

---

[1] N. van Rooijen and A. Sanders, *J. Leuk. Biol.* **62,** 702 (1997).
[2] N. van Rooijen and R. van Nieuwmegen, *Cell Tiss. Res.* **238,** 355 (1984).
[3] N. van Rooijen and A. Sanders, *J. Immunol. Meth.* **174,** 83 (1994).

molecules released in the circulation from dead macrophages will not enter cells again, because they are not able to cross cell membranes. Moreover, free clodronate molecules show an extremely short half-life in circulation and body fluids. They are removed by the renal system. The combination of low toxicity and short half-life of clodronate makes this drug the best choice for the liposome-mediated elimination of macrophages "*in vivo*". Clodronate in its free form is used widely as a drug for the treatment of malignant hypercalcemia[6] and painful bone metastasis caused by hormone-refractory prostate cancer,[7] emphasizing its nontoxic nature.

Clodronate Liposomes in Research

Clodronate Liposomes as a Tool to Investigate Macrophage Activities In Vivo

With the liposome-mediated macrophage "suicide" approach, functional aspects of macrophages have been established in hundreds of studies up to now. Many of the resulting publications are listed in the website http://www.ClodronateLiposomes.com.

Although liposomes can not cross vascular barriers such as the walls of capillaries, their meeting with particular macrophage populations can be achieved by choosing the right administration routes. Among the macrophages that might become useful targets for manipulation by liposomes are Kupffer cells in the liver and splenic macrophages (to be reached by way of intravenous injection),[8] alveolar macrophages in the lungs (to be reached by way of intratracheal instillation or intranasal administration),[9] phagocytic synovial lining cells (by means of intra-articular injection in the synovial cavity),[10] peritoneal macrophages (by means of intraperitoneal injection),[11] macrophages in the testis (by means of local injection),[12]

---

[4] N. van Rooijen, A. Sanders, and T. van den Berg, *J. Immunol. Meth.* **193,** 93 (1996).

[5] M. Naito, H. Nagai, S. Kawanao, H. Umezu, H. Zhu, H. Moriyama, T. Yamamoto, H. Takatsuka, and Y. Takkei, *J. Leuk. Biol.* **60,** 337 (1996).

[6] A. List, *Arch. Intern. Med.* **151,** 471 (1991).

[7] A. Heidenreich, R. Hofmann, and U. H. Engelmann, *J. Urol.* **165,** 136 (2001).

[8] N. van Rooijen, N. Kors, M. van den Ende, and C. D. Dijkstra, *Cell Tiss. Res.* **260,** 215 (1990).

[9] T. Thepen, N. van Rooijen, and G. Kraal, *J. Exp. Med.* **170,** 499 (1989).

[10] P. L. E. M. van Lent, A. E. M. Holthuyzen, L. A. M. van den Bersselaar, N. van Rooijen, L. A. B. Joosten, F. A. J. van de Loo, L. B. A. van de Putte, and W. B. van de Berg. *Arthritis Rheum.* **39,** 1545 (1996).

[11] J. Biewenga, B. van de Ende, L. F. G. Krist, A. Borst, M. Chufron, and N. van Rooijen, *Cell Tiss. Res.* **280,** 189 (1995).

[12] A. Bergh, J. E. Damber, and N. van Rooijen, *J. Endocrinol.* **136,** 407 (1993).

perivascular macrophages in the central nervous system (CNS, by means of intraventricular injection),[13] and lymph node macrophages (by means of injection in their draining areas).[14]

Depleted macrophages are replaced by new ones recruited from the bone marrow after various periods of time. In the liver, new Kupffer cells reappear after ~5 days, and repopulation of the liver with Kupffer cells is complete after ~2 weeks. In the spleen, red pulp macrophages, marginal metallophilic macrophages, and marginal zone macrophages reappear after ~1 week, 3 weeks, and 2 months, respectively.[15] The different repopulation kinetics can be used to study their functional specialization. The role of marginal zone macrophages in the antibody response to particulate antigens was established in mice repopulated by red pulp macrophages and marginal metallophilic macrophages 1 month after administration of clodronate liposomes; in such mice, only the marginal zone macrophages were still absent.[16] The repopulation kinetics of alveolar macrophages in the lung, testis macrophages, lymph node macrophages, peritoneal macrophages, and synovium lining macrophages in joints are available in the relevant literature.

Clodronate Liposomes in Immunodeficient Mice to Study Grafted Human Cells In Vivo

Immunodeficient mice are widely used to harbor xenogeneic grafts of human blood cells to study their role in host defense mechanisms, pathological disorders, and diseases. However, despite the absence of functional T and B lymphocytes as the effector cells of acquired immunity in mice bearing for example the *scid* mutation, elements of the innate immune system are still present. Macrophages are thought to form the core of the remaining resistance against the grafted human cells.

The effects of macrophage depletion in *scid* mice by administration of clodronate liposomes has been investigated in various models. Human peripheral blood lymphocytes injected into macrophage-depleted *scid* mice maintained a large proportion of human cells in the peripheral blood and spleen of the mice, whereas no human cells were detected in control mice within 72 h.[17] The minimum graft size of normal and leukemic human hemopoietic cells in *scid* mice, which results in an outgrowth of the human

[13] M. M. J. Polfliet, P. H. Goede, E. M. L. van Kesteren-Hendrikx, N. van Rooijen, C. D. Dijkstra, and T. K. van den Berg. *J. Neuroimmunol.* **116**, 188 (2001).
[14] F. G. A. Delemarre, N. Kors, G. Kraal, and N. van Rooijen, *J. Leuk. Biol.* **47**, 251 (1990).
[15] N. van Rooijen, N. Kors, and G. Kraal, *J. Leuk. Biol.* **45**, 97 (1989).
[16] F. G. A. Delemarre, N. Kors, and N. van Rooijen, *Immunobiology*, **182**, 70 (1991).
[17] C. C. Fraser, B. P. Chen, S. Webb, N. van Rooijen, and G. Kraal, *Blood*, **86**, 183 (1995).

cells in the mouse bone marrow seemed to be 10 times smaller in macro-phage-depleted *scid* mice than in normal *scid* mice.[18] This considerable reduction of the minimal graft size facilitates greatly studies on subsets of human hemopoietic cells, which are not easy to obtain in large numbers.

Mice lacking the elements of acquired immunity could be made suscep-tible to the development of the human malaria parasite *Plasmodium falci-parum* by depletion of macrophages, followed by substitution of mice erythrocytes by human red blood cells infected with *P. falciparum*.[19] This new model can be used for studies on host–parasite interactions and defense mechanisms against *P. falciparum*,[19] as well as in the development of anti-malarial drugs.[20] In view of the scarcity of animals able to harbor human parasites, this novel model offers new approaches for malaria research.

Clodronate Liposomes in Experimental Models of Therapy

Autoantibody-Mediated Disorders

Under normal circumstances, macrophages will not ingest the orga-nism's own particulate blood constituents. However, when autoantibodies are produced (e.g., against platelets in immune thrombocytopenic pur-pura [ITP] or against red blood cells [RBC] in autoimmune hemolytic anemia [AIHA]), macrophages are responsible for the clearance of large numbers of these autoantibody-coated platelets or RBC. As a con-sequence, macrophages may play a key role in the induction of these autoantibody-mediated disorders.

In a mouse model of ITP, depletion of splenic and hepatic macrophages by liposome-encapsulated clodronate inhibited the antibody-induced thrombocytopenia in a dose-dependent manner. Moreover, this treatment rapidly restored the platelet counts in thrombocytopenic animals to hemato-logical safe values, and despite additional antiplatelet antiserum treatment, mice were able to maintain this level of platelets for at least 2 days. The bleeding times in the treated animals were not different from those in con-trols, demonstrating that hemostasis was well controlled in these animals.[21]

[18] W. Terpstra, P. J. M. Leenen, C. van den Bos, A. Prins, W. A. M. Loenen, M. M. A. Verstegen, S. van Wijngaardt, N. van Rooijen, A. W. Wognum, G. Wagemaker, J. J. Wielenga, and B. Lowenberg, *Leukemia*, **11**, 1049 (1997).
[19] E. Badell, C. Oeuvray, A. Moreno, S. Soe, N. van Rooijen, A. Bouzidi, and P. Druilhe, *J. Exp. Med.* **192**, 1653 (2000).
[20] A. Moreno, E. Badell, N. van Rooijen, and P. Druilhe, *Antimicrob. Agents Chemother.* **45**, 1847 (2001).
[21] F. Alves-Rosa, C. Stranganelli, J. Cabrera, N. van Rooijen, M. S. Palermo, and M. A. Isturiz, *Blood*, **96**, 2834 (2000).

The possible application of clodronate liposomes in the treatment of autoantibody-mediated hemolytic anemia has also been shown in a mouse model.[22] Autoimmune hemolytic anemia (AIHA) is a disease in which autoantibodies against RBC lead to their premature destruction. Most clinically significant autoantibodies are of the IgG type, which lead primarily to the uptake and destruction of RBC by splenic and hepatic macrophages. In a mouse model of AIHA in which animals were given either anti-RBC antibodies or preopsonized RBC, liposomal clodronate substantially decreased RBC destruction. The treatment was rapidly effective within hours by first blocking and consecutively depleting macrophages, and its action lasted for 1 to 2 weeks.

Rheumatoid Arthritis

Macrophages play a key role in the production of inflammatory mediators such as cytokines, NO, and chemokines. Depletion of phagocytic lining cells in knee joints of mice by direct injection of liposome-encapsulated clodronate a few days before induction of arthritis with heterologous bovine type II collagen significantly reduced the inflammatory reaction compared with controls.[10] Cell influx into the synovium was decreased markedly, and expression of interleukin-1 mRNA in the synovium was reduced strongly. Also in the synovial washout samples, chemotactic activity was highly decreased. In addition, other experiments showed that cartilage destruction was reduced in the animals treated with clodronate liposomes.[23] Phagocytic synovial lining cells were also involved in acute and chronic inflammation after exacerbation of hyperreactive joints with antigen given either directly into the knee joint or intravenously in a mouse model of antigen-induced arthritis.[24] Further human studies revealed that a single intra-articular administration of clodronate liposomes resulted in macrophage depletion and decreased expression of adhesion molecules in the synovial lining of patients with long-standing rheumatoid arthritis.[25]

Transplantation

Corneal graft rejection is characterized by a massive infiltration of both T cells and macrophages. Macrophages are found in large numbers in rejected corneal grafts, suggesting a role for these cells in the rejection

[22] M. B. Jordan, N. van Rooijen, S. Izui, J. Kappler, and P. Marrack, *Blood*, **101**, 594 (2003).
[23] P. L. E. M. van Lent, A. E. M. Holthuyzen, N. van Rooijen, L. B. A. van de Putte, and W. B. van de Berg, *Ann. Rheum. Dis.* **57**, 408 (1998).
[24] P. L. E. M. van Lent, A. E. M. Holthuyzen, N. van Rooijen, F. A. J. van de Loo, L. B. A. van de Putte, and W. B. van de Berg, *J. Rheumatol.* **25**, 1135 (1998).
[25] P. Barrera, A. Blom, P. L. E. M. van Lent, L. van Bloois, G. Storm, J. Beijnen, N. van Rooijen, L. B. A. van de Putte, and W. B. van de Berg, *Arthritis Rheum.* **43**, 1951 (2000).

process. In rats treated postoperatively with subconjunctival injections of liposome-encapsulated clodronate at the time of transplantation and several times thereafter, grafts were not rejected during the maximum follow-up of 100 days. Cellular infiltration in these grafts was reduced, and there was a strong reduction in neovascularization of the cornea. Corneal grafts in rats that had received empty liposomes were rejected within the usual period of 17 days.[26] In additional experiments, treatment with clodronate liposomes was shown to down-regulate local and systemic cytotoxic T lymphocyte responses and to prevent the generation of antibodies. Depletion of macrophages in the initiation phase of the immune response seemed to induce a less vigorous attack on the grafted tissue and, therefore, to promote graft survival.[27]

Macrophage depletion by clodronate liposomes also prolonged survival and functioning of grafts after pancreas islet xenotransplantation,[28,29] as well as that of porcine neonatal pancreatic cell clusters contained in alginate macrocapsules and transplanted into rats.[30] Treatment with clodronate liposomes reduced markedly graft infiltration by macrophages and T cells, and evidence has been produced that macrophages play a role in graft rejection by promotion of T-cell infiltration.[28] Interestingly, recent evidence supports the idea that T-cell–activated macrophages themselves are capable of recognizing and rejecting pancreatic islet xenografts.[31] In the latter studies, it has been shown that $CD4^+T$ cells are required for macrophage activation in the presence of pancreatic islet xenografts. However, once activated, macrophages are capable of rejecting xenografts in the absence of any other effector cells; they are able to migrate to the graft site and to identify the graft independently of other signals from T cells. According to the authors, this suggests that in xenograft rejection, macrophages receive additional non-T-cell–mediated signals by way of the innate immune system. This could explain why immunosuppressive strategies that

[26] G. van der Veen, L. Broersma, C. D. Dijkstra, N. van Rooijen, G. van Rij, and R. van der Gaag, *Invest. Ophthalmol.* **35,** 3505 (1994).
[27] T. P. A. M. Slegers, P. F. Torres, L. Broersma, N. van Rooijen, G. van Rij, and R. van der Gaag, *Invest. Ophthalmol.* **41,** 2239 (2000).
[28] A. Fox, M. Koulmanda, T. E. Mandel, N. van Rooijen, and L. C. Harrison, *Transplantation*, **66,** 1407 (1998).
[29] G. Wu, O. Korsgren, J. Zhang, Z. Song, N. van Rooijen, and A. Tibell, *Xenotransplantation*, **7,** 214 (2000).
[30] A. Omer, M. Keegan, E. Czismadia, P. de Vos, N. van Rooijen, S. Bonner-Weir, and G. C. Weir, *Xenotransplantation* **10,** 240 (2003).
[31] S. Yi, A. M. Lehnert, K. Davey, H. Ha, J. Kwok Wah Wong, N. van Rooijen, W. J. Hawthorne, A. T. Patel, S. N. Walters, A. Chandra, and P. J. O'Connell, *J. Immunol.* **170,** 2750 (2003).

inhibit the alloimmune response are ineffective at suppressing T-cell–mediated xenograft rejection.[31]

Neurological Disorders

Depletion of blood-borne macrophages reduces strongly lesion formation and the development of clinical signs in experimental allergic encephalomyelitis (EAE),[32] an experimental model for multiple sclerosis (MS). Adoptive transfer of EAE with myelin basic protein–reactive CD4$^+$ T cells to SJL/J mice was abrogated by treatment with mannosylated clodronate liposomes. Invasion of the CNS by various cells was almost completely blocked by this treatment, and the myelin sheaths appeared completely normal, whereas marked demyelination was observed in the control groups.[33] These studies demonstrated a role for macrophages in regulating the invasion of autoreactive T cells and secondary glial recruitment that ordinarily lead to the demyelinating pathology in EAE and MS.[33] Recent studies demonstrated that perivascular macrophages and meningeal macrophages, constituting a major population of resident macrophages in the CNS, also contribute to the early stages of EAE development.[13]

Inflammatory mechanisms are believed to play an important role in hyperalgesia resulting from nerve injury. It was shown that intravenous injection of clodronate liposomes reduced the number of macrophages in an injured nerve, alleviated thermal hyperalgesia, and protected both myelinated and unmyelinated fibers against degeneration. Results confirmed the role of circulating monocytes and/or macrophages in the preservation of myelinated axons, decreased cavitation in the development of neuropathic hyperalgesia and Wallerian degeneration caused by partial nerve injury, and it was suggested that suppression of macrophage activity immediately after nerve injury could have some clinical potential in the prevention of neuropathic pain.[34]

Traumatic injury to the spinal cord initiates a series of destructive cellular processes that accentuate tissue damage at and beyond the original site of trauma. The cellular inflammatory response has been implicated as one mechanism of secondary degeneration. Within injured spinal cords of rats treated with clodronate liposomes, macrophage infiltration was significantly reduced at the site of impact. These animals showed marked improvement in hindlimb use during overground locomotion. Behavioral

[32] I. Huitinga, N. van Rooijen, C. J. A. de Groot, B. M. J. Uitdehaag, and C. D. Dijkstra, *J. Exp. Med.* **172,** 1025 (1990).

[33] E. H. Tran, K. Hoekstra, N. van Rooijen, C. D. Dijkstra, and T. Owens, *J. Immunol.* **161,** 3767 (1998).

[34] T. Liu, N. van Rooijen, and D. J. Tracey, *Pain,* **86,** 25 (2000).

recovery was paralleled by a significant preservation of myelinated axons, decreased cavitation in the rostrocaudal axis of the spinal cord, and enhanced sprouting and/or regeneration of axons at the site of injury. These data implicate blood-borne macrophages as effectors of acute secondary injury and suggest that clodronate liposomes may prove to be useful in therapy after spinal cord injury.[35]

Other Forms of T-cell–Mediated Tissue Damage

T cells seem to be responsible for liver damage in any type of acute hepatitis. T-cell–mediated liver injury is induced for example by several agents such as *Pseudomonas* exotoxin A (PEA), concanavalin A (ConA), and by a combination of subtoxic doses of PEA and the superantigen *Staphylococcus* enterotoxin B (SEB). Depletion of Kupffer cells (liver macrophages) by clodronate liposomes protected mice from PEA-, ConA-, or PEA/SEB–induced liver injury. In the absence of Kupffer cells, liver damage was restricted to a few small necrotic areas. These studies further indicated that Kupffer cells play an important role in T-cell activation-induced liver injury by contributing tumor necrosis factor.[36]

After administration of clodronate liposomes in nonobese diabetic (NOD) mice, it was shown that T cells lost their ability to differentiate into beta cell–cytotoxic T cells in a macrophage-depleted environment, resulting in the prevention of autoimmune diabetes. These T cells regained their beta cell–cytotoxic potential when they were returned to a macrophage-containing environment.[37] In these studies, the administration of IL-12 seemed to reverse the prevention of diabetes, which was conferred by macrophage depletion, in these NOD mice.

Gene Therapy

Replication-deficient recombinant adenovirus vectors are efficient at transferring genes to target cells. However, both the innate immune system and the acquired immune system may reduce the efficacy of this approach for gene transfer. By their activity as scavengers of foreign particulate material, macrophages may remove most of the injected gene carriers before they can reach their targets.

[35] P. G. Popovich, Z. Guan, P. Wei, I. Huitinga, N. van Rooijen, and B. T. Stokes, *Exp. Neurol.* **158,** 351 (1999).

[36] J. Schumann, D. Wolf, A. Pahl, K. Brune, T. Papadopoulos, N. van Rooijen, and G. Tiegs, *Am. J. Pathol.* **157,** 1671 (2000).

[37] H. S. Jun, C. S. Yoon, L. Zbytunik, N. van Rooijen, and J. W. Yoon, *J. Exp. Med.* **189,** 347 (1999).

It has been shown that depletion of Kupffer cells by liposome-encapsulated clodronate before intravenous administration of an adenovirus vector led to a higher input of recombinant adenoviral deoxyribonucleic acid (DNA) to the liver, an absolute increase in transgene expression, and a delayed clearance of both the vector DNA and transgene expression. One week after administration of the adenovirus vector, peak transgene expression was found to be enhanced about 10-fold in the macrophage-depleted animals. One month after administration, expression in the animals treated with clodronate liposomes was still 38% of this peak value, whereas control animals that got the adenovirus but not the liposomes showed no detectable expression after 2 weeks.[38] Significantly higher and stable expression levels resulting from high-capacity adenovirus vectors that were preceded by administration of clodronate liposomes have since been reported in various models of gene therapy.[39,40] Also in the lung, alveolar macrophages were shown to play an important role in the elimination of intratracheally administered adenovirus vectors, and their suppressing effect on adenovirus vector–mediated gene transfer could be eliminated by a preceding intratracheal administration of clodronate liposomes.[41] The nonlinear dose response, following the application of adenovirus vectors for gene therapy of Fabry disease in a mouse model, could be corrected by the preceding administration of clodronate liposomes. As a consequence, lower doses with strongly reduced toxicity are required.[42] These results also suggest that minimizing the interaction between the recombinant adenoviral vectors and the mononuclear phagocyte system may improve the therapeutic window of this vector system.[42]

Drug Targeting

Liposomes may be considered one of the most versatile and promising drug-carrier devices (see the present and accompanying volumes). However, the high phagocytic capacity of tissue macrophages prevents the bulk

[38] G. Wolff, S. Worgall, N. van Rooijen, W. R. Song, B. G. Harvey, and R. G. Crystal, *J. Virol.* **71,** 624 (1997).

[39] G. Schiedner, S. Hertel, M. Johnston, V. Dries, N. van Rooijen, and S. Kochanek, *Molec. Ther.* **7,** 35 (2003).

[40] M. K. L. Chuah, G. Schiedner, L. Thorrez, B. Brown, M. Johnston, V. Gillijns, S. Hertel, N. van Rooijen, D. Lillicrap, D. Collen, T. vanden Driessche, and S. Kochanek, *Blood*, **101,** 1734 (2003).

[41] S. Worgall, P. L. Leopold, G. Wolff, B. Ferris, N. van Rooijen, and R. G. Crystal, *Human Gene Ther.* **8,** 1675 (1997).

[42] R. J. Ziegler, C. Li, M. Cherry, Y. Zhu, D. Hempel, N. van Rooijen, Y. A. Ioannou, R. J. Desnick, M. A. Goldberg, N. S. Yew, and S. H. Cheng, *Human Gene Ther.* **13,** 935 (2002).

of all kinds of particulate carriers, including liposomes, to reach their targets. Several modifications of the original liposome formulations, such as the incorporation of amphipathic polyethylene glycol (PEG) conjugates in the liposomal bilayers, have been proposed to reduce the recognition of liposomes by macrophages. Nevertheless, a large percentage of these so-called long-circulating liposomes will still be ingested by macrophages, as has been shown in both the spleen[43] and lymph nodes.[44] Depletion of liver and splenic macrophages by clodronate liposomes significantly prolonged the circulation time of subsequently administered liposomes, even when the latter were long-circulating liposomes. The efficacy of drug targeting through the use of particulate carriers may thus be improved by the transient suppression of the phagocytic activity of macrophages by clodronate liposomes.

When drug targeting is considered, repeated injections of the drug carriers will often be necessary to achieve the required effects. It has been shown that long-circulating PEG-liposomes are cleared rapidly from the circulation when injected repeatedly in the same animal. However, when liver and splenic macrophages were previously depleted by clodronate liposomes, such an enhanced clearance of repeatedly injected liposomes was not observed,[45] emphasizing that suppression of phagocytic activity by clodronate liposomes may contribute to the success of drug-carrier–mediated therapy.

Materials and Methods

Preparation of Multilamellar Clodronate-Liposomes

1. Equipment and reagents

- Chloroform, analytical grade (Riedel-de Haën, Seelze, Germany)
- Argon gas (or other inert gas, e.g., nitrogen gas)
- Sterile phosphate-buffered saline (PBS) (Braun Melsungen AG Melsungen, Germany) containing 8.2 g NaCl, 1.9 g $Na_2HPO_4 \cdot 2H_2O$, 0.3 g $NaH_2PO_4 \cdot 2H_2O$ at pH 7.4 per liter.
- Stock solution of phosphatidylcholine (egg lecitin): 100 mg/ml phosphatidylcholine (Lipoid) in chloroform.[46] The solution is filtered

[43] D. C. Litzinger, A. M. J. Buiting, N. van Rooijen, and L. Huang, *Bioch. Bioph.* **1190,** 99 (1994).

[44] C. Oussoren, M. Velinova, G. Scherphof, J. J. van der Want, N. van Rooijen, and G. Storm, *Bioch. Bioph. Acta,* **1370,** 259 (1998).

[45] P. Laverman, M. G. Carstens, O. C. Boerman, E. Th. M. Dams, W. J. G. Oyen, N. van Rooijen, F. H. M. Corstens, and G. Storm, *J. Pharm. Exp. Ther.* **298,** 607 (2001).

   with a syringe-driven filter unit with 0.2-μm pores (Millex GN, Millipore, Bradford, MA) on a glass/Teflon syringe.

- Stock solution of cholesterol: 10 mg/ml cholesterol (Sigma) in chloroform.[47] The solution is filtered with a syringe-driven filter unit with 0.2-μm pores (Millipore, Millex GN) on a glass/Teflon syringe.
- 0.7 M clodronate solution: 50 g clodronate (Roche Diagnostics GmbH Mannhein, Germany) is dissolved in 150 ml Milli Q (or similar purified water). The pH is adjusted to 7.1 with 5 N NaOH. The final volume is brought to 200 ml with Milli Q. This solution is filtered with 0.2-μm pore bottle-top filter (Millipore, Steritop).
- Waterbath sonicator (Sonicor SC-200-22, 55 kHz; Sonicor Instr. Corp., Copiague, NY).
- High-speed centrifuge (Sorvall, RC 5B plus).
- Rotary evaporator (Büchi, Rotavapor).
- Sterile pipets (Cellstar, Greiner Bio-One, Frickenhausen, Germany).
- Polycarbonate centrifuge tubes (Sorvall).
- Bottle-top filter 0.2-μm pores (Millipore, Steritop).
- Autoclaved 3.0-μm pore polycarbonate membrane filter (Millipore, Isopore TSTP 2500) in filter holder (Millipore, Swinnex SX 2500).

2. Preparation of liposomes

- Forty-three milliliters of the phosphatidylcholine stock solution are added to 40 ml cholesterol stock solution in a 2-liter round-bottom flask.[48]
- The chloroform is removed by low-vacuum (120 mbar) rotary evaporation (150 rpm) at 40°. At the end, a thin phospholipid film will form against the inside of the flask. The condensed chloroform is removed, and the flask is aerated three times.
- The flask is vented by putting a pipet (without cotton-wool) at the end of the argon gas tube. The tip of the pipet can be used to vent deep into the flask to ensure ventilating the whole film and thus removing all remaining chloroform.
- The phospholipid film is dispersed in 200 ml PBS (for empty liposomes) or 0.7 M clodronate solution (for clodronate liposomes)

---

[46] This stock can be made in advance and stored at $-20°$ under argon gas. Argon gas is used to prevent oxidation of phosphatidylcholine.

[47] This stock can be made in advance and stored at $-20°$.

[48] Instructions are given for preparation of 200 ml liposome suspension as we usually do. However, smaller volumes can be made by reducing phosphatidylcholine, cholesterol, and dispersing liquid. Attention should be paid to choose the most suitable filter, because the amount of liquid loss depends on the diameter of the filter unit.

by gentle rotation (max., 100 rpm) at room temperature (RT) for 20–30 min (PBS) or 10–15 min (0.7 M clodronate solution).[49] Development of foam should be avoided.

- The milky white suspension is kept at RT for 1.5–2 h.
- The solution is shaked gently and sonicated in a waterbath for 3 min.
- The suspension is kept at RT for 2 h (or overnight at 4°) to allow swelling of the liposomes.[50]
- Before using the clodronate liposomes:
  — The nonencapsulated clodronate is removed by centrifugation of liposomes at 22,000g and 10° for 60 min. The clodronate liposomes will form a white band at the top of the suspension, whereas the suspension itself will be nearly clear.[51]
  — The clodronate solution under the white band of liposomes is carefully removed using a sterile pipet (about 1% will be encapsulated). The liposomes are resuspended in approximately 450 ml sterile PBS.
  — The nonencapsulated clodronate is recycled for re-use. The clodronate solution is centrifuged at 22,000g and 10° for 120 min. The remaining liposomes are discarded. This solution is filtered using a 0.2-μm bottle-top filter. This recycling procedure should not be repeated for more than five times.
- The liposomes should be washed four to five times using centrifugation at 22,000g and 10° for 25 min. The upper solution should be removed each time and the pellet resuspended in approximately 450-ml sterile PBS using a sterile pipet.
- The final liposome pellet is resuspended in sterile PBS and adjusted to a final volume of 200 ml. The suspension is shaken (gently) before administration to animals or before dispensing to achieve a homogeneous distribution of the liposomes in suspension.[52]

---

[49] Clodronate liposomes can be stored in the original clodronate solution at 4° under argon gas to prevent denaturation of phospholipid vesicles. This is particulary important in the case of clodronate liposomes, because they float on the aqueous phase after preparation. PBS liposomes form a pellet on the bottom of the tubes.

[50] To limit the maximum diameter of the liposomes for intravenous injection, the suspension can be filtered using membrane filters with 3.0-μm pores.

[51] There is no problem when the suspension is not completely clear, because the remaining liposomes will be very small. The relatively large clodronate liposomes are efficacious with respect to depletion of macrophages.

[52] Sterility can be tested by distributing 50-μm liposomes on a blood-agar plate.

Spectrophotometric Determination of the Amount of
  Liposome-Encapsulated Clodronate

  1. Equipment and reagents

- 3*1 ml of clodronate–liposome suspension to be tested (i.e., in triplicate).
- Milli Q or similar purified water.
- Chloroform, analytical grade (Riedel-de Haën).
- Sterile PBS (Braun Melsungen AG) containing 8.2 g NaCl, 1.9 g Na$_2$HPO$_4$·2H$_2$O, 0.3 g NaH$_2$PO$_4$·2H$_2$O at pH 7.4 per liter.
- Standard clodronate solution (10.0 mg/ml): to prepare this, 500 mg clodronate (Roche Diagnostics GmbH) is dissolved in 30-ml Milli Q. The pH is adjusted to 7.1 with 5 N NaOH. The solution is brought to a final volume of 50.0 ml with Milli Q.
- Phenol 90% (Riedel-de-Haën, 16018).
- 4 mM CuSO$_4$ solution: to prepare this, 0.64 g/liter CuSO$_4$ (Merck) is dissolved in Milli Q.
- HNO$_3$ solution: to prepare this, 65% HNO$_3$ (Merck, Darmstadt, Germany) is diluted 100 times in Milli Q.
- 16-ml glass tubes, caps with Teflon inlay (Kimble, Vineland, NY).
- 10-ml polystyrene tubes (Greiner).
- Spectrophotometer (UV-160A, Shimadzu, Kyoto, Japan).
- Pasteur pipets.
- Glass pipet 10 ml (piston pipet; Hirschmann).
- Pipets (P20, P200, and P1000, Gilson, Emeryville, CA).

  2. Extraction of clodronate from liposomes

- 1 ml of the clodronate–liposome suspension (in triplicate), 1 ml of standard clodronate solution, and 1 ml of the PBS solution is dispensed in separate glass tubes.[53]
- 8 ml of phenol/chloroform (1:2) is added to each tube.
- The tubes are vortexed and shaken extensively.
- The tubes are held at RT for at least 15 min.
- The tubes are centrifuged (1125 g) at 10° for 10 min.
- The aqueous (upper) phase is transferred to clean glass tubes using a Pasteur pipet.
- 6-ml chloroform per tube is added: the solution is reextracted by extensive vortexing.

---

[53] Attention should be paid to the right controls. If liposomes are suspended in PBS, PBS controls should be included. NB: Phosphate (depending on concentration) may disturb the assay.

- The tubes are held for at least 5 min at RT.
- The tubes are centrifuged (1125 g) at $10°$ for 10 min.
- The aqueous phase is transferred to 10-ml plastic tubes with a Pasteur pipet. These are the samples for determination of clodronate concentration.

3. Determination of clodronate concentration

- A standard curve using 0, 10, 20, 40, 50, 70, and 80 μl of the extracted standard clodronate solution adjusted with saline to a total volume of 1 ml per tube is prepared.
- The samples are diluted until they are within the range of the standard curve.[54]
- 2.25-ml 4 mM $CuSO_4$ solution, 2.20-ml Milli Q, and 0.05-ml HNO_3 solution is added to each tube, containing 1-ml sample or standard.
- All tubes are vortexed vigorously
- The samples are read at 240 nm using spectrophotometer and quartz cuvette.[55]

[54] A suspension of clodronate liposomes prepared according to protocol 1 contains about 6 mg clodronate per 1 ml suspension. Twenty microliters of extracted clodronate liposome suspension (thus diluting the sample 50 times) has an average absorption of 0.5 using a 1-cm quartz cuvette.

[55] J. Mönkkönen, M. Taskinen, S. O. K. Auriola, and A. Urtti, *J. Drug Targeting* **2**, 299 (1994).

[2] Trafficking of Liposomal Antigens to the *Trans*-Golgi Complex in Macrophages

By Mangala Rao, Stephen W. Rothwell, and Carl R. Alving

Introduction

The use of liposomes as potential carriers of antigens for vaccines in combination with a variety of adjuvants and mediators has been explored extensively,[1-4] and the first liposomal vaccine (for hepatitis A) was been licensed in Europe[5,6] The major justification and rationale for using

[1] C. R. Alving, V. Koulchin, G. M. Glenn, and M. Rao, *Immunol. Rev.* **145**, 5 (1995).

[2] C. R. Alving, *J. Immunol. Meth.* **140**, 1 (1991).

[3] G. Gregoriadis, *Immunol. Today* **11**, 89 (1990).

[4] L. F. Fries, D. M. Gordon, R. L. Richards, J. E. Egan, M. R. Hollingdale, M. Gross, C. Silverman, and C. R. Alving, *Proc. Natl. Acad. Sci. U S A* **89**, 358 (1992).

[5] L. Loutan, P. Bovier, B. Althaus, and R. Glück, *Lancet* **343**, 322 (1994).

METHODS IN ENZYMOLOGY, VOL. 373　　　　　　　　0076-6879/03 $35.00

liposomes as vehicles for vaccines has been the rapid uptake of liposomes by macrophages.[7-9] In this chapter, we describe methods for examining the intracellular fate of liposomes and liposomal antigens in macrophages.

Protein antigens are processed and presented either by the major histocompatibility complex (MHC) class I or class II pathways.[10,11] MHC class I molecules are expressed on the surface of all nucleated cells. In contrast, MHC class II molecules are expressed only on the surface of antigen presenting cells (APCs), such as macrophages, B cells, and dendritic cells. MHC class I and class II molecules are highly polymorphic membrane proteins that bind and transport peptide fragments of proteins to the surface of APCs. The MHC–peptide complex then interacts with either $CD8^+$ or $CD4^+$ T lymphocytes to generate a specific immune response.[10,12]

Endogenous antigens are presented by way of the MHC class I pathway, whereas exogenous antigens are presented by way of the MHC class II pathway. Therefore, most soluble antigens are relatively ineffective for priming MHC class I–restricted cytotoxic T lymphocyte responses because of the inability of the antigen to gain access to the cytoplasmic compartment. Several different methods have been used to channel antigens into the class I pathway.[1,2, 13-23] Among these methods, liposomes have proven

[6] R. Glück, *Vaccine* **17,** 1782 (1999).

[7] D. Su and N. Van Rooijen, *Immunology* **66,** 466 (1989).

[8] J. N. Verma, M. Rao, S. Amselem, U. Krzych, C. R. Alving, S. J. Green, and N. M. Wassef, *Infect. Immun.* **60,** 2438 (1992).

[9] J. N. Verma, N. M. Wassef, R. A. Wirtz, C. T. Atkinson, M. Aikawa, L. D. Loomis, and C. R. Alving, *Biochim. Biophys. Acta* **1066,** 229 (1991).

[10] R. N. Germain and D. H. Margoulies, *Ann. Rev. Immunol.* **11,** 403 (1993).

[11] T. J. Braciale, L. A. Morrison, M. T. Sweetser, J. Sambrook, M. J. Gething, and V. L. Braciale, *Immunol. Rev.* **98,** 95 (1987).

[12] A. Townsend and H. Bodmer, *Ann. Rev. Immunol.* **7,** 601 (1989).

[13] M. W. Moore, F. R. Carbone, and M. J. Bevan, *Cell* **54,** 777 (1988).

[14] K. Deres, H. Schild, K. H. Weismuller, G. Jung, and H. G. Rammensee, *Nature* **342,** 561 (1989).

[15] H. Schild, M. Norda, K. Deres, K. Falk, O. Rotzschke, K. H. Weismuller, G. Jung, and H. G. Rammensee, *J. Exp. Med.* **174,** 1665 (1991).

[16] C. V. Harding and R. Song, *J. Immunol.* **153,** 4925 (1994).

[17] M. Kovacsovics-Bankowski and K. L. Rock, *Science* **267,** 243 (1995).

[18] Y. Men, H. Tamber, R. Audran, B. Gander, and G. Corradin, *Vaccine* **15,** 1405 (1997).

[19] L. M. Lopes and B. M. Chain, *Eur. J. Immunol.* **22,** 287 (1992).

[20] R. Reddy, F. Zhou, S. Nair, L. Huang, and B. T. Rouse, *J. Immunol.* **148,** 1585 (1992).

[21] K. White, U. Krzych, T. D. Gordon, M. R. Porter, R. L. Richards, C. R. Alving, C. D. Deal, M. Hollingdale, C. Silverman, D. R. Sylvester, W. R. Ballou, and M. Gross, *Vaccine* **11,** 1341 (1993).

[22] W. I. White, D. R. Cassatt, J. Madsen, S. J. Burke, R. M. Woods, N. M. Wassef, C. R. Alving, and S. Koenig, *Vaccine* **13,** 1111 (1995).

[23] C. R. Alving and N. M. Wassef, *AIDS Res. Hum. Retrovir.* **10,** S91 (1994).

to be an efficient delivery system for entry of exogenous protein antigens into the MHC class I pathway because of their particulate nature.[24] A liposome formulation developed in our laboratory that contains dimyristoyl phosphatidylcholine, dimyristoyl phosphatidylglycerol, cholesterol, and an encapsulated protein antigen has been used in human clinical trials.[4] This formulation of liposomes has also been shown to be an effective vehicle for delivery of proteins or peptides to APCs for presentation by way of the MHC class I pathway in mice.[25,26]

By the use of fluorophore-labeled proteins encapsulated in liposomes, we have addressed the question of how liposomal antigens enter the MHC class I pathway in bone marrow–derived macrophages. After phagocytosis of the liposomes, the liposomal lipids and the liposomal proteins seem to follow the same intracellular route, and they are processed as a protein-lipid unit.[27] The fluorescent liposomal protein and liposomal lipids enter the cytoplasm where they are processed by the proteasome complex.[25] The processed liposomal protein is then transported into the endoplasmic reticulum and the Golgi complex by way of the transporter associated with antigen processing (TAP).[28] In these compartments, the peptides bind to the MHC class I molecules. Once bound, the antigenic peptides are transported to the cell surface to interact with receptors on T cells.[29,30] The procedures that we use to study the intracellular trafficking of liposome-encapsulated proteins are outlined in the following.

Experimental Design

We have developed an *in vitro* antigen presentation system consisting of bone marrow–derived macrophages as the APCs. Our system is well suited for studying intracellular trafficking, because we begin with precursor cells that can be differentiated into either dendritic cells or macrophages. Although B cells can also be used to study intracellular trafficking of antigens, we have used bone marrow–derived macrophages because of the ease of preparation, their inherent phagocytic properties, their ability to adhere to

[24] M. Rao and C. R. Alving, *Adv. Drug Deliv. Rev.* **41,** 171 (2000).

[25] S. W. Rothwell, N. M. Wassef, C. R. Alving, and M. Rao, *Immunol. Lett.* **74,** 141 (2000).

[26] R. L. Richards, M. Rao, N. M. Wassef, G. M. Glenn, S. W. Rothwell, and C. R. Alving, *Infect. Immun.* **66,** 2859 (1998).

[27] M. Rao, S. W. Rothwell, N. M. Wassef, A. B. Koolwal, and C. R. Alving, *Exp. Cell Res.* **246,** 203 (1999).

[28] M. Rao, S. W. Rothwell, N. M. Wassef, R. E. Pagano, and C. R. Alving, *Immunol. Lett.* **59,** 99 (1997).

[29] C. Bonnerot, V. Briken, and S. Amigorena, *Immunol. Lett.* **57,** 1 (1997).

[30] S. Joyce, *J. Mol. Biol.* **266,** 993 (1997).

plastic dishes, and their morphological characteristics that permit easy microscopic observation of living cells. It is important to realize that because these cultures are not synchronized, at any given time the cells will not be in the same state regarding phagocytosis and processing. Therefore, it is critical to observe and count cells from multiple fields to obtain representative results.

The MHC haplotype of the mouse strain is important in determining the MHC class I response. The peptide SIINFEKL is the cytotoxic T-cell epitope of ovalbumin that binds to H-2K^b class I molecules.[31] Therefore, C57BL/6 (H-2K^b) mice are the strain of choice for studies that use ovalbumin. Similarly, B10.Br (H-2K^k) mice are the strain of choice for studies that use conalbumin. The age of the mice is critical to the processing efficiency of the bone marrow–derived macrophages, and the mice should not be older than 3 months.

Materials and Reagents

Reagents and Sources

- All strains of mice (Jackson Laboratories, Bar Harbor, MA).
- C2.3 hybridoma cell line derived from C57BL/6 bone marrow macrophages (K. L. Rock, Harvard School of Medicine, Boston, MA).
- 25-D1.16 antibody (R. Germain, NIAID, NIH, Bethesda, MD).
- Texas Red sulfonyl chloride derivative of sulforhodamine 101, Texas red-1,2-dihexadecanoyl-sn-glycerol-3-phosphoethanolamine and N-(ε-NBD-aminohexanoyl)-D-erythro-sphingosine) (Molecular Probes, Eugene, OR).
- Chicken ovalbumin, conalbumin type I from chicken egg whites, trypan blue, Tris, glycine, glutaraldehyde, Triton X-100, and normal goat serum (Sigma-Aldrich, St. Louis, MO).
- Dimyristoyl phosphatidylcholine, dimyristoyl phosphatidylglycerol, and cholesterol (Avanti Polar Lipids, Alabaster, AL).
- Fluorescein–anti-rabbit and anti-mouse antibody (Boehringer Mannheim, Indianapolis, IN).
- Vectashield (Vector Laboratories, Burlingam, CA).
- Lactacystin (BIOMOL Research Laboratories, Inc., Plymouth Meeting, PA).
- RPMI-1640, fetal bovine serum, L-glutamine, penicillin, streptomycin, murine gamma-interferon, and phosphate-buffered saline (Gibco-BRL Life Technologies, Rockville, MD).

---

[31] K. Falk, O. Rotzschke, S. Faath, S. Goth, I. Graef, N. Shastri, and H. G. Rammensee, *Cell. Immunol.* **150,** 447 (1993).

- Dulbecco's phosphate-buffered saline without Ca^{2+}, Mg^{2+} (Bio-Whittaker Inc., Walkersville MD).

Equipment and Sources

- Pear-shaped glass flask, glass pipettes, vaccine vials with rubber stoppers (Kimble/Knotes, Vineland, NY).
- PD-10 (Sephadex G-25 columns) (BD-Pharmacia, San Jose, CA).
- Fluorescence-activated cell sorter (FACScan, Becton-Dickinson Immunosystems, San Jose, CA).
- VirTis Advantage Freeze Dryer (The VirTis Company, Gardiner, NY).
- Rotavapor Rotary evaporator (Brinkman/Buchi, Dumstat, Germany).
- Fluorescence microscope (Leitz Orthoplan, Leica, Deerfield, IL) with color digital camera (DEI-470, Optronics Engineering, Goleta, CA).
- Adobe Photoshop software (Adobe Systems, Inc., San Jose, CA).
- 4–20% Polyacrylamide Sodium dodecyl sulfate (SDS) gels, SDS-PAGE equipment (BioRad, Hercules, CA).
- CO_2 incubator (Forma Scientific, Inc. Marietta, OH).
- Refrigerated centrifuge (Sorvall RT 6000, RC-5B, Dupont Instruments, Newark, DE).
- Sterile tissue culture plasticware.
- Circular glass coverslips (no. 1), depression glass slides (VWR Scientific, West Chester, PA).
- Speed Vac SC100 (Savant Instruments, Holbrook, NY).
- Ultraviolet (UV) Visible-spectrophotometer (Jasco Inc, Easton, MD).
- UV transilluminator (FotoDyne, Inc., Hartland, WI).
- Solvent Recovery Still (Distilling equipment for organic solvents, Knotes/Martin, Vineyland, NY).

Methods

Labeling of Protein and Separation of Labeled Protein

Proteins can be labeled with many different fluorochromes. We have successfully labeled several different proteins, such as conalbumin, ovalbumin, recombinant malaria, and human immunodeficiency virus proteins in quantities ranging from 1–20 mg with Texas Red. The amino groups on proteins are covalently coupled to Texas Red when the protein and the reagent are mixed together.[32] The efficiency of labeling of the protein with

---

[32] J. A. Titus, R. Haugland, S. O. Sharrow, and D. M. Segal, *J. Immunol. Meth.* **50,** 193 (1982).

Texas Red is dependent on several factors: (1) Texas Red hydrolyzes rapidly in aqueous solution; therefore, Texas Red should be kept dry before addition to the protein solution. (2) The protein solution is kept on ice to enhance protein coupling. (3) Solid Texas Red is added directly to the chilled protein solution with rapid mixing. (4) The labeling efficiency depends on the pH of the buffer. Maximal conjugation is obtained at pH = 9.0. The procedure outlined in the following is the one that we have used in our studies.

Conalbumin (20 mg) is dissolved in 2 ml of 0.2 M carbonate–bicarbonate buffer (0.144 g of Na_2CO_3-H_2O and 0.272 g $NaHCO_3$, pH = 9.0). Once the protein goes into solution, it is filter sterilized. One milliliter (10 mg) of sterile conalbumin solution is transfered into a 12 × 75-cm sterile plastic tube containing a small stir bar. This is placed on ice, and 1 mg Texas Red sulfonyl chloride is added while stirring. The ice bucket is covered with aluminum foil, and it is stirred on ice for 3 h. During the 3-h reaction time, a Sephadex G-25 column is set up in the biological cabinet. A three-way stopcock is attached to the tip of the column. The column is washed with 30–35 ml sterile saline. After 3 h, the protein solution is loaded onto the column, and the protein is eluted with sterile saline. Until all the purple color is eluted, 0.5-ml fractions are collected. The purple dye is Texas Red conjugated to protein. The red dye is the unbound Texas Red. The labeled protein elutes typically starting at tubes 5–6 and is finished by tubes 8–9. Absorbance is measured at 280 nm (protein) and at 596 nm (Texas Red) in a spectrophotometer. The labeled protein fractions are pooled, filter sterilized, and kept in a 12 × 75-cm-sterile tube covered with foil at 4°. The Texas Red–labeled protein is now ready to be encapsulated in liposomes.

Conalbumin is released from liposomes with chloroform treatment.[33] The labeled proteins are analyzed for degree of degradation by SDS-PAGE electrophoresis. The proteins are separated by electrophoresis on 4% to 20% precast polyacrylamide gels using a Tris (25 mM)-glycine (192 mM)-SDS (20%) electrode buffer and viewed under UV light.

Preparation of Liposomes[34]

Stock solutions of lipids are prepared in chloroform and stored at −20° in glass-stoppered containers.[34] All glass containers and pipettes are wrapped in aluminum foil and heat-sterilized in an oven at 250° overnight.

[33] D. Wessel and U. I. Flügge, Anal. Biochem. 138, 141 (1984).

[34] C. R. Alving, S. Shichijo, I. Mattsby-Baltzer, R. L. Richards, and N. M. Wassef, Preparation and use of liposomes in immunological studies. In "Liposome Technology", 2nd ed. (G. Gregoriadis, ed.) Vol. 3, p. 317. CRC Press, Boca Raton, FL, 1993.

Because chloroform deteriorates on standing, it is important that the chloroform is redistilled every 3 months. After distillation, 0.7% ethanol is added as a preservative. Because of the potential carcinogenic nature of chloroform, all the distillation and rotary evaporation steps must be conducted in a fume hood.

Lipid solutions are mixed in molar ratios of 1.8:0.2:1 of dimyristoyl phosphatidylcholine, dimyristoyl phosphatidylglycerol, and cholesterol, in a pear-shaped flask that has a volume that is $10\times$ larger than the final volume of the resuspended liposome solution. To study the trafficking of liposomal lipids, fluorescent lipids (2 mol% with respect to the phospholipid concentration) are added. The fluorescent lipids can be purchased from Molecular Probes. We have used N-NBD-PE (nitrobenzoxadiazole phosphoethanolamine or TR-DHPE (Texas Red-1,2-dihexadecanoyl-sn-glycerol-3-phosphoethanolamine). Using a rotary evaporator, the solvent is removed at $40°$ under negative pressure provided by a filter pump aspirator attached to a water faucet. Lipids are dried under low vacuum ($<50 \mu$m Hg) for a minimum of 1 h in a dessicator. The dried lipids are stable under vacuum but are best used on the same day. The dried lipids are reconstituted in deionized water, followed by vigorous vortexing, and aliquots are placed into sterile vaccine vials with rubber stoppers. The vaccine vials are frozen at $-70°$, transferred to a lyophilizer, and the stoppers are loosened. The lipids are lyophilized, the vials are stoppered, and the tops are crimped. The lipids can now be stored indefinitely at $-70°$.

Encapsulation of Protein in Liposomes

Multilamellar liposomes are prepared by dispersion of lyophilized mixtures of lipids at a phospholipid concentration of 100 mM in Dulbecco's phosphate-buffered saline (PBS) containing either unlabeled conalbumin, Texas Red–labeled ovalbumin (TR-OVA), or Texas Red–labeled conalbumin (TR-conalbumin).[28,35] To encapsulate the proteins in liposomes, the vial containing lyophilized lipids at $-70°$ should be allowed to come to room temperature (RT). The required amount of protein is added into the vial in the biological cabinet and vortexed. No lumps should be present. Parafilm is wrapped around the cap, and foil is wrapped around the vial if either lipids or proteins are light sensitive; it is then kept in the refrigerator. After 48 h, the contents are transferred into a sterile centrifuge tube. The vial is washed well with 0.15 M NaCl (30 ml), and the contents are transferred to a sterile centrifuge tube. The tube in capped and centrifuged at RT 7500 g for 30 min in a Sorvall RC-5B refrigerated super speed centrifuge using an

[35] N. M. Wassef, C. R. Alving, and R. L. Richards, *J. Immunol. Meth.* **4,** 217 (1994).

SA600 rotor. At the end of the run, the supernatant is carefully decanted or transferred. The supernatant is then discarded, and the wash is repeated. The liposome pellet is resuspended in buffer to give a final phospholipid concentration of 30 mM and stored at 4° until used. The amount of antigen encapsulated in liposomes is determined by a modified Lowry procedure (see later).[34]

Modified Lowry Procedure

Aliquots (10–50 μl) of liposomes and protein standards (0–80 μg) are pipetted into 13 × 100 mm glass test tubes. To dissolve the lipids 0.5 ml chloroform is added to each tube. The tubes are placed in a Speed-Vac centrifuge to remove the chloroform and dry the samples. Two hundred microliter of deoxychlolate (15% w/v in deionized water) is added to each tube. The tubes are vortexed, and the normal Lowry procedure is followed for assaying the amount of protein present. The percent encapsulation as follows is calculated as follows:

% Encapsulation = Amount of protein measured by the Lowry assay/ amount of protein initially added to the lipids × 100

With ovalbumin and conalbumin, the percent encapsulation is between 45% and 50%.

Preparation of Macrophages

Media. The bone marrow macrophage growth media (BM media) consists of RPMI-1640 containing 10% fetal bovine serum (FBS) (heated at 56° in a water bath for 30 min to inactivate complement), 10% L-cell conditioned media, 100-U/ml penicillin and 100 μg/ml streptomycin (P/S), and 8 mM glutamine. The growth media is filter sterilized and store at 4°. Fresh media is prepared every 3 w.

To prepare L-cell conditioned media, L-929 cells (obtained from American Type Culture Collection) are grown in a tissue culture flask for a week in RPMI-1640 containing 10% FBS and 8 mM glutamine. The supernatant is collected and clarified by centrifugation in a Sorvall RT6000 refridgerated centrifuge at 800g for 10 min at 4°. The supernatant is aliquoted and stored at −70°.

Preparation of Single Cell Suspension from Mouse Femurs

Mice are euthanized by carbon dioxide inhalation followed by cervical dislocation. Typically, three mice are processed for each macrophage harvesting. Each mouse is dipped in 70% ethanol and placed on a sterile field. With sterile instruments, the skin and muscle of the hind legs is open to expose the femurs. The femurs are removed and placed in a 60-mm Petri dish containing 2 ml PBS (without Ca^{2+} and without Mg^{2+}). The dishes are transferred to a biological containment cabinet. No fat or cartilage should be attached to them, otherwise fibroblasts will overgrow the culture. The outside of femurs is scrubbed with sterile gauze. The ends of the bones are snipped off. The marrows are flushed with 10 ml PBS (without Ca^{2+} and without Mg^{2+}) using a 10-ml syringe and a 22-gauge needle into a sterile 50-ml conical tube. A single cell suspension is made by drawing up the cells with the syringe and passing them three times through the needle. Debris are allowed to settle ($\sim$1 min), and is transferred the cell suspension into a fresh sterile 50-ml conical tube, which is spun at $800g$ for 10 min at $4°$. The supernatant is discarded, the cell pellet is gently tapped to loosen the cells, and the cells are resuspended in 10 ml bone marrow (BM) media.

Cell Counting

Ten microliters of the cell suspension is transferred to a 12×75-mm polypropylene tube, and 90 μl trypan blue (2% solution made in PBS) is added to the tube; $10\mu l$ are counted in a hemocytometer. Several fields are averaged. The numbers of cells are calculated. (No. of live cells counted $\times 10^4$ Dilution of cells = Cells/ml). The cell concentration is adjusted to achieve 2×10^6 cells/ ml. The acid-washed coverslips are placed in 35-mm Petri dishes. One hundred-microliter cells/coverslip are seeded at a density of 2×10^5 cells, then the cells are spread over the area of the coverslip with a sterile pipette tip. The cells are allowed to adhere for 20 min at RT. After 20 min, 2 ml BM media/dish is carefully added to avoid disturbing the adhered cells, and the dishes are placed in a humidified, CO_2 incubator at $37°$. One milliliter of media is removed and replaced with 1 ml fresh BM media 48 h later. This process is repeated until day 9. On day 9, the macrophage cultures are supplemented with 10 U/ml murine IFN-γ and used for trafficking experiments the next day.[8,36]

Fluorescence Microscopy

The cells are examined with a Leitz Orthoplan (Leica, Deerfield, IL) microscope equipped with differential interference contrast objectives and a Leitz 63$\times$ oil immersion lens designed for fluorescence microscopy. Fluorescence signals are generated by use of fluorescence filters from Omega Optical that are optimized for Texas Red (excitation wavelength, 595; emission wavelength, 615) and fluorescein (excitation wavelength, 494; emission wavelength, 518) fluorochromes. Images are collected with a color digital camera (Model DEI-470, Optronics Engineering, Goleta, CA)

[36] M. Rao, N. M. Wassef, C. R. Alving, and U. Krzych, *Infect. Immun.* **63**, 2396 (1995).

coupled to an Apple Macintosh computer and are stored as Adobe Photoshop files (Adobe Systems, Inc., San Jose, CA).

Intracellular Trafficking of Liposomal Antigens

A detailed procedure for the examination of intracellular trafficking in macrophages is described below, with a general flow chart presented in Fig. 1 .

Coverslips containing macrophages from B10.BR mice are washed in Hanks balanced salt solution without phenol red (HBSS), pH 7.4, twice, and incubated in a total volume of 1 ml HBSS containing 30 μg of liposome-encapsulated TR-conalbumin [L(TR-CON)] at 37° in a CO_2 incubator for various time periods. After incubation, the coverslips are washed and mounted cell-side down on a depression slide containing a small quantity of buffer. The cells are viable under these conditions for at least

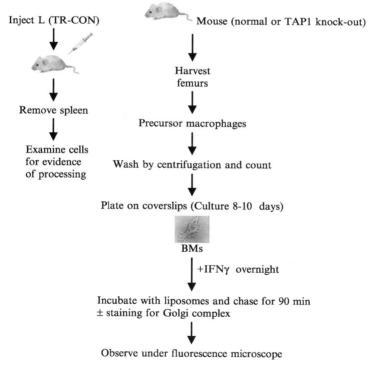

Fig. 1. Flowchart of basic procedures for studying the processing of liposomal antigens in murine macrophages.

2 h and can be put back in culture if needed. The uptake of L(TR-CON) can be observed as early as 5 min (Fig. 2). Areas with diffuse fluorescence can be seen within 15 min, suggesting the presence of protein in the cytoplasm of the macrophages. Internalization of the liposomal antigen continues, and by 45 min the liposomal antigen begins to concentrate in the perinuclear/Golgi area of the cells. After 90 min, the protein is mainly localized to a perinuclear region with some diffuse staining (Fig. 2). This localization is distinctly visualized by washing the cells in HBSS and incubating for a further 90 min chase at 37° in HBSS (Fig. 3A).

Colocalization of the antigen with the Golgi can be demonstrated in several ways.[25] At the end of the chase period, *trans*-Golgi is visualized by staining the cells with a green fluorescent analog of ceramide [N-(ε-NBD-aminohexanoyl)-D-erythro-sphingosine (C_6-NBD-ceramide).[37]

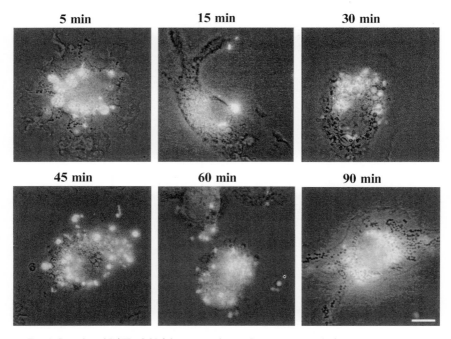

FIG. 2. Uptake of L(TR-CON) by macrophages. Bone marrow–derived macrophages from B10.BR mice were grown on coverslips and incubated with L(TR-CON) for different times (5 min–90 min) at 37°. The coverslips containing the cells were washed and mounted on depression slides. Live cells were observed under a Leitz-Orthoplan microscope with an oil immersion 63 × objective. Scale bar: 10 μm.

[37] R. E. Pagano, M. A. Sepanski, and O. C. Martin, *J. Cell Biol.* **109**, 2067 (1989).

Coverslips containing macrophages are incubated on ice with 2 nmoles/ml of C_6-NBD-ceramide for 30 min, then washed twice with HBSS, and transferred to 37° for 15 min. After washing twice with HBSS, cells are mounted and viewed as described previously. The localized liposomal conalbumin fluorescence (Fig. 3A) can be superimposed on the Golgi fluorescence (Fig. 3B).

For localization of conalbumin by immunofluorescence microscopy, macrophages are incubated with L(TR-CON) for 90 min followed by a 90-min chase in media. The cells are then fixed for 10 min in 2% formaldehyde

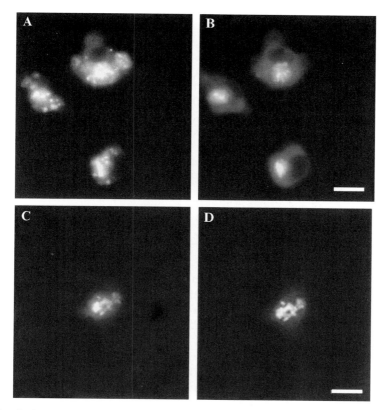

FIG. 3. Localization of L(TR-CON) to the *trans*-Golgi complex. Macrophages were incubated with L(TR-CON) for 90 min, followed by a 90-min chase at 37° and stained for *trans*-Golgi with C_6-NBD ceramide. L(TR-CON) concentrates in the perinuclear region (A) and colocalized with *trans*-Golgi as identified by the NBD-ceramide fluorescence (B). Fluorescent peptides (C) retain specificity for anticonalbumin antibodies after localization to a perinuclear area (D). Scale bar: 10 μm.

and 0.1% glutaraldehyde (final concentrations) and are permeabilized for 10 min with 0.5% Triton X-100 in PBS. The fixed and permeabilized cells are incubated with rabbit anticonalbumin antibody for 1 h and fluorescein-antirabbit antibody (5 μg/ml) for 1 h at 37° with three washes of PBS between each incubation. Before viewing, coverslips are mounted on slides using Vectashield to diminish photo bleaching. The conalbumin staining, as visualized by the Texas red fluorescence (Fig. 3C), colocalizes with the fluorescein staining observed using the conalbumin-specific antibody (Fig. 3D). This confirms that the conalbumin and/or its peptides still retains the Texas red labeling.

Effects of Proteasome Inhibitors

Peptides that are presented through the MHC class I pathway are generated by the degradation of cytoplasmic antigens through the proteolytic activities of the proteasome complex. Macrophages to be used in the proteasome inhibitor studies are incubated with the irreversible proteasome inhibitor, lactacystin (10 μM), for 30 min before incubation with L(TR-CON). Following the chase period, the cells are stained with the Golgi-specific stain, C_6-NBD ceramide. The cells are then washed in HBSS, mounted, and viewed. In contrast to localization of the TR peptides (Fig. 4A) in the Golgi region (Figure 4B), the lactacystin-treated cell shows a diffuse, granular pattern (Fig. 4C). However, the treatment with lactacystin does not affect the integrity of the Golgi itself, as shown by the ceramide staining of the Golgi in the same cell (Fig. 4D).

Demonstration of Transporter Associated with Antigen Processing Proteins in Trafficking

After the proteolytic degradation of the antigens by the proteasome complex, the antigenic peptides are translocated into the endoplasmic reticulum (ER) by means of the heterodimeric TAP proteins.[38] The TAP complex is composed of TAP1, TAP2, and tapasin.[39] Transport of peptides into the ER requires both TAP1 and TAP2 proteins.

To determine whether peptides derived from the liposomal proteins used TAP proteins for their transport, macrophages are obtained from TAP1 (−/−) knock-out mice (obtained from Jackson Laboratories) and from the corresponding TAP1 (+/+) wild-type mice. Because TAP1 knock-out is on a C57BL/6 background, the antigen of choice for trafficking

[38] M. J. Androlewicz, P. Cresswell, and K. S. Anderson, *Proc. Natl. Acad. Sci. U S A* **90**, 9130 (1992).

[39] S. Li, K. Paulsson, H. Sjogren, and P. Wang, *J. Biol. Chem.* **274**, 8649 (1999).

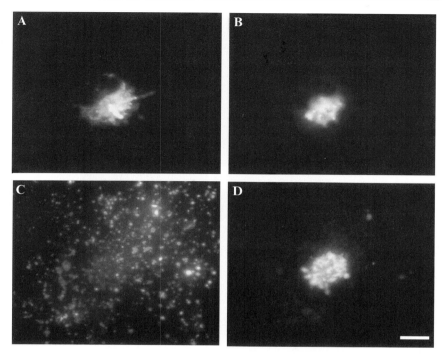

FIG. 4. Inhibition of processing of L(TR-CON) by lactacystin. Macrophages were preincubated with 10 μM lactacystin and then incubated with L(TR-CON), followed by staining the cells with C_6-NBD-ceramide. Cells incubated with L(TR-CON) in the absence of the inhibitor transported the fluorescent peptides into the Golgi area (A), TR-CON (B), (NBD-ceramide staining). In cells incubated with 10 μM lactacystin, the TR-CON remained widely distributed throughout the cells (C), TR-CON (D), (NBD-ceramide staining). Scale bar: 10 μm.

experiments should be ovalbumin. Experiments are performed with these macrophages as described previously.

As shown in Fig. 5A, OVA peptides derived from L(TR-OVA) are excluded from the area of the *trans*-Golgi (Fig. 5B). In macrophages derived from the wild-type mice, OVA peptides are localized exclusively in the *trans*-Golgi (Fig. 5C, D).

Detection of Major Histocompatibility C–Peptide Class I Complex on the Cell Surface

Once the peptides are transported into the Golgi complex by the TAP proteins, the peptides bind to the newly synthesized MHC class I molecules and are translocated to the cell surface. The expression of MHC–peptide

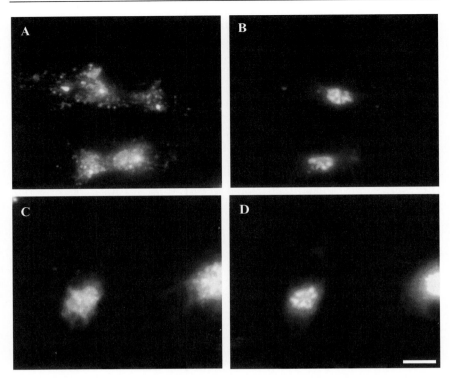

FIG. 5. Requirement of TAP proteins for the localization of L(TR-OVA) peptides to the *trans*-Golgi area. Macrophages from TAP1 knock-out mice (A and B) and C57BL/6 mice (C and D) were grown on coverslips and incubated with L(TR-OVA). Cells were washed and stained with C_6-NBD–ceramide (B and D). Peptides derived from L(TR-OVA) (A) were excluded from the *trans*-Golgi area (B) in macrophages from knock-out mice. Macrophages from wild-type mice concentrated the peptides (C) to the *trans*-Golgi area (D). Scale bar: 10 μm.

complexes on the cell surface can be visualized with fluorescence microscopy or can be measured quantitatively using flow cytometry. To detect the MHC–peptide complexes, an antibody that recognizes specifically the MHC–peptide complexes generated intracellularly is required. In the experiments described later, we have used a mouse monoclonal antibody, 25-D1.16, that binds to MHC class I–SIINFEKL complexes. This antibody, generated by Dr. Porgador[40] at NIH, was kindly provided to us by

[40] A. Porgador, J. W. Yewdell, Y. Deng, J. R. Bennink, and R. N. Germain, *Immunity* **6,** 715 (1997).

Dr. Germain (NIAID, NIH, Bethesda, MD). The positive control for these experiments is macrophages incubated with the ovalbumin peptide, SIIN-FEKL (500 μg) for 2.5 h at 37°. The negative control is either buffer alone or a liposomal antigen that is not recognized by this antibody.

Macrophages are incubated on coverslips in 35-mm dishes with L(OVA), as described previously, for 90 min, followed by a 90-min chase to generate MHC–peptide complexes. Cells are washed in PBS and Fc receptors are blocked by incubating with normal goat serum (1/100 dilution in 100 μl PBS) for 30 min on ice. After 30 min, the cells are not washed but the 25-D1.16 antibody (1 ml culture supernatant) is directly added to the cells. There are incubated overnight at 4°, washed three times with PBS, and then incubated with fluoresceinated goat-antimouse IgG (5 μg/ ml diluted in PBS containing 1/100 normal goat serum) for 1 h on ice. The cells are washed three times in PBS; coverslips are mounted on depression slides as described previously. These are observed with the fluorescence microscope.

For flow cytometry, macrophages are grown in Petri dishes. Cells are processed for detection of cell surface expression of MHC–peptides as described previously. After the final wash with PBS, the cells are scraped gently from the Petri dishes with a rubber policeman, and the cells are collected by centrifugation. The cell-associated fluorescence is measured using a FACScan flow cytometer, and the results are analyzed with Cell Quest software. Because more cells are required for flow cytometric analysis than for microscopy, the macrophage cell line, C2.3, is a suitable substitute for bone marrow–derived macrophages.

As shown in Fig. 6A , the antibody detects the expression of the H-2K^b–SIINFEKL complex when the C2.3 macrophage cell line is incubated with SIINFEKL peptide. Similar results are obtained when the cells are incubated with L(OVA) (Fig. 6B). The buffer control is shown in Fig. 6C . Antigen-specific fluorescence on the cell surface demonstrates that L(OVA) is processed intracellularly, and the peptide SIINFEKL generated binds to the MHC class I molecule, H-2K^b. The MHC–peptide complex is expressed on the cell surface and is recognized by the antibody. The expression of MHC–peptide complex is analyzed by flow cytometry (Fig. 6D). Incubation with L(OVA) or SIINFEKL result in 1–2 log increases in fluorescence.

In Vivo Processing of Liposomal Antigens

To determine whether the trafficking of liposomal antigen into the Golgi observed in vitro with bone marrow–derived macrophages also occurs in vivo, one can inject mice intravenously with the liposomal antigen and then examine spleen cells for evidence of antigen processing. Mice are

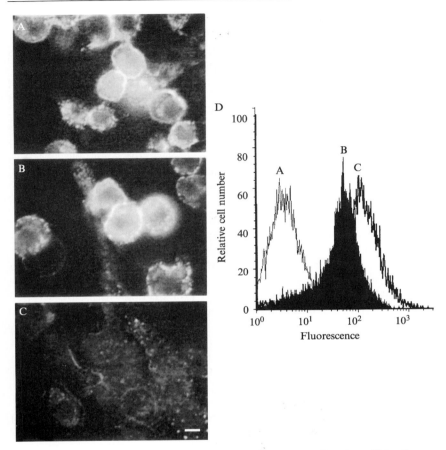

Fig. 6. Detection of MHC class I-peptide complexes on the cell surface. C2.3 cells were incubated with SIINFEKL peptide (A), L(OVA) (B), or buffer (C). Cells were processed for labeling with 25 D1.16 antibody and fluorescein-conjugated anti-mouse antibody and either examined by microscopy or analyzed by flow cytometry. Microscopy showed bright fluorescence on the outer surface of cells. Flow cytometry measurement (D) of MHC class I-peptide complexes following incubation with either SIINFEKL peptide or L(OVA) showed a 1–2 log increase in the surface staining. Peak A, buffer control; peak B, L(OVA); peak C, SIINFEKL. Scale bar: 10 μm

injected with liposomal antigen (150–200 μg in a total volume of 0.5 ml). After 60 min, the mice are euthanized, their spleens are removed, and the spleens are placed on a stainless-steel screen in a Petri dish containing HBSS. A single cell suspension is prepared by mashing the spleen with a syringe plunger. The cells are collected by centrifugation (800 g, 4°, 10 min).

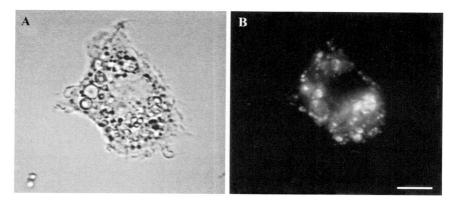

FIG. 7. Processing of L(TR-CON) *in vivo* by macrophages. Splenic macrophages were isolated from mice 60 min after intravenous injection of L(TR-CON). Single-cell suspensions were made, the cells were plated on glass coverslips for 1 h at 37° to allow adherence of macrophages, and the macrophages were then examined by fluorescence microscopy. The macrophages phagocytosed the liposomal antigen and localized the fluorescence to a perinuclear area consistent with the Golgi localization (A, light image) and (B, Texas Red fluorescence). Scale bar: 10 μm.

Cells (2×10^6) are plated on a glass coverslip and allowed to adhere for 1 h at 37°. The nonadherent cells are washed away and the coverslip is mounted on a glass slide, which is examined by fluorescence microscopy.

Intravenous injection of B10.BR mice with L(TR-CON) results in the uptake and processing of the liposomal antigen by the adherent macrophages. The splenic macrophages remaining on the coverslip had avidly phagocytosed L(TR-CON) (Fig. 7A). The fluorescence is localized to a perinuclear area (Fig. 7B), consistent with the Golgi localization seen in the *in vitro* experiments described previously. The fluorescence seen in macrophages is specific, because neither neutrophils nor lymphocytes phagocytose or concentrate the fluorescent label.

Conclusions

There have been several studies on the ability of macrophages and macrophage cell lines to present exogenous antigens. Liposomes have been used widely as carriers of protein or peptide antigens. Here we have presented detailed procedures to allow one to study the processing and trafficking patterns of liposomal antigens in living cells, because an understanding of the intracellular trafficking of liposomal antigens is essential for developing effective liposomal vaccines.

[3] Oil-in-Water Liposomal Emulsions for Vaccine Delivery

By GARY R. MATYAS, JEAN M. MUDERHWA, and CARL R. ALVING

Introduction

Liposomes containing lipid A are potent adjuvants for the induction of high-titer antibody and cell-mediated immunity.[1-8] To enhance further the adjuvant effect and to provide a depot vehicle for the slow release of liposomes to the immune system, we have developed an oil-in-water (o/w) emulsion containing liposomes.[9] Antigen is encapsulated in liposomes, and the liposomes are then used as an emulsifier for creating an o/w emulsion with light mineral oil. During the emulsification process, some liposomes are destroyed to stabilize the emulsion, but a large number of liposomes remain intact. As the body degrades the emulsion, or as the emulsion is ingested by phagocytes, liposomes are released to the cellular elements of the immune system. The rate of release of the liposomes can be controlled by designing emulsions with varying stabilities.

We have used these liposomal emulsions as vaccine carriers both in mice[9] and in phase I and phase II clinical trials with patients with prostate cancer.[10] Prostate cancer represents a unique challenge in that the patients have circulating levels of the immunogen—prostate-specific antigen (PSA)—and the goal is to create autoimmunity that breaks tolerance to

[1] R. L. Richards, M. Rao, N. M. Wassef, G. M. Glenn, S. W. Rothwell, and C. R. Alving, *Infect. Immun.* **66,** 2859 (1998).

[2] K. White, U. Krzych, D. M. Gordon, T. G. Porter, R. L. Richards, C. R. Alving, C. D. Deal, M. Hollingdale, C. Silverman, D. R. Sylvester, W. R. Ballou, and M. Gross, *Vaccine* **11,** 1341 (1993).

[3] D. G. Heppner, D. M. Gordon, M. Gross, B. Welde, W. Leitner, U. Krzych, I. Schneider, R. A. Wirtz, R. L. Richards, A. Trofa, T. Hall, J. C. Sadoff, P. Boerger, C. R. Alving, D. R. Sylvester, T. G. Porter, and W. R. Ballou, *J. Infect. Dis.* **174,** 361 (1996).

[4] C. R. Alving and N. M. Wassef, *AIDS Res. Hum. Retrovir.* **10**(suppl. 2), S91 (1994).

[5] C. R. Alving, V. Koulchin, G. M. Glenn, and M. Rao, *Immunol. Rev.* **145,** 5 (1995).

[6] L. F. Fries, D. M. Gordon, R. L. Richards, J. E. Egan, M. R. Hollingdale, M. Gross, C. Silveman, and C. R. Alving, *Proc. Natl. Acad. Sci. USA* **89,** 358 (1992).

[7] R. Glück, *in* "Vaccine Design: The Subunit and Adjuvant Approach" (M. F. Powell and M. J. Newman, eds.) p. 325, Plenum Press, New York, 1995.

[8] B. Mengiardi, R. Berger, M. Just, and R. Glück, *Vaccine* **13,** 1306 (1995).

[9] J. M. Muderhwa, G. R. Matyas, L. E. Spitler, and C. R. Alving, *J. Pharm. Sci.* **88,** 1332 (1999).

[10] D. T. Harris, G. R. Matyas, L. G. Gomella, E. Talor, M. D. Winship, L. E. Spitler, and M. J. Mastrangelo, *Semin. Oncol.* **26,** 439 (1999).

METHODS IN ENZYMOLOGY, VOL. 373   0076-6879/03 $35.00

PSA. Immunization with liposome-encapsulated PSA by the intramuscular or intravenous routes was ineffective in inducing strong immune responses to the self-antigen.[10] However, when the liposomes containing PSA and lipid A were used as an emulsifier with light mineral oil, potent immune responses were observed. This demonstrates clearly the enhanced adjuvanticity of an o/w emulsion containing liposomes, lipid A, and encapsulated antigen. In this chapter, we describe methods for the formulation and manufacture of Walter Reed liposomes[11,12] and use of the liposomes for subsequent emulsification with light mineral oil to form o/w emulsions (Fig. 1). The measurement of emulsion stability and the factors that affect emulsion stability also are described.

Preparation of Liposomes

Preparation of Lipid Stocks and Solvents

Highly purified lipids are essential for the preparation of stable liposomes that do not leak antigen. Oxidized or unsaturated lipids affect adversely the stacking of lipids within the liposomes, allowing leakage of antigen. We use 1,2-dimyristoyl-*sn*-glycero-3-phosphocholine (DMPC), 1,2-dimyristoyl-*sn*- glycero-3-phospho-*rac*-(1-glycerol) (DMPG), and cholesterol from Avanti Polar Lipids Inc. (Alabaster, AL). We have always found these lipids to be highly purified and with excellent lot-to-lot reproducibility. The adjuvant used in the liposomes is lipid A (either monophosphoryl or diphosphoryl), the active component of lipopolysaccharide. The lipid A, isolated from *Salmonella minnesota* R595, is predominately monophosphate. The lipid A can be obtained in the lyophilized, free acid form from Corixa Corporation (Seattle, WA), List Biological Laboratories (Campbell, CA), or Avanti Polar Lipids.

Chloroform (Burdick Jackson Laboratories Inc., Muskegon, MI) is used to solubilize the lipids. Chloroform is unstable and degrades with time. Chloroform used in the preparation of liposomes must be freshly manufactured and stabilized with 0.75% ethanol. If the chloroform was manufactured more than 60–90 days before use, it contains breakdown products, which do not evaporate and contaminate the liposomes. These breakdown products cause the liposomes to be somewhat leaky.[12] Chloroform can be distilled to remove the contaminants. The chloroform is placed in a

[11] N. M. Wassef, C. R. Alving, and R. L. Richards, *Immunomethods* **4**, 217 (1994).
[12] C. R. Alving, S. Shichijo, I. Mattsby-Baltzer, R. L. Richards, and N. M. Wassef, *in* "Liposome Technology," 2nd Edn, Vol. 3 (G. Gregoriadis, ed.) p. 317, CRC Press, Boca Raton, 1993.

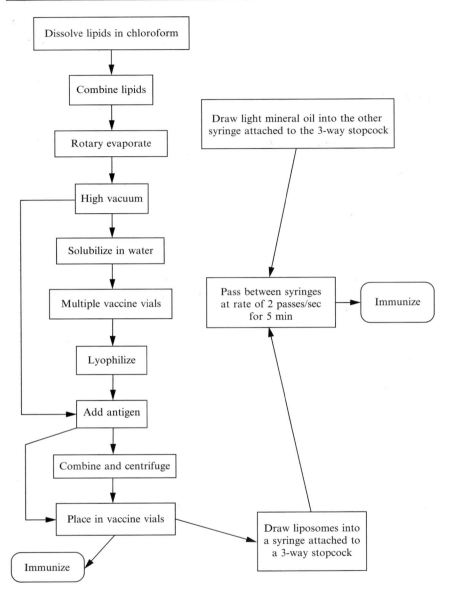

Fig. 1. Flowchart of the basic procedures for the manufacture of Walter Reed liposomes and the emulsification of the liposomes with light mineral to form an oil-in-water emulsion.

distillation flask, and a distillation column is attached. The distillation flask is heated using a heating mantle, and the distillation column is cooled by circulating water from a circulating water bath set at 4°. A thermometer is inserted into the distillation flask to monitor the boiling point of the chloroform solution. Chloroform boils at 61–62°. The first 5%–10% of the original volume that distills is discarded. The chloroform is collected, and absolute ethanol is added to a final percentage of 0.75. Distilled chloroform is stored in a brown bottle at room temperature.

To ensure sterility and to remove pyrogen, all glassware should be washed with detergents lacking phosphate. Residual amounts of detergent containing phosphate may interfere with the quantification of phospholipids as measured by an assay for phosphate. (The glassware is soaked in Crystal Simple Green, Sunshine Environmental Intelligence, Huntington Beach, CA). The opening of the glassware is covered with aluminum foil, and the glassware is baked in an oven at 220° overnight. All other materials should be autoclaved before use. If desired, the chloroform can be sterile-filtered by passing it through a previously autoclaved 0.22-μm GVWP filter (Millipore Corporation, Bedford MA). This is done with a solvent-filtering apparatus with ground-glass joints from Millipore Corporation. The filter is placed on a previously autoclaved cradle (baking chars the cradle). The funnel and clamp are attached, and the cradle is placed on a ground-glass Erlenmeyer flask. Vacuum is applied to filter the chloroform. The filtered chloroform is then transferred to a sterile brown bottle in a hood equipped with HEPA-filtered air.

The lipids can be dissolved in chloroform at any concentration. However, the stock solutions shown in Table I for ease of calculation of the volumes required for the preparation of liposomes. DMPC, DMPG, and cholesterol are weighed on an analytical balance, and the powder is transferred to a 500-ml volumetric flask. A wide-mouth funnel is placed in the flask, and the powdered lipid is added. The flask and funnel are tapped

TABLE I
LIPID STOCK SOLUTIONS

Lipid	Molecular weight	Stock concentration	Mass required	Volume	Solvent
DMPC	667.94	180 mM	61.0 g	500 ml	Chloroform
DMPG	688.86	20 mM	6.64 g	500 ml	Chloroform
Cholesterol	386.66	150 mM	29.0 g	500 ml	Chloroform
Lipid A	1955.81	1 mg/ml	10 mg	10 ml	Chloroform: Methanol (9:1, v/v)

until the powder is in the flask. The funnel is rinsed with chloroform to remove the remaining lipid. A 10-ml glass pipet is used to add the chloroform rinse. Disposable glass pipets containing blue graduations should not be used, because the blue graduations are solubilized by the chloroform, resulting in contamination of the lipid solutions. Chloroform is then poured into the flask until just below the neck. A ground-glass stopper is inserted, and the flask is shaken until the lipid dissolves. Chloroform is then added to the 500-ml mark. The DMPG flask may need to be warmed in a 37° water bath to ensure complete dissolution of the DMPG. Lipid A is the active component of endotoxin. To prevent contamination of the laboratory, we do not weigh it. A solution of chloroform:methanol (9:1, v/v) is added directly to the lipid A vial to give approximately a 2 mg/ml concentration. The lipid A solution is transferred to a 10-ml volumetric flask. The lipid A vials are rinsed with fresh solvent to remove residual lipid A, and this is added to the volumetric flask. Chloroform:methanol (9:1, v/v) is added to the volumetric flask to give a final volume of 10 ml. If sterility is a concern, the lipid solutions can be filtered through the 0.2-μm GVWP filter as described previously for chloroform. However, because chloroform is a sterilizing agent, we routinely do not filter our chloroform solutions. The lipid solutions should be stored at $-20°$ and are stable until the chloroform expires in approximately 90 days.

Combining Lipids

Walter Reed liposomes consist of DMPC, cholesterol, and DMPG in a molar ratio of (9:7.5:1). The phospholipid concentration (DMPC + DMPG) can range from 25–200 mM. Lipid A is added to give the desired dose in the final formulation. We use 20–25 μg in 0.05–0.1 ml per dose for a mouse. The procedure for the preparation of 10 ml of 100 mM phospholipid with a lipid A dose of 20 μg/0.05 ml will be described. The lipid solutions are allowed to warm to room temperature before use. Ten milliliters of 100 mM phospholipid is equal to 1 mmol of phospholipid. Because the DMPC:DMPG ratio is 9:1, 0.9 mmol of DMPC and 0.1 mmol of DMPG are needed. The ratio of cholesterol to DMPG is 7.5:1; therefore, 0.1 mmol of DMPG and 0.75 mmol cholesterol are needed. The desired lipid A concentration is 20 μg in 0.05 ml (400 μg/ml). For 10-ml liposomes, 4 mg lipid A is needed. Table II indicates the volume of each lipid solution required.

Rotary Evaporation and High Vacuum Treatment

The lipids are transferred to a 100-ml pear-shaped rotary evaporator flask using 5-ml glass pipets. The size of the pear-shaped flask should be at least 10 times the final volume of liposomes to be manufactured. The

TABLE II
AMOUNTS OF LIPIDS REQUIRED TO MANUFACTURE 10 ML OF 100 mM PHOSPHOLIPID LIPOSOMES

Lipid	Stock	Amount needed	Volume
DMPC	180 mM	0.9 mmol	5 ml
DMPG	20 mM	0.1 mmol	5 ml
Cholesterol	150 mM	0.75 mmol	5 ml
Lipid A	1 mg/ml	4 mg	4 ml

pear-shaped flask is placed on a rotary evaporator (Büchi, Model EL131, Brinkman Instruments Inc., Westbury, NY) fitted with a ground-glass solvent trap. The rotary evaporator water bath is preheated to 40°. The condenser coils are cooled with a circulating water bath set at 4°. The evaporator is attached to an oil-free vacuum pump (Büchi, Model B-171), which is set at 200 mbar. The flask is lowered into the water bath and is rotated at 80 rpm. The solvent will rapidly evaporate and collect in the collection reservoir. The flask contents will become viscous and the solvent stream, which collects in the reservoir, will stop flowing. The vacuum and flask rotation are then increased to 100 mbar and 150 rpm, respectively. Evaporation of the solution is allowed to continue until all the solvent is removed and a dry lipid layer coats the side of the flask. The evaporation process must be monitored for foaming. The solvent should boil, but not foam. If foaming occurs, the vacuum is released with the stopcock at the top of the distillation column. The vacuum pressure may be adjusted up, or the flask may be removed either partially or entirely from the water bath.

After the solvent is evaporated, the flask is removed from the rotary evaporator, and the mouth of the flask is immediately covered with autoclaved hardened ashless filter paper (No. 541, Whatman International Ltd., Maidstone, UK). The filter paper is secured in place with a rubber band. The flask is then placed in a desiccator (Dryseal, Wheaton, Millville, NJ), which has a silicon sealing ring (vacuum grease should be avoided). The desiccator is attached to the vacuum pump, and the pump is set as low as it goes (approximately 11 mbar). Because the pump cannot attain this pressure, the pump will run constantly. This high-vacuum treatment is continued for a minimum of 1 h for small volumes of liposomes (1–5 ml) to 12–24 h for large volumes of liposomes (>100 ml). The 10-ml sample described here should be under high vacuum for at least 2 h. This ensures that the residual solvent is removed. The lipids are stable for at least several days at room temperature under these conditions.

Lyophilization

After the flask is removed from the desiccator, it is transferred to a biological safety cabinet. The filter paper is removed. The dried lipids may be used directly for liposome formation and antigen encapsulation, but better encapsulation efficiency and lot-to-lot reproducibility is obtained by use of the lyophilization process described here. Nine milliliters of sterile water are added to the flask, and the flask is capped with a ground-glass stopper. The flask is shaken or vortexed until most of the lipid is removed from the side of the flask. A sterile 10-ml plastic pipet is used to transfer the contents to 5 depyrogenated (baked) 10-ml glass vaccine vials (Wheaton). The aqueous lipids are divided equally among the vials. The flask is rinsed two times with 5 ml of water until all the lipids are removed from the flask. Each rinse is divided equally among the vaccine vials. Two- or three-prong lyophilization stoppers (Wheaton) are inserted into the vials. The stoppers are autoclaved and inserted on the vials with sterile forceps. As an added precaution, the operator should be wearing sterile gloves during this process. The number and size of the vaccine vials are not critical. However, we have found that splitting the liposomes into multiple vials for lyophilization, and subsequent antigen encapsulation and pooling of all the liposome vials, ensures enhanced reproducibility of antigen encapsulation compared with single large-vial antigen encapsulation. The size of the vaccine vial affects the time required for lyophilization. Smaller vials have less surface area, requiring increased lyophilization time. We typically do not fill the vaccine vial more than half way.

The stoppered vials are either transferred to a $-80°$ freezer or frozen directly in the lyophilizer (The VirTis Company, Gardiner, NY, Advantage Model with computer control and a lyophilization chamber). Freezing time should be approximately 4–6 h but depends on the volume of the sample in the vials. It is critical that the samples are frozen completely before the application of vacuum in the lyophilizer. We prefer to freeze in the $-80°$ freezer. This allows the stoppers to remain inserted during the freezing process. If the vials are frozen directly in the lyophilizer, the contents are exposed to air during the entire freezing process. After freezing, the vials are transferred to the lyophilizer chamber. The stoppers are pulled up so that the slot in the stopper is above the lip of the vial. The lyophilizer is programmed as indicated in Table III . After the samples are dried, the vials are stoppered while under vacuum. The vials can be used directly for incorporation of antigen, or crimp seals can be placed on the vials, after which the vials can be stored at $\leq -20°$. We have stored lyophilized lipids under these conditions for 3 y without detectable degradation of the lipids.

TABLE III
PROGRAM FOR LYOPHILIZATION OF AQUEOUS LIPIDS

Step	Temperature (°)	Time (h)	Vacuum pressure (mbar)
Refreeze	−60	2	0
1	−40	2	100
2	−20	10	100
3	0	2	100
4	10	6	100
Hold	10	Until removed	100

Encapsulation of Antigen

The vials are sprayed with 70% alcohol to sterilize the outer surface of the vial and placed in a biological safety cabinet. If the vials have been frozen, they are allowed to warm to room temperature. Antigen is diluted to the desired concentration in buffer and sterile-filtered through a 0.2-μm low-protein binding syringe filter (Millex GV, Millipore Corp.). The antigen solution is added to the lyophilized lipids. This can be done with a needle and syringe by puncturing the stopper, or the stopper can be removed and the antigen pipetted into the vial. Sufficient antigen should be added to make a final phospholipid concentration of 200 mM. For the example described here, each vial would receive 1 ml of antigen. The vials are shaken to wet the lyophilized lipid cake. If the stoppers were removed, new stoppers are inserted and the vials sealed. The vials are then incubated at 4°. We routinely incubate 2 1/2 days. The minimum amount of incubation time required for maximal encapsulation of antigen depends on a number of variables, including the buffer and properties of the antigen. Although we have not rigorously studied this issue, at least an overnight incubation is required. The encapsulation efficiency of antigens varies greatly among antigens (Table IV).

Almost any buffer can be used for encapsulation, but a number of points must be considered: (1) If phosphate buffers are used, the phosphate assay described later cannot be used to determine the phospholipid concentration. (2) pH less than 6.0 should be avoided. The acid catalyzes hydrolysis of the fatty acid on the *sn*-2 position of the phospholipid, and liposomes will degrade with time.[13] (3) pH greater than 8.5 may also induce saponification of the phospholipids.[13] (4) Physiological concentrations of

---

[13] M. Grit and D. J. A. Crommelin, *Chem. Phys. Lipids* **64**, 3 (1993).

TABLE IV
PERCENT ENCAPSULATION OF DIFFERENT ANTIGENS IN LIPOSOMES

Antigen	Encapsulation (%)
R32NS1 (recombinant protein from *Plasmodium falciparum* circumsporite protein fused with NS1[6])	21
NKPKDELDYENDIEKKICKMEKCS (synthetic peptide from positions 367–390 of *Plasmodium falciparum* circumsporite protein[2])	33
Kallikrein (porcine)	40
Prostate-specific antigen	50
Conalbumin	55
Ovalbumin	60
Bovine serum albumin	77
Ricin A subunit	100

sodium chloride should be used to avoid differences in ionic strength when the liposomes are to be injected. Differences in osmolality may lead to leakage of the antigen from the liposomes. We typically use 50 mM TRIS-HCl–150 mM sodium chloride, pH 7.4; 0.9% saline; or, in cases in which quantification of the phospholipid concentration by phosphate determination is not required, Dulbecco's phosphate-buffered saline without calcium and magnesium.

After incubation to encapsulate antigen, the vials are sprayed with 70% alcohol to sterilize the outer surface of the vial and placed in the biological safety cabinet. After warming to room temperature, the stoppers are removed, and the liposomes are pooled and the vials rinsed with buffer. In the example described here, the final volume can be measured with a pipet, adjusted to 10 ml with buffer, and vialed for use. These liposomes will contain both encapsulated and unencapsulated antigen. Alternatively, for removal of unencapsulated antigen, the contents of each vial can be transferred to a autoclaved 50-ml polycarbonate screw-capped centrifuge tube (Nalgene Company, Rochester, NY). One tube is used for each vial. The vials are rinsed with buffer, which is added to the tube. Buffer is added to each tube to a final volume of 45 ml. The tubes are capped and shaken. They are centrifuged at 25,000–30,000g for 30 min. We use a Sorvall RC-5B (Dupont Company, Wilmington, DE) with a SA600 rotor set at 15,000 rpm. The tubes are removed and the supernatants removed by aspiration. The pellets are soft and will float away if supernatant is decanted or poured. Five milliliters of buffer is added to each tube, and contents are pooled in one tube. The tubes are rinsed with buffer, and the rinse is added to the tubes with liposomes. Buffer is added to almost fill the tube. The tube is capped and centrifuged as described previously. The supernatant is removed, and the final volume is adjusted to 10 ml. The liposomes are transferred to new vaccine vials, stoppered with normal stoppers (not split), and crimped with aluminum seals. The vials are stored at 4° and are stable for more than a year. The liposomes should not be frozen. Freezing causes fractures of the liposomal bilayer.

Characterization of the Liposomes

Phospholipids

Total phospholipids can be measured by use of a phosphate assay.[14] This assay will also measure the phosphate associated with lipid A, but the amount of lipid A (4 μmol) present in the liposomes is insignificant compared with the amount of phospholipid (1 mmol). The assay is linear from 0.05–0.3 μmol phosphate. For the liposomes to be within this range, they are diluted 1:10 in water, and 10–30 μl can be assayed. The phosphate is cleaved from the phospholipid by heating in sulfuric acid and perchloric acid. This perchloric acid hydrolysis must be performed in a fume hood specifically designed for perchloric acid use. This hood is made of stainless steel and is equipped with a system that washes the vent ducts with water to prevent the accumulation of potentially explosive perchloric acid residue. After cleavage of the phosphate from the phospholipid, the free phosphate is quantified with Fiske-Subbarow reducing agent (Sigma-Aldrich, St. Louis, MO) as described.[14] The assay should verify that the liposomes are approximately 100 mM phosphate.

The stability of the phospholipids can be monitored by thin-layer chromatography (TLC).[15] The most easily detectable breakdown products of phospholipids are lysophospholipids. The liposomes are diluted in 1:35 in chloroform to give approximately 2 mg/ml phospholipid concentration. LK6 silica gel 60 plates (Whatman) are cleaned by first running in acetone. The plates are dried and stored under vacuum. Twenty-five microliters (50 μg phospholipid) of liposomes dissolved in chloroform is spotted on the plate. Standards consisting of DMPC, DMPG, lysoPC (Avanti Polar Lipids), and lysoPG (Avanti Polar Lipids) are dissolved in chloroform at concentrations of 2 mg/ml, and 25 μl (50 μg) is spotted on the plate. The plate is developed in chloroform:methanol:water (65:25:4, v/v) in a filter paper–lined tank. Phospholipids are visualized by spraying with

[14] E. Gerlach and B. Deuticke, *Biochemische Zeitschrift* **337**, 477 (1963).
[15] J. M. Muderhwa, N. M. Wassef, L. E. Spitler, and C. R. Alving, *Vaccine Res.* **5**, 1 (1996).

molybdenum blue spray reagent (Alltech Associates Inc., Deerfield, IL Cat. No. 18213). Phospholipids appear blue on a white background. The R_f values for DMPC, DMPG, lysoPC, and lysoPG are 0.35, 0.39, 0.15, and 0.23, respectively.

Cholesterol

Cholesterol is measured by reaction with iron chloride.[16] The assay is linear from 100–500 μg of cholesterol. Five to 15 μl of liposomes is assayed without dilution. The assay should verify that the cholesterol content of the liposomes is approximately 29 mg/ml. The stability of cholesterol can be monitored by TLC.[15] The most easily detectable breakdown product of cholesterol is 25-hydroxycholesterol.[17] The liposomes are diluted 1:19.5 in chloroform, and 25 μl ($\sim$50 μg) is spotted on an acetone-washed plate. Fifty micrograms of cholesterol and 25-hydroxycholesterol (Sigma-Aldrich) standards are spotted on the plate. The plate is developed in benzene:ethylacetate (3:2, v/v) in a filter paper–lined tank. Cholesterol is detected by spraying 50% sulfuric acid or a solution containing 50 mg of ferric chloride in 90 ml of water, 5 ml glacial acid, and 5 ml of concentrated sulfuric acid and heating at 120° for 10 min.[18] The R_f for cholesterol and 25-hydroxycholesterol is 0.52 and 0.33, respectively.

Encapsulated Protein

We have conducted extensive studies that indicate that immune responses are generated by encapsulated antigen.[1–6] Immunization of free antigen mixed with preformed liposomes does not induce antigen-specific immune responses, but, in some cases, unencapsulated antigen may have induced immunosuppression.[19] For these reasons, it is important to determine the amount of encapsulated antigen. If the liposomes are unwashed, an aliquot should be washed as described earlier before protein determination. Encapsulated protein is quantified by use of modified Lowry procedure.[11,20] One-half milliliter of chloroform is added to 50–250 μl of liposomes in a 13 × 100-mm glass test tube. The solution is vortexed vigorously and dried in a Speed Vac (SC100, Savant Industries Inc.,

[16] A. Zlatkis, B. Zak, and A. J. Boyle, *Lab. Clin. Med.* **41,** 486 (1953).
[17] J. E. van Lier and L. L. Smith, *J. Org. Chem.* **35,** 2627 (1970).
[18] R. R. Lowry, *J. Lipid Res.* **9,** 397 (1968).
[19] C. R. Alving, R. L. Richards, M. D. Hayre, W. T. Hockmeyer, and R. A. Wirtz, *in* "Immunological Adjuvants and Vaccines (G. Gregoriadis, A. C. Allison, and G. Poste, eds.) p. 123, Plenum Publishing Corp., New York, 1989.
[20] O. H. Lowry, N. J. Rosebrough, A. L. Farr, R. J. Randall, *J. Biol. Chem.* **193,** 265 (1951).

Farmingdale, NY). The dried liposomes are solubilized in 200 μl of 15% sodium deoxycholic acid (Calbiochem-Novabiochem Corp., San Diego, CA) in 0.9% saline. Highly purified sodium deoxycholic acid is required for this assay. Other sources may not have as high a purity and may require recrystalization in acetone before use. The samples are vortexed vigorously until the lipid is released from the tube and solubilized. The assay is performed as described by Lowry et al.[20] for the low-volume assay, using antigen for the standard curve. Before reading the absorbance of the liposome samples, the samples are centrifuged at 1875g (Sorvall RT6000, 3000 rpm in an H-1000 rotor) at room temperature for 10 min. The supernatant is removed and the absorbance determined. Empty liposomes can be run as controls. The encapsulation efficiency is calculated by dividing the total antigen encapsulated by the amount of antigen added. Data are expressed as percent encapsulated antigen (compared with the amount added). Encapsulated antigen can also be quantified by HPLC or amino acid analysis. The lipids can be extracted as described later for sodium dodecyl sulfate–polyacrylamide gel electrophoresis (SDS-PAGE) analysis.

Encapsulated antigen can be monitored by SDS-PAGE analysis. The lipids in the liposomes cause smearing of the gel and must be removed before running the gel. This is achieved by use of modified method of chloroform:methanol extraction described by Wessel and Flügge.[21] One-tenth milliliter of liposomes is placed in a 1.5-ml screw cap microfuge tube (Bio-Rad Laboratories, Hercules, CA). Methanol (0.4 ml) is added, and the tube is capped, vortexed, and centrifuged in an Eppendorf microfuge (Brinkman Instruments Inc., Westbury, NY) for 1 min at 8800 g. Chloroform (0.2 ml) is added, and the tube is capped, vortexed, and centrifuged as described previously. Water (0.3 ml) is added, and the tube is capped, vortexed, and centrifuged as described earlier. The upper phase is removed with a 1-ml syringe fitted with a 271/2-gauge needle. An additional 0.3 ml of methanol is added, and the tube is capped, vortexed, and centrifuged as earlier for 2 min. The supernatant is removed and discarded. The tube is inverted on a paper towel and drained for at least 2 min. After blotting, the tube is inverted and placed under a nitrogen stream to remove the remaining solvent. Water (15 μl) and sample solubilizer (15 μl) are added, and the sample is processed for SDS-PAGE[22] and/or Western blotting.[23]

[21] D. Wessel and U. I. Flügge, *Anal. Biochem.* **138,** 141 (1984).
[22] U. K. Laemmli, *Nature (London)* **227,** 680 (1970).
[23] H. Towbin, T. Staehelin, and J. Gordon, *Proc. Natl. Acad. Sci. USA* **76,** 4350 (1979).

Preparation of Oil-in-Water Liposomal Emulsions

Materials

Light mineral oil (Sigma-Aldrich or Spectrum Chemical Manf. Corp., Gardena, CA) is filter-sterilized through a 0.2-μm polyethersulfone filter (Nalgene Company) and filled into depyrogenated 2-ml vaccine vials (Wheaton). Sterilized oil is stored at room temperature in the dark. Sterile, nontoxic, nonpyrogenic, 3-ml plastic Luer-lock syringes are from Becton Dickinson & Co (Franklin Lakes, NJ). Three-way nylon stopcock connectors with two female and one male Luer-lock adapter are from Kontes Glass Company (Vineland NJ, Cat. No. 420163-4503). For laboratory experiments, they are immersed in 70% ethanol for 10 min and transferred to autoclaved aluminum foil in a biological safety cabinet. After air drying, the stopcocks can be wrapped in foil. Alternately, the stopcocks are placed in sealable pouches and irradiated with 2500 cGy. Sterile 21-gauge needles are from Becton Dickinson & Co.

Procedure

Two syringes are placed on the female connectors of the stopcock, and the 21-gauge needle is placed on the male connector of the stopcock (Fig. 2). The liposomes (1.0 ml) are drawn up into the first syringe after inserting the needle through the stopper on the liposome vial. Light mineral oil (0.1 ml) is drawn into the other syringe. No air spaces should be

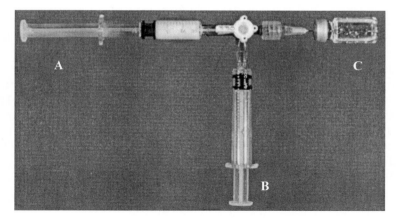

FIG. 2. Photograph of the setup for the emulsification of liposomes with light mineral oil. Syringe A contains the liposomes, and syringe B has drawn up the light mineral oil (C).

introduced into the syringes during the filling process. The needle is discarded. The switch key is turned to pass liquid between the two syringes. The syringes are pushed alternately to emulsify the samples. The pass rate should be 2 passes/s for 5 min. The emulsion is pushed to one syringe and the stopcock discarded. A needle is attached to the syringe, and the emulsion is injected. The preceding example is the procedure for making the liposomal emulsions used for our clinical trials in patients with prostate cancer.[10] The volumes of the liposomes and/or the oil can be modified.

During the emulsification process, a number of liposomes are broken, and the encapsulated antigen is released. The lipids stabilize the emulsion by coating the droplet of oil with lipids. The hydrophobic tail portions of the lipids are in the oil, and the hydrophobic head groups are on the surface of the oil interacting with the hydrophilic aqueous phase. However, a large number of liposomes remains intact (Fig. 3). Once injected, the emulsion serves as a depot for the slow release of the liposomes.

Emulsion Stability

Measurement of Emulsion Stability

The emulsion is placed in a 2-ml vaccine vial, the vial is stoppered, and the height is measured with a ruler from the bottom of the vial.[9] The emulsion is incubated at 4°, 25°, or 37°. At various intervals, the emulsions are

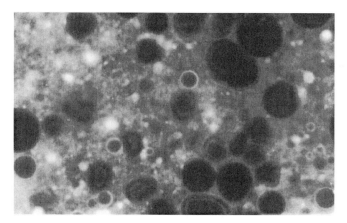

FIG. 3. Photomicrograph of a liposomal emulsion showing liposomes containing trace amounts of N-NBD-PE (bright areas) and mineral oil droplets (dark areas). The oil droplets are outlined by a fluorescent ring indicating that phospholipid from broken liposomes is coating the surface of the droplet to stabilize the emulsion.

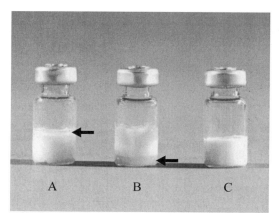

Fig. 4. Gross physical appearance of liposomal emulsions after storage. An emulsion containing 100 mM phospholipid liposomes and 10% light mineral oil has separated into a layer of light mineral oil on top (arrow) and liposomes on the bottom (A). An emulsion containing 125 mM phospholipid liposomes and 30% light mineral oil has separated into a layer of emulsion on top and a cream layer (arrow) on the bottom (B). An stable emulsion (for 3 y) containing 150 mM phospholipid and 40% light mineral oil (C).

removed from storage, and the height of the total sample and any separated phases are measured. There are two types of separations that can occur. The first is a clear water and oil separation, with the oil floating on top of the aqueous phase (Fig. 4A, arrow). The percent oil separated is calculated by the following formula:

$$\text{Height of oil (upper phase)} / \text{Total height of emulsion sample} \times F \quad (1)$$

where F stands for the initial fraction of oil in the emulsion (% oil divided by 100). The other type of separation that can occur is the separation of a lower creamed layer from the original emulsion (Fig. 4B, arrow). This may proceed eventually into the separation of an aqueous and oil layer. For these cream emulsion separations, the percent separation is calculated by the following formula:

$$100 \times \text{Height of the lower layer (cm)} / \text{Total height (cm)} \quad (2)$$

Factors Affecting Emulsion Stability

Two types of factors can affect emulsion stability. The first are those parameters that are purposely manipulated to change the stability of the emulsion, such as the phospholipid concentration of the liposomes and the amount of oil used to make the emulsion. The second type of parameter is other factors that are not inherent constituents of the emulsion, such as stopcock bore size, duration of the emulsification, rate of the passes during emulsification, and antigen.

Purposeful manipulations. We routinely manipulate the phospholipid concentration of the liposomes (Fig. 5) and amount of oil used in the emulsion (Fig. 6) to change the emulsion stability. The unstable emulsion that

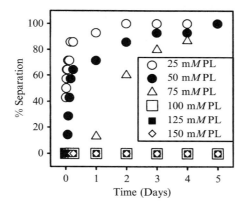

FIG. 5. Effect of liposomal phospholipid concentration on the stability of liposomal oil-in-water emulsions stored at room temperature. Liposomes containing various amounts of phospholipid were emulsified with 40% light mineral oil. The amount of separation was measured as a function of time.

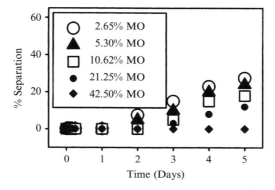

FIG. 6. Effect of different amounts of light mineral oil on emulsion stability. Liposomes containing 100 mM phospholipid were emulsified with the percent oil indicated and stored at room temperature. The amount of separation was measured as a function of time.

was used in our prostate cancer vaccine clinical trials[10] employed liposomes containing 100 mM phospholipid and 10% light mineral oil. It started to break after 8 h of incubation at 4° or 25°. We have made similar emulsions with liposomes containing 150 mM phospholipid and 40% light mineral that have been stable for 3 y (at the time of this writing) at both 4° and 25° (Fig. 4C). Emulsions that exhibit phase separation between these extremes can be made by altering the liposomal phospholipid concentration and percent light mineral oil (Fig. 5).

Other Factors Affecting Emulsion Stability. Other factors affect that emulsion stability have been investigated using the unstable emulsion used in the prostate cancer vaccine clinical trials. The emulsion contains liposomes with 100 mM phospholipid and 10% light mineral. Stopcock bore size has a dramatic effect on stability. We use stopcocks with a bore size of 0.1 cm. If a stopcock with a bore size of 0.22 cm is used, the emulsion stability is dramatically reduced. The emulsion starts to separate after 3 h with the 0.22-cm bore stopcock compared with 8 h with the 0.1-cm stopcock. If stopcocks with a bore size of 0.05 cm are used, the emulsions are stable for greater than 10 h but are still fully separated at 24 h. Although the use of a 0.05-cm stopcock produces more stable emulsions, it is difficult to push the syringes back and forth during the emulsion process. The 0.1-cm bore stopcock produces sufficiently stable emulsions with relative ease of manufacture.

Duration of emulsification and pass rate of the emulsification have a significant impact on the stability of the emulsion. At emulsification times of 3–5 min, the emulsion does not start to separate until 8 h. If the emulsification time is decreased to 0.5, 1, or 2 min, the time to start of separation is 0.5, 0.5 and 2 h, respectively. Similarly, if the pass rate is decreased from 2 passess/s to 0.5 or 1, the emulsion starts to break in 30 min. If the pass rate is increased to 4 passess/s, there is no significant change in emulsion stability. We also attempted to produce emulsions by vortexing but were unsuccessful. The emulsions made by vortexing break rapidly regardless of the concentration of the liposomes or the percent oil used in manufacture. It seems that vortexing does not supply sufficient shear force to make emulsions.

The antigen encapsulated within the liposomes also has an effect on the stability of the emulsion. In general, if the emulsion is made with liposomes lacking antigen, the resulting emulsion is more stable, requiring more time until it starts to separate. Interestingly, breakage of the emulsion to form a cream layer only appears when antigen is present. However, creaming only occurs with some antigens, and predominantly at higher liposome phospholipid and oil amounts.

[4] Synthetic Peptide–Based Highly Immunogenic Liposomal Constructs

By BENOÎT FRISCH, AUDREY ROTH, and FRANCIS SCHUBER

Introduction

Synthetic peptides corresponding to epitopes of proteins are potentially attractive in the design of vaccines.[1] They present many advantages (e.g., they are chemically defined and can be manufactured in rather large quantities in pure forms devoid of biological contaminants). The choice of the peptide epitope can also orient the immune response,[2] and multiepitope constructs can be envisaged that incorporate epitopes from the same and/ or different pathogens. Small peptides are, however, often poorly immunogenic, because they act as haptens that lack the necessary T-helper (Th) epitopes. For this reason these peptide haptens are generally coupled to carriers such as proteins (tetanus toxoid, ovalbumin, etc.) and administered in combination with adjuvants. Most of these approaches are not compatible with vaccine formulations, because the advantage of structurally well-defined peptides is largely offset by problems linked to the carriers such as carrier suppression,[3] to alterations of a residue essential to the antigenicity of the peptide during the conjugation step, or the scarcity of safe and efficient adjuvants.[4] An alternative to carrier proteins is the use of liposomes. These highly versatile phospholipid vesicles are characterized by a low toxicity and the absence of intrinsic immunogenicity. The same vesicles can carry antigens, either encapsulated or surface bound, and lipophilic immunoadjuvants, such as monophosphoryl lipid A (MPLA)[5] or lipopeptides, such as N^{α}-palmitoyl-S-(2,3-*bis*-palmitoyloxy-(2RS)-propyl)-(R)-cysteinyl-alanyl-glycine (Pam$_3$CAG),[6] associated with the bilayers. Moreover, their physicochemical properties such as size, pH sensitivity, and bilayer rigidity can be manipulated to influence the mode of antigen presentation. Liposomes have been used extensively to enhance the immunogenicity of poorly immunogenic proteins associated to them; this adjuvanticity has been ascribed to the tropism of the vesicles for antigen presenting cells

[1] T. Ben-Yedidia and R. Arnon, *Curr. Opin. Biotechnol.* **8,** 442 (1997).
[2] P. J. Delves, T. Lund, and I. M. Roitt, *Mol. Med. Today.* **3,** 55 (1997).
[3] L. A. Herzenberg and T. Tokuhisa, *Nature* **285,** 664 (1980).
[4] V. E. Schijns, *Curr. Opin. Immunol.* **12,** 456 (2000).
[5] E. Ribi, J. Cantrell, K. Takayama, N. Qureshi, and J. Peterson, *Rev. Infect. Dis.* **6,** 567 (1984).
[6] I. Fernandes, B. Frisch, S. Muller, and F. Schuber, *Mol. Immunol.* **34,** 569 (1997).

such as macrophages. In contrast, much fewer studies have been performed with liposomes as carriers of peptide antigens.[7]

In this chapter we give indications, based on our work, on the feasibility, using liposomes, to elicit an antibody response against small antigenic peptides. By use of a model peptide (IRGERA),[8] and varying several parameters of the vesicles (size, presence of MPLA) including the mode of association of the peptide (encapsulated versus surface-linked), an optimized chemically controlled preparation is defined (monoepitope construct) that yields an effective and relatively long-lasting immune response to the peptide and its parent protein (histone H3).[9] The application of this model system to a synthetic cyclic peptide (D-loop)[10] that mimics site A of influenza hemagglutinin was a successful test; it was found to impart protective immunity to OF1 mice challenged intranasally with influenza virus.[11] Finally, we also address the issue of the design of diepitope constructs (i.e., liposomes carrying at their surface both B and Th epitopes). The example given involves (1) a B epitope originating from a *Streptococcus mutans* cell surface adhesin;[12] this peptide was conjugated to a phospholipid anchor, and (2) a tetanus toxin–derived "universal" Th epitope[13] that was conjugated to the lipopeptide Pam$_3$CAG. This synthetic construct, when administered to mice, induces a highly intense, anamnestic, and long-lasting (mouse lifespan) immune response.[14]

Materials

Cholesterol (Chol) is recrystallized in methanol, egg yolk phosphatidylcholine (PC), phosphatidylglycerol (PG) (transesterified from egg PC), dicyclohexylcarbodiimide (DCC), *N*-hydroxysuccinimide (NHS), and bromoacetic acid *N*-hydroxysuccinimide ester (purchased from Sigma Chemical

[7] C. R. Alving, V. Koulchin, G. M. Glenn, and M. Rao, *Immunol. Rev.* **145,** 5 (1995).

[8] S. Muller, K. Himmelspach, and M. H. V. Van Regenmortel, *EMBO J.* **1,** 421 (1982).

[9] B. Frisch, S. Muller, J. P. Briand, M. H. V. Van Regenmortel, and F. Schuber, *Eur. J. Immunol.* **21,** 185 (1991).

[10] S. Muller, S. Plaue, J. P. Samama, M. Valette, J. P. Briand, and M. H. Van Regenmortel, *Vaccine* **8,** 308 (1990).

[11] M. Friede, S. Muller, J. P. Briand, S. Plaue, I. Fernandes, B. Frisch, F. Schuber, and M. H. V. Van Regenmortel, *Vaccine* **12,** 791 (1994).

[12] E. Lett, S. Gangloff, M. Zimmermann, D. Wachsmann, and J. P. Klein, *Infect. Immun.* **62,** 785 (1994).

[13] P. Panina-Bordignon, A. Tan, A. Termijtelen, S. Demotz, G. Corradin, and A. Lanzavecchia, *Eur. J. Immunol.* **19,** 2237 (1989).

[14] C. Boeckler, D. Dautel, P. Schelte, B. Frisch, D. Wachsmann, J. P. Klein, and F. Schuber, *Eur. J. Immunol.* **29,** 2297 (1999).

Co., St-Quentin Fallavier, France). Dipalmitoylphosphatidylethanolamine
(DPPE) was a gift from D3F (Doullens, France). Succinimidyl-4-(p-maleimi-
dophenyl)-butyrate (SMPB), maleimidobenzoyl succinimide (MBS), and
bisdiazotized benzidine (BDB) were obtained from Pierce Chemical Co.
(Rockford, MD). Monophosphoryl lipid A (MPLA, Table I) from *Salmonella
typhimurium* is obtained from Ribi Immunochem Research (Hamilton, MO).
Recombinant protein I/IIf is purified from pHB-*sr*-1 transformed *Escherichia
coli* cell extracts as described by Soell *et al.*[15] Alkaline phosphatase-goat anti-
mouse IgM, IgG, IgG1, IgG2a, IgG2b, and IgG3 are from Tebu (Le Perray-
en-Yvelines, France). Chicken erythrocyte histone H3 is isolated and purified
as described previously.[16]

Animals

Male and female BALB/c (H-2^d) mice (8–12 weeks old), noncongenic
3-week-old female OF1 mice, or randomly outbred male Swiss mice bred
in our own facilities, were obtained from IFFA Credo (L'Arbresle, France)
and used for immunization studies.

Synthetic Peptides (Table II)

CG-IRGERA (CG-IRG), CG-QYIKANSKFIGITEL (CG-QYI) and
C-TPEDPTDPTDPQDPSS (C-TPE) were obtained from Neosystem
(Strasbourg). D-loop is synthesized by Dr. Serge Plaué (Neosystem
Strasbourg). A cysteine residue is added to the *N*-terminus of the peptides
to permit their selective coupling to thiol-reactive functions. The purity of
the peptides, as assessed by high-performance liquid chromatography is at
least 80%.

Conjugation of Peptides to Carrier Proteins

The D-loop is coupled to ovalbumin (OVA) by means of a tyrosine residue by BDB as described previously.[10] Coupling of the other peptide (B epitopes) to bovine serum albumin (BSA) is obtained by reacting their cysteine
residue with maleimidobenzoyl succinimide (MBS)–derivatized BSA or
glutaraldehyde. The yield of coupling is obtained by the determination of
the amino acid composition of the final conjugate.[17]

---

[15] M. Soell, F. Holveck, M. Scholler, R. D. Wachsmann, and J. P. Klein, *Infect. Immun.* **62,**
1805 (1994).
[16] D. R. Van Der Westhuyzen and C. Von Holt, *FEBS Lett.* **14,** 333 (1971).
[17] J. P. Briand, S. Muller and M. H. V. Van Regenmortel, *J. Immunol. Methods* **78,** 59 (1985).

TABLE I
CHEMICAL STRUCTURE OF THE THIOL-REACTIVE ANCHORS AND ADJUVANTS

Structures	Abbreviations/roles
	DPPE-P$_3$-Mal Anchor
	DPPE-PB-Mal Anchor
	DPPE-Ac-Br Anchor
	MPLA Adjuvant
	Pam$_3$CAG-Mal Anchor/Adjuvant

TABLE II
CHEMICAL STRUCTURE OF THE PEPTIDES CITED

Peptides (1 letter code)	Peptides (3 letter code)	Abbreviations
CG-IRGERA	CysGlyIleArgGlyGluArgAla	CG-IRG
C-TPEDPTDPTDPQDPSS	CysThrProGluAspProThrAspPro ThrAspProGln AspProSerSer	C-TPE
CG-QYIKANSKFIGITEL	CysGlyGlnTyrIleLysAlaAsnSerLys PheIleGlyIle ThrGluLeu	CG-QYI
SKRGPGSGFDGGYC	*(cyclic peptide: Ser-Gly-Phe-Asp-...-Ser-Lys-Arg-Gly-Pro-Gly)* — Gly-Gly-Tyr-Cys-SH	D-loop

Synthesis of N-(4-(p-Maleimidophenyl)butyryl) Phosphatidylethanolamine (Fig. 1)

The lipid derivative is synthesized according to the method of Martin and Papahadjopoulos.[18] One hundred milligrams (155 μmol) of DPPE is dissolved in 1 ml of anhydrous CH_3OH and 10 ml of anhydrous $CHCl_3$ containing 20 mg of diisopropylethylamine (155 μmol) and 55 mg (155 μmol) of SMPB. The reaction is carried out overnight under argon at room temperature, and a near-quantitative conversion of DPPE to a faster running product is generally observed by thin-layer chromatography (TLC) analysis (Rf = 0.6, solvent: $CH_2Cl_2/CH_3OH/ H_2O$, 65/25/4). Solvents are removed under reduced pressure, and the products are redissolved in $CHCl_3$. This organic phase is then extracted twice with 1% NaCl to remove water-soluble by-products. After purification by silica gel chromatography (solvent: $CHCl_3/CH_3OH$; 40/1 to 10/1), DPPE-PB-Mal (Table I) is concentrated under reduced pressure and gives a colorless oil: yield, 60 mg (42%). Analysis by TLC (solvent: $CH_2Cl_2/CH_3OH/ H_2O$, 65/25/4) should indicate a single phosphate-positive and ninhydrine-negative spot.

Synthesis of N-(Bromoacetyl)phosphatidylethanolamine

N-(Bromoacetyl)phosphatidylethanolamine (Table I) is synthesized according the same procedure as DPPE-PB-Mal. Succinimidyl-4-(p-maleimidophenyl) butyrate is replaced by bromoacetic acid N-hydroxysuccinimide ester.

[18] F. J. Martin and D. Papahadjopoulos, *J. Biol. Chem.* **257**, 286 (1982).

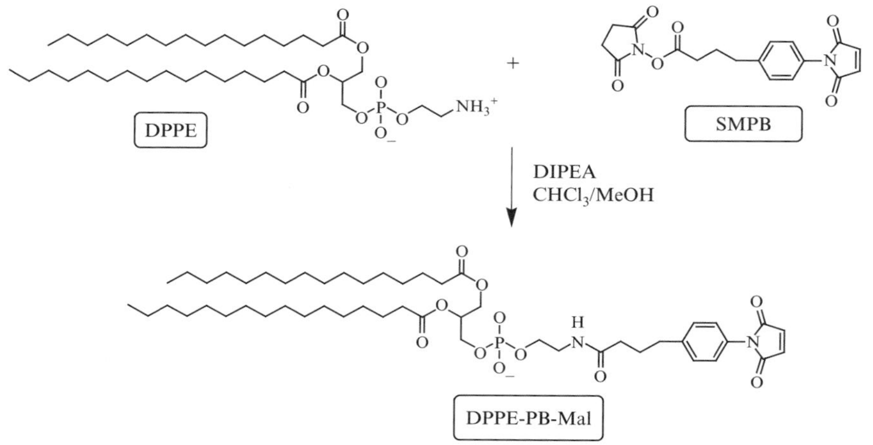

FIG. 1. Design and synthesis of the thiol-reactive functionalized phosphatidylethanolamine (DPPE-PB-Mal).

Synthesis of N-(Maleimido-ethoxy-ethoxy-acetamido)phosphatidylethanolamine

N-(Maleimido-ethoxy-ethoxy-acetamido)phosphatidylethanolamine (Table I) is synthesized according the same procedure as DPPE-MB-Mal.

Succinimidyl-4-(p-maleimidophenyl)butyrate is replaced by the mixture of 1 eq. {2-[2-(2,5-Dioxo-2,5-dihydro-pyrrol-1-yl)-ethoxy]-ethoxy}-acetic acid (Mal-P$_3$-COOH),[19] 1 eq. DCC, and 1 eq. NHS.

Synthesis of Thiol-Reactive Lipopeptide (Fig. 2)

Synthesis of Pam$_3$CAG-Mal requires first the production of the lipopeptide Pam$_3$CAG.[20] This is achieved by solid-phase synthesis on a Wang resin. N^{α}-palmitoyl-S-(2,3-bis-palmitoyloxy-(2RS)-propyl)-(R)-cysteine (Pam$_3$C-COOH), synthesized according to Wiesmuller et al.[20] or obtained commercially (Novabiochem, Laufelfingen, Switzerland), is coupled to the dipeptide AlaGly, previously prepared as a resin conjugate, according to conventional N-α-Fmoc amino acid method. Pam$_3$CAG is finally isolated, after cleavage from the solid phase by a simple treatment with trifluoroacetic

[19] B. Frisch, C. Boeckler, and F. Schuber, *Bioconj. Chem.* **7**, 180 (1996).
[20] K. H. Wiesmuller, W. Bessler, and G. Jung, *Hoppe Seylers Z. Physiol. Chem.* **364**, 593 (1983).

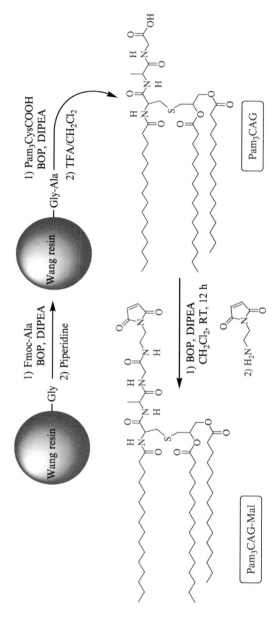

Fig. 2. Design and synthesis of thiol-reactive lipopeptide (Pam₃CAG-Mal).

acid (TFA/H_2O/anisol, 95/4/1), by precipitation with a mixture of CHCl$_3$/CH$_3$OH (1:5) at $-20°$ (total yield: 40%–60%). Compared with the liquid-phase synthesis, this strategy gives much better yields of pure compounds, mostly because it allows a straightforward removal of reagents and of Pam$_3$C-COOH, which is used in a three-fold excess in the coupling step. The last step consists in the conjugation of the maleimido group to the lipopeptide. This is accomplished by coupling to the terminal carboxylic group of Pam$_3$CAG to 1-(2-aminoethyl)-pyrrole-2,5-dione (a bifunctional molecule that contains a free amino group and a maleimide group) in the presence of 1.1 eq. benzotriazol-N-oxytrisdimethylaminophosphonium hexafluoro-phosphate and 1.1 eq. diisopropylethylamine.[21] The final product (Pam$_3$CAG-Mal) is obtained in about 70–80% yield.

Characteristics of the Liposomes

Liposomes are prepared from egg PC, PG (i.e., phospholipids with low phase transition temperatures), Chol, adjuvants, and modified lipids (Table I). A net negative charge is maintained to minimize the aggregation of the vesicles, and the proportion of Chol, which amounts to one third of the total lipid content, is selected to ensure good stability of the vesicles *in vivo*. Because of their amphiphilicity, Pam$_3$CAG-Mal, MPLA, and the other modified lipids (Table I) can be easily incorporated into the bilayers of the liposomes during their formation (the adjuvanticity of Pam$_3$CAG-Mal and MPLA is preserved when these compounds are incorporated into liposome bilayers).[6,22] The liposomes used are either large unilamellar vesicles (LUV) obtained by the reverse-phase evaporation technique (REV)[23] or small unilamellar vesicles (SUV) prepared by sonication/extrusion.

Small Unilamellar Vesicles

Lipids (PC, PG, and Chol at 65/25/50 molar ratio) in chloroform (6.5 μmol/ml) are mixed in a 1 × 10-cm glass tube; depending on the construct, the following compounds are added: MPLA (1 mol% of the phospholipids) and/or Pam$_3$CAG-Mal or DPPE-Ac-Br or DPPE-BP-Mal (10 mol% of the phospholipids) (i.e., the thiol-reactive functionalized lipopeptide or phosphatidylethanolamine). The solvents are then evaporated to dryness under high vacuum. SUV are prepared from a suspension of multilamellar vesicles obtained by rehydration, under mechanical stirring

[21] C. Boeckler, B. Frisch, and F. Schuber, *Bioorg. Med. Chem. Lett.* **8**, 2055 (1998).
[22] C. R. Alving and E. C. Richardson, *Rev. Infect. Dis.* **6**, 493 (1984).
[23] F. C. Szoka and D. Papahadjopoulos, *Proc. Natl. Acad. Sci. USA* **75**, 4194 (1978).

(2 min; Vortex), of the dried lipid film in 1 ml of 10 mM HEPES buffer (pH 7.4) containing 145 mM NaCl. The mixture is pulse-sonicated at 25° (5 s cycles interrupted for 1.25 s) with a 3-mm diameter probe sonicator (Vibra Cell, Sonics and Material Inc, Danbury, CT) at 300 W for 60 min under a continuous flow of nitrogen. SUV preparations are finally centrifuged at 10,000 g for 10 min to eliminate the probe fragments. When needed, the liposomes are extruded through 100-nm pore size polycarbonate filters (Nuclepore). Vesicles uniformly distributed in size are obtained, with a mean diameter of 100 ± 10 nm ($n = 3$) for the standard preparations, as determined by a photon correlation spectroscopy technique, using a Coulter (model N4SD) submicron particle analyzer (Coulter Electronics, Hialeah, FL). The liposomes are analyzed for phosphorous content.[24] Usually, the yield of vesicles is about 90% for phosphate content.

Large Unilamellar Vesicles

LUV are prepared by the REV technique according to Szoka and Papahadjopoulos.[23] Lipids (PC, PG, and Chol at 65/ 25/50 molar ratio) in chloroform (6.5 μmol/ ml) are mixed in a 1×10-cm glass tube with MPLA (1 mol% of the phospholipids) and DPPE-BP-Mal (10 mol% of the phospholipids) (i.e., the thiol-reactive functionalized phosphatidylethanolamine). The solvent is evaporated under vacuum, and the lipid film is dissolved in 1 ml diethylether. A 10 mM HEPES buffer (pH 6.5), containing 145 mM NaCl, is then added (0.33 ml), and the mixture is emulsified by sonication for 3 min in a bath-type sonicator (Laboratory Supply Co., Hicksville, NY). Diethylether is removed under partial vacuum (550 mm Hg) until a gel phase is obtained. The gel is repeatedly broken up by vortexing, and evaporation is resumed. Finally, the homogeneous suspension of liposomes is placed under full vacuum (water pump) to eliminate residual ether. Vesicles uniformly distributed in size are obtained with a mean diameter of 300 ± 120 nm ($n = 3$) for the standard preparations. Usually, the yield is about 90% for phosphate content.

Quantification of Thiol-Reactive Functions with Ellman's Reagent

Thiol functions are quantified according to Riddles et al.[25] An aqueous solution of HS-peptide (40 μl, 0–200 nmol) is added to 1 ml Ellman's reagent (5,5'-dithiobis{2-nitrobenzoic acid} 2 mM, EDTA 1 mM in 0.2 M sodium phosphate buffer, pH 7.27) and the absorbance is read at

---

[24] G. Rouser, J. Fleischer, and A. Yamamoto, *Lipids* **5**, 494 (1970).
[25] P. W. Riddles, R. L. Blakeley, and B. Zerner, *Anal. Biochem.* **94**, 75 (1979).

412.5 nm ($\varepsilon = 14, 150$ M^{-1} cm^{-1}). Calibration curves are established with 2-mercaptoethanol (0–200 nmol).

Quantification of Thiol-Reactive Functionalized Phosphatidylethanolamines and Lipopeptides Incorporated into Liposomes

Thiol-reactive functions are reacted with a two-fold molar excess of 2-mercaptoethanol, and the residual thiols are quantified with Ellman's reagent (see earlier). *In brief*, To a liposome suspension (700 μl, 140 nmol of surface-accessible thiol-reactive functions), maintained at 25°, is added an aqueous solution of 2-mercaptoethanol (30 μl, 280 nmol). After 60 min, min, the thiols are measured. Calibration curves are established with 2-mercaptoethanol (0–200 nmol) under the same experimental conditions but in the absence of liposomes.

Conjugation of the Peptides to Preformed Liposomes: Monoepitope Constructs (Fig. 3)

Conjugation of the peptide (D-loop, CG-IRG or C-TPE) is performed by reacting, at 4° for 12 h under argon, freshly prepared liposomes with an equimolar quantity of peptide with respect to the thiol-reactive anchors (i.e., approximately 2:1 excess peptide vs surface-accessible thiol-reactive functions) in a 10 mM HEPES buffer (pH 6.5) containing 145 mM NaCl when using the maleimide derivatives or in a 50 mM sodium borate buffer (pH 9.0) containing 145 mM NaCl when using the bromoacetyl derivatives. This step is followed by the addition of a 10-fold excess of 2-mercaptoethanol to derivatize the unreacted thiol-reactive groups. After 1 h of this treatment, the liposomes are dialyzed extensively against 10 mM HEPES buffer (pH 7.4) containing 145 mM NaCl to eliminate unconjugated peptides and excess reagents.

Diepitope Constructs (Fig. 4)

Small unilamellar vesicles are prepared as previously, from PC, PG, Chol, Pam$_3$CAG-Mal, and DPPE-Ac-Br at a molar ratio of 55/25/50/10/10 in a 10 mM HEPES buffer (pH 6.5) containing 145 mM NaCl. The subsequent procedures are all performed at 4°. Conjugation of CG-QYI is achieved by mixing freshly prepared liposomes with an equimolar quantity of peptide (with respect to Pam$_3$CAG-Mal) for 12 h under argon in the same buffer. The liposomes are then dialyzed extensively against 50 mM sodium borate buffer (pH 9.0) containing 145 mM NaCl to eliminate the

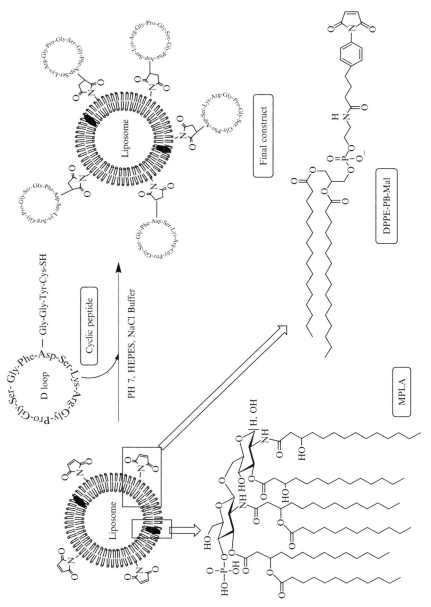

Fig. 3. Conjugation of the peptide D-loop to preformed liposomes: monoepitope construct.

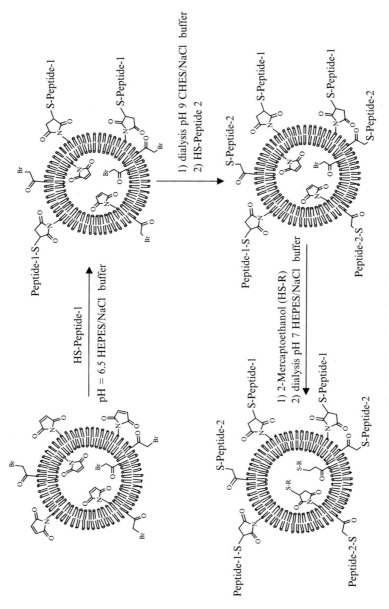

FIG. 4. Design of diepitope liposomal constructs.

unconjugated peptide. The coupling of C-TPE is obtained by addition of an equimolar quantity of peptide (with respect to DPPE-Ac-Br), at 4° for 12 h under argon at pH 9.0. This step is followed by an addition of a 10-fold excess of 2-mercaptoethanol. After 1 h of this treatment, the liposomes are dialyzed extensively against 10 mM HEPES buffer (pH 7.4) containing 145 mM NaCl. Alternate constructs are prepared according to the same method using liposomes prepared from PC, PG, Chol, DPPE-P$_3$-Mal, and DPPE-Ac-Br at 65/25/50/10/10 molar ratios.

Quantitation of the Peptides Associated to Liposomes

An aliquot of the liposomal preparation (50 μl) is mixed with an equal volume of 12 N HCl and hydrolyzed for 12 h at 110° in a sealed vial. The resulting amino acids are quantitated using a fluorometric assay with fluorescamine.[26] Ten microliters of the hydrolysis solution is added to 1.5 ml of 50 mM sodium borate buffer, pH 9.0, followed by the addition of 500 μl of a fluorescamine solution in dioxane (300 μg/ml). After mixing, the fluorescence is read at 475 nm (excitation: 390 nm). Calibration curves are established by hydrolysis of known amounts of the peptides, and the contribution of the liposomes to the total fluorescence is estimated by hydrolysis of vesicles containing 2-mercaptoethanol–derivatized thiol-reactive anchors. It is advised that control experiments in which liposomes lacking the thiol-reactive anchors—or containing thiol-reactive anchors that are reacted with an excess of 2-mercaptoethanol—are incubated with free peptides and worked up as previously; this yields information on the proportion of peptides spontaneously adsorbed to vesicles. In the examples given here, this is very low (<2% peptide bound).

Immunization for Diepitopes Construct

Mice (six animals per group) are injected intraperitoneally on days 1, 21, and 41 with 150 μl/injection of the appropriate liposome preparation, containing 1 μmol lipid and about 60 μg peptide/injection per animal. A booster immunization is given on day 151 with the same amount of antigen. Blood samples are harvested periodically from day 7 to day 150 and from day 155 to day 350 or 700 by retroorbital puncture. All the samples are tested individually for the presence of anti-peptide or anti-parent proteins (S. mutans adhesin (I/IIf) or protein H3 from which the peptides are derived) antibodies.

[26] P. Böhlen, S. Stein, W. Dairman, and S. Udenfriend, Arch. Biochem. Biophys. 155, 213 (1973).

Enzyme-Linked Immunosorbent Assay

The presence of antibodies (IgM, IgG), and their isotype distribution (IgG1, IgG2a, IgG2b and IgG3), against B epitopes is determined in sera by an indirect enzyme-linked immunosorbent assay (ELISA) procedure. Microtiter plates (Nunclon, Poly Labo, Strasbourg, France) are coated with 50 μl of solutions containing the peptides conjugated to BSA (1 μg/ml) or the parent proteins (1 μg/ml) in 0.1 M sodium carbonate buffer (pH 9.6). After incubation overnight at 4°, the plates are treated with phosphate-buffered saline (PBS) containing 0.05% Tween 20 (PBST) and 0.5% gelatin. Thirty microliters serial dilutions of each individual serum is added to the wells (1 h, 37°). After washing with PBST, antibody binding is detected with 50 μl (1/5000) of alkaline phosphatase-goat anti-mouse isotype antibodies (1 h, 37°). After incubation of the plates (1 h, 37°) with 50 μl of alkaline phosphatase substrate (p-nitrophenylphosphate) solution, the A_{405} is read with an spectrophotometer. Each assay is run in triplicate in parallel with preimmune sera. Antibody titers are expressed as the reciprocal of the highest serum dilution that gives an absorbance twice that of the preimmune control.

Protection Experiments

Noncongenic 3-week-old female OF1 mice (10 mice per group) are injected intraperitoneally three times, with 3-week intervals between injections, with 60 μg of peptide coupled either to proteins carriers (OVA) or SUV. In the case of peptide–protein conjugates, the first injections are performed in complete Freund's adjuvant and subsequent injections in incomplete Freund's adjuvant. The liposomes (1 μmol phospholipid containing 2 μg MPLA per injection) are injected in PBS. As a control, liposomes containing MPLA but no peptide or liposomes containing MPLA and an unrelated peptide (CG-IRG) are injected. Five days after the last injection, the mice are bled from the retroorbital plexus. The mice are then challenged intranasally with influenza virus in allantoic fluid using a dose predetermined to give 50%–60% mortality in a mouse lot immunized with saline only. The doses required are of the order of 60 μl of undiluted allantoic fluid (HA titer 5120) per mouse. Protection is calculated by the formula:

$$\text{Protection (\%)} = \frac{\left(\frac{N \text{ dead in } Cont}{\text{Animals in } Cont}\right) - \left(\frac{N \text{ dead in } Exp}{\text{Animals in } Exp}\right)}{(N \text{ dead in } Cont/\text{Animals in } Cont)} \times 100$$

where N = number, $Cont$ = control group, and Exp = experiment.

Peptides (Table II)

Two B epitopes are selected for the preparation of monoepitope liposomal constructs; they are synthesized with a thiol (Cys)-containing linker
either at their N- or C-terminus to permit their conjugation to preformed
vesicles containing thiol-reactive anchors (see later). The first one is
a model peptide IRGERA corresponding to the C-terminal residues
130–135 of histone H3.[8] To increase the accessibility of this peptide when
bound to carriers, two additional spacer residues are added to its amino-
terminus to produce the peptide CG-IRGERA (CG-IRG). The second B
epitope is a cyclic peptide (D-loop) corresponding to the region 139–147
(site A) of influenza virus hemagglutinin (strain X31), where Cys 139 in
the original sequence is replaced by a Ser residue; the peptide is cyclized
by formation of an amide bond between the N-terminus of Ser 139 and
the γ-carboxylate of an additional C-terminal Asp. A linker comprising
the sequence GGYC is added to the C-terminus of the cyclic peptide.
GG is necessary to allow the cyclization step in good yields, and the Tyr,
added at the penultimate position, is used for the conjugation of this peptide to carrier proteins using the BDB technique. The B epitope selected
for the diepitope constructs is the peptide TPEDPTDPTDPQDPSS
(TPE) comprising the residues 1495–1510 of S. *mutans* surface adhesin
(protein I/IIf) that mediates the attachment of this bacteria to the host
cells and to tooth surfaces. This synthetic peptide, which has been shown
previously to be poorly immunogenic and strictly helper-dependent,
becomes immunogenic only after being coupled to a carrier such as
OVA.[12] To overcome MHC-restriction, we have chosen the "universal"
Th epitope QYIKANSKFIGITEL (QYI), which belongs to the tetanus
toxin (residues 830–844) and which is characterized by its promiscuous recognition by Th cells.[13] For conjugation purposes, both peptides are also
functionalized at their N-terminus, (i.e., C-TPE and CG-QYI).

Design of Liposomal Monoepitope Constructs

The peptides are covalently linked to the surface of the vesicles. Several
techniques are available to that end;[27] we have adapted to small ligands the
method first described by Martin and Papahadjopoulos[18] for the conjugation
of antibodies to the surface of liposomes. It involves the reaction under mild
temperature and pH conditions between preformed liposomes (REV or
SUV) containing an electrophilic lipid derivative, such as a maleimido moiety
linked through a spacer-arm to DPPE (DPPE-PB-Mal or DPPE-P$_3$-Mal;

---

[27] T. D. Heath and F. J. Martin, *Chem. Phys. Lipids* **40,** 347 (1986).

see Fig. 1 and Table I), and a peptide carrying a thiol group (Table II). The conjugated peptide is thus anchored to the outer lipid bilayer of the liposomes by means of a thioether linkage and a relatively long spacer arm; it is thus expected to have enough degrees of freedom to be accessible to receptors present at the surface of immunocompetent cells. As assessed by amino acid analysis, the yield of conjugation of the peptides used to the maleimido groups exposed at the outer surface of the vesicles is essentially quantitative.

Important Parameters for the Immunogenicity of the
  Monoepitope Constructs

To define which parameters are of influence on the immunogenicity of the model peptide IRGERA, the relative importance of carriers (protein and liposomes) and of adjuvants was analyzed.[9] The results indicate that the peptide is able to trigger an antibody response only when associated to a protein carrier such as OVA, injected in the presence of an adjuvant such as Freund's adjuvant. In contrast to the situation prevailing with proteins associated to liposomes, no immunoadjuvant effect is detectable when IRGERA is associated (surface-conjugated or entrapped) to simple phospholipid vesicles.[9] These observations led us to associate the peptide to liposomes containing a chemically well-defined amphiphilic adjuvant, MPLA (Fig. 1). This compound has been described as a nontoxic, nonpyrogenic, and potent immunostimulant.[5] MPLA can be incorporated into bilayers of liposomes while retaining its adjuvanticity and B-cell mitogenicity.[22] The highest antibody titers are obtained when IRGERA is covalently bound to the outer surface of vesicles that contain MPLA. In contrast, when the peptide is encapsulated free in the same type of liposomes, no response is detected.

The size is an important parameter that greatly influences the fate of liposomes *in vivo*. Small vesicles have a longer circulation time than large vesicles of similar composition, and, importantly, whereas LUV can only be engulfed by phagocytic cells, SUV can also be endocytosed by cells such as B lymphocytes.[28] The antibody production obtained after injecting mice with LUV constructs (average diameter, 300 nm surface-conjugated peptide containing MPLA) gradually decreases and vanishes after the last injection (Fig. 5). A subsequent boost (day 97) with the same preparation is necessary to restimulate the immune response. Moreover, both IgM and IgG are produced with a persistence in time of high levels of IgM reacting

---

[28] P. Machy and L. D. Leserman, *Biochim. Biophys. Acta* **730**, 313 (1983).

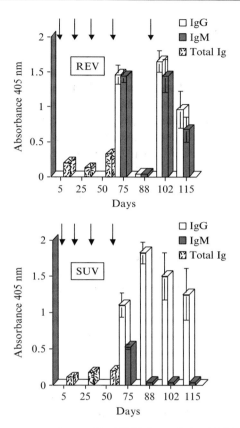

FIG. 5. Antibody response to the peptide CG-IRG coupled to REV and SUV containing MPLA. Arrows indicate the schedule of mice immunization. The antibody (total Ig, IgG, and IgM) levels are measured by ELISA using histone H3 as antigen and antisera diluted 1:1000. The histograms represent the average absorbance at 405 nm obtained in each group of mice ($n = 4$). The bars correspond to SD.

with H3 protein. The epitope density (ranging from 5–25 mol% of the phospholipid composition), the charge, and the presence of Chol do not greatly affect the immune response. In sharp contrast, when mice are given the peptide CG-IRG conjugated to the surface of SUV (average diameter, 90 nm) containing MPLA (Fig. 5), a strong antibody response is observed after the boost, which remains high without further injections for at least 50 days after the last injection. This pattern is typical of a T-dependent response with memory cell induction. The isotypes of the antibodies are essentially IgG (reacting with H3 protein), and the IgM, which are initially present, gradually disappear. It seems, therefore, that the formation of memory cells is strongly influenced by the size of the liposome bearing the antigenic peptide.

In Vivo Protection by Cyclic Peptides Coupled to Liposomes in a Murine Influenza Model

Antibodies raised against short linear peptides corresponding to the region 139–146 (site A) of influenza virus hemagglutinin (HA) do not react with the intact virus, nor do they confer protection against the virus,[29] suggesting that the free peptide does not mimic the three-dimensional structure of this antigenic site of HA. Muller et al.[10] have shown that cyclization of the 139–147 peptide by insertion of an amide bond between the N-terminus and an additional C-terminal Asp results in peptide (D-loop), which was shown by energy minimization and nuclear magnetic resonance studies to resemble very closely the native conformation of site A. It was also found that the D-loop in solution is recognized by anti-virus antibodies and that it inhibits the binding of these antibodies to NT 60/68 virus. This D-loop is able to confer an 83% protective immunity in an outbred OF1 mice model as measured by survival after viral challenge (influenza virus strain A/NT/60/68, which is a lethal strain for OF1 mice).[10] This conformationally restrained peptide, when properly derivatized with a Cys residue (Table II) and coupled to the surface of small liposomes (mean diameter, 100 nm) containing MPLA and a maleimido-derivatized phospholipid (according to the protocol optimized with IRGERA, see earlier), is also able to induce a 70% protective immunity against the virus in the same murine model (Table III). In contrast, much lower protection is observed with liposomes containing MPLA and carrying an unrelated peptide (CG-IRG) used as a control (17% protection). Because such a liposomal conjugate is synthetic and does not rely on carrier proteins or Freund's adjuvant, it represents a successful model for a synthetic influenza vaccine.[11]

Design of Liposomal Diepitope Constructs

To design diepitope constructs (i.e., liposomes carrying on their surface two distinct and structurally independent peptides, such as B and Th epitopes, each on a specific anchor), we have developed the conjugation strategy outlined in Fig. 6 . It is based on earlier observations from our group

[29] M. Shapira, M. Jibson, G. Muller, and R. Arnon, Proc. Natl. Acad. Sci. USA **81,** 2461 (1984).

TABLE III
PROTECTION OF IMMUNIZED MICE FROM INFLUENZA VIRUS INFECTION

Preparation used for immunization (adjuvant used)	N dead/N tested		Protection (%)
	Immunized	Control	
OVA-D loop (Freund's adjuvant)	1/10	6/10	83
SUV-D loop (MPLA)	4/20	13/20	70
SUV-CG-IRG (MPLA)	5/10	6/10	17

1 mice (10 per group) are immunized by intraperitoreal injection of peptides coupled to OVA, liposome, or saline (control). After three injections, the mice are challenged with the intranasally. Protection is calculated as described in "methods." When the experiments are repeated several times independently, the sum of the mice used is shown. Less than 10% variation is found among independent experiments in the same group.

indicating the existence of a marked difference in reactivity of maleimide and bromoacetyl functions toward thiol groups at neutral pH.[19,30] Compared with the maleimide moiety in a phospholipid derivative such as DPPE-P$_3$-Mal, which readily reacts at pH 6.5 with an HS-peptide, the bromoacetyl moiety in DPPE-Ac-Br reacts very sluggishly. Consequently, preformed liposomes that incorporate both thiol-reactive phospholipid derivatives react, at neutral pH, with an HS-peptide overwhelmingly by means of the maleimide moiety. A different HS-peptide can subsequently be conjugated to the same vesicle by means of the bromoacetyl function, but the conjugation must be performed at higher pH values (about 8.5–9.0) to observe appreciable rates. A complete analysis of the differences in reactivity of the maleimide and bromoacetyl groups has been reported elsewhere.[30] Several thiol-reactive lipophilic anchors have been synthesized (Table I) that differ by their thiol-reactive functions, (i.e., maleimide and bromoacetyl) and by the nature of their amphiphilic moieties (i.e., DPPE and Pam$_3$CAG). To adapt this latter amphiphilic immunoadjuvant (the properties of this molecule are discussed later) to our coupling strategy, we have synthesized its thiol-reactive derivative Pam$_3$CAG-Mal (Fig. 2).[21] Small unilamellar vesicles are prepared from egg PC, PG, and Chol (55/25/50 molar ratio) containing the different thiol-reactive anchors at given concentrations. The diepitope constructs are prepared by coupling separately the B (C-TPE) and Th (CG-QYI) epitopes to the surface of these preformed vesicles (Fig. 3). As assessed by amino acid analysis, the yields of conjugation of the peptide Th to maleimido groups and the

[30] P. Schelté, C. Boeckler, B. Frisch, and F. Schuber, *Bioconj. Chem.* **11,** 118 (2000).

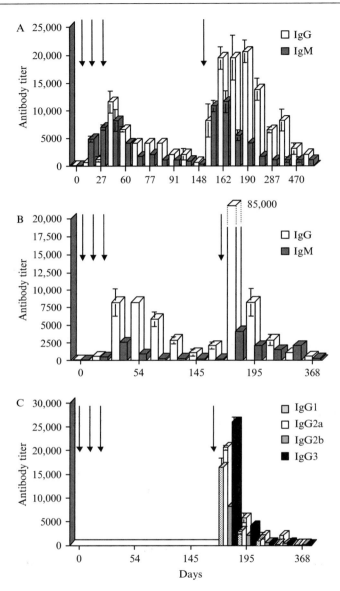

FIG. 6. Profiles of the antibody responses against peptide TPE in BALB/c (A) and outbred Swiss (B) mice immunized with diepitope constructs. Arrows indicate the schedule of mice immunization. Sera were tested individually at serial dilutions for the presence of specific antibodies: IgG and IgM. Sera (outbred Swiss mice) are tested individually for the presence of IgG1, IgG2a, IgG2b, and IgG3 (C). Antibody titers are expressed as the reciprocal of the highest serum dilution that give an absorbance twice that of the control. The data represent the means ± SD (error bars) of triplicate determinations from sera of six mice.

peptide B to bromoacetyl group, exposed at the outer-surface of the vesicles, are essentially quantitative.

Immunogenicity of Liposomal B and Th Diepitope Constructs

For this construct, liposomes are prepared containing Pam_3CAG-Mal and DPPE-Ac-Br. The Th epitope (CG-QYI) is coupled in the first step to Pam_3CAG-Mal, followed by the coupling of the B epitope (C-TPE) to DPPE-Ac-Br in the second step (Fig. 4). BALB/c (H-2^d haplotype) mice are immunized with this diepitope construct, which is found to be a very effective immunogen; a single injection elicits strong serum IgM and IgG responses (Fig. 6A) against the peptide TPE and the parent protein. Immunized mice, allowed to rest for 110 days after the last injection, are then boosted with a single dose containing the same amount of peptides. This results in higher titers of both IgM and IgG anti-TPE antibodies, which peak at about 12,000 and 20,000, respectively, and then decline very slowly over an 18-month period; thus even after 700 days, titers of about 1000–2000 are still measurable. The IgG subtypes determined in the same series of boosted animals consist of IgG1, IgG2a, IgG2b, and IgG3. The IgG1 response, which is found to be more pronounced than the IgG2a, IgG3, and IgG2b ones, clearly indicates that this type of construct allows observation of the occurrence of a Th-dependent response to the B epitope TPE of both Th1 and Th2 types. The antibodies obtained with the diepitope constructs when examined for their cross-reactivity with the native protein I/IIf indicate similar IgM and IgG patterns. To test the importance of the haplotype on such a response, outbred Swiss mice are also immunized with the same vesicles under identical conditions. Anti-TPE antibody titers are similar to those obtained in BALB/c mice (i.e., comparable high titers of IgM and IgG after a single injection, as well as an intense IgG response [titers about 85,000] after the boost injection are observed) (Fig. 6B). The subtypes IgG1, IgG2a, IgG2b, and IgG3 parallel those in the H-2^d–restricted mice (Fig. 6C). Taken together, these results provide further support for the utility of such diepitope constructs in inducing anamnestic responses in both H-2 and H-2^d–restricted mice.

Important Parameters in the Immunogenicity of Liposomal B and Th Diepitope Constructs

The preceding experiments indicate that liposome populations that carry B and Th epitopes on the same vesicles are effective in generating strong antibody responses against the B epitope, which are characterized by high titers and by a long duration. We have tested the importance of this response on different parameters: (1) the association of the two epitopes with a single vesicle, (2) the chemical nature of the B and Th epitope anchors. To test the first point, we have prepared a "split" version of the diepitope construct consisting of two populations of B and T monoepitope liposomes that were mixed in equal proportions and injected intraperitoneally into BALB/c mice. The first population consists of preformed vesicles to which TPE is coupled by means of DPPE-Ac-Br, whereas the second group contains liposomes conjugated to QYI by means of Pam$_3$CAG-Mal. This combination does not elicit an immune response even after multiple injections, exemplifying the need for both epitopes to encounter simultaneously the same cells.

So far we have conjugated the "universal" Th epitope QYI to Pam$_3$-CAG-Mal at the surface of the diepitope liposomes. This was done on purpose, because this type of anchor has distinct properties. The lipopeptides such as Pam$_3$CAG are adjuvants for peptide haptens that are coupled to them.[31] From the results obtained in our study on diepitope constructs, we can conclude that the lipopeptide/peptide coupling chemistry is not of paramount importance for the activity of the Th epitope. In our constructs, the peptide is coupled, by means of a thioether linkage, to a maleimide moiety carried by the C-terminus of the lipopeptide (Fig. 2), whereas in the conventional lipopeptide/peptide antigen conjugates, both moieties are linked by a peptide bond. The goal of coupling the Th epitope to a lipopeptide such as Pam$_3$CAG is to be able to deliver the QYI peptide to cell compartments that carry MHC class II molecules and to elicit a Th-dependent immune response. To investigate this point further, we have prepared another diepitope construct in which both the B and the Th epitope are coupled to DPPE-P$_3$-Mal, (i.e., a thiol-reactive phospholipid derivative that differs from Pam$_3$CAG-Mal by carrying a net negative charge). The magnitude of the immune response induced by such a diepitope construct without adjuvant is much less important; for example, after the boost, the titers are much lower, and the immunoglobulin isotypes show a prevalence for IgM. In sharp contrast to the results obtained with the first diepitope construct (with Pam$_3$CAG anchor), the IgG titers decline quite rapidly after the boost and among the IgG subtypes, IgG3 are the highest, pointing to a T-independent (TI-2) response. These results indicate that the nature of the anchor of the Th epitope is very important in these liposomal diepitope constructs and that, compared with a lipopeptide, a charged phospholipid is much less potent to induce a powerful immune response and to elicit a vigorous Th-cell response.

[31] W. G. Bessler, W. Baier, U. V. D. Esche, P. Hoffmann, L. Heinevetter, K. H. Wiesmuller, and G. Jung, *Behring Inst. Mitt.* **98,** 390 (1997).

Conclusions and Prospects

In this chapter we have described a simple liposome preparation that associates well-defined B and "universal" Th peptide epitopes, both covalently linked to the surface of the same vesicle by means of specific anchors.[32] These totally synthetic liposomal diepitope constructs, which mimic the context of an *in vivo* antigenic challenge, elicit humoral responses that are characterized by an immunological memory and by particularly intense and long-lasting T-dependent secondary responses.

An extension of this approach and a new challenge will be the development of anticancer therapeutic vaccines that trigger intense and cell-specific immune responses by generating a large number of tumor antigen-specific cytotoxic T lymphocytes (CTL) and the production of an appropriate cocktail of cytokines. This new generation of vaccines could make use of the liposome diepitope construct strategy in which the B epitope is replaced by an epitope-specific CTL (epitope Tc) response that mimics tumor-associated antigens. which is much more water-soluble and not plagued by such limitations.

Acknowledgments

The work described in this chapter has involved several coworkers whose names are given in the references; their contribution is gratefully acknowledged. We are also much indebted to the laboratories headed by Drs. M. Van Regenmortel, S. Muller, J.-P. Klein, and Professor D. Wachsmann, with whom we have collaborated fruitfully over the past years, and who were in charge of the immunochemical and the *in vivo* studies. We also thank the Centre National de la Recherche Scientifique and the Région Alsace for their financial support.

[32] We have experienced some difficulties in reproducing the conjugation of the CG-QYI peptide (Th epitope). Depending on the batch, this peptide was found to be marginally soluble in the buffers used, and the conjugation step, which gave in these particular cases poor yields, was also accompanied by some aggregation of the vesicles. To circumvent this problem, the peptide was dissolved in HEPES buffer (pH 6.5) in the presence of 5% (w/v) sorbitol and the conjugation performed in the same medium. More recently, we have also used another "universal" Th epitope PKY (PKYVKQNTLKLAT, HA 307–319).[33]

[33] D. O'Sullivan, T. Arrhenius, J. Sidney, M.-F. Del Guercio, M. Alberson, M. Wall, C. Oseroff, S. Southwood, S. M. Colon, F. C. A. Gaeta, and A. Sette, *J. Immunol.* **147,** 2663 (1991).

[5] Influenza Virosomes in Vaccine Development

By Anke Huckriede, Laura Bungener, Toos Daemen, and
Jan Wilschut

Introduction

Membrane-enveloped viruses, like influenza virus, carry spike glycoproteins on their surface that mediate binding of the virus to cellular receptors and fusion of the viral membrane with cellular target membranes. Ever
since the key role of the spike proteins in virus cell entry became apparent,
there has been considerable interest in reconstitution of these proteins in a
liposomal background. Clearly, reconstituted viral spike proteins represent
a useful model system for the study of biochemical and biophysical aspects
of virus–membrane interactions. Moreover, viral spike proteins are suitable
ligands for targeting of liposomes to cells. The first successful reconstitution
of viral spike proteins in a liposomal background was reported by Almeida
et al. in 1975.[1] With preformed liposomes and hemagglutinin (HA) and
neuraminidase (NA) purified from influenza virus, these authors succeeded
in generating membrane vesicles with spike proteins protruding from the
vesicle surface. Because these liposomal structures resembled native influenza virions when investigated by electron microscopy, they were called
"virosomes." Since then, reconstituted virus envelopes have been generated from various other viruses, including Sindbis virus, human immunodeficiency virus, Sendai virus, Epstein-Barr virus, and vesicular stomatitis
virus.[2-6] However, influenza virus has remained one of the viruses most
widely used for the generation of virosomes.

Influenza Virus

Influenza virus, well known as the causative agent of yearly flu epidem-
ics, is a membrane-enveloped RNA virus. The 10 influenza proteins are
encoded on 8 negative-strand RNA segments. Together with tightly bound

[1] J. D. Almeida, C. M. Brand, D. C. Edwards, and T. D. Heath, *Lancet* **2**, 899 (1975).
[2] R. K. Scheule, *Biochemistry* **25**, 4223 (1986).
[3] B. Cornet, M. Vandenbranden, J. Cogniaux, L. Giurgea, D. Dekegel, and J. Ruysschaert, *Biochem. Biophys. Res. Commun.* **167**, 222 (1990).
[4] S. Bagai and D. P. Sarkar, *Biochim. Biophys. Acta* **1152**, 15 (1993).
[5] S. Grimaldi, D. Pozzi, A. Lisi, N. Santoro, and G. Ravagnan, *J. Liposome Res.* **3**, 663 (1993).
[6] M. Paternostre, M. Viard, O. Meyer, M. Ghanam, M. Ollivon, and R. Blumenthal, *Biophys. J.* **72**, 1683 (1997).

nucleoprotein (NP) and viral polymerases (PA, PB1, PB2), the RNA segments form ribonucleoproteins. These are surrounded by the viral membrane the inside monolayer of which is covered with a layer of the matrix protein M1.[7] The viral membrane carries two major proteins, HA and NA, and a small number of copies of the M2 protein. Three HA monomers, each consisting of an HA1 and an HA2 polypeptide chain linked by a disulfide bridge, form a trimer in the membrane.[8,9] The major function of HA is binding of the virus to its cellular receptor, sialic acid, and mediation of fusion of the viral with the cellular membrane. Neuraminidase is inserted in the membrane as a tetramer.[10] Its major function is the cleavage of sialic acid residues from the cell membrane. Thereby, NA prevents binding of newly formed virus particles to the membrane of the virus-producing cell and enables the release of the new generation of virions.

Infection of a cell by influenza virus starts by interaction of the viral HA with sialic acid residues on the cell surface and subsequent internalization of the virus by receptor-mediated endocytosis (recently reviewed by Skehel & Wiley[11]). At the mildly acidic pH of the endosome, HA undergoes a conformational change.[12,13] A small hydrophobic peptide, buried earlier in the interior of the molecule, becomes exposed and mediates fusion of the viral membrane with the membrane of the endosome.[14] By this process, the viral genome is delivered into the cytosol, and virus replication can start.

Hemoglutenin not only plays a very important role in the infection process, it is also the most important antigen of influenza virus and target of strong antibody responses. Antibodies directed against HA can prevent virus binding to the cellular receptor and can thus provide neutralizing immunity.[15]

Virosomes

Influenza virosomes are membranous vesicles that carry the viral spike proteins, particularly HA, in their membrane. When properly reconstituted, the receptor-binding and membrane fusion–mediating properties of

---

[7] R. A. Lamb and R. M. Krug, in "Virology" (D. M. Knipe and P. M. Howley, eds.), p. 1487. Lippincott Williams & Wilkins, Philadelphia, 2001.

[8] D. C. Wiley and J. J. Skehel, J. Mol. Biol. 112, 343 (1977).

[9] I. A. Wilson, J. J. Skehel, and D. C. Wiley, Nature 289, 366 (1981).

[10] J. N. Varghese, W. G. Laver, and P. M. Colman, Nature 303, 35 (1983).

[11] J. J. Skehel and D. C. Wiley, Annu. Rev. Biochem. 69, 531 (2000).

[12] C. M. Carr and P. S. Kim, Cell 73, 823 (1993).

[13] P. A. Bullough, F. M. Hughson, J. J. Skehel, and D. C. Wiley, Nature 371, 37 (1994).

[14] K. S. Matlin, H. Reggio, A. Helenius, and K. Simons, J. Cell Biol. 91, 601 (1981).

[15] T. Bizebard, B. Gigant, P. Rigolet, B. Rasmussen, O. Diat, P. Bosecke, S. A. Wharton, J. J. Skehel, and M. Knossow, Nature 376, 92 (1995).

the viral HA are retained.[16] Virosomes are therefore very useful tools for the *in vitro* study of the role of HA in receptor binding and in the process of membrane fusion.[17,18]

When incubated with cultured cells, virosomes seem to bind to cellular sialic acid receptors, enter the cell by endocytosis, and fuse with the endosomal membrane just like native virus. Compounds encapsulated in the virosomal lumen therefore eventually end up in the cytosol. This was demonstrated conclusively using the membrane-impermeable protein synthesis inhibitors gelonin and the A chain of diphtheria toxin.[19,20] Although both compounds are ineffective when added to cultured cells in soluble form, they block completely protein synthesis when delivered encapsulated within virosomes. Cytosolic delivery of the toxins depends strictly on virosomal fusion activity: Virosomes that are fusion inactivated by preexposure to low pH are unable to deliver the toxins. Moreover, treatment of target cells with ammonium chloride, which prevents acidification of the endosome lumen, also inhibits delivery of the virosome-encapsulated inhibitors to the cytosol.

Interaction of Virosomes with the Immune System

Antibodies elicited against HA can neutralize the virus and prevent infection. Because virosomes present HA to the immune system in a natural conformation, their potential use as vaccines was recognized as soon as the virosome concept was conceived.[1] Early studies concentrated mainly on the induction of antibody responses against HA reconstituted in a membranous environment.[21,22] However, influenza virosomes have also been used to potentiate antibody responses against unrelated antigens bound to the virosomal surface.[23,24]

The virosomal capacity to deliver material to the cytosol of target cells offers the possibility to use virosomes for the induction of cytotoxic

[16] T. Stegmann, H. W. M. Morselt, F. Booy, J. F. L. van Breemen, G. Scherphof, and J. Wilschut, *EMBO J.* **6**, 2651 (1987).

[17] V. Chams, P. Bonnafous, and T. Stegmann, *FEBS Lett.* **448**, 28 (1999).

[18] S. Günther-Ausborn, P. Schoen, I. Bartoldus, J. Wilschut, and T. Stegmann, *J. Virol.* **74**, 2714 (2000).

[19] P. Schoen, R. Bron, and J. Wilschut, *J. Liposome Res.* **3**, 767 (1993).

[20] R. Bron, A. Ortiz, and J. Wilschut, *Biochemistry* **33**, 9110 (1994).

[21] N. El Guink, R. M. Kris, G. Goodman-Snitkoff, S. J. P. A., and R. J. Mannino, *Vaccine* **7**, 147 (1989).

[22] R. Glück, R. Mischler, B. Finkel, J. U. Que, B. Scarpa, and S. J. Cryz, Jr., *Lancet* **344**, 160 (1994).

[23] R. Glück, R. Mischler, S. Brantschen, M. Just, B. Althaus, and S. J. Cryz, Jr., *J. Clin. Invest.* **90**, 2491 (1992).

[24] B. Mengiardi, R. Berger, M. Just, and R. Glück, *Vaccine* **13**, 1306 (1995).

T-lymphocyte (CTL) responses. Activation of CTLs requires interaction of the T-cell receptor on CD8-positive lymphocytes with antigenic peptides presented by major histocompatibility (MHC) class I molecules on antigen-presenting cells (APCs). The peptides are derived from proteins that are present in the cytosol of the APC, usually either cellular proteins or proteins derived from intracellular pathogens like viruses. The proteins are degraded by proteasomes, and the produced peptides are transported into the endoplasmic reticulum. Here the peptides bind to MHC class I molecules that are transported subsequently to the plasma membrane.[25,26] Inactivated vaccines do not have access to the cytosol and thus cannot enter this route of antigen presentation. Consequently, they are incapable of activating CTLs. Virosomes have the capacity to actively transport material to the cytosol and can thus be used to deliver antigen to the MHC class I presentation route.

The various ways in which virosomes are likely to interact with the immune system, in particular with B cells and dendritic cells (DCs), are depicted schematically in Fig. 1. Hemogglutinin and NA spikes protruding from the virosomal membrane (as well as material bound to the surface of the virosomes) can be recognized by membrane-associated antibodies on B lymphocytes (1). The repetitive arrangement of the antigens on the virosomal surface probably enables cross-linking of surface antibody molecules on the B cells, which is known to be an exceptionally strong activation signal.[27] Virosomes can also bind to the cellular receptor for HA and sialic acid, thereby initiating receptor-mediated endocytosis (2). At the mildly acidic pH in the lumen of endosomes, the HA spikes undergo conformational changes and take on a fusion-active conformation. This triggers fusion of the virosomal membrane with that of the endosome, thus, the release of the virosome contents, for example encapsulated antigen, into the cytosol of the APC (3). The antigen can then be processed and presented by means of the MHC class I presentation route, resulting in CTL activation (3). Hemogglutinin remaining in the endosomal membrane after the fusion process but also virosomes that do not undergo fusion will be degraded in the endosome. This process will generate peptides (derived from HA or NA or from encapsulated proteins) that can eventually bind to MHC class II molecules. After transport to the plasma membrane, the peptide/MHC class II complexes are presented to CD4$^+$ T cells (4).

Thus, the structure and the composition of virosomes, as well as their capacity to enter cells, can be very useful in activating various immune

[25] J. W. Yewdell and J. R. Bennink, *Adv. Immunol.* **52**, 1 (1992).
[26] R. N. Germain and D. H. Margulies, *Annu. Rev. Immunol.* **11**, 403 (1993).
[27] M. F. Bachmann, R. M. Zinkernagel, and A. Oxenius, *J. Immunol.* **161**, 5791 (1998).

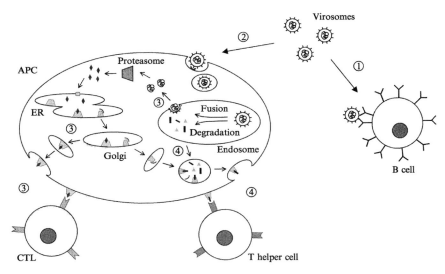

FIG. 1. Schematic depiction of the putative interaction of virosomes with the immune system. See text for further explanation.

mechanisms. Virosomes are therefore interesting devices to elicit humoral and cellular immune responses against the virus they have been derived from but also against unrelated antigens.

Preparation and Characterization of Influenza Virosomes

Background

Preparation of virosomes involves three steps: (1) solubilization of the viral membrane, (2) removal of the nucleocapsids and (3) reconstitution of the membrane proteins in a lipid environment. Solubilization of influenza virus membranes can be achieved with various detergents. However, the choice of the detergent is very important when the fusogenic properties of the reconstituted HA are to be retained (see later).[16,28] The nucleocapsids can be removed conveniently by ultracentrifugation. Reconstitution of the membrane proteins, particularly HA, is induced by extraction of the detergent from the solubilized membrane.[16,21] Some protocols involve addition of extra lipid before this step.[23] Alternately, solubilized viral

[28] R. Bron, A. Ortiz, J. Dijkstra, T. Stegmann, and J. Wilschut, *Meth. Enzymol.* **220,** 313 (1993).

membrane proteins can be incorporated in the membrane of preformed liposomes in a process called postinsertion.[1,29]

With respect to presentation of HA to the immune system in the context of influenza vaccines, any reconstitution of HA into a membranous environment is probably beneficial for vaccine efficacy. The cell-binding and fusion-mediating activity of HA, although possibly having additional positive effects, is not crucial for HA immunogenicity. However, when delivery of antigens to the cytosol and thus to the MHC class I antigen presentation route is envisaged, fusion activity of the virosomes, and thus retention of the functional properties of HA, is essential.

Our group was the first to develop a method for functional reconstitution of HA and to demonstrate fusion activity of virosomes thus generated.[16] Below, we will describe this method of preparation for virosomes, as well as for virosomes with encapsulated antigens. Biochemical, morphological, and functional characterization of these virosomes will also be presented. We will then give some examples for application of virosomes for vaccination purposes.

Procedure for Production of Fusion-Active Virosomes with or without Encapsulated Antigen

Virus, Reagents, Miscellaneous. Influenza virus, grown in the allantoic cavity of 10-day-old embryonated chicken eggs or in cell culture systems,[30] can be obtained from manufacturers of influenza vaccine or can be prepared in the laboratory as described.[28] The virus material has to be of high quality and has to be purified from excess chicken egg or serum proteins by sucrose density centrifugation. Protein content can be determined according to Peterson.[31] After quantitative lipid extraction, the phospholipid content can be measured as described by Böttcher *et al.*[32,33] The purified and characterized virus can be stored in small aliquots at −80°.

The detergent $C_{12}E_8$ (octaethyleneglycol mono(*n*-dodecyl)ether) has been found to be suitable for functional reconstitution of HA.[16] It has to be of the highest quality and can be purchased from Calbiochem (San Diego, CA) or other suppliers. The detergent is dissolved in HNE buffer (5m*M* HEPES, 150 m*M* NaCl, 1 m*M* EDTA, pH 7.4) at a concentration

[29] I. Babai, S. Samira, Y. Barenholz, Z. Zakay-Rones, and E. Kadar, *Vaccine* **17,** 1223 (1999).
[30] A. M. Palache, R. Brands, and G. J. van Scharrenburg, *J. Infect. Dis.* **176** Suppl. 1, S20 (1997).
[31] G. L. Peterson, *Anal. Biochem.* **83,** 346 (1977).
[32] E. G. Bligh and W. J. Dyer, *Can. J. Biochem. Biophysiol.* **37,** 911 (1959).
[33] C. J. F. Böttcher, C. M. van Gent, and C. Fiers, *Can. J. Med. Sci.* **24,** 203 (1961).

of 100 mM. Pyrene-labeled C_{10}-phosphatidylcholine (pyrene-PC), used for determination of the fusion activity of virosomes, can be obtained from Molecular Probes (Leiden, The Netherlands). BioBeads SM2 (BioRad, Richmond, CA), necessary for controlled detergent removal, have to be washed once in methanol and four times in HN buffer (5mM HEPES, 150 mM NaCl, pH 7.4) before use. They can be stored in HN supplemented with 0.05% sodium azide and have to be drained on filter paper before weighing. Sucrose solutions for gradient centrifugation are prepared in HNE on a weight per volume basis. Optiprep for gradient purification of antigen-loaded virosomes can be obtained from Axis-Shield PoC AS (Oslo, Norway).

Antigens to be encapsulated should be in solid or lyophilized form or should be dissolved at high concentration in HNE. The antigen of interest may be purchased, isolated from natural sources, purified from prokaryotic or eukaryotic expression systems, or, in the case of peptides, produced synthetically. In the examples described in the following, we used ovalbumin (OVA; grade VII, purchased from Sigma Chemicals Co, St Louis, MO) or peptides representing CTL epitopes of NP (produced synthetically). To allow determination of antigen recovery in the virosome preparation, it is convenient to include trace amounts of fluorescently or radioactively labeled antigen homologues. We used fluorescein isothiocyanate–labeled OVA (Molecular Probes) or iodinated peptide for this purpose.

Reconstitution and Encapsulation Procedure. Sucrose-gradient–purified influenza virus (approximately 1.5 μmol phospholipid, corresponding to about 5 mg viral protein) is diluted twofold in HNE and pelleted by ultracentrifugation (30 min, 50,000g, 4°). The pellet is resuspended in 700 μl 100 mM $C_{12}E_8$ in HNE and is left on ice for 1 h or overnight to allow for complete solubilization of the viral membrane. After careful homogenization using a syringe equipped with a 25-gauge needle, the suspension is centrifuged for 30 min at 85,000g, 4° to remove the viral nucleocapsids. Efficient solubilization of the virus material can be checked by determination of phospholipid and protein amounts in the supernatant; at least 90% of the phospholipids and about 30% of the protein (representing essentially all of the viral membrane proteins) should be recovered at this stage.

Empty virosomes can be prepared directly from the supernatant by detergent extraction (see later). To allow for determination of the fusion activity of the virosomes, phosphatidylcholine pyrene-(PC) can be incorporated in the virosomal membrane by adding the supernatant to a dry film of the probe (5–10 mol% with respect to the viral phospholipid) and mixing gently. If antigen is to be encapsulated in the virosome lumen, the desired amount is added to the solubilized viral membrane in solid or lyophilized form, or from a concentrated stock solution. We have successfully used NP peptide at 1 mg/ml final concentration and OVA at 3–100 mg/ml.[34,35] Addition of the antigen in diluted form should be avoided, because it would increase the total volume, and thus decrease the concentration of HA, which may interfere with successful HA reconstitution.

To induce virosome formation, a volume of 650 μl of the suspension containing the viral membrane lipids and proteins (and possibly added material such as pyrene-PC and/or antigen) is transferred to a microvial. Bio-Beads (275 mg wet weight), pretreated as described previously, are added, and the suspension is shaken at 1400 rpm for 60 min at room temperature in a Vibrax-VXR shaker (IKA Labortechnik, Staufen, Germany). Subsequently, an additional amount of 138 mg BioBeads are added, and shaking is continued for 10 min at 1800 rpm. By this latter treatment, the clear solution becomes turbid, indicating formation of vesicular structures.

The virosome suspension is purified by centrifugation on a discontinuous sucrose gradient consisting of 1 ml 40% (w/v) sucrose in HNE and 3 ml 10% (w/v) sucrose in HNE. Centrifugation is performed for 90 min at 130,000 g in a swing-out rotor. The virosomes appear as an opalescent band at the interface of the sucrose layers and are collected in a volume of about 0.5 ml. Empty virosomes are dialyzed directly against HNE at this point. Virosomes with encapsulated antigen are further purified by an Optiprep flotation gradient to separate loosely associated material from antigen encapsulated in the virosomes. For this purpose, the 60% Optiprep stock solution is mixed with the virosome suspension to give a final Optiprep concentration of 35%. This mixture is transferred to a centrifugation vial and topped with 3 ml 30% Optiprep in HNE and 0.5 ml HNE. On centrifugation (160,000 g, 2 h, 4°), the virosomes with encapsulated antigen will float to the top of the gradient, while the nonencapsulated antigen will remain in the Optiprep solution. The purified virosomes are harvested from the HNE layer and are dialyzed against HNE to remove any residual Optiprep solution. When used for immunization purposes, the virosomes are sterilized by filtration through a 0.45-μm filter. Centrifuge tube filters (Corning Incorporated; Corning, NY) can be used for this purpose to keep loss of material as small as possible.

Characterization of Virosome Properties In Vitro

Composition. Influenza virosomes prepared by the method described previously consist solely of the lipids (phospholipids and cholesterol at a molar ratio of about 2:1) and the major membrane proteins HA and NA

---

[34] A. Arkema, A. Huckriede, P. Schoen, J. Wilschut, and T. Daemen, *Vaccine* **18**, 1327 (2000).
[35] L. Bungener, K. Serre, L. Bijl, L. Leserman, J. Wilschut, T. Daemen, and P. Machy, *Vaccine* **20**, 2287 (2002).

of the virus. The ratio of protein to phospholipid in the viral membrane is approximately 1–1.4 μg protein/μmol lipid. Recovery of phospholipid in virosomes is about 20%–25%, and recovery of membrane protein is about 20%–30%, such that the protein/phospholipid ratio in the virosome preparation is virtually the same as that in the viral membrane. This indicates that the surface density of the spike proteins in virosomes is similar to that in the viral membrane. However, because the orientation of the proteins in the membrane is random (see later), the number of spike proteins protruding to the virosome exterior is usually lower than in the virus.

Analysis of influenza virosomes by sodium dodecyl sulfate–polyacrylamide gel electrophoresis (SDS-PAGE) shows that virosomes contain HA (split into HA1 and HA2 under reducing conditions) and NA (Fig. 2A, lane 2). Viral core proteins like NP and M1, the latter being the most prominent protein of influenza virions (Fig. 2A, lane 1), are absent from virosomes, indicating quantitative removal of nonmembranous material.

Encapsulation of substances into the virosome lumen is in principle a passive process. Consequently, encapsulation efficiency depends on the volume enclosed in the virosomes. On the basis of an average diameter of the virosomes of 100–200 nm, the volume trapped in the virosome lumen

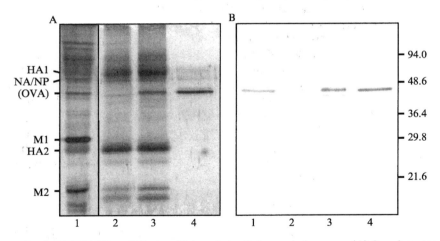

FIG. 2. SDS-PAGE and Western blot analysis of virus and virosomes. (A) Samples were separated by gel electrophoresis on a 12.5% acrylamide gel. Proteins were detected by silver staining. Lane 1, influenza virus (strain A/Panama/2007/99); lane 2, empty virosomes; lane 3, OVA-loaded virosomes; lane 4, OVA. (B) OVA was localized on a blot derived from a gel as in (A) with the help of an OVA-specific antibody and a secondary antibody conjugated with alkaline phosphatase. Note that small amounts of OVA are present in the virus sample because of growth of the virus in chicken eggs.

can be calculated to be about 3–5 μl/μmol of phospholipid.[36] When measuring recoveries of NP peptide (starting concentration, 1 mg/ml), we found encapsulation of 2.1 μg peptide/μmol phospholipid, a value close to the theoretically expected value of 3–5 μg. This amount corresponds to 0.5% of the initially used material. When encapsulating OVA (starting concentration, 3 mg/ml), recoveries were found to be higher (about 50 μg/μmol phospholipid or 3.6% of the initially added protein). Obviously, encapsulation of OVA is not purely a passive process; the protein accumulates in or on virosomes because of the interaction with membrane components. Encapsulated OVA can be detected easily on silver-stained SDS gels of virosome preparations (Fig. 2A, lane 3) and on Western blots (Fig. 2B, lane 3).

Morphology. When investigated by negative-stain electron microscopy, virosomes appear as spherical vesicles with a diameter of 100–200 nm (Fig. 3B). Spike proteins protruding from the virosome surface or into the lumen of the virosomes are clearly visible. The random orientation of the spikes indicates that virosomes have indeed been reconstituted from solubilized constituents. Except for spike orientation, the overall appearance of virosomes is similar to that of intact virus (Fig. 3A). The virosome preparation may also contain a minor fraction of small starlike structures that represent rosettes formed by aggregated HA molecules. Their presence indicates that reconstitution of membrane proteins is not necessarily always quantitative. Virosomes with encapsulated material, like OVA-containing virosomes depicted in Fig. 3C, show the same overall appearance as empty virosomes.

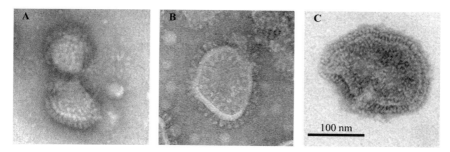

FIG. 3. Morphology. Purified influenza virus (A), empty virosomes (B), and OVA-loaded virosomes (C) were negatively stained by incubation with ammonium molybdate and evaluated by electron microscopy. Adapted from reference 35, with permission.

[36] J. Wilschut, *in*: "Methodologie des Liposomes/Liposome Methodology" (L. D. Leserman and J. Barbet, eds.), p. 9. Editions INSERM Vol. 107, 1982.

Fusion Activity. When virosomes are to be used for delivery of encapsulated foreign antigen into the cytosol of APCs, such that it can be processed and presented by means of the MHC class I route, fusion activity of the virosomes is important. For assessing fusion activity we use an assay based on the fluorescent probe pyrene-PC.[37] When excited at 343 nm, pyrene emits fluorescent light with a maximum of 377 nm for pyrene monomers and a maximum of 480 nm for pyrene excimers (excited dimers). The number of excimers, and thus excimer fluorescence, depends directly on the concentration of the probe in the membrane. On fusion of pyrene-labeled virosomes with an unlabeled target membrane, the pyrene-labeled lipids are diluted, and excimer density decreases. By monitoring the concomitant change in excimer fluorescence, the fusion process can be followed online.

Examples of fusion of empty and of OVA-loaded virosomes with erythrocyte ghosts, used as target membranes, are given in Fig. 4A. During a 60-s preincubation at neutral pH, the virosomes bind to sialic-acid receptors on the erythrocyte ghost surface. On acidification of the reaction buffer to a pH of 5.5 (the pH at which the viral HA assumes the fusion-active conformation), the fusion process starts immediately. Maximal fusion of 70%–75% is reached after about 120 s. Fusion of empty virosomes and virosomes loaded with a concentration of 3 mg/ml OVA follow

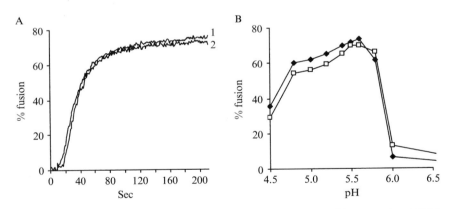

Fig. 4. Fusion activity. Empty virosomes (line 1) and virosomes with encapsulated OVA (line 2) were labeled during preparation with 10 mol% pyrene-PC. Virosomes were incubated with erythrocyte ghosts at neutral pH. Fusion was started by acidification of the incubation buffer to pH 5.5 and was followed online for 200 sec (A). Maximal fusion of empty (♦) and OVA-virosomes (□) was determined for various pH values demonstrating the strict pH dependence of fusion activity (B).

[37] T. Stegmann, P. Schoen, R. Bron, J. Wey, I. Bartoldus, A. Ortiz, J. L. Nieva, and J. Wilschut, *Biochemistry* **32**, 11330 (1993).

the same kinetics and reach approximately the same final extent (Fig. 4A). This indicates that proteins encapsulated in the virosome lumen or on the virosomal surface do not interfere with fusion activity. The conformational change of HA required for fusion is triggered by changes in pH. As can be deduced from Fig. 4B, optimal exposure of the fusion peptide takes place in a small pH range from pH 5.5–5.8 for virosomes derived from A/Panama virus used in this example. Again, virosomes loaded with 3 mg/ml OVA showed the same pH dependence of fusion as empty virosomes, further underlining that the presence of OVA does not interfere with the conformational change of HA. Exposure of virosomes to low pH in the absence of target membrances irreversibly inhibits the fusion mediating activity of HA (not shown). Virosomes thus treated can therefore be used conveniently to assess the relevance of the fusion process for virosome function.

Application of Influenza Virosomes for Immunization Purposes

Induction of Antibodies

Responses against Influenza Virus. The first application of influenza virosomes for the induction of immune responses against influenza virus was reported by EL Guink et al.[21] Using virosomes reconstituted by detergent extraction from octylglucoside-solubilized virus, these authors succeeded in inducing hemagglutination-inhibiting antibodies in mice. Antibodies of various subclasses were elicited by intramuscular (IgM, IgG) but also by intranasal (IgM, IgG, IgA) immunization. The induced immune response provided protection against virus challenge.

Virosomes have been investigated extensively as influenza vaccines in humans. In fact, a virosomal influenza vaccine has been commercially available for a number of years (Inflexal V, Berna Biotech, Berne, Switzerland). This vaccine is produced by mixing PC and phosphatidylethanolamine (PE) (3:1 w/w) with $C_{12}E_8$-solubilized viral membrane.[22,38–40] Because a large excess of extra lipid (10–100 × the amount of viral lipids) is added during generation of these virosomes, the surface density of the reconstituted spike proteins is much lower than that in the viral membrane. Nevertheless, intramuscular administration of these virosomes induced in humans similar or even higher hemagglutination-inhibition (HI) titers than conventional whole virus or subunit vaccines. Results were, however,

[38] P. Conne, L. Gauthey, P. Vernet, B. Althaus, J. U. Que, B. Finkel, R. Glück, and S. J. Cryz, Jr., *Vaccine* **15**, 1675 (1997).
[39] U. Glück, J.-O. Gebbers, and R. Glück, *J. Virol.* **73**, 7780 (1999).
[40] R. Glück, R. Mischler, P. Durrer, E. Fürer, A. B. Lanf, C. Herzog, and S. J. J. Cryz, *J. Infect. Dis.* **181**, 1129 (2000).

somewhat variable and dependent partly on the vaccine strain.[22,38] The virosomal vaccine has also been used for intranasal vaccination.[39,40] Induction of protective immune responses by way of this route required coadministration of a strong mucosal adjuvant (heat-labile enterotoxin of *Escherichia coli*) and two successive immunizations.

We have used virosomes prepared by the previously described $C_{12}E_8$/Biobead reconstitution method for the induction of anti-HA responses in mice. At all doses tested, virosomes injected intramuscularly induced higher antibody and HI titers as high or higher than a subunit vaccine derived from the same virus strain (A/Panama/2007/99). Although all antibody subclasses were higher after virosome immunization, IgG2a titers were particularly enhanced after virosome immunization. The resulting increased IgG2a/IgG1 ratio indicates that virosomes elicit a response that is skewed to a T helper 1-type reaction when compared with the response obtained with the subunit vaccine. Interestingly, influenza-specific T cell activity as measured by the extent of T cell proliferation as well as by the number of cytokine-releasing cells was substantially higher in virosome-immunized mice compared to subunit-immunized animals (A. Huckriede, unpublished observations).

Virosomes as a Carrier System for Unrelated Antigens. Influenza virosomes have also been used to increase antibody responses against noninfluenza antigens. Intact virions of inactivated hepatitis A virus (HAV) have been noncovalently coupled to the surface of virosomes prepared from solubilized viral membrane and added extra lipids as described previously.[23] Adsorption of HAV, presumably by means of electrostatic interaction with PE molecules included in the preparation, was strong enough to allow separation of virosome-associated HAV from unbound HAV by sucrose density centrifugation. The virosome-formulated HAV vaccine induced seroconversion in 100% of vaccinees with high and long-lasting HAV-specific antibody titers.[23,41] The vaccine is marketed in several countries under the name Epaxal Berna.

A number of small protein antigens have been coupled covalently to virosomes. For this purpose, PE was derivatized with a linker allowing covalent coupling of proteins by means of a disulfide bridge.[42,43] The antigens (diphtheria toxoid, tetanus toxoid, and the malaria peptide vaccine SPf66) were bound covalently to the lipid, and the lipid/protein constructs were added during virosome preparation. All three formulations induced rather poor antibody responses against the coupled antigens. However, responses

[41] L. Loutan, P. Bovier, B. Althaus, and R. Glück, *Lancet* **343,** 322 (1994).

[42] R. Zurbriggen and R. Glück, *Vaccine* **17,** 1301 (1999).

[43] F. Pöltl-Frank, R. Zurbriggen, A. Helg, F. Stuart, J. Robinson, R. Glück, and G. Pluschke, *Clin. Exp. Immunol.* **117,** 496 (1999).

could be increased substantially when mice were preimmunized with influenza vaccines before immunization with the modified virosomes. Possibly, influenza-specific antibodies target the derivatized virosomes to APCs, thereby facilitating antigen capture and presentation.

We have immunized mice with virosomes with encapsulated OVA prepared by the method described previously (Bungener *et al.*, manuscript in preparation). Two immunizations with OVA-containing virosomes (2.5 μg OVA, 50 nmol viral lipid) induced antibody responses against the viral HA but also against the encapsulated OVA. Anti-OVA responses were highest on intraperitoneal injection, whereas intramuscular application was less efficient, and subcutaneous application failed to induce any significant response. Because OVA present inside the lumen of the virosomes would probably not be accessible for recognition by B lymphocytes, these results are in line with OVA being present also at the virosomal surface as discussed previously.

Induction of Cytotoxic T-Lymphocyte Responses

Activation of CTLs requires processing of antigens in the cytosol and presentation of antigenic peptides by surface MHC class I molecules. Virosomes have the capacity to deliver material to the cytosol and are therefore suitable to channel antigen into the MHC class I processing route. The antigen has to be encapsulated in virosomes, and the virosomes have to be reconstituted such that the fusion activity of HA is retained.

Delivery of NP Peptides. Making use of the method described previously, we have prepared virosomes with encapsulated peptides derived from influenza NP. These peptides represent known CTL epitopes of influenza virus. The capacity of the virosomes to deliver the encapsulated peptides for MHC class I presentation was demonstrated in two ways. *In vitro*, we established that peptide-loaded virosomes can sensitize target cells for recognition by influenza-specific CTLs. For this purpose, P815 cells (H-2K^d) were incubated with different amounts of the virosome preparation for 1 h at 37°. The washed cells were labeled with ^{51}Cr and were exposed to influenza-specific CTLs generated from Balb/c mice (H-2K^d) that had been immunized with live influenza virus. At all concentrations tested, incubation with NP-containing virosomes was as effective as infection with influenza virus to sensitize P815 cells for lysis by influenza specific CTLs (measured as ^{51}Cr in the medium). Empty virosomes consistently failed to sensitize target cells, indicating that they do not contain residual NP.[34]

In vivo, we investigated whether NP peptide-containing virosomes can induce NP-specific CTL responses. Mice immunized intraperitoneally twice (with a 2-week interval) with the peptide-loaded virosomes developed strong CTL responses against the peptide.[34] An amount of 0.5 μg of virosome-encapsulated peptide was sufficient to give a similarly strong response as immunization with active influenza virus (Fig. 5). In contrast, immunization with 100 μg of free peptide did not result in CTL induction. Fusion activity of the virosomes was essential for activation of the immune response, because fusion-inactivated virosomes or peptide-loaded liposomes failed to elicit a CTL response.

Delivery of Protein Antigens. To explain the capacity of influenza virosomes to introduce unrelated intact protein antigens into the MHC class I processing and presentation route, we used virosomes with encapsulated OVA.[35] In contrast to NP peptides, OVA requires processing by cytosolic proteasomes before presentation of OVA-derived peptides on MHC class I molecules can occur. Cultured DCs were incubated with OVA-containing virosomes overnight, washed, and coincubated with $CD8^+$ T cells specific for the OVA peptide SIINFEKL (OT-1 cells). After 48 h of incubation, the medium was harvested, and the concentration of IL-2 was determined as a measure for activation of the T cells. Picomolar concentrations of virosome-encapsulated OVA were sufficient to induce T-cell activation (Fig. 6). Fusion activity was an essential prerequisite, because fusion-inactivated

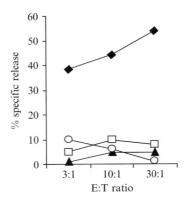

FIG. 5. Induction of CTL activity by vaccination with virosomes. Mice were immunized twice (with a 2-week interval) by intramuscular injection. Spleen cells were harvested 2 weeks after the second immunization and were stimulated by incubation with influenza-infected syngeneic spleen cells for 5 days. The bulk spleen cells (E, effector cells) were then incubated at different ratios with ^{51}Cr-labeled, influenza-infected P815 cells (T, target cells). Radioactivity in the supernatant after a 4-h incubation was taken as a measure for lytic activity of the effector cells. Immunization was performed with ($\blacklozenge$) virosomes (prepared from influenza strain A/X47) with 0.5-μg encapsulated NP peptide (TYQRTRALV), ($\square$) free peptide (100 μg), ($\blacktriangle$) fusion-inactivated virosomes with encapsulated peptide, or ($\bigcirc$) liposomes with encapsulated peptide, respectively. Adapted from reference 34, with permission

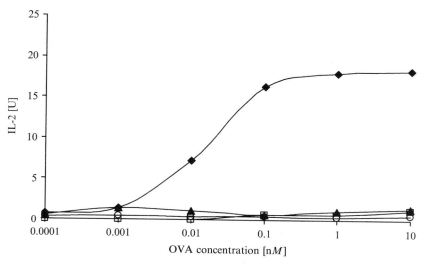

FIG. 6. Virosome-mediated delivery of OVA into the MHC class I presentation route. Bone marrow–derived dendritic cells (cultured for 7 days in the presence of GM-CSF) were incubated overnight with (♦) OVA-loaded virosomes (from influenza A/Johannesburg 33), (□) free OVA, (▲) fusion-inactivated OVA-virosomes, or (○) liposomes with encapsulated OVA targeted to the Fcγ receptor, respectively. After washing, the DCs were coincubated with OT-1 transgenic T cells. IL-2 produced during a 48-h period was determined using an IL-2 bioassay. Adapted from reference 35, with permission.

virosomes and liposomes (even when targeted to the Fc-gamma receptor on DCs) were incapable of delivering protein to the MHC class I presentation route at all concentrations tested.

Ovalbumin-loaded virosomes were also capable of delivering OVA for MHC class I presentation *in vivo* (Bungener *et al.*, manuscript in preparation). Again fusion-active virosomes were superior to fusion-inactivated virosomes and nonfusogenic liposomes in inducing CTL responses. These results underline the potential usefulness of virosomes for the induction of CTL responses against a variety of antigens.

Delivery of Antigen to the MHC Class II Presentation Route

Uptake of virosomes by APCs either by means of the sialic acid receptor (DCs) or interaction of surface immunoglobulins with the viral spike proteins (B cells) results in transfer of the particles to endosomes. Although many virosomes might fuse with the endosomal membrane at this stage, others will be degraded in the endosome, leading to the generation of peptides derived from the viral spike proteins and from encapsulated antigen. These peptides may eventually associate with MHC class II molecules. Presentation of the MHC class II/peptide complexes on the surface of the APC to CD4[+] T cells will induce strong T-helper responses that are essential for the support of antibody-forming B cells, as well as for the stimulation of CTLs.

The effect of virosomes on CD4[+] T cells *in vitro* was evaluated by incubating spleen cells from mice that had been immunized with live influenza virus with different amounts of influenza virosomes. The cells were stimulated efficiently, proliferated *in vitro*, and produced various cytokines, including interferon-gamma and IL-4 (A. Huckriede, unpublished observations). *In vivo*, subcutaneous injection of virosomes in influenza-immunized mice induced a strong delayed-type hypersensitivity reaction. These results indicate that the viral spike proteins present in virosomes are processed efficiently for presentation on MHC class II molecules, and thus for stimulation of CD4[+] T cells *in vitro* as well as *in vivo*.

To study *in vitro* presentations of an unrelated virosome-encapsulated antigen to T-helper cells, we used the OVA-loaded virosomes described previously.[35] DCs were incubated with OVA-carrying virosomes overnight, washed, and then coincubated with a CD4[+] T-cell hybridoma

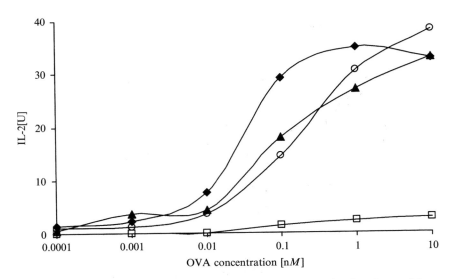

FIG. 7. MHC class II presentation of OVA. Dendritic cells incubated overnight as described previously were cocultured for 48 h with OT4H. 1D5 T-cell hybridomas. IL-2 in the medium was then determined using an IL-2 bioassay. Adapted from reference 35, with permission.

(OT4H.1D5) specific for an unidentified OVA epitope presented in the context of the MHC class II molecule I-A^b. Measurement of IL-2 in the cell supernatant after 48 h of coincubation indicated that virosome-encapsulated OVA is processed and presented to the T-helper cell hybridoma (Fig. 7). Because processing within the endosome is independent of fusion activity, fusion-inactivated virosomes and Fc receptor-targeted liposomes performed as well in this assay. Free OVA, on the other hand, was ineffective up to concentrations of 10 nM.

Summary and Conclusions

Influenza virosomes can be regarded as unilamellar liposomes carrying the spike proteins of influenza virus on their surface. Vaccination with influenza virosomes elicits high titers of influenza-specific antibodies, indicating that HA (and NA) reconstituted into a membranous environment exhibit strong immunogenicity. Moreover, virosomes can be used as presentation systems for unrelated antigens bound to the virosome surface. Because of their intrinsic adjuvant activity, virosomes support antibody formation and induction of T-helper cell responses against such surface-associated antigens. Provided that the fusogenic properties of the reconstituted HA are retained, virosomes can also be used to elicit cytotoxic T-cell responses against encapsulated antigens. Vaccines capable of activating the cellular branch of the immune response can be very important for protection against acute virus infections, especially for viruses with rapidly changing envelope glycoproteins like HIV and influenza virus. Moreover, virosomes can suit as powerful carriers in the development of prophylactic and immunotherapeutic strategies against cancer and premalignant disease.

The use of virosomes as commercial influenza vaccine and as commercial adjuvant for a hepatitis A vaccine demonstrates that production of virosomes on an industrial scale is feasible, both technically and economically. The industrial production procedure currently followed has not been designed to retain the functional properties of HA. In fact, several steps in the procedure are probably incompatible with retention of fusion activity. As mentioned previously the fusogenic properties of virosomes are important for CTL activation and might also play a role in the induction of T-helper cell and antibody responses. Therefore, a number of key adaptations in the virosome production protocol will be necessary. Thus improved, virosomes are very attractive devices for the development of highly efficacious vaccines against a range of antigens.

[6] Liposomal Cytokines and Liposomes Targeted to Costimulatory Molecules as Adjuvants for Human Immunodeficiency Virus Subunit Vaccines

By Bülent Özpolat and Lawrence B. Lachman

Introduction

The recent insights into the nature of antigen uptake by antigen-presenting cells (APC) and induction of immunoreactive B and T cells have led to novel approaches for vaccination of different diseases.[1,2] The use of adjuvants is critical for the development of successful subunit vaccines,[3] because most peptide antigens are either nonimmunogenic or weakly immuniogenic.[4,5] Adjuvants provide a mechanism for antigen persistence at the injection site and enhance the immune response to immunogens by prolonging their release and time of interaction with professional APC (i.e., macrophages and dendritic cells).[4] Cytokines have significant potential as adjuvants, because most adjuvants function by inducing production of cytokines from cells of the immune system.[6] Cytokines such as IL-1, IL-2, IL-6, IL-12, and IFN-γ have been used as adjuvants.[4,7] In addition to their carrier-adjuvant function, liposomes can be used for their depot effect to release antigens and other incorporated adjuvants, to increase antigen uptake by targeting antigens, and to generate strong immune responses.[8,9]

Identification of B7-1 (CD80) and B7-2 (CD86) as the prototypic costimulatory molecules has led to new strategies for immunotherapy and vaccination.[10–13] Full T-cell activation requires signals from the antigen-specific T-cell receptor (TCR) and the antigen-independent CD28

[1] T. R. Mosmann, *Ann. N. Y. Acad. Sci.* **664,** 89 (1992).
[2] J. Sprent, *Curr. Biol.* **5,** 1095 (1995).
[3] C. H. Allan, D. L. Mendrick, and J. S. Trier, *Gastroenterology* **104,** 698 (1993).
[4] A. C. Allison and N. E. Byars, *Biotechnology* **20,** 431 (1992).
[5] N. Bhardwaj, *Trends Mol. Med.* **7,** 388 (2001).
[6] A. W. Heath and J. H. Playfair, *AIDS Res. Hum. Retroviruses* **8,** 1401 (1992).
[7] L. B. Lachman, B. Ozpolat, X. M. Rao, *Eur. Cytokine Netw.* **7,** 693 (1996).
[8] G. Gregoriadis, *J. Drug Targ.* **2,** 351 (1994).
[9] N. C. Phillips and J. Dahman, *Immunol. Lett.* **45,** 149 (1995).
[10] J. A. Bluestone, *Immunity* **2,** 555 (1995).
[11] C. A. Chambers and J. P. Allison, *Curr. Opin. Cell Biol.* **11,** 203 (1999).
[12] S. E. Townsend and J. P. Allison, *Science* **259,** 368 (1993).
[13] B. Özpolat, X. M. Rao, M. F. Powell, and L. B. Lachman, *AIDS Res. Hum. Retroviruses* **14,** 409 (1998).

receptor. T-cell activation leads to the secretion of proinflammatory cytokines and effector T-cell functions. CTLA4, which is expressed on T cells, binds to B7-1 and B7-2 with an affinity greater than for CD28.[10,11] Cross-linking of CTLA4 with antibodies inhibits T-cell proliferation and IL-2 release, whereas simply blocking with CTLA4 Fab increases proliferation of T cells.[14,15] These findings suggest that manipulation of the TCR-costimulatory receptor network offers therapeutic opportunities for control of hyperesponsiveness, such as autoimmune diseases, and hyporesponsiveness as exhibited by a poor response to subunit vaccine antigens and tumor antigens.

Materials and Methods

Antigen, Antibodies, and Cytokines

Recombinant gp120 from HIV-1 (MN strain, CHO-derived) and recombinant murine IFN-γ (specific activity, 1.0×10^7 U/mg) are provided by Genentech, Inc. (South San Francisco, CA). Biotinylated monoclonal anti-mouse B7-1 (CD80), anti-mouse B7-2(GL1), anti-mouse CTLA4, anti-mouse CD28, anti-mouse CD11a, and control rat IgG$_{2a}$ are used to create immunoliposomes for vaccination purposes. Monoclonal antibodies to murine CD3(01082A) and FcγII/III are used as controls to demonstrate *in vitro* interactions between liposomes and target cells. All antibodies were purchased from Pharmingen, Inc. (San Diego, CA). Some antibodies are biotinylated using a biotinylation kit from Pierce Chemical Co. (Rockford, IL). Monoclonal antibodies were found to contain less than 12 pg of endotoxin per immunization dose.

Preparation of Liposomes Containing gp120 and Cytokines

The type of liposomes used are dehydration-rehydration vesicles (DRV) created from phosphatidylcholine (PC), cholesterol (CHO), and biotinylated phosphatidylethanolamine (PE-B) (Avanti Polar Lipids Inc., Pelham, AL) according to the procedure of Gregoriadis *et al.*[16] We prefer DRV to traditional liposomes prepared by adding the solute to a dried lipid film, because DRV entrap greater quantities of solute.[13] When calculating the amount of phospholipids required for an experiment, one extra dose is always included for syringe retention. Thus, we routinely prepare 0.6 ml of liposomes for a group of 5 mice (0.1 ml/mouse) or 1.1 ml for a group of 10

[14] P. S. Linsley, J. L. Greene, P. Tan, J. Bradshaw, J. A. Ledbetter, C. Anasetti, and N. K. Damle, *J. Exp. Med.* **176**, 1595 (1992).
[15] M. F. Krummel and J. P. Allison, *J. Exp. Med.* **182**, 459 (1995).
[16] G. Gregoriadis, D. Davis, and A. Davies, *Vaccine* **5**, 145 (1987).

mice. All lipids and used from stock solutions prepared in chloroform (CHCl$_3$). PC (500 nmole/mouse dose), CHO (500 nmole/mouse dose), and PE-B (100 nmole/mouse dose) at a 5:5:1 molar ratio are dissolved in anhydrous chloroform and placed in 25-ml sterile round-bottom screw-capped glass centrifuge tubes (Corex, Fisher Scientific, Pittsburgh, PA) (Table I). The tubes are rotated continuously at a 45° angle while being dried by a stream of N$_2$ gas. The residual chloroform is removed by vacuum desiccation for 1 h.

DRV are prepared by first creating standard liposomes from a dry lipid film and sequentially adding a small volume (50–100 μl) of distilled, autoclaved water until the total amount of water required has been added. The gp120 (15 μg/mouse), INF-γ (1.5 μg/mouse), and/or IL-6 (5U, R&D Systems, Minneapolis, MN) are added to the water suspension of liposomes with vigorous vortexing for 30 s. The liposome suspension is frozen in a dry ice–acetone bath ($-20°$) by rotating the tubes slowly at a 45° angle to create a thin lipid film. The tubes containing the frozen lipid film are lyophilized to dryness with the screw caps loosely open. The dried liposomes are hydrated with 100-μl aliquots of water added slowly with continuous vortexing. The lyophilization and hydration steps are repeated two more times to increase antigen trapping by the liposomes. Unincorporated antigen and adjuvant are removed from the final liposome preparation by centrifugation and washing three times with 10 ml of ice-cold phosphate-buffered saline (PBS) at 13,000g (10,000 rpm in a Sorvall RC-5B centrufuge with SS34 rotor) at 4° for 30 min. Following the third wash, the liposomes are resuspended in water (100 μl per mouse dose).

To create immunoliposomes, DRV prepared as described previously, are treated with 2.5 μg/per mouse dose of avidin (Pierce, Rockford, IL) dissolved in water and incubated at room temperature for 30 min with gentle mixing of the tubes every 10 min.[9,13] After washing two times with ice-cold PBS, the liposomes are incubated with biotinylated monoclonal antibodies. In our experiments, we use either monoclonal anti-mouse B7-1, anti-mouse B7-2, anti-mouse CTLA4, anti-mouse CD28 (2 μg/mouse), anti-mouse CD11a (all at 2 μg/mouse dose), or rat IgG$_{2a}$ (2 μg/mouse) as a control. Following

TABLE I
DRV FORMULATION

Lipid used in DRV	One mouse dose
Phosphotidylcholine	380 μg
Cholesterol	193 μg
Phosphatidylethanolamine (biotinylated)	10 μg

incubation for 1 h at 37°, immunoliposomes are washed twice with 10 ml of PBS at 10,000 rpm at 4° for 30 min to remove unbound antibodies. Immunoliposomes are resuspended in PBS (100 μl/mouse) and used immediately for immunization of animals.

All glassware is treated at 180° for 4 h to inactivate endotoxin. Sterile, pyrogen-free water and PBS are used to prepare all solutions, and liposomes are prepared in a sterile tissue culture cabinet.

Entrapment Capacity of Dehydration–Rehydration Vesicles

We previously performed a series of experiments to determine the percentage of the antigen and cytokines trapped by DRV-type liposomes using radiolabeled gp120 and cytokines.[17] The liposomes are prepared in the standard manner from PC and phosphatidylserine (PS) (PC/PS, 1:1), or PC and cholesterol (PC/CHO, 7:3). As shown in Table II, PC/CHO trap larger amounts of cytokines, although both types of liposomes entrap about the same amount of gp120. The trap ratios probably differ because of differences in charge and the ratio of hydrophobic amino acids in the proteins.

Slow Release of Trapped Proteins by Dehydration–Rehydration Vesicles

The liposomes containing radiolabeled cytokines and gp120 are kept in RPMI-1640 medium containing 5% fetal bovine serum (FBS) at 37° for up to 3 days to determine the level of retention of cytokines and antigen in

TABLE II
ENTRAPMENT OF CYTOKINES AND GP120 IN DRV

Protein	Percent entrapment[a]	
	PC/CH	PC/PS
IL-1α	13.5 ± 0.4	5.2 ± 1.1
IL-1β	24.7 ± 2.6	20.2 ± 0.2
IL-6	34.5 ± 3.3	17.5 ± 2.1
TNFα	24.5 ± 2.8	7.2 ± 1.1
INFδ	22.7 ± 3.8	14.5 ± 0.3
gp120	16.9 ± 0.2	16.3 ± 1.6

[a] Trapping ratio is the presence of added radiolabeled recombinant cytokine or gp120 associated with the DRV following three washes in PBS.

[17] L. B. Lachman, L. C. Shih, X. M. Rao, X. Hu, C. D. Bucana, S. E. Ullrich, and J. L. Cleland, *AIDS Res. Hum. Retroviruses* **11,** 921 (1995).

DRV. The percentage of radiolabeled cytokine associated with the DRV is in the range of 87%–94% in PC/CHO liposomes after 3 days incubation in the media.[17] The percentage of radiolabeled gp120 associated with DRV is about 80% for the same period of time, suggesting that liposomes are able to release the proteins slowly to the environment.

In Vitro *Binding Capability of Immunoliposomes to Target Cells*

To demonstrate that immunoliposomes were able to bind target cells expressing costimulatory molecules, we performed analysis by flow cytometry. Flourescent immunoliposomes are prepared by incorporating NBD (7-nitro-2-1,3 benzoxadiazol-4-yl)-PC at 0.1% of the total PC content. Immunoliposomes containing MAb to B7-1 are able to bind K-1735 cells transfected with B7-1 but do not bind to parental K-1735 cells after 2 h incubation at 4° (Fig. 1). Immunoliposomes containing either MAb to CD3 or no MAb do not bind the cells, suggesting that the binding of liposomes to the cells is mediated by specific MAb to B7-1. Similarly, immunoliposomes containing MAb to B7-2 are incubated with the murine macrophage cell line, J774, which express B7-2. As shown in Fig. 1, immunoliposomes containing MAb to B7-2 bind to J744 cells, whereas those containing MAb to CD3 or no MAb do not bind. The presence of cell surface expression of B7-1 and B7-2 molecules is confirmed using soluble MAb to both molecules.

Immunization of Mice with Soluble gp120 and Immunoliposomes Containing Cytokines and gp120

Groups of five mice are immunized subcutaneously 3 times at 14-day intervals. The immunoliposomes are injected in 0.1 ml of PBS. Each injection contains 380 μg PC, 193 μg CHO, 10 μg of PE-B, 2.0 μg MAb, and 10 μg gp120, as previously determined from trapping experiments. Blood samples are collected 2 weeks after the last vaccine treatment on day 42. Blood samples taken from the tail vein are kept at 25° for 30 mins and centrifuged for 10 minute at 5000 rpm using an Eppendorf microcentrufuge. The supernatant sera are separated into aliquots and frozen for future assays. Sera are tested for gp120-specific IgG, IgG$_1$, and IgG$_{2a}$ by enzyme-linked immunosorbent assay (ELISA). For ELISA, flat-bottom 96-well plates are coated with 20 μg recombinant gp120. Plates are blocked with 100 μl of 1% bovine serum albumin (BSA) plus 1% polyvinylpyrrolidone (PVP-40, Sigma, St. Louis, MO) and are incubated at 25°, for 30 min. After three washes with PBS-Tween (6 mM sodium phosphate, pH 7.2, 0.15 M NaCl, 0.05% Tween 20, Sigma), serum samples are added to the plates. The blocking buffer is used as a diluent for antibodies and

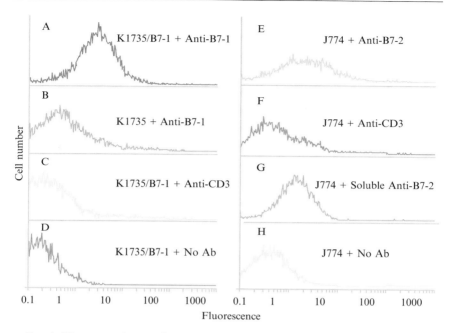

Fig. 1. Fluorescent immunoliposomes containing mAb to B7-1 bound to K1735 cells transfected with B7-1 and immunoliposomes containing mAb to B7-2 bound to J774 cells. Flow cytometric analysis of NBD-containing immunoliposomes prepared with anti-B7-1 demonstrate tight binding to K1735 cells transfected with B7-1 (A) but not to *neo*-transfected control K1735 cells (B). Immunoliposomes containing anti-CD3 (C) or lacking mAb (D) do not bind to B7-1–transfected K1735 cells. Immunoliposomes containing mAb to B7-2 bind to J774 cells (E), whereas immunoliposomes containing mAb to CD3 do not. Soluble anti-B7-2 demonstrates the presence of B7-2 on the surface of the J774 cells (G) and immunoliposomes lacking a mAb do not bind to J774 cells (H).

conjugates for all subsequent steps. To measure gp120-specific IgG response, the serum is tested at a 1:50 and 1:100 dilution. For IgG, diluted samples are loaded in wells as duplicates and incubated for 2 h at room temperature. After washing with buffer, 100 μl of a 1:2,000 dilution of anti-mouse IgG conjugated with horseradish peroxidase (HRP) (Sigma), 1:1,000 dilution of anti-mouse IgG$_1$ (Serotech, Oxford, England), or IgG$_{2a}$ (Pharmingen, San Diego, CA) conjugated with HRP is added to wells and incubated for 30 min. Plates are washed five times with wash buffer and once with only PBS, and the color is developed by adding 100 μl of the substrate solution *o*-phenylenediamine dihydrocholride (Sigma) in phosphate citrate buffer. The reaction is stopped at 30 min by adding 3 M sulfuric acid.

Optical density is measured in an ELISA plate reader (Dynatech MR5000) at 450 nm.

Delayed-Type Hypersensitivity and Humoral Immune Response in Mice after Vaccination with Liposomes and Immunoliposomes Containing Cytokines and gp 120

To determine whether immunoliposomes induce an immune response to gp 120 compared with liposomes lacking MAb, delayed-type hypersensitivity (DTH) responses are determined 2 weeks after the last immunization. DTH is measured as footpad swelling 24 h after injection of 1 μg of rgp 120 into both hind feet of mice. As shown in Fig. 2, mice vaccinated

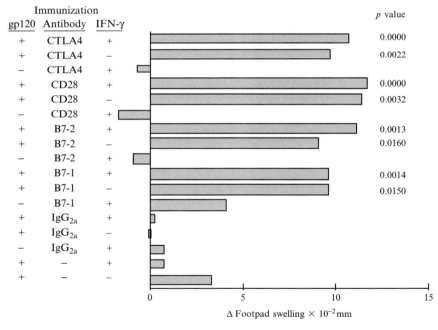

FIG. 2. Mean DTH responses of groups immunized three times with immunoliposomes containing rgp120 and mAbs to either CTLA4, CD28, B7-2, or B7-1 and in the presence or absence of IFN-γ. Fourteen days following the final immunization, the hind footpads are injected with 1 μg of soluble rgp120, and the amount of swelling is measured 24 h later. Control mice are immunized with immunoliposomes containing rat IgG$_{2a}$ or no mAb. The p values are determined by comparing the experimental group with the control group containing a mAb but lacking rgp120. Experimental groups containing IFN-γ are compared with the control group containing IFN-γ but lacking an mAb.

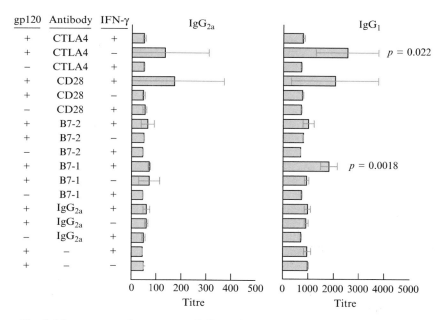

Fig. 3. Measurement of mean serum IgG_{2a} and IgG_1 Ab levels to gp120 in the serum of immunized mice. The end-point titer for each animal is determined by ELISA as described in the text.

with immunoliposomes containing MAb to B7-1, B7-2, CD28, and CTLA4 have significantly increased foodpad swelling 24 h after injection of the antigen. However, immunoliposomes containing control rat Ig_{2a} antibodies or liposomes lacking MAb do not result in significant footpad swelling. Similarly, immunoliposomes that do not contain gp 120 are also not able to produce significant footpad swelling. Interestingly, we do not find any additional effect in the increase of footpad swelling when IFN-γ is included in the liposomes. Our findings demonstrate clearly that the strategy of immunization by using liposomes targetted to costimulatory molecules induces a DTH response, and immunization with immunoliposomes does not induce statistically significant IgG_1 and IgG_{2a} titers in the mice (Fig. 3).

Cytokine Production by Lymph Node Cells from Mice Vaccinated with Immunoliposomes Containing Cytokines and gp 120

To determine whether a T-helper lymphocyte 1 (Th1) vs Th2 response to antigen results from vaccination with immunoliposomes, lymph nodes from mice vaccinated 6–8 wk earlier are isolated and incubated with liposomes containing only gp120 for 48 h.[13] Interleukin-4 and IFN-γ levels in the culture medium are determined by ELISA (Pharmingen, San Diego, CA). Interleukin-4 was undetectable in all samples, and IFN-γ was detected only in groups immunized with immunoliposomes containing gp120 + anti-B7-2 + IFN-γ (293 pg/ml), gp120 + anti-CTLA4 (334 pg/ml), and gp120 + anti-CTLA4 + IFN-γ (415 pg/ml), demonstrating that these immunoliposomes induce a Th1 immune response.

Conclusion

The purpose of the experiments just described was to determine whether attachment of MAb to costimulatory molecules made liposomes more effective for the induction of a primary immune response to antigen. Presumably, the MAb to B7-1 and B7-2 would direct the liposomes to APC and the MAb to CD28 and CTLA4 toward T lymphocytes. Our assumption that immunoliposomes containing MAb to CD28 and CTLA4 could bind to T lymphocytes must be considered in light of the fact that liposomes are cleared by phagocytic cells, some of which are APC. Our findings support our hypothesis that immunoliposomes do in fact induce a greater immune response to gp120 than liposomes not containing the MAb.

Acknowledgments

The authors thank Michael F. Powell, formerly of Genentech, Inc., for providing the gp120 used in our studies. This research was supported by an Advanced Technology Program grant (ATP-15-100) from the Texas Higher Education Coordinating Board and Cancer Center Core Grant CA16672 from the National Cancer Institute.

[7] Liposomes Targeted to Fc Receptors for Antigen Presentation by Dendritic Cells *In Vitro* and *In Vivo**

By Karine Serre, Laurent Giraudo, Lee Leserman, and Patrick Machy

Introduction: Why and How We Study Antigen Presentation to T Cells

Adaptive immune responses occur in jawed fishes and their descendents. In responses of this type, cells previously stimulated by antigen (Ag) react with enhanced speed to subsequent exposure to the Ag, and these responses are generally of higher magnitude. Adaptive immune responses of mammals depend on cells specialized in the acquisition of

Ag, called Ag presenting cells (APC), and cells that respond to that Ag, T and B lymphocytes. The two major arms of the adaptive immune response are antibody (Ab) production and cytotoxicity. T lymphocytes are divided into $CD4^+$ and $CD8^+$ T cells. $CD4^+$ T cells are necessary to induce the immunoglobulin gene rearrangements necessary for production of somatically mutated IgG Ab by B cells. This Ab may have a higher affinity for Ag than nonmutated IgM Ab. Cytotoxicity is primarily a property of $CD8^+$ T cells, but these cells require $CD4^+$ T-helper cells for complete differentiation and acquisition of effector functions.

Induction of the immune response is thus a complex series of sequential interactions involving Ag contact with APC, APC contact with lymphocytes, and interaction between lymphocytes. The goal of our research is to increase understanding of the processes of Ag acquisition by APC and of the stimulation of $CD4^+$ lymphocytes by these APC. We want to do so in sufficient detail to be able to optimize immunization, not only against microorganisms but also against tumors that are generally considered poorly immunogenic. We are particularly interested in the use of liposomes as Ag carriers. Liposomes are excellent for this purpose, because they have been well studied in man and may be made to contain a large variety of Ag. Furthermore, they can be radiolabeled or fluorescent, so that they may be followed *in vivo*. The usefulness of liposomes as Ag carriers for the generation of Ab responses *in vivo* has been known for almost 30 years.[1] By use of liposomes as Ag carriers, alone or in conjunction with well-defined adjuvants, we can hope to increase our ability to induce desired immune responses while limiting undesirable consequences. As opposed to microbial Ag carriers, liposomes are very simple and lack elements such as endotoxin, repeated carbohydrates, nucleic acids, and other determinants that stimulate so-called innate immune responses.[2] They also lack microbial elements designed to inhibit immune responses. As a consequence, they permit a more precise determination of the elements of the Ag carrier that affect adaptive immune responses and will hopefully help us to design better vaccines.

$CD4^+$ T cells do not recognize protein Ag as such but recognize Ag after uptake, partial proteolysis, and reexpression of Ag-derived peptides

---

* These studies were supported by institutional grants from l'Institut National de la Santé et de la Recherche Médicale and the Centre National de la Recherche Scientifique, by a grant from the Association Pour la Recherche sur le Cancer (ARC), and by contracts QLG1-1999-00622 and BMH4-CT97-2503 from the European Community. K.S. was supported by fellowships from the Ministère de l'Education Nationale, de la Recherche et de la Technologie and from the ARC.
[1] A. Allison and G. Gregoriadis, *Nature* **252,** 252 (1974).
[2] R. Medzhitov and C. A. Janeway, Jr., *Cell* **91,** 295 (1997).

at the surface of APC in association with major histocompatibility complex (MHC)-encoded class II molecules. The noncovalent association of peptides and MHC molecules constitute the ligands that actually stimulate the Ag-specific receptor of these T cells, called the T-cell receptor (TcR). Class II molecules are normally expressed only on APC. Antigen-presenting cells include dendritic cells (DC), macrophages, and B cells. Dendritic cells are APC that are widely distributed in tissues. It is currently thought that only DC and not macrophages or B cells are capable of acting as APC to stimulate naive T cells. Macrophages and B cells may lack some critical cell-surface molecules called costimulatory molecules or secreted cytokines required for this process. However, the role of B cells is subtler, as discussed later. B cells present Ag to Ag-experienced T cells by virtue of the clonally distributed Ab molecules they express as cell-bound receptors (BcR). Contact with Ag and CD4$^+$ T cells permits their proliferation and secretion of a modified form of that Ag receptor, as soluble Ab. This Ab can interact with Ag at a distance. Both DC and macrophages may be able to use Ab secreted by B cells to bind Ag for enhanced presentation to naive CD4 T cells. DC and macrophages have receptors for Ag–Ab complexes, the immunoglobulin Fc receptors, which are a family of molecules, expressed at the surface of DC and several other cell types.

We and others have established that, by virtue of their Fc receptors, DC may acquire and efficiently present to T cells Ag recognized by specific immunoglobulin G (IgG).[3–7] Immunoglobulin G is the most prevalent soluble form of the receptor for Ag of B cells. IgG circulates before Ag exposure in young mammals, because it is transported from the mother to the fetus across the placenta or transferred to the neonate through colostrum. This Ab may bind to its corresponding Ag, not only permitting its rapid clearance by binding to Fc receptors expressed on macrophages but also permitting enhanced presentation by DC, allowing the naive T cells of the infant to rapidly respond to the Ag. This is a fascinating process, permitting the neonate to profit from the adaptive immune response that was generated in the previous generation. In the case of liposomes bearing a ligand for which Ab exists, the high degree of specificity of circulating Ab has the possibility of increasing the capture of these liposomes compared

[3] F. Sallusto, M. Cella, C. Danieli, and A. Lanzavecchia, *J. Exp. Med.* **182,** 389 (1995).
[4] K. Serre, P. Machy, J.-C. Grivel, G. Jolly, N. Brun, J. Barbet, and L. Leserman, *J. Immunol.* **161,** 6059 (1998).
[5] A. Regnault, D. Lankar, V. Lacabanne, A. Rodriguez, C. Théry, M. Rescigno, T. Saito, S. Verbeek, C. Bonnerot, P. Ricciardi-Castagnoli, and S. Amigorena, *J. Exp. Med.* **189,** 371 (1999).
[6] P. Machy, K. Serre, and L. Leserman, *Eur. J. Immunol.* **30,** 848 (2000).
[7] P. Machy, K. Serre, and L. Leserman, *J. Immunol.* **168,** 1172 (2002).

with other circulating opsonins known to associate with liposomes. Most of these are older in evolution than Ab, and their receptors have not been well characterized. These include albumin, fibronectin, complement, beta-2 glycoprotein I, and various apolipoproteins.[8] However, the IgG Fc receptor is a family of at least three molecules with some functional redundancy and different tissue expression, so a role for the Fc receptor in Ag presentation in vivo has been difficult to demonstrate formally.

The ability of DC to stimulate naive T cells depends on (1) the capacity of DC to acquire Ag and to present it in association with class II molecules; (2) the expression by DC of other molecules (cytokines or costimulatory molecules) necessary for the optimal stimulation of T cells; (3) *in vivo*, the capacity of DC to localize in or to migrate to an environment in which they can interact with T cells specific for the Ag in question.[9] For our studies on Ag presentation by DC, we use mice transgenic (Tg) for the expression of a TcR of known Ag specificity. With normal mice, the contribution of the few Ag-specific clones of T cells is difficult to measure at early times in the response to the Ag, because they are submerged in the mass of non-Ag–specific and thus nonresponding clones. With Tg technology, it is possible to derive mice in which all T cells express the same TcR, thus confining the response of those mice to a single antigenic peptide. Notwithstanding the homogeneity of their TcR, the physiology of Tg T cells is considered to be the same as that of normal T cells. We have used DC derived from cultured bone marrow as a source of reasonably homogeneous DC *in vitro*. We have studied Ag acquisition and its presentation by these DC to homogeneous T cells, whose responses may be quantified easily by virtue of their Ag-dependent proliferation or production of secreted proteins, called interleukins. Recently, we have extended these *in vitro* studies to studies *in vivo*. In this chapter we describe methods to evaluate the capacity of tissue DC to acquire and present Ag to the same Tg T cells, which are adoptively transferred and whose proliferation may be evaluated by fluorescence techniques.

In most of our experiments, Tg T cells are specific for peptides derived from hen egg lysozyme (hEL) or ovalbumin (OVA). These proteins are readily available, well characterized, and immunogenic in mice. We can compare the response of hEL or OVA-specific T cells (called 3A9[10] and OT-II cells,[11] respectively) to peptides derived from these proteins in

[8] S. C. Semple, A. Chonn, and P. R. Cullis, *Adv. Drug Deliv. Rev.* **32**, 3 (1998).

[9] P. Guermonprez, J. Valladeau, L. Zitvogel, C. Thery, and S. Amigorena, *Annu. Rev. Immunol.* **20**, 621 (2002).

[10] W. Ho, M. Cooke, C. Goodnow, and M. Davis, *J. Exp. Med.* **179**, 1539 (1994).

[11] M. J. Barnden, J. Allison, W. R. Heath, and F. R. Carbone, *Immunol. Cell. Biol.* **76**, 34 (1998).

different contexts or physical forms. In particular, we have evaluated the response to these Ag encapsulated in liposomes. Here we present some of the methods used in our laboratory to evaluate the consequences of targeting liposomes to APC. We also present an overview of ongoing experiments using liposomes as carriers of Ag to DC to stimulate T cell *in vivo*. The questions posed are: How are liposomes recognized by APC? Can we identify a role for Fc-receptor uptake of liposomes *in vivo*? By targeting liposomes to a particular type of APC or receptor on APC, can we optimize the response to liposome-associated Ags or favor development of a particular type of response?

Results presented in this chapter are partly based on previous publications from our laboratory.[4,6,7] They include *in vitro* experiments in which we used hEL or OVA-containing liposomes bearing Dinitro-phenyl-hapten determinants[12] on their surfaces. These liposomes can be opsonized by anti-DNP Ab for binding to FcR.[13] Ag encapsulated in liposomes opsonized by IgG Ab to the hapten is taken up by DC by means of immunoglobulin FcR (FcγRI, II and/or III) and presented efficiently to T cells. Free Ag or Ag in nonopsonized liposomes is taken up and presented inefficiently. We also present new data showing that uptake of and presentation of Ag-containing liposomes by DC *in vivo* seems not to be mediated by the FcR. This should not be considered as a definitive conclusion, because a role for the FcR is being evaluated in ongoing experiments.

Materials and Methods and Experimental Plan

Liposomes

Liposomes (80 μmol with respect to lipids) are made from 65% (mol/mol) dimyristoyl phosphatidylcholine, 34.5% cholesterol (Sigma), and 0.5% DNP-caproyl-phosphatidylethanolamine (DNP-cap PE) (Molecular Probes Eugene, OR). Liposomes are formed by exposing lipids evaporated from chloroform/methanol (9:1 v/v) to an aqueous solution containing hEL (20 mg/ml, 1.4 μM), or OVA (60 mg/ml, 1.4 μM) (both from Sigma) and 10 mM carboxyfluorescein (CF)[14] (Molecular Probes) in phosphate-buffered saline (PBS). Following repeated cycles of freezing and thawing, liposomes are formed by extrusion (Extruder, Lipex Biomembranes,

[12] H. R. Six, K. I. Uemura, and S. C. Kinsky, *Biochemistry* **12,** 4003 (1973).
[13] L. Leserman, J. Weinstein, R. Blumental, and W. Terry, *Proc. Natl. Acad. Sci. USA* **77,** 4089 (1980).
[14] J. N. Weinstein, S. Yoshikami, P. Henkart, R. Blumenthal, and W. A. Hagins, *Science* **195,** 489 (1977).

Vancouver)[15] through polycarbonate filters of 200 nm pore size at 45°, followed by gel filtration to eliminate unencapsulated solute. Lipid vesicles are sterilized by filtration through 0.45-μm polycarbonate filters. Different Ag concentrations in lipid vesicles are obtained by dilution of this stock solution. Ag levels are confirmed by fluorescence of dilutions of free and liposome-entrapped CF with reference to the stock solution of Ag and CF used for liposome preparation. The internal volume of the liposomes used is about 4.2×10^{-18} liters, so at the 1.4-μM Ag used, each liposome contains about 3500 molecules of hEL or OVA.

Antibodies

Monoclonal IgG2a anti-DNP Ab is produced from hybridoma cells kindly provided by Zelig Eshhar (Weizmann Institute of Science, Rehovot, Israel). Phycoerythrin-labeled anti-B220, CD11c, CD4, and CD8 are obtained from BD-Pharmingen. Other Ab are from the ATCC or the laboratories that published them.

Mice

3A9 Tg mice, which express the same TcR as 3A9 hybridoma specific for the immunodominant hEL 46-61 peptide in the context of IA^k, are maintained on a CBA/J (CBA) background. OT-II mice, in which the Tg TcR is specific for OVA peptide 323–339 in the context if IA^b are maintained on the C57BL/6 (B6) background. Animals are treated according to national and institutional guidelines.

Cells

Dendritic cells are obtained from femur bone marrow of normal CBA/J (for hEL) or C57BL/6 mice (for OVA). Cells are cultured in Iscove's modified Dulbecco's medium with 10% fetal bovine serum (FBS), antibiotics, 2×10^{-5} M 2-mercaptoethanol (2-ME) glutamine and supernatants from NIH3T3 cells supplemented with 10–20 ng/ml murine granuclocyte-macrophage colony-stimulating factor (GM-CSF).[16] After 3 days of culture, cells are diluted 1:1 in the same medium and, after an additional 4–6 days of culture, plastic nonadherent cells are used as APC. At this time, most of these cells have immature dendritic cell morphology. For *in vitro* experiments, T cells from the spleens of TcR Tg mice are purified by passage over nylon wool columns as described.[17]

---

[15] M. J. Hope, M. B. Bally, G. Webb, and P. R. Cullis, *Biochim. Biophys. Acta* **812**, 55 (1985).
[16] P. Rovere, V. S. Zimmermann, F. Forquet, D. Demandolx, J. Trucy, P. Ricciardi-Castagnoli, and J. Davoust, *Proc. Natl. Acad. Sci. USA* **95**, 1067 (1998).

The IL-2–dependent line cytotoxic lymphoid line (CTLL)[18] is cultured in RPMI 1640 medium supplemented with 5% FBS, 2×10^{-5} M 2-mercaptoethanol (2-ME), 2 mM glutamine and antibiotics, and 10 U/ml of recombinant IL-2 (Boerhinger Mannheim).

Tests of Antigen Presentation

Twenty thousand DC are distributed in wells of 96-well flat-bottom microtiter plates (Costar) in RPMI medium supplemented with 5% FBS, 2×10^{-5} M 2-ME, glutamine, and antibiotics. Tenfold dilutions from stock solutions of free Ag or DNP-liposomes containing Ag are added to the wells, with or without the IgG anti-DNP Ab at a final concentration of 5 μg/ml. Cells and Ags are incubated overnight at 37° in a total volume of 100 μl, after which 20,000 T cells from relevant Tg mice are added and incubated for an additional 48 h in a total volume of 110 μl. At this time, the concentration of IL-2 in supernatant fluids is determined by adding them to IL-2–dependent CTLL cells[18] (10,000 cells/well). Following 16 h incubation, proliferation of CTLL cells is assessed by measurement of radiolabeled thymidine incorporation into DNA at the end of an additional 8 h incubation in the presence of 0.5 μCi [^{3}H] thymidine.[4]

Immunization Protocols

Mice are immunized subcutaneously in hind footpads with different doses of free Ag or Ag encapsulated in liposomes in 50 μl total volume. Control mice are not injected or receive empty liposomes. In some experiments, mice are injected intravenously with 100 μg of IgG2a anti-DNP 3 days before liposome injection.

Ex Vivo Antigen Presentation Assay

Draining popliteal lymph nodes are harvested 24 h after immunization with fluorescent Ag-containing liposomes. Cell suspensions are subjected to mild collagenase type 1 digestion (Sigma) at 37° for 20 min. Unseparated lymph node cells are used as APC, and various dilutions of cells are distributed in 100 μl in wells of 96-well flat-bottom microtiter plates (Costar) in RPMI medium. T cells for ex vivo Ag presentation are obtained from the spleens of Tg mice as indicated previously. Twenty thousand 3A9 cells (or OT-II Tg T cells, not shown) are added for 48 h in supplemented

---

[17] P. Machy, J. Barbet, and L. Leserman, *Proc. Natl. Acad. Sci. USA* **79,** 4148 (1982).

[18] S. Gillis, M. M. Ferm, W. Ou, and K. A. Smith, *J. Immunol.* **120,** 2027 (1978).

RPMI, at which time undiluted supernatant fluids are harvested for determination of IL-2 secretion as previously described.

FACS® Analysis and Purification of Antigen-Presenting Cells from Lymph Nodes

To characterize APC that had taken up Ag-containing liposomes, mice are injected with CF-containing liposomes. Draining popliteal and distal brachial lymph nodes are harvested 24 h after immunization, and cell suspensions are subjected to collagenase digestion as described previously. For *ex vivo* analysis of APC that had taken up CF-containing liposomes, cell suspensions from lymph nodes are collected 24 h after immunization and are sorted on the basis of forward angle scatter (which detects size differences) of cells that took up fluorescent liposomes and are therefore detected in the FL-1 channel in a FACScan® cytofluorometer (Becton-Dickinson).

Preparation of Antigen-Specific Transgenic T Cells for Adoptive Transfer

Popliteal, inguinal, brachial, axillary, cervical, mesenteric, and paraaortic lymph nodes are harvested from TcR Tg mice in RPMI medium. Cells not of $CD4^+$ T cell lineage are depleted by incubating lymph node cell suspensions with rat anti-B220 (RA3-6B2), anti-pan MHC class II (m5/114), anti-CD11b (M1/70), anti-FcγRII/III (24G2), and Ab H59.101.2 specific for $CD8^+$ T cells. Cells are incubated with Ab at 4° for 30 min, washed twice in cold RPMI medium, and then the Ab-binding cells are removed with anti-rat Ab-coupled magnetic beads. The TcR Tg T cells at this stage are usually about 95% pure. Labeling with the intracellular fluorescent dye carboxyfluorescein diacetate succinimidyl ester (CFSE; Molecular Probes) is performed as described.[19] T cells are washed twice in PBS containing 0.1% BSA and resuspended at 10^7 cells/ml in PBS/0.1% BSA with 10 μM CFSE for 10 min at 37°. Cells are washed twice with cold RPMI medium followed by two washes in PBS. Carboxyfluorescein diacetate succinimidyl ester-labeled T cells (2×10^6) in 200 μl of PBS are injected intravenously. Recipient mice are unimmunized or immunized with 50 μl of OVA-containing liposomes in footpads 3 days after adoptive T-cell transfer. After an additional 3 days, T cells are retrieved from draining popliteal and nondraining brachial lymph nodes, and cell division in analyzed by flow cytofluorometry. Cells are gated for T lymphocytes on the basis of forward angle and side-scatter profiles, and 500,000 to 1,000,000 events are collected.

[19] A. B. Lyons, *J. Immunol Methods* **243,** 147 (2000).

Use of Hapten-Bearing Liposomes for Antibody Targeting

The strategy for the use of hapten-bearing liposomes incubated together with Ab for targeting to the FcR of DC is shown schematically in Fig. 1.

Interleukin-2 Production by Transgenic T Cells Stimulated by DC In Vitro

An important question is whether DC could bind liposomes and present Ag encapsulated within them in the absence of a ligand permitting their association with an FcR. To answer this question DC are incubated with

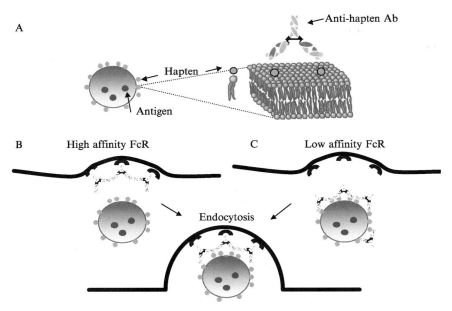

FIG. 1. Schematic presentation of hapten-bearing liposomes and their association with Fc receptors. (A) Schema of liposomes and the composition of the lipid bilayers surrounding an aqueous space containing passively entrapped hEL or OVA. DC express FcγRI (B), which will bind monovalent IgG2a Ab with high affinity. These prearmed receptors may then bind DNP-bearing liposomes. DC also express the low-affinity receptors FcγRII and FcγRIII (C), which do not bind uncomplexed Ab but will bind liposomes opsonized by Ab, because multivalent binding increases the avidity of the interaction between the FcR and the Ab. For reviews on Fc receptor biology, see footnotes 24 and 25. In both instances, after binding, liposomes are taken up by endocytosis, and the Ag they contain is released in endocytic vesicles where peptides derived from it may associate with class II molecules that are subsequently expressed at the cell surface (bottom panel). It remains to be established whether these different FcR deliver their contents into the same compartments or whether DC are signaled differentially by binding of multivalent ligands to these different receptors.

[24] M. Daeron, *Annu. Rev. Immunol.* **15,** 203 (1997).
[25] J. V. Ravetch and S. Bolland, *Annu. Rev. Immunol.* **19,** 275 (2001).

OVA-containing, DNP-bearing liposomes in the absence of cell-targeting Ab or in the presence of IgG2a monoclonal anti-DNP Ab, together with OVA-specific OT-II T cells. Production of IL-2 by these T cells is evaluated after 48 h. Results of a typical experiment are presented in Fig. 2. Neither free OVA nor liposome-encapsulated OVA incubated without anti-DNP Ab were taken up sufficiently well to be presented at concentrations below about 100 nM. In the presence of IgG2a anti-DNP Ab, DNP-bearing liposome-encapsulated OVA is efficient for T-cell stimulation at 1 pM. This is at least 10,000-fold less than the amount of free OVA or non-targeted liposomes required for the same level of IL-2 production. The anti-DNP Ab does not induce IL-2 production when OVA-containing liposomes are made without DNP on their membranes or when DNP-bearing liposomes are made without encapsulated OVA (data not shown). We have already reported similar results for responses against hEL, as well as the fact that the binding of liposomes to the DC depends on the opsonizing Ab.[4] The enhanced presentation of opsonized Ag is explained by the enhanced internalization kinetics of the Ag and activating signals transmitted to the DC through the FcR. In addition to showing that the FcR is efficient

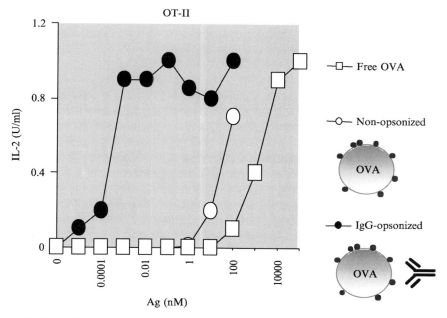

FIG. 2. Antigen in IgG-opsonized liposomes efficiently stimulates T cells. This experiment shows the response of OT-II T cells to free and liposome-encapsulated OVA in the presence or absence of opsonizing anti-DNP Ab.

for inducing liposome uptake and Ag presentation, these data indicate that in the *in vitro* environment studied, protein Ag taken up by fluid-phase endocytosis is not very efficient at inducing IL-2 synthesis by T cells. They also indicate that there is no marked affinity of receptors on DC for other lipid components expressed by liposomes, nor are there significant levels of opsonins in the FBS-containing medium in which liposomes are cultured that permits their uptake by receptors expressed by these bone marrow–derived DC.

Efficiency of Antigen Stimulation by Cells That Take Up Liposome-Encapsulated Antigen In Vivo

CBA mice are injected in footpads with 5–10 μg free or liposome-encapsulated hEL. After 24 h and 48 h, *ex vivo* Ag stimulation of 20,000 3A9 Tg T cells is tested using as APC different numbers of cells from the draining lymph node. These APC were used without additional Ag *in vitro*, so presentation depended on the amount of Ag they acquired *in vivo*. Data for individual mice are presented in Fig. 3. A response was barely detectable in either of two mice injected with free hEL at the highest number of APC used (100,000 cells), whereas responses were seen in 6 of 6 mice for 25,000 APC when mice were injected with liposomes.

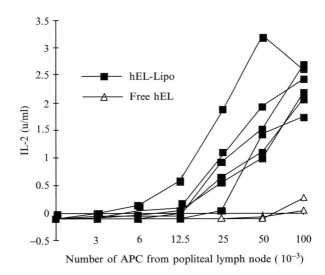

FIG. 3. The immunogenicity of an antigen is increased for *ex vivo* presentation when the antigen is encapsulated in liposomes. Results of an experiment showing hEL presentation by APC in draining lymph nodes of mice injected with free hEL or hEL-containing liposomes.

Determination of the Cell Types That Take Up Liposomes in Lymph Nodes

It is important to establish which cells take up and present efficiently. Ag injected *in vivo*. The fate of liposomes injected subcutaneously in mice is determined by means of fluorescent liposomes and flow cytofluorometry of cells in draining lymph nodes. Twenty four h after injection of CF-containing liposomes in hind footpads of mice, we determined by FACS® analysis that the popliteal lymph nodes contained essentially two fluorescent major cell types, which represented about 4% of the total cells in the node (Fig. 4A). More than half of these fluorescent cells expressed the DC-specific marker CD11c (Fig. 4C), plus high levels of MHC class II and class I, and costimulatory molecules (CD80, CD86, ICAM-1, CD40), suggesting that these were activated DC. The other fluorescent cell type that represents 2% of the total cells expressed the B220 specific–B-cell marker (Fig. 4D) and low levels of MHC class II and costimulatory molecules. The number of DC that had captured liposomes peaked in the node at 24 h and subsequently declined. No fluorescent DC were seen at day 5 following liposome injection. No fluorescent cells could be detected in the nondraining brachial node at any time (Fig. 4B). These results indicate that DC are the primary cells in lymph nodes that captured liposomes (Fig. 4) and are presumably responsible for the Ag presentation of APC *ex vivo* (Fig. 3).

Efficient Activation of Specific CD4 T Cells After In Vivo *Injection of Antigens Contained in Liposomes*

To assess the efficiency of Ag in liposomes to induce T-cell proliferation *in situ*, B6 mice are adoptively transferred with 2×10^6 CFSE-labeled OT-II Tg T cells before immunization. After 3 days, OVA in liposomes is injected in the hind footpads of these mice, and control mice are not immunized. Three days later, recipients are sacrificed, and cell suspensions of the draining popliteal lymph node are monitored for the presence and proliferation of adoptively transferred T cells. As seen in Fig. 5, division of cells is not seen in the absence of immunization with OVA (see explanation in the legend to Fig. 5). Furthermore, when OVA is injected, the extent of division depends on the amount of Ag injected and on its form. Antigen stimulated OT-II T cell proliferation at doses as low as 0.15 ng when encapsulated in liposomes but required 10,000-fold more Ag (1500 ng) for equivalent proliferation when the Ag is injected in free form (K. Serre *et al.*, unpublished observations). Thus, the marked advantage for liposome-encapsulated compared with free Ag observed *in vitro* is duplicated *in vivo*.

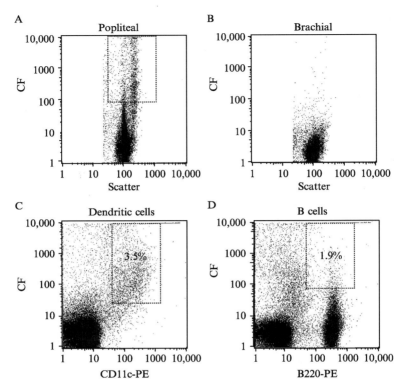

FIG. 4. Flow cytofluorometric data of cells that took up fluorescent liposomes in lymph nodes *in vivo*. The FACS® histogram shows populations of larger and smaller cells in the draining node (box with dotted lines) that are positive for CF, whereas these cells are not seen in the nondraining node (B). The fluorescent populations were sorted on the basis of size. The larger population is positive with a phycoerythrin (PE)-marked Ab specific for the DC marker CD11c (C, box with dotted lines), whereas the smaller population is positive for PE-marked anti-B220 (D, box with dotted lines). The number indicated as percent refers to the percentage of cells with the selected size and fluorescence parameters among the total cells present in the lymph node.

Determination of the Role of Fc Receptors in Liposome Uptake *In Vivo*

In the experiment presented in Fig. 5, Tg T cells specific for a liposome-encapsulated Ag were induced to divide *in vivo* even though liposomes were injected without intentional opsonization, that is, in the absence of added anti-DNP Ab. Furthermore, the titers of naturally occurring or cross-reactive anti-DNP Ab susceptible to opsonize the liposomes in the absence of added Ab were undetectable in these mice. We do not have a clear understanding of the mechanism of the uptake of fluorescent liposomes responsible for staining the DC in Fig. 4. The B cells that bind liposomes are probably DNP-specific (K. Serre et al., unpublished data). To evaluate a possible role for opsonizing Ab in the acquisition of liposomes by DC, CFSE-labeled OT-II cells are injected intravenously in mice and immunized with DNP-liposomes containing OVA 3 days later, similar to the experiment presented in Fig. 5. In some mice, 100 μg of anti-DNP Ab are injected at the same time as the OT-II cells. At the time of immunization with liposomes, circulating levels of anti-DNP Ab are still high.

Data for CFSE fluorescence for both the draining popliteal and the nondraining brachial lymph node are presented in Fig. 6. Although there is some variability between the groups, there is no clear difference in the number of T cells induced to divide in the draining popliteal node in mice in which anti-DNP Ab is introduced (right panel of Fig. 6A) compared with those mice not receiving Ab (left panel of Fig. 6A). This indicates that the uptake of liposomes and presentation of the Ag encapsulated in them are unaffected by the presence of opsonizing Ab.

We have not demonstrated that the Ab we passively administered was present in tissues through which the liposomes passed on their way to the node, but in a limited series of experiments we have incubated liposomes with subaggregating amounts of anti-DNP Ab before their subcutaneous injection, without changes in the preceding results. The aggregation of liposomes in tissues draining an injection site has been reported to increase their uptake in draining nodes, as shown by scintigraphy studies.[20] However, in that report there was no evaluation whether this aggregation increased Ag presentation in the target lymph node.

In contrast to the lack of an effect in the draining lymph node, the administration of passive anti-DNP Ab decreases Ag presentation to T cells in the nondraining brachial node, where the Ag presumably arrives through transport in the blood (Fig. 6B). In the absence of passive Ab (6B, left panel), the two highest doses of Ag clearly stimulate T cells at this distal site, indicating that a number of liposomes insufficient to be detected by fluorescence (Fig. 4) arrive in the node and could stimulate proliferation of Ag-specific T cells, presumably after uptake by DC there. In contrast, passive Ab eliminates presentation in the node at all Ag doses tested (right panel). This suggests that, in the absence of Ab, the Ag arrives in the distal node still encapsulated in liposomes and neither as free Ag that had leaked from liposomes nor as Ag already associated with some migrating cell population. If this were the case, we would expect no effect of the

[20] W. T. Phillips, R. Klipper, and B. Goins, *J. Pharmacol. Exp. Ther.* **295,** 309 (2000).

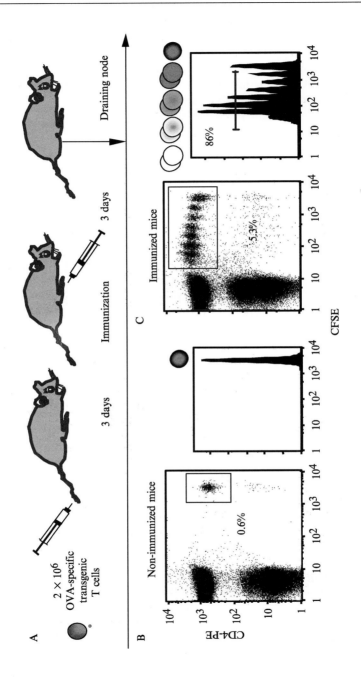

anti-DNP Ab. Depletion of circulating DNP liposomes by passive and induced anti-DNP Ab has been reported.[21] These liposomes are cleared presumably by low-affinity FcR, because their action is partially blocked by an Ab (2.4G2) that binds to these two receptors but not the high-affinity FcR.

We have not investigated whether the passive Ab enhances presentation of liposome-encapsulated Ag in the spleen or liver, which are the major sites for clearance of circulating Ab-opsonized liposomes. Further evaluation of the fate of opsonized liposomes in mice deficient for one or more FcR, as well as Ab-deficient mice, are in progress. In preliminary experiments, we have determined that presentation of Ag in liposomes to Ag-specific T cells in draining nodes is at least not markedly affected in mice in which transmembrane IgM was genetically deleted (K. Serre *et al.*, unpublished results). These mice lack the B cells that make Ab, because signals through this molecule are required for the survival of B cells.[22] Because these mice have no circulating Ab, this means that DC

---

[21] D. Aragnol and L. Leserman, *Proc. Natl. Acad. Sci. USA* **83**, 2699 (1986).
[22] D. Kitamura, J. Roes, R. Kühn, and K. Rajewsky, *Nature* **350**, 423 (1991).

Fig. 5. Measurement of adoptively transferred antigen-specific CFSE-labeled CD4$^+$ T-cell proliferation. (A) The design of this experiment is shown schematically. CFSE-labeled OT-II T cells (shown as uniformly labeled spheres) are injected intravenously into B6 mice. Three days later, mice are immunized in footpads with OVA in liposomes or not immunized. After 3 additional days, mice are killed and cells prepared from the draining popliteal lymph nodes. (B) FACS® data for unimmunized mice. On the ordinate of the left panel is shown staining with a phycoerythrin-labeled anti-CD4 Ab. About 30% of the cells in the node stain with this Ab; the other cells in the node are negative, because most of these are either CD8$^+$ T cells or B cells, and neither of these populations express the CD4 molecule. A small percentage of cells in the node are positive both for CD4 and for CFSE (positive on the abscissa). These correspond to the fraction of the injected CFSE-labeled OT-II cells that can be found in the popliteal node at this time. The staining of this population is uniform and similar to the fluorescence of the input cells when the data for CFSE fluorescence are presented alone (B, right panel). (C) Data for mice that were immunized with OVA in liposomes 3 days previously, after transfer of the OT-II cells. The CD4$^+$ staining profile in the left panel is similar to B, except that the percentage of cells that are CD4$^+$ and are CFSE positive has increased, and the level of CFSE fluorescence of that population is also heterogeneous. In the right panel of C, which shows only cells stained with CFSE, a series of peaks with fluorescence diminishing by about 50% is seen, symbolized by the spheres above the histogram. This is explained as follows: OT-II cells stimulated by Ag in the lymph node are induced to divide. When the cells divide, each daughter cell receives one half of the cytoplasm. Because CFSE stains primarily proteins in the cytoplasm, the amount of CFSE decreases 50% with each division. In this instance, dividing cells represent at least 86% of the CFSE$^+$/CD4$^+$ T cells. The sensitivity of the technique is limited by the ability to distinguish cells that have divided more than about six times from cells that were never CFSE labeled. Nevertheless, this is a powerful technique for evaluation of early events in responding T-cell populations after Ag presentation *in vivo*.

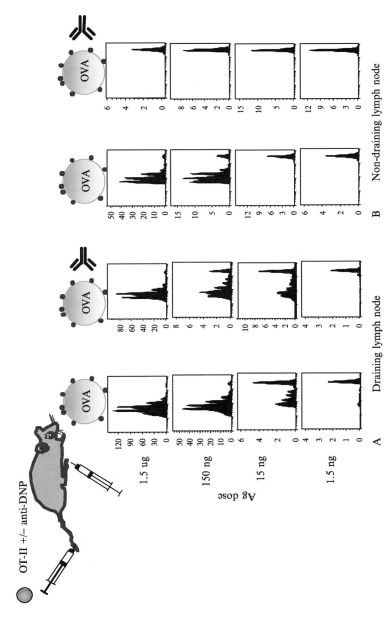

FIG. 6. Are Fc receptors implicated in liposome uptake *in vivo?* Responses in draining and nondraining lymph nodes of mice injected with liposomes in the presence or absence of passively administered opsonizing Ab.

do not require Ab, Ab-mediated complement fixation, or immunoglobulin Fc receptors, which require Ab binding to mediate liposome recognition and uptake of liposomes *in vivo*.

Conclusions and Perspectives

The data presented indicate that Ag encapsulated within liposomes is taken up and presented to Ag-specific T cells in a manner markedly enhanced by binding of the liposomes to an FcR expressed by bone marrow–derived DC *in vitro*. The data also indicate that liposomes are taken up primarily by cells expressing DC markers *in vivo* (Fig 4). It is presumably these DC that are primarily implicated in Ag presentation by APC *ex vivo* (Fig. 3) and for presentation *in situ* (Figs. 5 and 6). Liposomes are thus excellent carriers for Ag presentation, even if we do not completely understand how they are recognized by APC. This would be consistent with the evolutionary importance of a population of APC capable of recognizing microbes or particles that resemble microbes. The role for DC is supported by our data showing that purified DC, corresponding to 3% of the cell population in the lymph node, are largely or entirely responsible for the activity of the node in Ag presentation (K. Serre *et al.*, manuscript in preparation). We cannot, however, identify clearly the mechanism(s) or receptor(s) on DC required for this process *in vivo*; there may be many of these mechanisms, and they may be redundant. At this point, at least two possibilities may account for our data. The first is that the uptake of liposomes occurs primarily at the site of injection by cells that migrate to the nodes or that DC in draining nodes are placed strategically to take up the bulk of incoming particles by a mechanism not involving the FcR. Our data in Fig. 6, showing uptake of liposomes and presentation of their contents in distal nodes, presumably after passage through the blood, would argue against this, because lymph nodes do not manifestly filter the circulation. The second possibility is that the DC present in lymph nodes *in vivo* differ in their properties from bone marrow–derived DC that we studied *in vitro*. Among these differences, DC present *in vivo* may have some means of recognizing injected liposomes, for example, by direct recognition of liposome phospholipids or by binding by the liposomes of an opsonin or opsonins other than IgG present in tissue. Evaluation of this possibility requires accurate methods of identifying DC in lymph nodes and of culturing them *in vitro*. It has been determined only recently that there are at least five[23]

---

[23] S. Henri, D. Vremec, A. Kamath, J. Waithman, S. Williams, C. Benoist, K. Burnham, S. Saeland, E. Handman, and K. Shortman, *J. Immunol.* **167,** 741 (2001).

phenotypically different populations of DC in lymph nodes. It is entirely possible that these phenotypic differences are associated with different interactions with liposomes, as well as with potential functional differences in immunity. Experiments on isolated lymph node populations *in vitro* are thus of interest. Finally, despite our inability to demonstrate differences in the initial steps of stimulation of CD4[+] T cells by Dc *in vivo* in the presence or absence of added Ab as an opsonin, we do not know whether having acquired an Ag that is opsonized or not may play a role later in the response. We have already reported that the form of the Ag (that is, in opsonized liposomes or not), plays an important role in the ability of CD4[+] T cells to stimulate Ag presentation in the context of MHC class I molecules by DC[7], and analogous roles for the manner of Ag acquisition are possible *in vivo*.

[8] Liposomes Composed of Reconstituted Membranes for Induction of Tumor-Specific Immunity

By Larry E. Westerman and Peter E. Jensen

Introduction

This chapter deals with methods to generate experimental cell-free cancer vaccines consisting of reconstituted tumor membrane liposomes with incorporated immunomodulatory molecules. By use of a detergent dialysis technique, tumor cell membranes can be reassembled with the addition of purified membrane proteins included to promote tumor-specific immune responses.[1,2] A major benefit of this approach is that immunostimulatory proteins can be presented on a membrane like structure that resembles the tumor cell plasma membrane. Tumor-specific antigens and preformed major histocompatibility molecules (MHC)–peptide complexes containing relevant tumor-associated peptide antigens may be preserved in the liposomes. The immunogenicity of antigens and MHC molecules in liposomes has been demonstrated.[3–5]

[1] L. E. Westerman and P. E. Jensen, *J. Immunol. Meth.* **236,** 77 (2000).
[2] L. E. Westerman S. C. Sund, P. Selvaraj, and P. E. Jensen, *J. Immunother.* **23,** 456 (2000).
[3] A. A. Brian and H. M. McConnell, *Proc. Natl. Acad. Sci. USA* **81,** 6159 (1984).
[4] P. E. Jensen, *J Exp. Med.* **174,** 1111 (1991).
[5] M. A. Sherman, H. A. Runnels, J. C. Moore, L. J. Stern, and P. E. Jensen, *J. Exp. Med.* **179,** 229 (1994).

Our expanding knowledge of the mechanisms involved in the activation of T lymphocytes is being exploited to develop strategies for enhancing the immune response to tumors. Considerable effort has been focused on increasing the immunogenicity of tumors by the introduction of immunomodulatory molecules into tumor cells by gene transfer. Promising results have been obtained in experimental models in which tumors are engineered to express genes encoding cytokines, MHC, or costimulatory molecules such as CD80, CD86, and CD40L.[6–10] Transfection and viral transduction have been used to induce protein expression in viable tumor cells. However, these techniques can be difficult and time consuming, particularly in the clinical setting.

Direct protein transfer techniques provide an alternative to gene transfer. Purified integral membrane proteins with immunostimulatory activity can be added to detergent-solubilized tumor membranes. Selective removal of the detergent by dialysis results in the reassembly of proteins and lipids to form reconstituted membrane structures containing additional components not present in the original tumor membrane.

General Strategy

Successful generation of reconstituted tumor membrane liposomes is comprised of three principal aspects: the isolation of tumor membranes, the partial or complete solubilization of membrane protein and lipid components with detergent, and the reassembly of solubilized proteins and lipids to form unilamellar vesicles. The strategy to generate these liposomes is outlined in Fig. 1. Also illustrated in Fig. 1 is a one-step procedure to introduce additional purified transmembrane proteins into the liposomes.

Isolation of Tumor Membranes

General Considerations

Membrane preparations can be obtained from tumor cells by various lysis methods followed by centrifugation to isolate the cell membranes. This chapter focuses on tumor cells that are from established tissue-cultured

[6] S. Baskar, S. Ostrand-Rosenberg, N. Nabavi, L. M. Nadler, G. J. Freeman, and L. H. Glimcher, *Proc. Natl. Acad. Sci. USA* **90,** 5687 (1993).
[7] G. Dranoff, E. Jaffee, A. Lazenby, P. Golumbek, H. Levitsky, K. Brose, V. Jackson, H. Hamada, D. Pardoll, and R. C. Mulligan, *Proc. Natl. Acad. Sci. USA* **90,** 3539 (1993).
[8] G. Yang, K. E. Hellstrom, I. Hellstrom, and L. Chen, *J. Immunol.* **154,** 2794 (1995).
[9] S. E. Townsend and J. P. Allison, *Science* **259,** 368 (1993).
[10] M. F. Mackey, J. R. Gunn, P. P. Ting, H. Kikutani, G. Dranoff, R. J. Noelle, and R. J. Barth, *Cancer Res.* **57,** 2569 (1997).

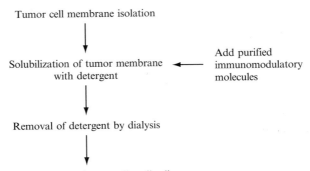

Tumor cell membrane isolation

Solubilization of tumor membrane ◄——— Add purified
with detergent   immunomodulatory
                                            molecules

Removal of detergent by dialysis

Reconstituted tumor membrane unilamellar liposomes
with incorporated immunomodulatory molecules

FIG. 1. Flowchart for the generation of tumor membrane liposomes, incorporating additional purified proteins to increase immunogenicity.

cell lines. The techniques described here can also be adapted to obtain membranes from tumor cells isolated from solid tumor samples. Methods to isolate tumor cells from solid masses are described elsewhere.[11–13]

Several lysis techniques can be used as a first step in the isolation of cell membranes. These include N_2 cavitation, solubilization with detergents, hypotonic lysis, and shearing with homogenizers or passage through small needle apertures. With each of these methods, it is important not to disrupt the integrity of cell nuclei, releasing viscous DNA, which interferes with membrane isolation. It is wise to monitor the extent of cell breakage by microscopic examination so that excessive lysis can be avoided. Pilot experiments should be done to identify a technique that will provide the desired degree of lysis while preserving important tumor membrane proteins. Different tumor cell lines may require different conditions for optimal lysis. Hypotonic conditions can be used effectively to lyse murine B-cell lymphomas such as A20, CH27, and M12.4 cell lines. However, hypotonic conditions alone will not result in effective lysis of the murine melanoma cell line K1735 or the murine fibrosaroma cell line AG104. The combination of hypotonic conditions and shearing techniques can effectively lyse K1735 and AG104 tumor cells.

[11] J. J. Penning and J. H. LeVan, *J. Natl. Cancer Inst.* **66,** 85 (1981).

[12] E. Kedar, B. L. Ikejiri, G. D. Bonnard, and R. B. Herberman, *Eur. J. Cancer Clin. Oncol.* **18,** 991(1982).

[13] J. F. Ensley, Z. Maciorowski, H. Pietraszkiewicz, G. Klemic, M. KuKuruga, S. Sapareto, T. Corbett, and J. Crissman, *Cytometry* **8,** 479 (1987).

All lysis techniques will subject membrane proteins to potential proteolysis. The inclusion of protease inhibitors in the solubilization buffer is usually indicated. A good inexpensive cocktail of protease inhibitors would have a final concentration of 2 mM iodoacetamide, 1 mM phenylmethylsufonyl fluoride (PMSF), and 1% aprotinin.

Following cell lysis, nuclei are removed by centrifugation. The membranes are then isolated from the supernatants by high-speed or ultracentrifugation. Pelleted membranes can be resuspended in a saline-buffer solution and solubilized with detergent for immediate use or they can be stored at $-70°$. Typically, 10^8 tumor cells will yield 0.75–1.25 mg total protein as determined by the method of Bradford[14] and 0.8–3.0 μmol phospholipid as determined by malachite green inorganic phosphate assay.[15] Determining the protein or phospholipid quantities will become important for optimizing the detergent concentration needed for membrane solubilization and preservation of critical membrane proteins in the liposomes.

Cell Lysis Techniques

Hypotonic Lysis. All step are preformed in the cold or on ice. Tumor cells are washed with a cold saline–buffer solution (0.15 M NaCl, 10 mM Tris-HCl, pH 7.4). To the pelleted cells, cold hypotonic buffer (25 mM Tris-HCl, pH 8.0) is added at a concentration of $1–5 \times 10^7$ cells per ml buffer. The buffer should include freshly added protease inhibitors. The mixture is incubate for 15–60 min. The extent of cell lysis is determined by microscopic examination with trypan blue (see later). The optimal cell concentration and incubation time should be determined for each cell line to minimize lysis time. If cells do not lyse under these conditions, a shearing technique can be used additionally.

Shearing Techniques. Following the procedure for hypotonic lysis, cells are incubated with hypotonic buffer plus protease inhibitors for 30 min. The tumor cells should appear swollen. The cells are aspirated through a 27-gauge needle attached to a 10-ml syringe. The cells are dispensed back through the needle to the original container. The procedure is repeated two to four times until cells are lysed. The extent of cell lysis is determined by microscopic examination with trypan blue. Alternately, the cells can be sheared in a hand-held homogenizer. The number of strokes needed to lyse ~90% of the tumor cells is monitored. A motor-driven homogenizer, should not be used, because it can cause frothing.

N_2 Cavitation. Washed tumor cells are resuspended in 10 mM Tris-HCl, pH 7.4, containing 0.25 M sucrose and fresh protease inhibitors at a

[14] M. M. Bradford, *Anal. Biochem.* **72**, 248 (1976).
[15] H. H. Hess and J. E. Derr, *Anal. Biochem.* **63**, 607 (1975).

concentration of $1-2 \times 10^8$ cells per ml. Cells are placed in a small nitrogen cavitation bomb, and the bomb is pressurized to 30 psi with N_2 for 15 min on ice. The cell suspension is collected, and the extent of lysis is examined by trypan blue dye exclusion.

Trypan Blue Dye Exclusion to Determine Cell Lysis. A $1-2$ μl aliquot of the cell suspension is diluted with 25 μl of a saline-buffer solution. Twenty five microliters of a 0.4% trypan blue solution are added. A small amount of trypan blue–cell suspension mixture is introduced into the chamber of a hemacytometer. The cells are examined for intact nuclei, which appear deep blue. Unlysed cells will have a deep blue nucleus with a light blue cytoplasm. The cell membrane isolation procedure is initiated when greater than 85% of the cells are lysed.

Centrifugation to Isolate Tumor Membranes. After cell lysis, nuclei are removed by centrifugation at 3600g for 10 min, and the supernatant is centrifuged at 22,000g for 30 min. Membrane pellets are washed two times with a cold saline-buffer solution. The pellets are resuspended to 5×10^8 cell equivalents per ml in a 0.15 M NaCl, 10 mM Tris-HCl pH 8.0 solution with protease inhibitors. The membrane preparation is solubilized immediately or aliquots are prepared and stored at $-70°$.

Solubilization of Tumor Membranes and Liposome Generation

General Considerations

The detergent used to generate reconstituted membrane liposomes is of considerable importance. Some of the essential criteria to consider in choosing a detergent include the ability to solubilize membrane proteins and lipid components, the preservation of important tumor antigens or MHC proteins, and the ease of detergent removal by dialysis. Knowing the physical properties of the detergent and its effect on the efficiency of solubilizing membrane proteins and lipids will help in designing a successful strategy for the generation of reconstituted membrane liposomes. There are excellent reviews that deal extensively with the physical properties of detergents and their use in the solubilization of membrane proteins.[16,17,17a]

The detergent of choice for our studies is an octyglucoside, n-octyl-β-D-thioglucoside (SOG). Octyglucosides are mild nonionic detergents, causing minimal denaturation of membrane proteins.[17]

When solubilizing cell membranes, the detergent must be in sufficient quantity to disrupt the lipid bilayer and solubilize the protein and lipid

---

[16] J. R. Silvius, *Ann. Rev. Biophys. Biomolec. Struct.* **21,** 323 (1992).
[17] J. M. Neugebauer, *Methods Enzymol.* **182,** 239 (1990).

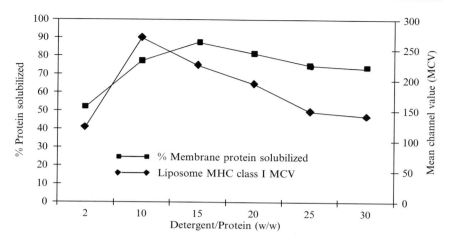

FIG. 2. Effect of the detergent/protein ratios on the ability to solubilize cell membrane components and preserve antigens in reconstituted liposomes. K1735 tumor membranes were solubilized with a 1% SOG solution at varying detergent/membrane protein ratios (w/w). The percentage of membrane protein solubilized was determined by the Bradford assay. Liposomes were generated by detergent dialysis, and samples were supported on polystyrene beads and analyzed for MHC class I expression by flow cytometry with the monoclonal antibody M1/42 (anti-murine MHC class I).

components. However, excessive quantities of detergent will result in protein–detergent and lipid–detergent mixed micelles instead of liposome structures.[17,18] Figure 2 illustrates the importance of having the appropriate detergent ratio in relation to the quantity of membrane protein in the sample. Experiments were preformed in which varying amounts of K1735 tumor cell membranes were solubilized in a 1% SOG solution. At low detergent membrane-protein ratios, the percentage of total membrane protein that is solubilized is low. At high detergent membrane-protein ratios, the liposomes generated from the solubilized membranes can have reduced quantities of important membrane constituents, in this example MHC class I molecules.

It is also necessary to choose a final detergent concentration that should be greater than the critical micelle concentration. A useful range of concentrations for SOG is 0.2% (w/v) to 2.0% (w/v). In our experiments, a 1% SOG solution is used. When tumor membranes are solubilized

[17a] J. L. Rigand and D. Levy, *Methods Enzymol.* **373,** 465 (2003).
[18] L. M. Hjelmeland and A. Chrambach, *Methods Enzymol.* **104,** 305 (1984).

at a detergent membrane-protein ratio of 10:1 (w/w), the final protein concentration is approximately 1 mg/ml.

Unilamellar liposomes prepared by octylglucoside dialysis have been described previously.[19] SOG has a relatively high critical micelle concentration, so it can be removed easily by dialysis. In addition, liposomes formed by SOG detergent dialysis are fairly uniform in size.[19] K1735 tumor membrane liposomes generated by this technique were determined to have diameters between 130 and 250 nm by dynamic light-scattering measurements. A disadvantage of the detergent dialysis technique is that residual detergent may accumulate in the liposomes, which can affect the stability of the liposomes.[20]

Protocol for Membrane Solubilization and Liposome Generation

All steps are preformed in the cold or on ice. The isolated tumor membranes are solubilized in 0.15 M NaCl, 25 mM Tris-HCl, pH 7.4, containing 1% SOG at 10^8 cell equivalents per ml. Purified immunomodulatory molecules can be added to the solubilized membrane solution. After equilibration for 30 min, insoluble material is removed by centrifugation (10,000 rpm for 10 min in a microcentrifuge).

The supernatant is dialyzed for 24 h at 4° against three changes of phosphate-buffered saline (PBS), using dialysis membranes with a molecular mass cutoff of 12,000 Da to remove detergent and generate unilamellar liposomes. The liposomes can be stored at 4° for up to 1 week.

Incorporation of Immunomodulatory Molecules in Reconstituted Tumor Membranes

General Considerations

A major advantage of this method to generate reconstituted tumor membranes is the ease in which immunomodulatory molecules can be transferred into tumor membranes. These molecules can be added directly to detergent solubilized tumor membranes and incorporated into the liposome bilayers during detergent removal by dialysis.

Most immunomodulatory molecules that have been incorporated into liposomes have been transmembrane proteins. Transferred proteins used thus far include MHC molecules,[4,5,21] adhesion molecules lymphocyte

---

[19] L. T. Mimms, G. Zampighi, Y. Nozaki, C. Tanford, and J. A. Reynolds, *Biochem.* **20,** 833 (1981).

[20] M. Ueno, C. Tanford, and J. A. Reynolds, *Biochem.* **23,** 3070 (1984).

function-associated ontigen-3 and intracellular adhesion molecule-1,[22] and costimulatory molecules such CD86 and CD40L (L. E. Westerman and P. E. Jensen, unpublished data, 2000).[1,2] All these molecules have been purified by immunoaffinity chromatography techniques. During purification, octylglucoside is included in the elution buffer to maintain solubility of membrane proteins. As noted previously, octylglucoside is removed readily by dialysis. Detergents with low critical micelle concentrations should be avoided. Functional assays to determine the activity of molecules to be incorporated in liposomes should be developed. This is helpful in ensuring that the protein will be functional after transfer into reconstituted tumor membranes. We demonstrated in a T-cell costimulation assay that purified CD86 could stimulate T-cell proliferation and that this activity was preserved after incorportion in tumor membrane liposomes.[1,2]

When gene transfer techniques are used to introduce immunomodulatory molecules into tumor cells, the level of expression by the tumor cells is difficult to regulate. By use of the liposome protein transfer technique, known quantities of purified molecules can be incorporated into liposomes. A range of 0.25–2.5 μg of purified CD86 glycoprotein has been demonstrated to incorporate into 10^7 cell equivalents of K1735 membranes without loss of tumor cell–derived MHC Class I expression in the liposomes.[2]

Evaluation of Reconstituted Tumor Membranes

General Considerations

In any vaccination strategy, the consistency of the immunogen is important to ensure reproducible results. It is important to characterize and quantify the liposomes. Physical and chemical properties such as liposome size and protein and phospholipid content should be measured. It is also important to confirm that relevant tumor-specific antigens or MHC molecules are preserved in the liposomes and to quantify the incorporation of immunostimulatory components and confirm that their function is preserved.

Reconstituted tumor membrane liposomes can be coated on 10-μm polystyrene beads. This serves two purposes, to provide a means to evaluate the contents of the liposomes and to determine the functionality of

[21] B. Prakken, M. Wauben, D. Genini, R. Samodal, J. Barnett, A. Mendivil, L. Leoni, and S. Albani, *Nat. Med.* **6,** 1406 (2000).
[22] G. Ganpule, R. Knorr, J. M. Miller, C. P. Carron, and M. L. Dustin, *J. Immunol.* **159,** 2685 (1997).

tumor membrane proteins and transferred molecules incorporated in the liposomes.

Liposome-coated beads can be handled in a manner similar to cells. The beads can be stained with fluorescent monoclonal antibodies and evaluated by flow cytometry. The presence and quantity of tumor membrane proteins can be measured using this technology. This is helpful in optimizing the lysis and solubilization steps and in monitoring the incorporation of additional components.

It has been demonstrated that liposomes containing MHC molecules need to be supported on glass or other surfaces to stimulate T-cell responses *in vitro*.[3] No responses are observed when unsupported liposomes are added directly to culture media. In some situations, it is desirable to determine whether reconstituted tumor membranes can be recognized by appropriate antigen-specific T cells *in vitro*. Polystyrene beads provide a convenient support for evaluating liposomes in this kind of functional assay.

Protocol for Coating of Liposomes on Polystyrene Beads

Polystyrene beads, 10 μm in size (Polysciences, Inc., Warrington PA), are added to a liposome preparation at a ratio of 10:1 cell equivalents to beads. It has been determined that at this ratio, optimal coating of the beads is achieved. The mixture is incubated for 1 h at 4° on a rocker. The beads are then pelleted by centrifugation for 3 min at 3000 rpm in a microcentrifuge, washed two times with PBS, and resuspended in PBS. The liposome-coated beads can be evaluated by flow cytometry or used in functional assays.

Concluding Remarks

The protocols described in this chapter illustrate an approach that can be used to generate reconstituted tumor membrane liposomes. Methods are described to isolate membranes from tumor cell lines, solubilize protein and lipid components, and reassemble solubilized components in the form of unilamellar vesicles. Also described is a general method to enhance the immunogenicity of tumor membrane liposomes by incorporating purified immunostimulatory transmembrane proteins. The presence and function of these molecules can be examined by coating polystyrene beads with the liposomes.

The potential use of this protein transfer technique to generate tumor vaccines is attractive because of its simplicity. It is relatively easy to generate consistently a high level of expression of immunomodulatory molecules in reconstituted vesicles. The detergent dialysis technique allows for the incorporation of more than one type of membrane protein into a single liposome preparation. This provides opportunities to enhance immunogenicity through incorporation of multiple complementary adhesion and costimulatory molecules, as well as molecules with other activities in promoting immune responses.

The use of this technique to generate tumor liposomes has several advantages as an immunotherapeutic approach for generating experimental tumor vaccines. In particular, this is a cell-free system that avoids the use of live tumor cells and circumvents difficulties associated with gene transfer techniques. Like genetically engineered tumor vaccines, tumor liposomes have the potential to be generated from primary tumor tissue isolated from individual patients, providing a vaccine that is specialized for each individual. However, many questions remain in evaluating this approach. Key antigens may be lost during liposome preparation, and preformed MHC–peptide complexes may dissociate. In addition, the MHC and costimulatory molecules have the potential to orient in either direction in the membrane bilayer, limiting availability on the external surface of the liposome. The fate of liposome *in vivo* remains to be examined. It is not clear whether tumor antigen components of the liposomes are efficiently internalized by professional antigen presenting cells and whether they can be presented to T cells through "cross-presentation" mechanisms. Candidate tumor antigens derived from cytoplasmic and nuclear proteins are only present in the liposomes as peptides bound in stable preformed MHC–peptide complexes derived from the tumor cells. Nevertheless, this is an interesting experimental approach that merits further investigation.

[9] pH-Sensitive Liposomes as Adjuvants for Peptide Antigens

By JIN-SOO CHANG and MYEONG-JUN CHOI

Introduction

Liposomes have been studied as an effective drug delivery system for many years, and pH-sensitive liposomes were introduced as a tool to deliver encapsulated drugs to tumors, whose environment is slightly lower in pH than those of normal tissues.[1] pH-sensitive liposomes can deliver

---

[1] M. B. Yatvin, W. Kreutz, B. A. Horwitz, and M. Shinitzky, *Science* **219**, 1253 (1980).

encapsulated substances into the cytosol following endocytosis.[2,3] Liposomes are taken up by cells through receptor-mediated endocytosis and degraded through the endosome–lysosome pathway, in which a small percentage of contents can avoid degradation and reach their nuclear or cytoplasmic target. In the case of pH-sensitive liposomes, however, the slightly acidic pH of early endosomes can trigger the destabilization of the liposomal and endosomal membranes and mediate the release of their aqueous contents into the cytosol.[1,2,4] Although a recent report showed that acidification of the endosome is not the only mechanism of the destabilization of liposome inside the cell and the unknown mechanism also helps the translocation of liposomal compartments into the cytosol,[5] the translocation efficiency of the aqueous substance of pH-sensitive liposomes is higher than that of conventional liposomes. If the entrapped substances of pH-sensitive liposomes are proteins or antigenic epitope peptides, they are presumably transferred to the endoplasmic reticulum to associate with class I MHC molecules.[6,7] Some pH-sensitive liposomes are retained in endosomes and pass through the endosome–lysosome pathway, and the compartments are presented on class II MHC molecules. Thus the antigens encapsulated in pH-sensitive liposomes can be presented on both class I and class II MHC molecules and induce both cellular and humoral immune responses.

Cytotoxic T lymphocytes (CTL), one of the major cells for the cell-mediated immune response, play a major role in destroying cancer or virus-infected cells. The peptide-specific CTL response has been studied for the development of peptide-based therapeutic vaccines against cancer and viral infectious disease. Because the immunogenicity by the 8-9mer epitope peptide is very low, however, an adjuvant is required to induce antigen-specific CTL responses with epitope peptide. The pH-sensitive liposome is one of the candidates. Conventional liposomes are endocytosed on contact with antigen-presenting cells and degraded, along with the entrapped molecules, inside the endosome (endosome–lysosome pathway), whereas pH-sensitive liposomes after endocytosis release their contents into the cytosol because of their fusion capacity with the endosomal membrane at low pH, in which the pH range of early and late endosome is 5.5–6.5. The released peptides from liposome behave like endogenous antigens in the cytosol and are presented by major histocompatibility complex

[2] R. M. Straubinger, N. Düzgüneş, and D. Papahadjopoulos, FEBS Lett. **179,** 148 (1985).
[3] C. J. Chu, J. Dilkstra, M. Z. Lai, K. Hong, and F. C. Szoka, *Pharm. Res.* **7,** 824 (1990).
[4] M. Marsh and A. Helenius, *Adv. Virus Res.* **36,** 107 (1989).
[5] S. Simoes, V. Slepushkin, N. Düzgüneş, and M. C. Pedroso de Lima, *Biochim. Biophys. Acta* **1515,** 23 (2001).
[6] M. W. Moore, F. R. Carbone, and M. J. Bevan, *Cell,* **54,** 777 (1988).
[7] J. W. Yewdell and J. R. Bennick, *Adv. Immunol.* **52,** 1 (1992).

(MHC) class I restriction pathway. The peptide antigen presented with MHC class I molecules can induce effective CTL responses without any side reactions. The immunogenicity of CTL epitope peptides drived from tumor-associated antigens or viral proteins is too low to induce immune responses with conventional adjuvants, because of its short length of 8 to 10 amino acids. Furthermore, the peptide must be presented on MHC class I molecules to induce a CTL response and thus, the peptide antigen must be delivered into the cytosol. The pH-sensitive liposomes can therefore transport these short epitope peptides into the cytosol where they act as an endogenous antigen and are presented on the MHC class I molecule to induce an effective CTL response.

Inclusion of phosphatidylethanolamine as the main lipidic component of pH-sensitive liposomes determines the pH sensitivity and the capacity to be destabilized and/or fusion at acidic pH.[1,2] The incorporation of cholesterol or ganglioside, or cholesterol hemisuccinate (CHOHS) can enhance the stability of pH-sensitive liposomes in plasma and serum to move the pH-sensitive liposomes to the target cell safely.[1,2] These modified pH-sensitive liposomes can deliver the encapsulated material more safely and efficiently than conventional liposomes. In some studies, pH-sensitive liposomes composed of palmitoyloleoyl phosphatidylethanolamine (POPE) CHOHS were used as a delivery system for epitope peptides to induce CTL responses[8,9] and show effective immune response to destroy cancer cells. In this chapter we describe the preparation and characterization of pH-sensitive liposomes composed of POPE/CHOHS and methods evaluate the induction of peptide-specific CTL responses. In this chapter, we demonstrate the effectiveness of parameters in using pH-sensitive liposomes to induce peptide-specific CTL response *in vivo*, such as liposome preparation methods, size of individual liposomes, ratio of lipid to peptide, and the effect of immune supplements. Several methods in effective liposome preparation are introduced, and pH-sensitive liposomes prepared by these methods can induce effective CTL responses with little difference among them.

Lipids and Chemicals

Phosphatidylethanolamine-β-oleoyl-γ-palmitoyl, CHOHS, and monophosphoryl lipid A (MPL) are purchased from Sigma Chemical Co (St. Louis, MO). Only one spot for each of the lipids is observed on

[8] J. S. Chang, M. J. Choi, T. Y. Kim, S. Y. Cho, and H. S. Cheong, *Vaccine* **17,** 1540 (1999).
[9] J. S. Chang, M. J. Choi, H. S. Cheong, and K. Kim, *Vaccine* **19,** 3608 (2001).

thin-layer chromatography, and they are used without further purification. 8-Aminonaphthalene 1,3,6-trisulfonic acid disodium salt (ANTS), and *p*-xylene bis (pyridinium) bromide (DPX) are purchased from Molecular Probes (Eugene, OR). Aminomethylcoumarin acetate (AMCA) is purchased from Pierce Chemical Co (Rockford, IL).

Mice and Cells

C57BL/6 or Balb/c mice (5-to 6-week-old female mice) are purchased from Charles River Laboratories (Wilmington, MA). EL4 and P815 cells are obtained from American Type Culture Collection (ATCC; Rockville, MD). EL4, P815, and mouse splenocytes are cultured at 37° in RPMI 1640 medium (Life Technologies, Grand Island, NY) supplemented with 10% fetal bovine serum (FBS), 2 mM L-glutamine, 20 mM HEPES, 1 mM sodium pyruvate, 10^{-5} M β-mercaptoethanol, and gentamicin (all from Life Technologies).

Peptides

Peptides to be encapsulated in liposome must have a purity of more than 97%. Peptides are characterized by reverse-phase C-18 high-pressure liquid chromatography (HPLC) before liposome preparation. Stocks of peptides in lyophilized form at −20° are preferred to that in solution form because of the low stability of peptides in solution. In the case of peptide solutions used in cell culture, the peptide solution is sterilized by filtration through a 0.22-μm membrane and kept at 4° (peptide names from various origin and their sequences for CTL induction are shown in Table I.).

Methods for the Encapsulation of Peptides in pH-Sensitive Liposomes

Freezing-Thawing Method. POPE, CHOHS, and MPL (mole ratio 7:3:0.01; total phospholipid) are dissolved in 1 ml chloroform in a glass tube and dried under a stream of nitrogen gas while rotating the tube, to form a thin film. If the thin film on the wall of the glass tube is not homogeneous, a little more chloroform is added, and the drying step is repeated to make a homogenous thin film. After formation of the film, the tube is flushed with N$_2$ gas to minimize the oxidation of the lipid. The peptide dissolved in phosphate-buffered saline (PBS) (1–0.5 mg/ml) is poured into the glass tube coated with the lipid film. The peptide solution to be encapsulated in pH-sensitive liposomes must not exhibit any precipitation before addition to the lipid film. The peptide–lipid mixture is vortexed vigorously to form multilamellar vesicles (MLV). The average liposome size is about

TABLE I
CTL Epitope Peptides

Name	Origin	MHC	Sequence
Inf	Influenza NP	H-2K^d	TYQRTRALV
LCMV	LCMV NP	H-2K^d	PQASGVYMG
Sendai	Sendai NP	H-2K^b	FAPGNYPAL
R15K	HIV-1 env	H-2K^d	RIQRGPGRAFVTIGK
T26K	HIV-1 env	H-2K^d	TRPNNNTRKRIQRGPGRAFVTIGK
M6	HTN NP(221–228)	H-2K^b	SVIGFLAL
E7	HPV E7(44–57)	H-2K^b	QAEPDRAHYNIVTF

2 μm for MLV. Freezing-thawing vesicles are prepared as described by Zhou et al.[10] Small unilamellar vesicles (SUV) from MLV are produced by bath-type or tip-type sonication above the phase transition temperature of the lipids in aqueous solution.[11] The SUV solution is transparent with an average size of 70 $\pm$ 20 nm and further incubated at room temperature (RT) for 1 h. The SUV are briefly sonicated to enhance the peptide encapsulation efficiency, then rapidly frozen in liquid nitrogen and slowly thawed at RT. The freezing–thawing process is repeated three or four times. The prepared FTV have an average size of 800 $\pm$ 200 nm, and are stored in liquid nitrogen.

Extrusion Method. A lipid film is prepared in a round-bottom flask by thoroughly evaporating the chloroform in a rotary evaporator. N_2 gas is flushed for about 20 min, and then the peptide solution is poured into the flask. After hydration with the peptide solution, the lipid film is suspended by vortexing vigorously. The liposome suspension is sonicated briefly in a bath-type sonicator until the lipid film is completely suspended. This solution is extruded serially through 3.0-, 1.0-, and 0.4 μm polycarbonate membrane filters using Extruder (Northern Lipids Inc., BC, Canada).[11,12] The average liposome size is 350 $\pm$ 50 nm for extrusion liposomes (EXT) after filtered with 0.4-μm membrane filter.

Dehydration-Rehydration Method (Freeze-Drying method). Freeze-dried liposomes are prepared by the same process as the extrusion method, except that PBS containing trehalose (at four times the amount of lipid (w/w)) is used. After extrusion through the polycarbonate membranes, the liposome solution is lyophilized. Dried liposomes are stored at 4° and rehydrated with distilled water before use.

---

[10] F. Zhou, B. T. Rouse, and L. Hang, *J. Immunol. Meth.* **145**, 143 (1991).
[11] N. Düzgüneş, *Methods Enzymol.* **367**, 23 (2003).
[12] J. Hernandez, J. Estelrich, and R. Pouplana, *J. Microencapsul.* **4**, 315 (1987).

pH Sensitivity of the Liposomes

ANTS/DPX encapsulating liposomes have a very low have fluorescence intensity, but when entrapped ANTS is diluted in the external medium, the fluorescence intensity increases.[13] The pH sensitivity of the pH-sensitive liposomes is determined by the degree of leakage of the ANTS/DPX. They are then diluted in HEPES buffer at different pH to 30 μM of final lipid concentration. After 1 h incubation at 37°, the fluorescence intensity is measured at 360 nm for excitation and 535 nm for emission, using a JASCO FP-777 spectrofluorometer. The 0% leakage is set by the fluorescence intensity of the liposomes immediately after passage through a Sephadex G-50 column (1 cm × 20 cm) at RT,[14] and the 100% leakage is obtained after disruption of the liposomes with Triton X-100. Liposomes are stable at neutral pH, and as the pH is lowered, the leakage is gradually increased. At pH 6.0 and 5.0, 30% and 60% to 70% of encapsulated ANTS/DPX are released during incubation, respectively.

Encapsulation Efficiency and Size Determination of pH-Sensitive Liposomes

Peptide encapsulation efficiency can be determined by HPLC. To separate the unencapsulated free peptide from the peptide encapsulated in pH-sensitive liposomes, the liposome suspension is centrifuged at 100,000g for 30 min, and the supernatant is collected. This supernatant and the peptide solution before liposome preparation are analyzed with the same analytical HPLC column chromatography. The areas of peptide peaks from these two fractions are compared with each other.

Another method to determine the encapsulation efficiency uses AMCA fluorescence intensity. The N-terminus of the peptide is labeled with AMCA as described.[15] The AMCA-labeled peptide is purified and characterized by reverse-phase C-18 HPLC and spectrofluorometry. The AMCA-labeled peptides encapsulated in liposomes are separated from unencapsulated AMCA peptides by passing through a Sepharose CL-4B column (1 × 20 cm). The encapsulation efficiency is calculated by comparing the fluorescence of AMCA measured before and after passing through the column. The fluorescence intensity of AMCA is measured at 450 nm with the excitation wavelength set at 350 nm.

Liposome size is determined by dynamic light scattering with a Zetasizer 3000 (Melvern, UK).

[13] N. Düzgüneş, *Methods Enzymol.* **372,** 260 (2003).
[14] F. Zhou and L. Hang, *Vaccine* **11,** 1139 (1993).
[15] R. L. Richard, M. Rao, N. M. Wassef, G. M. Glenn, S. W. Rothwell, and C. R. Alving, *Infect. Immunol.* **66,** 2859 (1998).

TABLE II
PHYSICAL PROPERTIES OF pH-SENSITIVE LIPOSOMES

Criteria	Remark
pH sensitivity at pH 6.0	<30% leakage
Stability	<10% at 4° for 300
(Leakage %)	<30% at 37° for 45
Encapsulation efficiency	30%–35%

The physical properties of the pH-sensitive liposomes prepared in this study are summarized in Table II.

Biological Characterization of pH-Sensitive Liposomes: Peptide-Specific Cytotoxic T Lymphocyte Assay

C57BL/6 mice or Balb/c mice are immunized with 50 μg/mouse of peptide encapsulated in pH-sensitive liposomes and boosted after a week. The mice are killed, and the spleens are isolated 2 or 4 weeks after boosting. The splenocytes (5×10^6 cells/ml) in RPMI 1640 culture medium are cultured *in vitro* in the presence of epitope peptide (10 μg/ml) for 6–7 days. The EL4 or P815 cells (ATCC) are incubated with 20 μCi Cr^{51} (Dupont; Boston, MA)/10^6 cells for 2 h and washed with RPMI 1640 medium twice. Cr^{51}-labeled target cells are incubated with 10 μg of epitope peptide for 1 h and washed twice. The peptide-loaded target cells (2×10^4 in 100 μl) are incubated with varying numbers of effector cells (100 μl) in a U-bottomed microwell plate. The Cr^{51} release is measured in 100 μl of supernatant harvested after a 4-h incubation at 37°. The specific lysis is calculated by the following equation: Specific lysis (%) = [(sample release − spontaneous release)/(maximum release − spontaneous release)] × 100. The Cr^{51}-labeled target cells are incubated in 100 μl of 4% Triton X-100 solution for maximum release or in 100 μl of the culture medium for spontaneous release.

Induction of Antigen-Specific Cytotoxic T Lymphocyte Response In Vivo by peptides Encapsulated in pH-Sensitive Liposomes

Immunization of mice with epitope peptides encapsulated in pH-sensitive liposomes induces an epitope peptides specific CTL response. The splenocytes are obtained 2 weeks after boosting and cultured with appropriate peptide to generate effector CTL. These mice effectively mount a CTL response, although there is little difference in the CTL activity between the peptides (Fig. 1).

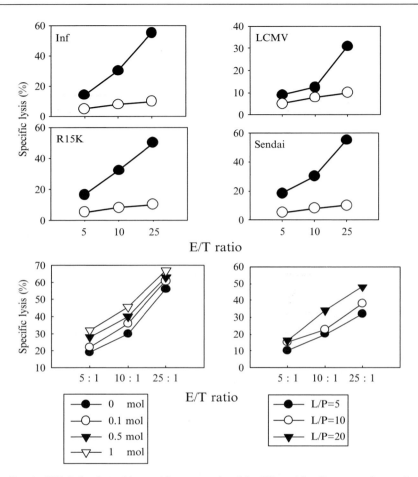

FIG. 1. CTL induction with peptides encapsulated in pH-sensitive liposomes (upper four panels), and the effects of MPL content (lower left) and L/P ratio (lower right) on the peptide-specific CTL response. E/T ratio = effector-to-target ratio.

In addition to the CTL response, the pH-sensitive liposomes can also induce humoral immune responses when an epitope peptide contains a B-cell epitope. When R15K and T26K, both derived from the HIV V3 loop containing B-cell and T-cell epitopes, are encapsulated in pH-sensitive liposomes, they can induce both CTL and humoral immune responses.[8] The humoral immune responses can be measured by enzyme-linked immunosorbent assay and Western blot.

Effect of Liposome Size and Encapsulation Efficiency on the Cytotoxic T Lymphocyte Response In Vivo

Liposomes prepared by different methods induce similar CTL response In this experiment, the concentration of the lipid and peptide was fixed at 500 μg for lipid and 50 μg for peptide/mouse, with the Sendai peptide used as a control antigen. Although the intermediate, MLV, shows lower encapsulation efficiency and CTL inductivity, the FTV, EXT, and DRV show similar encapsulation efficiencies and CTL inductivities. SUV can induce the CTL response regardless of low encapsulation efficiency. Table III summarizes the size of liposomes, encapsulation efficiency, and percent specific lysis of induced CTL.

The Role of the Lipid/Peptide Ratio and Monophosphoryl Lipid Content

Because L/P ratio and MPL are other parameters that may affect the induction of an effective CTL response with pH-sensitive liposomes, peptides must be encapsulated in pH-sensitive liposomes at variable L/P ratios and with varying MPL content to select optimal conditions. For example, pH-sensitive liposomes may be prepared with L/P ratios of 20, 10, or 5. The dependence of the peptide-specific CTL response on MPL content (0–1.0 mol%) is determined with a constant L/P ratio. Figure 1 shows the effects of MPL contents and L/P ratio on the peptide-specific CTL response. In general, higher levels of CTL response can be obtained from mice immunized with more MPL. In the case of pH-sensitive liposomes, antigen-specific CTL responses can be induced without MPL. The L/P ratio regulates nonspecific immune responses. Although the different L/P ratio and MPL contents show a little difference among them, the decrease of nonspecific immune responses is noticeable.

TABLE III
BIOLOGICAL ACTIVITY OF SEVERAL LIPOSOMES REPRESENTED BY DIFFERENT PREPARATION METHODS

Liposomes	SUV	MLV	FTV	EXT	DRV
Size	70 ± 20 nm	2000 ± 500 nm	800 ± 200 nm	350 ± 50 nm	1000 nm
Encapsulation efficiency	12%	12%	35%–40%	35%	25%–30%
CTL activity	40%	15%	65%	60%	50%

Conclusion

The pH-sensitive liposomes are one of the most readily available adjuvants for peptide antigen. Because of low immunogenicity with peptide, a high dose of peptides is required for immunization. Since the pH-sensitive liposomes can deliver the encapsulated antigen into the cytosol, the antigen can be processed and presented in the same manner as endogenous antigens and can induce CTL responses more effectively than other adjuvants. Because of the easy peparation process of pH-sensitive liposomes, minimal difference in adjuvancity among the preparation methods, and the small amount of peptide antigen necessary to induce CTL response, they seem to be a good candidate adjuvant for peptide-based CTL vaccines. Now, our pH-sensitive liposomes are being used as an adjuvant system for peptide-based CTL vaccine against hepatitis B and papillomavirus.

[10] The Interaction of Liposomes with the Complement System: *In Vitro* and *In Vivo* Assays

By Janos Szebeni, Lajos Baranyi, Sandor Savay, Janos Milosevits, Michael Bodo, Rolf Bunger, and Carl R. Alving

Introduction

Liposomes can interact with all four "arms" of the immune system: the cellular and humoral arms of acquired immunity and the corresponding parts of native immunity. As shown in Table I the first two interactions underlie the attempts to use "immune" or "adjuvant" liposomes as antigen carriers to augment T-cell and antibody responses to liposomal vaccines.[1,2] The third interaction, manifested in nonspecific uptake of liposomes by phagocytic cells, has been studied widely as the major clearance mechanism of most "nonstealth" drug carrier liposomes.[3-5] However, this

[1] C. R. Alving, *Ann. N. Y. Acad. Sci.* **754,** 143 (1995).
[2] C. R. Alving, V. Koulchin, G. M. Glenn and M. Rao, *Immunol. Rev.* **145,** 5 (1995).
[3] H. M. Patel, *Crit. Rev. Ther. Drug Carrier Syst.* **9,** 39 (1992).
[4] N. Van Rooijen and A. Sanders, *J. Immunol. Methods* **174,** 83 (1994).
[5] N. van Rooijen, J. Bakker and A. Sanders, *Trends Biotechnol.* **15,** 178 (1997).

TABLE I
CATEGORIZATION OF LIPOSOMES ACCORDING TO THEIR INTERACTION WITH THE IMMUNE SYSTEM

| Liposome type | Provoked immune response | | | |
| | Specific (acquired) immunity | | Nonspecific (innate) immunity | |
	Cellular (T cells)	Humoral (antibodies)	Cellular (macrophages)	Humoral (complement)
Immunogenic	+/−	+	+	+
Drug carrier	−	−	+	+
Stealth	−	−	−	+

interaction is also essential for the immunogenicity of antigen-carrying liposomes, because phagocytic uptake is the first step in antigen presentation.[6,7] The fourth type of interaction of liposomes with the immune system, referred to as nonspecific humoral, is manifested in liposome-induced complement (C) activation with consequent opsonization of vesicles by C split products, mainly C3b and its by-products. Importantly, C activation seems to be a common characteristic of almost all kinds of liposomes, including some of the Stealth® vesicles, which have a long circulation time. The goal of the this chapter is to describe some basic principles, as well as the details, of the most widely used experimental methods in studying the interaction of liposomes with the C system.

The fact that liposomes can interact with the C system was described some 34 years ago by Kinsky *et al.*, who used liposomes as a model membrane to study the mechanism of action of the membrane attack complex (MAC, C5b–9).[8–11] Since then, dozens of studies have dealt with various details of this interaction.[12–14]

[6] M. Rao and C. R. Alving, *Adv. Drug Deliv. Rev.* **41,** 171 (2000).

[7] M. Rao, S. W. Rothwell, and C. R. Alving, *Methods Enzymol.* **373,** 16 (2003).

[8] J. A. Haxby, C. B. Kinsky and S. C. Kinsky, *Proc. Natl. Acad. Sci. USA* **61,** 300 (1968).

[9] J. A. Haxby, O. Gotze, H. J. Muller-Eberhard, and S. C. Kinsky, *Proc. Natl. Acad. Sci. USA* **64,** 290 (1969).

[10] S. C. Kinsky, J. A. Haxby, D. A. Zopf, C. R. Alving, and C. B. Kinsky, *Biochemistry* **8**(10), 4149 (1969).

[11] C. R. Alving, S. C. Kinsky, J. A. Haxby, and C. B. Kinsky, *Biochemistry* **8,** 1582 (1969).

[12] A. J. Bradley, D. V. Devine, S. M. Ansell, J. Janzen and D. E. Brooks, *Arch. Biochem. Biophys.* **357,** 185 (1998).

[13] J. Szebeni, *Crit. Rev. Ther. Drug Carrier Syst.* **15**(1), 57 (1998).

[14] J. Szebeni, *Crit. Rev. Ther. Drug Carr. Syst.* **18,** 567 (2001).

The C system is composed of some 30 plasma and membrane proteins. Along with the three other proteolytic cascades in blood (coagulation, fibrinolytic, and kallikrein-kinin systems), it plays an essential role in maintaining life. In particular, the C system provides the first line of defense against infection in recognizing and killing foreign cells, orchestrating their clearance, and augmenting the body's second specific response.[15–17] As illustrated in Fig. 1, the central step in C activation, conversion of C3, can

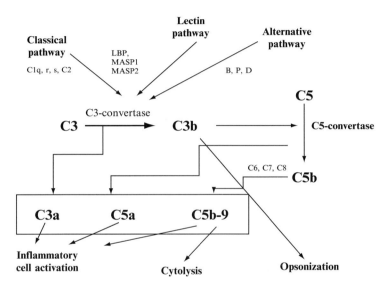

FIG. 1. Scheme of complement activation and its biological consequences. Classical pathway activation involves the binding of antibodies to the vesicles with subsequent activation of the C1 complex, C2, C4, and C3, leading to the formations of the C3 and C5 convertases. In the case of alternative pathway activation, formation of the C3 convertase is triggered by covalent attachment of C3b to the membrane. The anaphylatoxins C3a and C5a activate mast cells, basophils, platelets, and other inflammatory cells with resultant liberation of inflammatory mediators (histamine, PAF, prostaglandins, etc.). These, in turn, set in motion a complex cascade of respiratory, hemodynamic, and hematological changes, promoting inflammation and leading to numerous adverse clinical effects.

[15] Rother, K., Ed., "The Complement System," Springer Verlag, 1998.
[16] J. D. Lambris, and U. M. Holers, Eds., "Therapeutic Interventions in the Complement System," Humana Press, 2000.
[17] J. Bernard and M. J. Morley, Eds. "The Complement Factsbook," Walport Paperback, 2000.

proceed by way of three pathways, each involving different C proteins. The resulting production of C3b, MAC, and anaphylatoxins (C3a, C5a) exert their protective functions by opsonization, direct lysis of invader cells, and activation of inflammatory cells, respectively. The latter effect underlies the immediate clearance and specific immunity-triggering functions of C activation, but it also leads to mobilization of numerous biologically active mediators that assist indirectly in these effector mechanisms.

Bacteria, viruses, yeasts, and essentially all particulate material foreign to the body can activate the C system. However, C is also activated by a variety of activators that do not threaten the host, whereupon the activation itself can cause harm. Such conditions include intraversus administration of certain liposomal drugs, resulting in hypersensitivity reactions in susceptible individuals because of anaphylatoxin production.[14] It is this recently recognized untoward side effect of liposome therapy, referred to as "C activation-related pseudo-allergy" or "CARPA,"[14,18,19] that lends clinical significance to some of the tests of C activation by liposomes described in the following.

Occurrence of Complement Activation by Liposomes

Considering that liposomes represent particulate substances with individual particles having sufficiently large surfaces to bind substantial amounts of different plasma proteins, and that liposomes do not carry natural C inhibitors that prevent C activation on host cells (e.g., sialic residues, C receptor type I [CR1], decay accelerating factor [DAF], membrane cofactor protein [MCP]), it is perhaps not surprising that almost all liposomes can activate the C system when exposed to plasma under appropriate conditions for sufficient time. Thus, in the authors' experience, incubation of liposomes (final phospholipid concentration: 5–10 mM) with undiluted human or animal serum for 20–30 min at 37° results in significant C activation relative to phosphate-buffered saline (PBS) control (i.e., adding PBS instead of liposomes for volume adjustment), regardless of the liposome characteristics. However, there are great differences in the degree of C activation. Small unilamellar, neutral vesicles (DMPC/Chol 55:45 mole ratio) and large multilamellar, negatively charged liposomes with high cholesterol content (DMPC/DMPG/Chol 45:5:71), respectively, represent

[18] J. Szebeni, J. L. Fontana, N. M. Wassef, P. D. Mongan, D. S. Morse, D. E. Dobbins, G. L. Stahl, R. Bünger, and C. R. Alving, *Circulation* **99**, 2302 (1999).
[19] J. Szebeni, B. Baranyi, S. Savay, M. Bodo, D. S. Morse, M. Basta, G. L. Stahl, R. Bunger, and C. R. Alving, *Am. J. Physiol.* **279**, H1319 (2000).

the least and most activating liposome species, at least in the authors' experience (examples will be shown later).

Mechanisms of Complement Activation by Liposomes

As shown in Fig. 1, the central step of C activation, formation of the C3 convertase, can proceed by way of three pathways, with involvement of multiple proteins in each. These activation sequences lend substantial variation and redundancy to the process by which C3 is activated. Taken together with the liposome variables; i.e., lipid composition, size, surface charge, amount and impact of encapsulated material on vesicle properties, which all seem to influence C activation, one may conclude that the mechanism of C activation by liposomes is as diverse as these complex systems are. Various reported mechanisms include classical pathway activation triggered by the binding of specific or natural IgG, IgM, C1q, and C-reactive protein (CRP), and alternative pathway activation triggered by the binding to the activator surface of C3b, IgG, or C4b2a3b.[13]

Measuring the Interaction of Liposomes with the Complement System

Table II lists the various methods available for quantitative analysis of the interaction of liposomes with the C system. One of the two major approaches measures the changes in serum or plasma C levels caused by liposomes (Table IIA), whereas the other analyzes the changes of liposomes occurring as a result of C activation (Table IIB). Here we focus on the first approach and describe in detail later those methods in Table IIA with which the authors have had experience.[18–24] As for C activation–related changes of liposomes, Table IIB provides references containing detailed descriptions of the methods.

[20] J. Szebeni, N. M. Wassef, H. Spielberg, A. S. Rudolph, and C. R. Alving, *Biochem. Biophys. Res. Comm.* **205,** 255 (1994).

[21] J. Szebeni, N. M. Wassef, A. S. Rudolph, and C. R. Alving, *Biochim. Biophys. Acta* **1285,** 127 (1996).

[22] J. Szebeni, N. M. Wassef, K. R. Hartman, A. S. Rudolph, and C. R. Alving, *Transfusion* **37,** 150 (1997).

[23] J. Szebeni, H. Spielberg, R. O. Cliff, N. M. Wassef, A. S. Rudolph, and C. R. Alving, *Art. Cells Blood Subs. Immob. Biotechnol.* **25,** 379 (1997).

[24] J. Szebeni, B. Baranyi, S. Savay, L. U. Lutz, E. Jelezarova, R. Bunger, and C. R. Alving, *J. Liposome Res.* **10,** 347 (2000).

TABLE II

METHODS FOR STUDYING THE INTERACTION OF LIPOSOMES WITH THE COMPLEMENT SYSTEM

Analyte	Species	Assay method	Assay source/procedure
A. Measurement of complement changes			
CH_{50}, a measure of total functional C level in plasma/serum	Human, rat, porcine	Hemolytic activity on sensitized sheep red cells	For procedure see text
	Human	CH_{50} ELISA measuring TCC formation following classical pathway activation	Quidel Corp., San Diego, CA
C3a, one of the direct measures of anaphylatoxin production	Human	C3a-desArg ELISA	Quidel Corp., San Diego, CA, Pharmingen, San Diego, CA
		C3a-desArg RIA (^{125}I)	Amersham Biosciences, Inc., Piscataway, NJ
	Porcine, rat	Bioassay	For procedure see text
C4a, one of the direct measures of anaphylatoxin production	Human	C4a-desArg RIA (^{125}I)	Amersham Biosciences, Inc., Piscataway, NJ
C5a, one of the direct measures of anaphylatoxin production	Human	C5a-desArg ELISA	Quidel Corp., San Diego, CA, Pharmingen, San Diego, CA
		C5a-desArg RIA (^{125}I)	Amersham Biosciences, Inc., Piscataway, NJ
	Porcine	ELISA	For procedure see text
	Porcine, rat	Bioassay	For procedure see text
SC5b-9, a measure of terminal C complex (TCC) formation	Human	SC5b-9 ELISA	Quidel Corp., San Diego, CA
Bb, a measure of alternative pathway C activation	Human	Bb ELISA	Quidel Corp., San Diego, CA
C4d, a measure of classical pathway C activation	Human	C4d ELISA	Quidel Corp., San Diego, CA
B. Measurement of liposome changes			
MAC-caused membrane damage	Any species	Chemical measurement of glucose release	Refs. 25–28
Surface-bound C3b and/or iC3b	Human	ELISA or PAGE analysis of C3b or iC3b deposition on liposomes	Refs. 22,24

ELISA, enzyme-linked immunoassay; PAGE, polyacrylamide-gel electrophoresis; RIA, radioimmune assay; TCC, Terminal C complex.

Measurement of Liposome-Induced Changes in Serum/Plasma
  Complement Levels

Rationale and General Notes

   To quantify liposome-induced changes in serum or plasma C levels
in vitro, one needs first to incubate the vesicles with serum, then to measure
C consumption, or the production of scission products, as indirect or direct
indices of C activation, respectively. (Table III.)
   Complement Activation by Liposomes. Liposomes are incubated with
human or animal serum with constant shaking, typically by adding 50-μl
vesicles from a 40-mM phospholipid stock to 200 μl undiluted serum in Eppendorf tubes, followed by through vortex mixing and incubation at 37° for
30–45 min while shaking at 80 rpm. For negative and positive controls,
serum is incubated with PBS, pH 7.4, and 5 mg/ml zymosan, respectively.
After incubation, the liposomes are separated from the serum by centrifugation (e.g., in a desktop centrifuge at 14,000 rpm for 10 min), and the
serum is either tested immediately, or it is stored at −70° for later tests.
Centrifugal separation of unilamellar liposomes from serum may not be
possible as described previously. We have found, however, that the CH_{50}
or other C changes can be determined in the presence of such liposomes
as well (i.e., the liposomes do not interfere with any of the assays described
in the following).

TABLE III
STANDARD CONDITIONS FOR MEASURING LIPOSOME-INDUCED C ACTIVATION BY CH_{50} ASSAYS

- Use of fresh serum or stored serum that had been stored at −20° to −70° and not thawed
  more than twice.
- Use of Alsever's solution (71 mM NaCl, 114 mM dextrose, 27 mM Na-Citrate, pH 6.1, set
  with 1 M citric acid) to store SRBC.
- Use of veronal-buffered saline, pH 7.4, containing 0.15 mM Ca^{2+}, 0.5 mM Mg^{2+}, and 0.2%
  gelatine (VBS^{2+}-gel) for the incubation of SRBC with serum.
- Use of a constant number of SRBC ($\sim$7 × 10^7/ml)

[25] C. R. Alving, R. L. Richards, and A. A. Guirguis, *J. Immunol.* **118**, 342 (1977).
[26] C. R. Alving, K. A. Urban, and R. L. Richards, *Biochim. Biophys. Acta*, **600**, 117 (1980).
[27] S. Shichijo, G. Toffano, and C. R. Alving, *J. Immunol. Methods* **85**, 53 (1985).
[28] R. L. Richards, R. C. Habbersett, I. Scher, A. S. Janoff, H. P. Schieren, L. D. Mayer, P. R.
    Cullis, and C. R. Alving, *Biochim. Biophys. Acta* **855**(2), 223 (1986).

Measurement of Total C (CH$_{50}$)

Rationale and General Notes. These assays measure functional C level in serum or plasma, a composite measure of all factors contributing to terminal C complex (TCC, C5b–9) formation. TCC can be measured either indirectly, by means of its hemolytic activity on heterologous sheep red blood cells (SRBC) sensitized with specific antibodies (hemolysins), or directly, by TCC (ELISA). In actuality, these assays measure the level of the C factor that is rate limiting to TCC formation, which, in human plasma, can be either C2 (the least abundant C protein in plasma with normal levels in the 11–35 μg/ml range[17]) or one of the TCC components, the mean levels of which range between 45 and 90 μg/ml[17]. Although higher initially, the levels of the latter C proteins may decline faster during C activation than that of C2 because of the alternative pathway amplification.

A major advantage of the hemolytic CH$_{50}$ assays is that they can be used with most mammalian species' sera and that they require commonly available instruments and inexpensive chemicals. The limitations include their labor intensity (particularly the pipetting of small volumes), the short shelf-life, and the variable interlot performance of SRBC, and the low sensitivity of the assay relative to measuring C scission products.[29] The CH$_{50}$ ELISA is less labor intensive, and it does not use SRBC; thus, its shelf-life and consistency are superior to the hemolytic assays. It also avoids the need for multiple dilutions of test samples, the method can be automated for use in large, fully automated laboratories, as well as in semiautomated smaller laboratories. Unlike the hemolytic assays, however, the CH$_{50}$ Eq EIA can be used only for human serum samples; animal or ethylenediamine tetraacetic acid (EDTA)-anticoagulated human plasma are unusable.

As mentioned, there are numerous versions of the hemolytic CH$_{50}$ assay, whose arbitrary standardization with respect to concentration of sensitized SRBC (10^8/1.5 ml), concentration and type of sensitizing antibody (heterophilic Forssmann, i.e., rabbit antisheep RBC glycolipid), Ca^{++} (0.15 mM), Mg^{++} content (0.5 mM), and pH (7.4) of the solvent traces back to Osler *et al.*,[30] and Mayer.[31] The procedures described here represent its adaptations to small volumes, thereby conserving serum and test materials.

[29] D. Labarre, B. Montdargent, M.-P. Carreno, and F. Maillet, *J. Appl. Biomat.* **4,** 231 (1993).
[30] A. G. Osler, J. H. Strauss, and M. M. Mayer, *Am. J. Syph. Gonorrhea Vener. Dis.* **36,** 140 (1952).
[31] M. M. Mayer, Complement and complement fixation, in "Kabat and Mayer's Experimental Immunochemistry," 2nd ed., E. A. Kabat and M. M. Mayer Eds., p. 133, Springfield, IL, Charles C. Thomas, 1961.

A Simplified Tube Assay for Hemolytic C (CH_{50}). This assay measures CH_{50} in absolute terms (i.e., by the number of CH_{50} units present in 1 ml serum), where CH_{50} unit denotes the volume of serum that hemolyses approximately 3.3×10^7 sensitized SRBC (50% of 6.7×10^7 cells) in 1 ml veronal-buffered saline containing 0.2% gelatin (VBS^{2+}-gel).[26,27] The assay is carried out as follows.

1. Sheep RBCs (suspended in 3 volumes of Alsever's solution) are washed three times and suspended in VBS^{2+}-gel at 6.7×10^7/ml. A simple way to adjust this cell count is to measure the hemoglobin (Hb) content of washed SRBC stock (e.g., SRBC suspended in VBS^{2+}-gel after the third wash) following hemolysis with Na_2CO_3. In particular, we establish the volume of SRBC necessary to add to 1 ml of 0.1% Na_2CO_3 to give an OD of 0.7, and then add this volume to each ml of VBS^{2+}-gel to prepare the final test suspension. For example, to prepare 50 ml SRBC test suspension, we add 50 times the volume of washed SRBC that gives an OD_{540} of 0.7 after lysing in 1 ml of 0.1% Na_2CO_3.

2. Sensitization of SRBCs is done by mixing anti-sheep erythrocyte antiserum (hemolysin) to the preceding cell suspension, typically at a 1:1000 hemolysin to SRBC volume ratio.

3. Cells are allowed to stand for 30 min at room temperature and then aliquoted in Eppendorf tubes, placing 1 ml in each.

4. From each serum sample 4–5 aliquots are added to different 1-ml SRBC samples, in volumes increasing in the 2–20 μl range. Thus, if we wish to test the C activating capabilities of three different liposomes in one human serum, we will have four serum samples to test: the PBS baseline and the three liposome-exposed sera. Taking 5 aliquots from each sample (e.g., 3, 6, 9, 12, and 15 μl), we will require 24–26 1-ml sensitized SRBC, $4 \times 5 = 20$ for the four serum samples, two to three for spontaneous hemolysis control (e.g., to which 15 μl PBS is added), and two to three for 100% hemolysis control (e.g., to which 1% Tritox X-100 is added).

5. The tubes are incubated for 1 h at $37°$ while shaking at a rate of 80 rpm.

6. SRBC are sedimented by centrifugation at 14,000g for 5 min, using a benchtop (Eppendorf) centrifuge.

7. OD_{540} is read in the supernatant and used to compute the percentage of erythrocytes lysed by the liposome-exposed serum using the formula: % = $(OD_L - OD_{PBS})/OD_{MAX} - OD_{PBS}) \times 100$, where OD_L, OD_{PBS}, and OD_{max} are OD_{540} readings in the liposome-exposed test samples, PBS baseline, and Triton X-100 control, respectively.

8. For each liposome, samples plot the percent hemolysis values against the volume of serum added (V) (e.g., % vs 3, 6, 9, 12, and 15 μl serum). The

test is validated by establishing that this dose-response curve is sigmoid. If so, it follows von Krogh's equation: $x = k\ y/(100-y)^{1/n}$, or its logarithmically transformed form, $\log x = \log k + 1/n\ \log\ [y/(100-y)]$, where x, k, y and n denote V, CH_{50}, percent and a constant, respectively.

9. The log V values are plotted on the y axis against $\log \%/(100-\%)$ on the abscissa, and regression analysis is carried out using the points that lie between 10% and 90% hemolysis. These lines cross the y axis at $\log CH_{50}$; thus, from the equation of the regression line ($y = ax + b$), the constant (b) will give $\log CH_{50}$, whereas the slope, (a), gives $1/n$. CH_{50} will then be derived as 10^b.

Notes. The preceding assay introduces some trivial changes relative to other CH_{50} methods that simplify the calculation of CH_{50}/ml, such as the use of 1-ml assay volumes both for determining the SRBC count and for the incubations of SRBC with serum. In the authors' experience,[19,20,22] regression lines constructed as described previously from three to five points in the effective dynamic range of the assay have R^2 in the 0.97–1.00 range. Notwithstanding the relative technical simplicity and accuracy of CH_{50}/ml determinations, in studies in which the C-activating capabilities of liposomes are compared in different sera, it is more straightforward to compare the changes relative to baseline than giving the absolute CH_{50}/ml. This practice is illustrated in Table IV, which presents data from a recent study from our laboratories comparing the C-activating capabilities of different liposomes in pig serum.[19] We gave mean C consumption ($\pm$ SE) in terms of percent which was calculated from measurements in three to six different sera, using the formula: $100-(CH_{50-L}/CH_{50-PBS})$. CH_{50-L} denotes CH_{50}/ml in the presence of liposomes, whereas CH_{50-PBS} denotes CH_{50}/ml in the PBS controls, which corrects for spontaneous C consumption because of the incubation. In addition to ranking the liposomes in terms of C-activating power, Table 4 also highlights the time-dependence and lipid concentration–dependence of C consumption in some preparations.

Plate Assay for Hemolytic C (CH50). Osler and Mayers' CH_{50} assay has also been adapted to 96-well plates, allowing further reduction of assay volume and, at the same time, simplifying the assay procedure. Such assay was used recently, for example, by Plank *et al.*[32] to compare C activation by different cationic peptides. The assay developed in our laboratory consists of the following steps:

After incubation with liposomes, serum samples are diluted five-fold in VBS^{2+}-gel, and 2- to 20-μl aliquots are added to quadruplicate wells of 96-well ELISA plates.

[32] C. Plank, M. X. Tang, A. R. Wolfe, and F. C. Szoka, *Hum. Gene Ther.* **10**, 319 (1999).

TABLE IV

COMPLEMENT ACTIVATION BY DIFFERENT LIPOSOMES IN PIG SERUM *In Vitro*

Liposome	Complement consumption (% relative to baseline) 1 ml/ml[a]		8 mg/ml
	10 min[b]	30 min	5 min
DMPC/DMPG/Chol (50:5:45) LMV	18.9 ± 4.2 (5)	32.5 ± 7.6 (6)	59.6 ± 14.7 (3)
DMPC (100) LMV	8.9 ± 4.9 (5)[c]	16.0 ± 5.8 (6)[c]	
DMPC/Chol (55:45) LMV	9.4 ± 9.9 (6)[d]	16.4 ± 5.7 (6)[d]	
DMPC/DMPG (95:5) LMV	13.9 ± 7.2 (5)	24.7 ± 8.7 (6)[c]	
DMPC/DMPG/Chol (24:5:71) LMV	100 ± 0 (3)[d]		
DMPC/DMPG/Chol (50:5:45) LUV	-4.1 ± 4.5 (5)[d]	17.6 ± 9.6 (6)[d]	33.2 ± 3.1 (3)[d]

Dimyristoyl and distearoyl phosphatidylcholines (DMPC and DSPC), dimyristoyl phosphatidylglycerol (DMPG), and cholesterol (Chol) were purchased from Avanti Polar Lipids (Alabaster, AL). Entries are mean C consumption $\pm$ S.E., with (n) independent tests with different pig sera. Complement consumption was calculated by the formula: $100-(CH_{50L}/CH_{50PBS})$, where CH_{50L} denotes CH_{50}/ml values measured in the presence of liposomes, and CH_{50PBS} denotes CH_{50}/ml values measured in PBS-containing controls. Baseline CH_{50}/ml at 10 min: 48.8 ± 4 ($n = 6$).

[a] Final lipid concentration.
[b] Incubation time. ANOVA of C consumption values at each incubation time, followed by Student-Newman-Keuls test indicated significant difference relative to DMPC/DMPG/Chol (50:5:45) LMV at
[c] $P < 0.05$
[d] $P < 0.01$.

1. The volumes are adjusted with VBS^{2+}-gel to match the highest added volume (e.g., 20 μl). Positive (maximum hemolysis) and negative (spontaneous hemolysis) controls are wells to which 20 μl of 5% Triton X-100 or 20 μl VBS^{2+}-gel are added, respectively.

2. To each well 200 μl sensitized SRBC (10^9 cell/ml VBS^{2+}-gel) is added. The suspension is made as described previously for the tube assays.

3. The plate is then incubated at $37°$ for 1 h with shaking, followed by centrifugation at 4000 rpm for 10 min, using a plate centrifuge. The supernatants are transferred to another plate, and A_{540} is read in a plate reader.

4. Computation of CH_{50} is similar to that described for the tube assay, except that the readings need to be corrected (multiplied) by a factor of 1.1 (220/200) (i.e., the dilution of SRBC by the samples). (This factor is 1.02 with the tube assay, which is neglected).

Notes. The preceeding CH_{50} plate assay decreases the test volume (the five-fold dilution of serum/plasma compensates for the five times less volume of SRBC). The procedure is simpler and shorter than the tube assay owing to the feasibility for serial pipetting and one-step OD reading, and the use of quadruplicate wells for each plasma dilution ensures increased accuracy. The assay also offers relative analysis ($A_{540-L}/A_{540-PBS}$) rather than computing the absolute CH_{50}/ml.

CH_{50}-Equivalent Enzyme Immunoassay. An alternative approach to assess total C levels in blood is an ELISA that measures formation of SC5b-9 in serum following C activation by a standardized activator. Like hemolysis in the hemolytic assays, here the level of SC5b-9 correlates inversely with prior C activation in serum causing C consumption. The ELISA (Quidel, CH_{50} Eq EIA) uses human gamma globulins and murine monoclonal antibodies as C activators, whereas the monoclonals capturing and detecting SC5b-9 are the same as in the SC5b-9 kit (discussed later). The details of this assay are specified in the kit.

Measurement of Complement Scission Products

Rationale and General Notes. As shown in Fig. 1, C activation generates an array of proteolytic scission products displaying various biological activities. The most important ones and their derivatives include the anaphylatoxins, C3a, C5a, and SC5b-9, a stable, nonlytic form of the TCC. These activation products, as well as many others discussed later, can be measured in human serum or plasma with commercially available ELISA or radioimmunoassay kits, listed in Table 2. The specificity of these immunoassays for C activation lies in the fact that they use proprietary monoclonal antibodies to capture only the cleavage products, by means of neo-epitopes, but not the parent, unclipped C molecules. HRP-labeled or radiolabeled antibodies raised against other epitopes on the product then detect the trapped C fragments.

One important common practice with these assays is the preservation of antigens in plasma until the assay. Thus, if C activation is studied in whole blood (e.g., *in vivo*), one needs to block C activation and to separate blood cells from plasma immediately after sampling to prevent further *ex vivo* activation and uptake of anaphylatoxins by leukocytes. The addition of 10–20 mM EDTA or other Ca^{++} chelators (citrate) represents effective measures to block both coagulation and C activation, although they do not prevent further activation by means of the lectin pathway. The latter process can be inhibited by adding the serine protease inhibitor, futhan, to the samples in addition to EDTA.[33] If C activation is studied *in vitro*, in serum, the reaction may need to be stopped by EDTA. Regardless of

[33] P. H. Pfeifer, M. S. Kawahara, and T. E. Hugli, *Clin. Chem.* **45,** 1190 (1999).

the activation process, if the C assay cannot be done immediately, it is mandatory to store the plasma or serum samples at $-20°$, or, for extended periods, at $-70°$.

Anaphylatoxin Immunoassays. These kits measure C3a-desArg and C4a-desArg, which are stable end-products arising from the scission of a terminal Arg from the short-lived C3a and C4a molecules by the omnipresent carboxypeptidase N. Positivity in Quidel's C3a-desArg and/or C5a-desArg ELISAs, or by Amersham's C3a-desArg, C4a-desArg, and C5a-desArg RIAs, can therefore be taken as direct indicators of C activation and anaphylatoxin production.

TCC ELISA. Quidels SC5b-9 ELISA measures the cleavage of C5 and subsequent terminal pathway activation. Specifically, SC5b-9 is generated by the assembly of C5-C9 as a consequence of C activation via all three pathways and subsequent binding to the naturally occurring regulatory serum protein, the S protein (vitronectin). Actually, the S protein binds to nascent C5b–9 complexes at the C5b–7 stage of assembly. Regarding this assay, it should be pointed out that during C activation C5b–9 is also deposited on the activator surface as the membrane attack complex (MAC), the alternative form of TCC that mediates irreversible membrane damage. Thus, SC5b–9 represents only a part of all activated C5b and terminal complex molecules, and although its formation is proportional to C activation, SC5b-9 does not quantify total TCC.

Other Miscellaneous ELISAs. Quidel C4d and Bb ELISAs are specific for classical and alternative pathway C activations, respectively. Both analytes are late by-products of C activation: C4d is a scission derivative of C4b, whereas Bb rises in blood as a consequence of spontaneous dissociation, or factor H- or C receptor type I (CR1)-catalyzed scission from the alternative pathway C3 convertase (C3bBb). Quidel iC3b ELISA measures iC3b, a cleavage product of C3b, it is an additional marker for C activation via all three pathways.

Examples for C Scission Product Immuno Assays. Figure 2 illustrates a characteristic feature of liposome-induced C activation in humans, the substantial individual variation in both the extent and the spectrum of changes of different activation markers. In the experiment presented,[22] Hb plus albumin-containing large unilamellar liposomes (LEHA) were incubated with 10 different human sera, and the changes in SC5b-9, Bb, and C4d were measured. Consistent with C activation, LEHA caused significant increases of serum C4d, Bb, and SC5b-9 in most, but not all, sera, and the changes of different markers were not necessarily parallel with each other. The increment in serum levels of these C fragments was smallest with C4d, up to fourfold, and greatest with C5b–9, up to 10-fold in some subjects.[22] Another similar study,[34] using Taxol as C activator, suggested a tendency

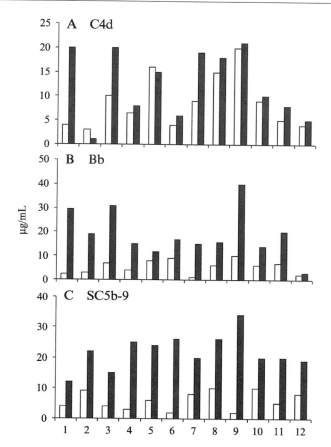

Fɪɢ. 2. Effect of liposomes on serum levels of C4d (A), Bb (B), and SC5b-9 (C) in normal human sera. Sera were incubated with PBS (empty bars) or with microfluidized (large unilamellar) liposomes containing cross-linked $(\alpha\alpha)$ Hb and human serum albumin (filled bars). Vesicles were separated by centrifugation and the above split products were determined in the supernatant by ELISA kits described in the text. Data obtained from Ref. 22 with permission.

for increased individual variation of activation markers that are multiple steps downstream in the proteolytic cascade from C3 conversion. In particular, the individual variation of different scission products increased in the following order: Bb < C3a-desArg < iC3b < Bb < SC5b–9.

[34] J. Szebeni, C. R. Alving, S. Savay, Y. Barenholz, A. Priev, D. Danino, and Y. Talmon, *Intern.* **1,** 721 (2001).

In Vivo Assays of Liposome-Induced Complement Activation

Rationale and General Notes

Complement activation has been well known to cause major physiological changes in animals and humans, with symptoms listed in Table V.[13] Importantly, some liposomal anticancer and antifungal drugs most importantly liposomal doxorubicin (Doxil), have been reported to cause acute allergic symptoms in patients, which can be explained with C activation.[24,35] We reported that these same liposomal drugs, as well as other liposomes, cause physiological changes in pigs[18,19,24] and rats[20] by means of C activation, some of which correspond to the human symptoms of hypersensitivity.[14] We found that pigs are particularly sensitive to liposomes, developing easily quantifiable hemodynamic changes following intravenous injection of minute (milligram) amounts of C activating liposomes.[18,19,24] In particular, intravenous injection of certain liposomes was shown to cause significant rises in pulmonary arterial pressure (PAP), rises or falls in systemic arterial pressure (SAP) and heart rate, and falls in cardiac output (CO). The electrocardiography changes include arrhythmia with ventricular fibrillation and cardiac arrest, the latter being lethal unless the animal is resuscitated with epinephrine with or without CPR and electroconversion. This cardiopulmonary distress is also associated with transient declines in blood oxygen saturation, reflecting pulmonary dysfunction (dyspnea),

TABLE V

COMPLEMENT-MEDIATED PHYSIOLOGICAL CHANGES IN ANIMALS AND CLINICAL SYMPTOMS OF
COMPLEMENT ACTIVATION IN HUMANS

Physiological changes in animals	Human symptoms
Arrhythmia, bradycardia coronary vasospasm, hypertension, hypoxemia, hypotension, leukocytosis, tachycardia, metabolic acidosis, rash, pulmonary marginalization of leukocytes, respiratory distress, thrombocytopenia	Bronchospasm, chest pain, chill, choking, confusion, coughing, cyanosis, diaphoresis, dispnea, erythema, facial edema, facial rash, fever, flush, headache, hypertension, hypotension, hypoxemia, low back pain, lumbar pain, nausea, metabolic acidosis, pruritus, rash, respiratory distress, skin eruptions, sneezing, tachypnea, urticaria, wheezing

[35] A. Chanan-Khan, J. Szebeni, S. Savay, L. Liebes, N. M. Rafigue, C. R. Alving, F. M. Muggia, *Ann. Oncol.* **14,** 1430 (2003).

and with transient skin reactions (rash, erythema, flushing), thus truly mimicking many aspects of the human hypersensitivity responses to liposomes. These changes were shown to be due to C activation, because C-activating substances mimicked, whereas specific C inhibitors inhibited the reaction.[18,19]

The reason underlying the high sensitivity of pigs to liposome reactions has not been explained to date. Our current hypothesis is that the phenomenon may be due to the presence of macrophages in the lungs of pigs, which, like Kupffer cells in the liver, are directly exposed to the blood. These macrophages may have a low threshold for anaphylatoxin-induced activation and promptly secrete thromboxane, histamine, and other vasoactive mediators that mediate the reaction.[18,19]

Although liposome-induced C activation entails major physiological changes in guinea pigs, rats, dogs, and probably many other animals, we describe here solely the use of pigs as an *in vivo* assay for liposome-induced C activation. We believe this model is particularly useful for quantitative assessment of the C-activation potential of liposomal drugs in humans.

Porcine Bioassay of Liposome-Induced Complement Activation

Animals. Adolescent Yorkshire swine (20–40 kg).

Drugs and Chemicals used were halothane or isoflurane, ketalar, and infusion media (0.9% NaCl).

Equipment: anesthesia machine, cardiac output monitor, pressure transducers with monitoring system and software, electrocardiograph, routine surgical instruments for catheter implantation, infusion equipment, syringes, and needles.

Data Recording and Analysis Software. The long-term, simultaneous, digital recording and signal analysis of several hemodynamic and ECG parameters can be performed, for example, using the NI 6011E multifunctional data acquisition board (National Instruments, Austin, TX). It can digitize signal from up to 16 single-ended analog channels (ECG, pco_2, blood pressure, body temperature). Data are recorded at 200 Hz at 16-bit resolution, using software for recording and analyzing physiological data.

Surgical Procedures. Pigs are sedated with intramuscular ketamine (Ketalar) and then anesthetized with halothane or isoflurane through a nose cone. The subsequent steps are as follows. The trachea is intubated to allow mechanical ventilation with an anesthesia machine, using 1%–2.5% halothane or isoflurane. A pulmonary artery catheter equipped with a thermodilution-based continuous cardiac output detector (TDQ CCO, Abbott Laboratories, Chicago, IL) is advanced through the right internal jugular vein through the right atrium into the pulmonary artery to measure PAP, central venous pressure (CVP) and CO. A 6 Fr Millar Mikro-Tip catheter (Millar Instruments, Houston TX) is inserted into the right femoral artery and advanced into the proximal aorta for blood sampling and to measure SAP. A second 6 Fr pigtail Millar Mikro-Tip catheter is inserted through the left femoral artery and placed into the left ventricle to monitor left ventricular end-diastolic pressure (LVEDP). Systemic vascular resistance (SVR) and pulmonary vascular resistance (PVR) are calculated from SAP, PAP, CO, CVP, and LVEDP using standard formulas.[18] Blood pressure values and lead II of the ECG are obtained continually.

Liposome Injections. Liposome stock solutions contain between 5 and 40 mM phospholipid (approximately 5–40 mg/ml lipid) from which 50–200 μl is diluted to 0.5–1 ml with PBS or saline and injected using 1 ml tuberculin syringes, either into the jugular vein through the introduction sheet or through the pulmonary catheter directly into the pulmonary artery. Injections are performed relatively fast (within 10–20 s) and are followed by 10 ml PBS or saline injections to wash in any vesicles remaining in the void space of the catheter.

Hemodynamic and Electrocardiographic Monitoring. Monitoring of hemodynamic parameters (PAP, SAP, LVEDP, CVP), heart rate, and ECG starts 3–5 min before the injections and continues until all hemodynamic parameters return to baseline, usually within 15–30 min. Then, baseline monitoring is started for the next injection.

Blood Sampling. Five- to 10-ml blood samples are taken from the femoral artery into heparinized tubes before each injection (baseline) and at the top of liposome reactions, usually between 4 and 10 min after the injections. Blood is centrifuged immediately at 4°, and the plasma is stored at −20° until the various assays are conducted.

Typical Results. Figure 3 demonstrates the hemodynamic responses of three different pigs to introvenous injections of 5 mg (5 μmol phospholipid in 1 ml PBS) large multilamellar liposomes prepared from dimyristoyl phosphatidylcholine (DMPC), dimyristoyl phosphatidylglycerol (DMPG), and cholesterol (mole ratio: 50:5:45). These injections caused substantial, although transient, hemodynamic changes, including a 50%–250% increase in PAP (Fig. 3A), 0% to 80% decline in CO (Fig. 3B), twofold to sixfold increase in PVR (Fig. 3C), 5%–10% increase in heart rate (Fig. 3D), 20%–40% fall, or rise, or biphasic changes in SAP (Fig. 3E), and 0%–400% rise of SVR (Fig. 3F). These responses were observed within the first minute, reached their peak within 5–6 min, and returned to baseline within 10–15 min.

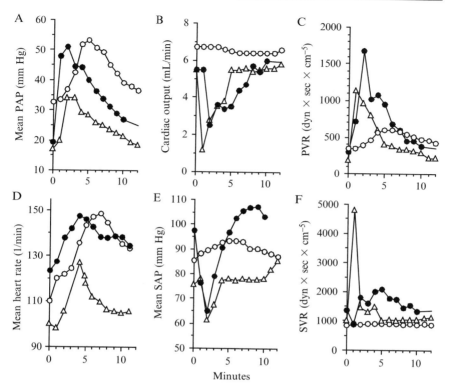

FIG. 3. Hemodynamic changes induced by liposomes in pigs. Typical curves from three pigs injected with the liposome boluses. PAP, pulmonary arterial pressure; PVR, pulmonary vascular resistance; SAP, systemic arterial pressure; SVR, systemic vascular resistance. Different symbols designate different pigs. Data reproduced from Ref. 18 with permission.

Notes. The unique advantage of using pigs to measure liposome-induced C activation is the capability to assess potential reactogenicity of various liposomes intended for therapeutic use in humans. With certain liposomes, namely those with a large multilamellar structure (d = 0.4–10 μm) or with unsaturated egg lecithin components, numerous injections can be given to the same animal at 20 to 40 min intervals over many hours without tachyphylaxis (i.e., attenuation of response). This phenomenon is illustrated in Fig. 4, showing the changes of PAP after eight consecutive injections of the same DMPC/DMPG/Chol–containing multilamellar liposomes (MLV). These MLVs are also "omnipotent", inasmuch as essentially all pigs react to them (in our experience, none of ~70 pigs tested to

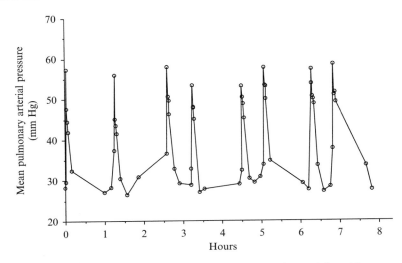

Fig. 4. PAP responses to repetitive liposome injections. A pig was injected introvenously with 5-mg liposomes consisting of DMPC/DMPG/cholesterol (45:5:50 mole ratio) eight times at the indicated time points, and the changes of PAP were recorded. Data reproduced from Ref. 18 with permission.

date failed to react to MLV). With other liposomes, such as Doxil or other pegylated large unilamellar vesicles (d = 100–150 nm), the animals may lose sensitivity after the first or second injection, possibly because of the consumption of natural antibodies whose reaction with liposomes triggers the physiological changes. Interestingly, sensitivity is maintained to MLV or other unilamellar preparations differing in phospholipid composition (unpublished data).

Limitations. The main limitation of the preceding pig bioassay is the need for special instruments and skills to conduct the surgery.

[11] Archaeobacterial Ether Lipid Liposomes as Vaccine Adjuvants

By G. Dennis Sprott, Girishchandra B. Patel, and Lakshmi Krishnan

Introduction

Archaeosomes, liposomes composed of the polar lipids unique to the domain *Archaea*, have unexpected adjuvant properties, first reported in 1997 for the humoral response in mice.[1] Mixtures of total polar lipids extracted from each of several archaea were tested, and, although all formed archaeosomes that promoted an enhanced humoral response to the associated protein antigen, there were differences noted in the degrees of adjuvant activity. These differences undoubtedly reflect subtle differences in the structural properties of the lipids.

The polar lipids of all *Archaea* share the constant-length phytanyl chains, usually fully saturated, and bonded by ether linkages to a glycerol backbone in a mirror image *sn*-2,3-configuration to the glycerolipids of the domains *Bacteria* and *Eukarya*.[2] However, different archaea express their own unique pattern of polar lipids, which is well conserved within the genus level of classification, caused by variations in both the polar head groups and, in some cases, to the core lipid structure itself. In the case of core lipids, the *sn*-2,3-diphytanylglycerol is ubiquitous to all strains, but some have the ability to form additional lipid cores (Fig. 1). Those polar head groups exposed to the outer surface of archaeosomes have the potential to interact with mammalian receptors, whereas the type and proportion of lipid cores markedly influence the stability[3] and permeability[4] of the archaeosome structure. *In vitro* studies indicate that a controlling factor for the stability of archaeosomes is the proportion of caldarchaeol membrane–spanning core lipids.[3,5] *In vivo* stability studies comparing various archaeosome types have yet to be reported.

Archaeosomes are mixed adjuvants capable of promoting strong humoral (antibody and Th2),[1,6] cell-mediated (Th1),[6] and cytotoxic T-cell

---

[1] G. D. Sprott, D. L. Tolson, and G. B. Patel, *FEMS Microbiol. Lett.* **154,** 17 (1997).

[2] M. Kates, *Biochem. Soc. Symp.* **58,** 51 (1992).

[3] C. G. Choquet, G. B. Patel, and G. D. Sprott, *Can. J. Microbiol.* **42,** 183 (1996).

[4] J. C. Mathai, G. D. Sprott, and M. L. Zeidel, *J. Biol. Chem.* **276,** 27266 (2001).

[5] C. G. Choquet, G. B. Patel, T. J. Beveridge, and G. D. Sprott, *Appl. Microbiol. Biotechnol.* **42,** 375 (1994).

[6] L. Krishnan, C. J. Dicaire, G. B. Patel, and G. D. Sprott, *Infect. Immun.* **68,** 54 (2000).

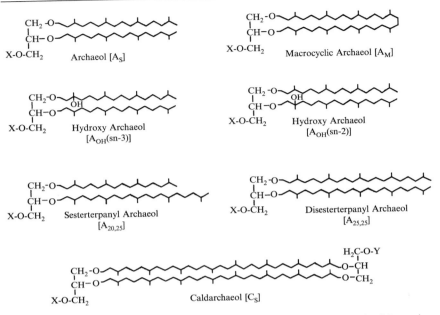

FIG. 1. Structures of the most commonly found polar lipid cores synthesized by various members of the *Archaea*. Symbols X and Y represent the various headgroups of polar lipids or protons in the case of the lipid cores. Standard caldarchaeol (C_s) is 2,2',3,3'-tetra-*O*-dibiphytanyl-*sn*-diglycerol; standard archaeol (A_s) is 2,3-di-*O*-phytanyl-*sn*-glycerol; hydroxyarchaeol is (A_{OH}), and archaeol with an additional C5 unit to form a sesterterpanyl chain ($A_{20,25}$ or $A_{25,25}$). *M. smithii* total polar lipids are caldarchaeols, archaeols, and minor amounts of hydroxyarchaeols.[18]

responses[7] to entrapped protein antigens. Physical association of the protein antigen with the archaeosome seems to be important for the induction of a strong humoral response, and all major antibody isotypes (IgG1, IgG2a, IgG2b) are induced.[6] Immunization of mice with archaeosome-encapsulated bovine serum albumin, ovalbumin (OVA), or lysozyme results in strong cell-mediated immune responses as well, measured as antigen-dependent proliferation of splenic cells, and substantial production by these cells of gamma interferon (a Th1 cytokine) and interleukin-4 (a Th2 cytokine).[6] In addition to major histocompatibility complex (MHC) class II presentation, archaeosomes induce a strong MHC class I presentation of associated protein antigens by antigen presenting cells (APCs), both *in vitro* and *in vivo*.[7] The induced cytotoxic T-cell response is CD8[+] T–cell dependent, because killing of target cells does not occur on removal of

[7] L. Krishnan, S. Sad, G. B. Patel, and G. D. Sprott, *J. Immunol.* **165**, 5177 (2000).

effector CD8$^+$ T cells and is primarily perforin mediated, because killing was not induced in perforin-deficient mice.[7] Processing of antigen is Brefeldin A sensitive, suggesting that archaeosome-antigen presentation is through the classical cytosolic pathway with peptides being transported through the endoplasmic reticulum and presented by MHC class I molecules.[7] Finally, adjuvant activity is sufficiently potent to bypass a requirement for CD4$^+$ T-cell help.[7] Comparisons with conventional liposomes and alum were made throughout all these adjuvant activity studies and indicate clearly the superior adjuvant activities of archaeosomes.[6,7] Evidently, archaeosome-antigen is taken up and released into both cytosolic and phagolysosome compartments for processing, although the exact mechanism of antigen delivery has yet to be determined. To date, total polar lipid extracts from all archaea tested form archaeosomes capable of antigen delivery and subsequent presentation by both MHC pathways.

Initially, we believed that the mechanism of adjuvant activity with archaeosomes was primarily attributable to a superior antigen-carrier property, featured by enhanced phagocytosis of archaeosomes by APCs compared with other liposome types.[8] Discovery of immune modulating activity by archaeosomes soon changed this view.[9] The addition of archaeosomes to cultures of J774A.1 macrophages resulted in up-regulation of costimulatory molecules B7.1 and B7.2, and MHC class II molecules on the cell surface. Similar results were found for mouse bone marrow–derived dendritic cells. Importantly, no such up-regulation of genes occurred in response to liposomes made from dimyristoylphosphatidylcholine(DMPC), dimyristoylphosphatidylglycerol(DMPG), and cholesterol. Furthermore, archaeosomes enhanced both the recruitment and activation of antigen presenting cells (APC) (dendritic cells and macrophages) to the injection site in mice. Finally, an enhanced production of tumor necrosis factor by APCs exposed to archaeosomes was observed. The ability of archaeosomes to activate APCs correlated with an enhanced ability of archaeosome-treated APCs to stimulate allogenic T-cell proliferation.[9]

Induction of immunological memory can be striking following a primary archaeosome-antigen adjuvanted immune response.[6] However, different archaeosome lipid types seem to give varied levels of memory response, and these aspects critical for vaccine development are under active investigation. Recently we demonstrated that an archaeosomal peptide vaccine conferred long-term protection against infection of mice by the intracellular pathogen *Listeria monocytogenes*.[10] Protection against

[8] D. L. Tolson, R. K. Latta, G. B. Patel, and G. D. Sprott, *J. Liposome Res.* **6**, 755 (1996)

[9] L. Krishnan, S. Sad, G. B. Patel, and G. D. Sprott, *J. Immunol.* **166**, 1885 (2001).

[10] J. W. Conlan, L. Krishnan, G. E. Willick, G. B. Patel, and G. D. Sprott, *Vaccine* **19**, 3509 (2001).

this pathogen is known to require an effective cytotoxic T-cell response, implicating archaeosomes as an effective means of conferring CD8$^+$ T-cell memory and corroborating earlier cytotoxic T-cell assays performed up to 5 months subsequent to vaccination.[7] Published reports of archaeosome action have been limited to our own laboratory[11] and feature the polar lipids from *Methanobrevibacter smithii*, *Thermoplasma acidophilum*, and *Halobacterium salinarum*. Because much of our data is for *M. smithii* archaeosomes, we restrict the protocols herein to demonstrate the production of antigen (OVA) loaded *M. smithii* archaeosomes and to assessment of their capability to promote Th1, Th2, and cytotoxic T-cell (CTL) responses in mice.

Growth of *Methanobrevibacter smithii*

M. smithii ALI (DSM 2375) is a mesophilic, methanogenic archaeon isolated from human feces. *M. smithii* derives energy for growth by using 4 moles of hydrogen to reduce 1 mole of carbon dioxide, resulting in the production of 1 mole of methane gas. Hydrogen is the preferred electron donor, with formate as a less than optimal substitute. Methanogenic bacteria are fastidiously anaerobic, requiring an initial growth E_h of below -115 mV at neutral pH. To achieve such low E_h, specialized oxygen-free gas mixtures are required, along with devices to scrub the gases of any contaminating oxygen. Special gassing stations are used to flush vessels free of oxygen, to perform the anaerobic techniques required to reduce the redox potential of the medium, and to protect the culture from exposure to oxygen during culture transfer.[12] The use of sealed culture vessels that are pressurized with a H_2/CO_2 (4:1) gas mixture and use of sulfide require the exercise of extreme caution and safety procedures.[13] These safety precautions must be followed rigorously to avoid injury and damage. By use of the safety measures outlined,[12,13] we have grown methanogens successfully in 58-l batches of media contained in a 75-l fermenter for more than a decade without adverse incident.

M. smithii ALI is grown in a medium modified from Balch *et al.*[14] of the following composition (mg/l): NH_4Cl, 1000; K_2HPO_4, 200; CH_3COO-Na·$3H_2O$, 2500; Bacto yeast extract, 2000; Bacto tryptone, 2000; $NaHCO_3$,

[11] G. B. Patel and G. D. Sprott, *Critical Rev. Biotechnol.* **19**, 317 (1999).
[12] K. R. Sowers and K. M. Noll, *in* "Methanogens" (K. R. Sowers and H. J. Schreier, eds.), p. 15, *in* "Archaea—A Laboratory Manual" (F. T. Robb, ed.-in chief). Cold Spring Harbor Laboratory Press, Plainview, NY, 1995.
[13] L. Daniels, *in* "Methanogens" (K. R. Sowers and H. J. Schreier, eds.), p. 63, *in* "Archaea—A Laboratory Manual" (F. T. Robb, ed.-in chief). Cold Spring Harbor Laboratory Press, Plainview, NY, 1995.

2500; KH_2PO_4, 300; $(NH_4)_2SO_4$, 300; NaCl, 610; $MgSO_4 \cdot 7H_2O$, 160; $CaCl_2 \cdot 2H_2O$, 8; nitrilotriacetic acid, 15; $MnSO_4 \cdot H_2O$, 5; $FeSO_4 \cdot 7H_2O$, 3; $CoCl_2 \cdot 6H_2O$, 1; $ZnSO_4 \cdot 7H_2O$, 1; $CuSO_4 \cdot 5H_2O$, 0.1; $AlK(SO_4)_2 \cdot 12H_2O$, 0.1; H_3BO_3, 0.1; $Na_2MoO_4 \cdot 2H_2O$, 0.1; resazurin, 0.2; pyridoxine-HCl, 0.1; thiamine-HCl, 0.05; riboflavin, 0.05; nicotinic acid, 0.05; p-aminobenzoic acid, 0.05; lipoic acid, 0.05; biotin, 0.02; folic acid, 0.02; vitamin B_{12}, 0.005; 2-mercaptoethanolsulfonic acid-Na salt (HS-CoM-Na salt), 0.008; $Na_2S \cdot 9H_2O$, 500; L-cysteine-HCl, 500; and the following volatile fatty acids (ml/l medium): isobutyric acid, 0.49; 2-methylbutyric acid, 0.55; n-valeric acid, 0.55; and isovaleric acid, 0.55. All ingredients are mixed, except the cysteine-HCl and $Na_2S \cdot 9H_2O$, and the pH is adjusted to 7.6. The medium is boiled briefly, the headspace is flushed with H_2/CO_2, and cysteine-Na_2S solution is added to reduce the medium (resazurin must turn colorless). The reduced medium is dispensed into glass serum vials under H_2/CO_2 gas phase, and the vials are capped with butyl rubber stoppers crimped in place by aluminum seals.[12] The medium is autoclaved (121°, 20 min) in sealed vials contained within a stainless steel metal container with 8-mm diameter holes in the bottom to allow steam entry, and a lid secured by metal clasps. After autoclaving, the medium is allowed to cool to room temperature and is removed from the metal container. The postautoclave pH of this medium after equilibration at room temperature for 24 h is 6.9 ± 0.1. This medium is used to maintain working stock cultures and to prepare inoculum for the fermenter vessels. The organism is inoculated by transferring aseptically and anaerobically using syringe techniques,[12] or modifications thereof, and grown at 35° with agitation.

The same medium is used for fermenters, except that the initial amount of cysteine-HCl and Na_2S is decreased to 60 mg/l of each. All ingredients except the vitamins, volatile fatty acids, HS-CoM-Na salt, $FeSO_4 \cdot 7H_2O$, and cysteine-Na_2S are mixed and sterilized *in situ* in the fermenter (the fermenter is steam sterilized, in place), with the vessel fully vented to the atmosphere (under aerobic conditions) as is the normal practice. After sterilization, the fermenter is sparged with 80% H_2:20% CO_2 to maintain a positive pressure (5–10 psi) as the vessel is cooling (50 l medium in a 75 l Chemap AG fermenter). When the temperature is 35°, the remaining ingredients are added as separately autoclaved solutions, and the pH is adjusted to 6.9. The pH during growth is maintained using sterile 5 N NaOH or 5 N HCl (kept under N_2). Just before inoculation with a mid-logarithmic phase culture (5%–10%, v/v), the medium is supplemented

[14] W. E. Balch, G. E. Fox, L. J. Magrum, C. R. Woese, and R. S. Wolfe, *Microbiol. Rev.* **43**, 260 (1979).

with 16% Na_2S solution (autoclaved, under N_2) to achieve a dissolved sulfide concentration of 0.1–0.2 mM. The sulfide is maintained in that range during growth by further additions, as needed. The 80% H_2:20% CO_2 sparge rate is adjusted upward during growth in response to the increase in biomass. The culture is harvested in mid to late logarithmic growth phase (A_{660} of > 2.0 between 48 and 72 h) using a tangential flow cell separator, and the concentrated cells are centrifuged. In addition to the safety precautions described earlier, additional steps must be adhered to as specified for fermenter-growth of methanogens on H_2/CO_2.[13]

Lipid Extraction

Total polar lipids are prepared from frozen and thawed pastes of *M. smithii* by the method of Bligh and Dyer,[15] as described previously.[16] Thirty grams (dry weight) of cell paste are thawed by adding water to a volume of 420 ml. One to 2 mg of DNase I is included, and the mixture stirred magnetically until cells are dispersed. Solvents are added to obtain a volume of 2 l in a ratio of $CH_3OH/CHCl_3/H_2O$ (v/v) of 2:1:0.8, and the mixture is stirred for 16 h at 23°. The suspension is centrifuged at 4080g for 15 min in 150-ml glass centrifuge bottles. The cell pellet is extracted twice more with 1-l volumes of solvent, and the supernatants containing the lipids are combined. The total volume of extract is divided by 3.8 to obtain the volume of $CHCl_3$ and water needed to achieve phase separation. The bottom $CHCl_3$ phase is removed and combined with a $CHCl_3$ wash of the methanol/water phase to obtain the total lipid extract. The volume of $CHCl_3$ is depleted to dryness by rotary evaporation. The lipid residue is dissolved into $CHCl_3$ adding the minimum volume of methanol required, and the polar lipids are precipitated by adding 20 volumes of ice-cold acetone. After storage at −20° for several hours, the white precipitate of total polar lipids is collected by centrifugation. The precipitate is washed once by redissolving in $CHCl_3$, and acetone-precipitated as before. Any acetone remaining in the pellet is dried with a stream of N_2, and the total polar lipids are dissolved in $CHCl_3$. Lipid extracts dissolved in $CHCl_3$ are stable indefinitely but are placed at 4° to minimize solvent evaporation over long storage periods and the resultant precipitation of lipids from concentrated extracts.

[15] E. G. Bligh and W. J. Dyer, *Can. J. Biochem.* **37**, 911 (1959).
[16] G. D. Sprott, C. G. Choquet, and G. B. Patel, *in* "Methanogens" (K. R. Sowers and H. J. Schreier, eds.), p. 329, *in* "Archaea—A Laboratory Manual" (F. T. Robb, ed.-in chief). Cold Spring Harbor Laboratory Press, Plainview, NY, 1995.

Fast atom bombardment mass spectrometry (FAB MS), a technique in which molecular fragmentation is minimal, and thin-layer chromatography, combined with color development for glyco, amino, phospho, and total lipids,[17] provide convenient checks on batch-to-batch consistency of lipid composition. In our laboratory, this variation has been minimal when using the following protocols, but quality control checks are still recommended. An FAB MS spectrum of a typical polar lipid extract from *M. smithii*[18] is shown in Fig. 2. Note that signal height is not necessarily indicative of relative abundance, as shown in Table I. Indeed, the mol% of caldarchaeol lipids is much higher than predicted from the FAB MS spectrum and accounts for about 40 mol% of the polar lipids, as can be seen from the assignments and quantitative data shown in the Table. The remaining 60 mol% are primarily archaeol lipids, in particular archaetidylserine (30 mol%), β-Glc$p$-(1,6)-β-Glc$p$-(1,1)-archaeol (12 mol%), archaetidic acid (8 mol%), and archaetidylinositol (<8 mol%).

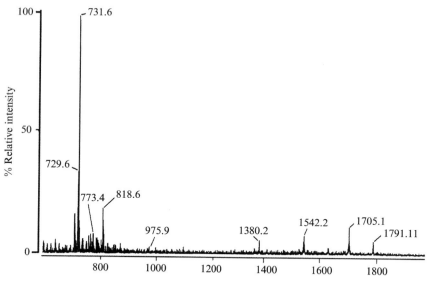

FIG. 2. Negative ion FAB MS analysis of total polar lipids from *M. smithii*. Data[18] are shown with permission from Elsevier Science, Inc.

[17] M. Kates, "Techniques of Lipidology. Isolation, Analysis and Identification of Lipids" (R. H. Burdon and P. H. van Knippenberg, eds.). Elsevier, New York, 1986.
[18] G. D. Sprott, J.-R. Brisson, C. J. Dicaire, A. K. Pelletier, L. A. Deschatelets, L. Krishnan, and G. B. Patel, *Biochim. Biophys. Acta* **1440,** 275 (1999).

TABLE I
QUANTIFICATION OF *M. SMITHII* POLAR LIPIDS

Structure of main lipids[a]	$[M^-]$ m/z[b]	Mol%[b]
$(Glc)_2$-C_s-P	1705	34 ± 2
$(Glc)_2$-C_s-PS[c]	1791	
A_{OH}-PG	821	2 ± 0.3
Unknown	1705	3 ± 0.05
A_s-PI	893	8 ± 0.6
A_s-PS	818	30 ± 2
Glc-C_s-P	1542	3 ± 0.2
A_s-$(Glc)_2$	975	12 ± 0.4
A_s-P	731	8 ± 0.6

[a] C_s and A_s represent caldarchaeol and archaeol lipid cores; P, phosphate; S, serine; G, glycerol; and I, inositol.
[b] The [^{14}C] total polar lipids were separated by two-dimensional thin-layer chromatography. Lipid spots were removed and counted. FAB MS was performed on unlabeled lipid spots obtained from duplicate plates. Some spots contained minor ether lipids not shown here. Further details for these data may be found in Sprott *et al.*[18]
[c] Presence of $(Glc)_2$-C_s-PS is deduced from the Rf value and the ninhydrin positive nature of this spot, which also contains lipid $(Glc)_2$-C_s-P.
Data are shown with permission from Elsevier Science, Inc.

Preparation of Antigen (OVA)-Loaded Archaeosomes

Archaeosomes may be prepared from the total polar lipids of *M. smithii* by simple hydration in the presence of antigen, size reduction, lyophilization, and rehydration. Archaeal lipids are in a liquid crystalline-like state at ambient temperature and, hence, vesicles may be prepared at a range of temperatures including ambient. All glassware are prebaked for 6 h at 180° to render them pyrogen-free and sterilized by autoclaving. Pyrogen-free water is used to prepare autoclaved water and phosphate-buffered saline (PBS contains 160 m*M* NaCl and 10 m*M* potassium phosphate, pH 7.1). When possible, manipulations are performed aseptically in a biohood. Conventional liposomes are prepared from DMPC:DMPG:cholesterol (1.8: 0.2:1.5, molar ratio; Sigma Chemical Co., St. Louis, MO) by the same procedure described in the following for archaeosomes.

1. Thirty milligrams of total polar lipids are transferred with a glass-tipped pipette into a 50-ml round-bottomed flask. A 20-ml scintillation vial may be used instead. Plastic tips should not be used when pipeting chloroform or plasticizer will be transferred.

2. The lipid is dried with an N_2 stream. If solvent is trapped, the lipid film is pierced with a needle. Alternatively, to obtain a thin uniform film of

dried lipid, the solvent is removed by rotary evaporation. The dried lipid is then placed under vacuum on a lyophilizer for 1 h. It is essential that all traces of solvent be removed.

3. Three milliliters of water and 10 glass beads (3-mm diameter) are added, then the antigen. To achieve high loading, OVA (stock 10 mg/ml water) is used at a ratio of 5 mg/10 mg lipid). The mixture is vortexed and left to hydrate on the bench for 2–3 h, and shaken/vortexed a few times during that time to aid in hydration. The hydrated lipid is then sonicated for 2 min in a sonic bath (Branson 3200) and incubated in a 35° shaker for at least an hour.

4. The average vesicle diameter is reduced to about 100 nm. This may be achieved with a Sonic Dismembrator 550 (Fisher) at setting #4 for at least 5 min. Pressure extrusion through 100-nm filters is also effective (sequentially using 800-, 400-, and 100-nm pore filters). The average diameter is assessed with a particle sizer such as a 5-mW He/Ne laser (Nicomp 350, Santa Barbara, CA).

5. Lipid vesicles are frozen in the round-bottomed flask by immersing in dry ice–alcohol, rotating to obtain a thin layer, and dried under vacuum.

6. The dried vesides are rehydrated by adding 0.5 ml of water. The flask is rotated gently to humidify all powder and incubated for 20–30 min. Half a milliliter more of water is added, and the vesicles are sonicated at low energy for 2–5 min in a Branson 3200 bath, or equivalent, to ensure homogeneity. An additional 1.0 ml of water is added, and the vesicles are incubated in a 35° shaker for 1–2 h.

7. The archaeosomes are annealed by incubating at 4° overnight.

8. The archaeosomes are transferred to sterile 15-ml conical tubes and centrifuged at 400g for 5 min in a clinical centrifuge. The small white pellet is discarded.

9. Nonentrapped OVA is removed by centrifugation for 1 h at 327,000g R_{max}. For this step, archaeosomes are transferred to alcohol-sterilized Ti 65.31 plastic tubes and the volume brought up to about 7 ml with water. After centrifugation, the archaeosome pellet is resuspended, into water by vortex mixing and washed twice. Finally, the pellet is re-suspended in 1–2 ml of water.

10. The preparation is filtered using a 0.45-μm, 25-mm diameter, syringe-driven sterilizing filter (Millex-HV, nonpyrogenic, low-protein binding). The archaeosomes are combined with a 0.5-ml water rinse also passed through the filter.

11. The size distribution and dry weight of OVA-loaded archaeosomes are determined. The suspension is diluted to achieve 15 μg OVA/0.1 ml PBS. Typically, the average diameter is 150 ± 100 nm, and lipid yields are 18 to 20 mg.

Quantification of Antigen Loading

The amount of protein antigen encapsulated in archaeosome formulations is determined by an sodiumdodecyl sulfate (SDS)-Lowry colorimetric assay.[19] Interfering color from amino lipids is avoided by lipid depletion before the assay according to Wessel and Flugge,[20] with modifications.

Methanol (0.4 ml) is added to 0.1 ml of the archaeosomal formulation contained in a 1.5-ml microcentrifuge tube, the mixture is vortexed, and 0.2 ml $CHCl_3$ is added. The mixture is vortexed again, 0.3-ml deionized water is added, followed by vigorous vortexing and centrifugation at 9000g for 3 min. The upper phase is removed with a Pasteur pipet and discarded, taking care not to disturb the precipitated protein at the interface. Methanol (0.3 ml), is added to the tube, and the mixture is vortexed and centrifuged for 3 min. The liquid is removed carefully, making sure not to displace any of the precipitated protein pellet. The protein pellet is dried under a stream of nitrogen and resuspended in 0.1 ml of 10% SDS for assay by the SDS-Lowry method.[19] Typical loading is 80–100 μg/mg archaeosomes. Filter-sterilized *M. smithii* archaeosomes loaded with OVA may be stored at 4° in PBS for at least 12 months without loss of immunological responses.

Antigen loading and stability may be assessed also by SDS-polyacrylamide gel electrophoresis (PAGE). Results typical for OVA entrapped in archaeosomes and conventional liposomes reveal some minor interference in band migration by the presence of lipids (Fig. 3).

Immune Responses in Mice

Inbred C57BL/6 female mice, 6–8 wk of age, may be purchased from Charles River Breeding Laboratories (St. Constant, Quebec) or The Jackson Laboratory (Bar Harbor, ME). A minimum of five mice per group are immunized at 0 and 3 wk by subcutaneous injection at the base of the tail with 0.1-ml volumes of PBS containing 15 μg OVA (no adjuvant) or 15 μg OVA entrapped in lipid vesicles. For comparison, groups of mice are also inoculated with 15 μg OVA either entrapped in conventional liposomes or mixed according to the manufacturer's instructions with Imject® Alum (Pierce, Rockford, IL). A naïve group receives no injections. The animals are test-bled periodically from the tail vein to measure the systemic antibody titers. Spleens from euthanized mice are removed aseptically in a biohood and prepared for proliferative and cytotoxic T-cell assays as described in the following:

[19] G. L. Peterson, *Anal. Biochem.* **83,** 346 (1977).
[20] D. Wessel and U. I. Flugge, *Anal. Biochem.* **138,** 141 (1984).

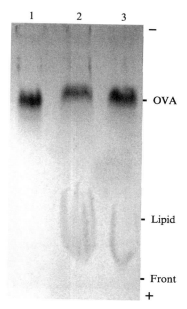

Fig. 3. SDS-PAGE analysis showing that SDS Lowry provides a valid, comparable estimate of OVA entrapment in *M. smithii* archaeosomes and conventional liposomes. SDS Lowry analysis was performed on OVA entrapped in archaeosome and conventional liposome preparations. On the basis of these SDS Lowry data, lanes were loaded with 0.7 μg OVA (lane 1), 0.7 μg OVA entrapped in 37 μg of archaeosomes (lane 2), or 0.7 μg OVA entrapped in 23 μg of conventional liposomes (lane 3). Data from Krishnan *et al.*[7] are shown with permission from *The Journal of Immunology*.

1. The spleens are removed from duplicate mice and placed in a Petri dish, with about 10–15 ml RPMI medium (Gibco-BRL, Life Technologies Inc., Grand Island, NY), and a single cell suspension is produced by grinding between the frosted ends of two sterile glass slides.

2. The cell suspension is filtered into a 15-ml centrifuge tube, using a 70-μm nylon membrane filter (both from Falcon, Becton Dickinson, Franklin Lakes, NJ), and the cells are harvested in a refrigerated centrifuge (4°, 10 min at 470g).

3. The supernatant is discarded, and the cell pellet is tapped gently to loosen any clumps. Five milliliter of TRIS-buffered ammonium chloride (RBC lysing buffer, Sigma Chemical Co.) is added to resuspend the pellet by drawing the mixture into and out of a 10-ml pipette for about 1 min. When the cell suspension turns slightly yellow, indicating RBC lysis, 10 ml of RPMI 1640 containing 8% FBS (HyClone, Logan, Utah) is added and mixed thoroughly. The cells are harvested by centrifugation at 470g.

4. The spleen cells are washed once or twice with RPMI 1640. Each wash will include gently tapping the cell pellet and resuspending the cells in RPMI 1640 medium and spinning at 470g to harvest the cells. After the final wash, the pellet is resuspended in 5 ml of RPMI + 8% FBS medium.

5. The cells are counted in a hemocytometer. The recommended dilution is 10 μl of spleen cell solution diluted with 90 μl of RPMI (dilution 1). Ten microliters of dilution 1 are removed and 10 μl of trypan blue (0.4%, Sigma Chemical Co., St. Louis, Mo.) is added. The number of cells in the 4 sets of 16 corner squares (WBC chamber) of the hemocytometer are counted

$$\text{No of cell/ml} = \text{No of cells per 16 squares} \times 20 \times 5 \times 10^4$$

6. The cells are resuspended to 5 $\times$ 10^6/ml in RPMI + 8% FBS.

Anti-Ovalbumin Antibody Titrations

Solutions

To prepare PBS for ELISA, NaCl (8.0 g), KH$_2$PO$_4$ (0.2 g), and KCl (0.2 g) are mixed, and distilled water is added to 1 l.

PBS-Tween is PBS for ELISA containing 0.5-ml Tween 20/l.

Skim milk powder (Difco, Detroit, MI.) at 0.5% and 3% (wt/vol) is dissolved in PBS for ELISA.

The blood is allowed to clot at 4° in Microtainer serum separator tubes (Becton Dickinson) and centrifuged at 10,000g for 5 min to obtain the serum. Antibody titers are determined by an indirect antigen-specific ELISA. Enzyme immunoassay microtitration plates (96 wells, flat bottomed, from ICN Biomedicals, Inc., Aurora, OH) are coated with antigen in water (10 μg/ml) by adding 0.1 ml/well and incubating at 37° until dry. The wells are washed four times with PBS-Tween, and all liquid is removed. To block nonspecific binding sites on the plastic 0.2 ml of 3% skim milk is added to each well, and incubated for 1 h at 37°. The wells are washed four times with PBS-Tween. Sera from individual mice are diluted 1:100 in 0.5% skim milk and further diluted serially (2x) in duplicate as the test antibody. The plate is incubated for 1 h at 37° and then washed four times. An appropriate dilution of horseradish peroxidase–conjugated goat anti-mouse immunoglobulin-revealing antibody (Caltag, San Francisco, CA) is made in 0.5% skim milk to measure total antibody titers. The incubation and washes are repeated. Color development is initiated with the ABTS Microwell peroxidase system (Kirkegaard and Perry Laboratories, Gaithersburg, MD), and the absorbance is determined at 415 nm after 15 min at 23°.

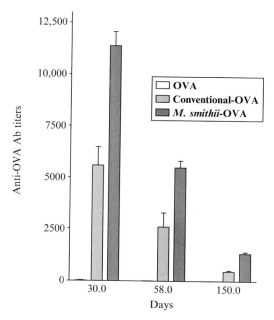

FIG. 4. Comparison in anti-OVA antibody titers in sera of C57BL/6 mice immunized with 15 μg of OVA (OVA, no adjuvant), *M. smithii*-OVA archaeosomes (15 μg OVA in 0.46 mg archaeosomes), and conventional-OVA liposomes (15 μg in 1.44 mg liposomes). Mice were bled at the times shown after first injections.

Titers are calculated as the end point dilutions exhibiting an optical density of 0.3 U above background. Results shown in Fig. 4 reveal higher anti-OVA antibody titers in sera of mice immunized with *M. smithii* OVA-archaeosomes compared with conventional OVA-liposomes.

Proliferative Responses of Splenic Cells

Antigen-induced proliferation assays are performed in 96-well round-bottom tissue culture plates. Spleen cells (0.1 ml, 5×10^5/well) are cultured, including 0.1 ml OVA to achieve 0, 5, 25, and 100 μg/ml (final concentrations) in triplicate wells, in RPMI 1640 + 8% FBS. Dilutions for 24-, 48-, and 72-h incubations are set up in different plates. After incubation at 37° for the appropriate length of time, the supernatant (150 μl) is collected without disturbing the cell pellet. If needed, the plates are centrifuged before collecting supernatants. Supernatants are stored at −70° for cytokine assays. The recommended length of cell culture time before collection of supernatants is 48–72 h for assays of IFN-γ and IL-4. Fresh RPMI 1640 (50 μl) + 8 % FBS medium is added to the cell pellet, and then 50 μl [³H]thymidine (1 μCi/well) is added. The cells are incubated for an additional 18–20 h at 37° and then harvested onto 90 × 120 mm glass fiber filters (Wallac, Oy Turku, Finland), and the [³H]thymidine incorporated is quantified by scintillation counting. Typical proliferative

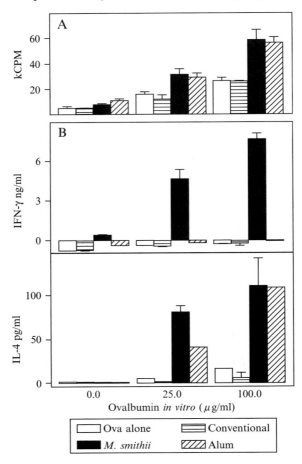

FIG. 5. OVA-specific spleen cell proliferation (A) and induction of both Th1 and Th2 cytokines (B) by *M. smithii* archaeosomes after immunization. Groups of female BALB/c mice were immunized twice at 0 and 3 weeks with 15 μg OVA either without an adjuvant, adsorbed to alum, entrapped in archaeosomes, or entrapped in conventional liposomes. Spleens were harvested on day 28. Cytokines were quantified in 72-h culture supernatants. Data from Krishnan *et al.*[6] are shown with permission from the American Society for Microbiology.

responses of splenocytes are similar for OVA-archaeosome and OVA-alum adjuvanted mice, and lesser for OVA-conventional liposome immunized animals (Fig. 5A).

Cytokine Sandwich Enzyme-Linked Immunosorbent Assay

The following solutions are required to quantify cytokines in the culture supernatants prepared previously.[21]

Solutions

$25 \times$ PBS: 33 g NaH_2PO_4, 188 g K_2HPO_4, and 1000 g NaCl taken to 5-l with distilled water, pH 7–7.5.

Sample diluent: 8% FBS in RPMI 1640 medium.

Secondary antibody and conjugate diluent: To prepare 1% BSA in PBS-Tween, 20 ml of $25 \times$ PBS, 0.25 ml Tween 20, and 5 g BSA are mixed and diluted to 400 ml with H_2O. This is stirred until BSA dissolves, and the volume adjusted to 500 ml with distilled H_2O.

Na_2HPO_4: 26.8 g in 500 ml H_2O.

Citric acid: 10.5 g in 500 ml H_2O (make sure citric acid is high purity, Fisher brand recommended, and check to see that the pH is strongly acidic).

ABTS substrate in a 50 ml total volume: 50 mg ABTS, 11 ml Na_2HPO_4, 14 ml citric acid, 25 ml H_2O, and 5 μl H_2O_2.

All final reagents are filter-sterilized through 0.2-μm filters and stored at $4°$ for long-term use.

Enzyme immunoassay microtitration plates (96 wells, flexible, U-bottom, from Falcon, Beckton and Dickinson, Franklin Lakes, N. J.) are coated for 1 h at room temperature ($23°$) with 50 μl/well of first antibody diluted in PBS. The wells are washed twice with PBS-Tween. All liquid is discarded and the plates are patted dry. Fifty microliters of sample are added and standards/well diluted in RPMI containing 8% FBS and incubated at room temperature for 1 h, followed by two washes in PBS-Tween. Usually two standard curves composed of several doubling dilutions, running in opposite directions along both edges, are included on each plate. Samples are usually tested at 1:2 dilution for cell culture supernatants from antigen-induced stimulations. The sample step is followed by 50-μl/well biotinylated-second antibody diluted in PBS-Tween, incubated for 30 min at room temperature, and washed twice. Next, 50 μl/well of appropriately diluted streptavidin-HRPO conjugate is added (Jackson

[21] T. R. Mosmann and T. A. Fong, *J. Immunol. Methods* **116,** 151 (1989).

ImmunoResearch Laboratories, Inc., West Grove, PA) in PBS-Tween. This is incubated for 30 min at room temperature and washed twice. Color development is initiated with 100 μl/well of fresh ABTS substrate prepared as described previously, or the ABTS Microwell peroxidase system as described for evaluation of antibody titers. Color development occurs optimally when incubations are done in the dark. The absorbance at 415 nm is determined when the color development reaches optimum for the standard curve ($\sim$ 15–20 min). Antibody pairs used for these assays include RA-6A2 (ATCC HB170) + XMG1.2-biotin[22] for IFN-γ, and 11B11[23] + BVD6-24G2-biotin (Pharmingen Canada Inc, Mississauga, Canada) for IL-4. Each pair of first and second antibody is titrated, using a range of standards from 20 ng onward, to establish appropriate concentrations. IFN-γ and IL-4 standards may be purchased from ID Labs (London, Canada). We suggest that samples that are to be compared against each other always be tested in the same assay. Results shown in Fig. 5B reveal substantial IFN-γ production only from splenic cultures of *M. smithii* archaeosome-OVA–immunized mice. On the other hand, both archaeosomes and alum induce IL-4 production.

Cytoytic T-Cell Assay

The CTL assay to follow requires that antigen-specific effector cells be restimulated *in vitro* to increase populations to easily measure specific killing of a target cell presenting a target peptide(s). Direct measurement of numbers of antigen-specific effectors in splenic populations requires other techniques, such as Elispot, not described here.

1. Spleen cell suspensions are prepared as described for assay of the proliferative response in the previous section. In a 25-cm^2 tissue culture flask (Falcon) 30 $\times$ 10^6 spleen cells and 5 $\times$ 10^5 irradiated target cells (10,000 rad) are cultured. The target cells are EG.7, a subclone of EL-4 (T cell lymphoma) stably transfected with the gene encoding OVA (ATCC TIB 39). Interleukin-2 (0.1 ng/ml) (ID labs, London, Ont.) is added 10 μl/flask from a 100 ng/ml solution. The total volume in the flask is made up to 10 ml with RPMI + 8% FBS medium. The flask is placed upright in the incubator at 37°, 8% CO$_2$, and 95% humidity, and incubated for 5 days.

2. ^{51}Cr labeled targets are prepared. EG.7 target cells (5 $\times$ 10^6), and EL-4 control cells not expressing OVA are centrifuged in 15-ml tubes at 470g for 10 min. As much of the supernatant as possible is removed, and

---

[22] H. M. Cherwinski, J. H. Schumacher, K. D. Brown, and T. R. Mosmann, *J. Exp. Med.* **166**, 1229 (1987).
[23] J. Ohara and W. E. Paul, *Nature* **315**, 333 (1985).

the cell pellet is tapped gently. ^{51}Cr (75 μCi) is added to the tube and mixed well with a pipette tip. The targets are inculated at 37° for 45 min. The cap of the 15-ml tube is loosened during incubation. RPMI medium (10 ml) is added to the ^{51}Cr-loaded cells and spun at 470g, 10 min. All the supernatant is discarded carefully without disturbing the cell pellet. The cell pellet is tapped to loosen the cells and 10 ml of fresh RPMI medium is added. The cells are reincubated at 37° for 10 min. The cells are centrifuged and again the supernatant is discarded as before. The cells are resuspended in 5 ml of RPMI + 8% FBS medium.

3. Effector cells are prepared. The cells are retrieved from the restimulation flasks into a 50-ml tube. The cells are counted using a hemocytometer as described previously. The effector cells are spun at 470g for 10 min and resuspended to obtain a final concentration of 100×10^6/ml.

4. The CTL assay is performed in a 96-well, round-bottomed tissue culture plate (Falcon). The killing potential of each splenic sample is tested at various effector/target ratios, in triplicate. RPMI + 8% FBS medium (25 μl) is added to wells A3–A12, numbered from right to left. In wells A1–A3 37.5 μl of effector cells is added from the original stock of 100×10^6 cells/ml. 12.5 μl is removed from wells A1–A3, transferred to wells A4–A6 and aspirated well using the same set of tips. Again 12.5 μl is transfered from wells A4–A6 into wells A7–A9. This procedure is repeated for the last set of three wells of rows A9–A12. Here, after aspirating, the extra 12.5 μl is discarded, and 25 μl of labeled target cells (from stock of 10^6/ml) is added to all the wells. Thus, effector/target ratios of 100:1, 33.3:1, 11.1:1 and 3.7:1 are achieved. Finally, 50 μl of RPMI + 8% FBS medium is added to all wells to achieve a volume of 100 μl. In triplicate wells measure spontaneous release (SR) by including 25 μl of labeled targets and 75 μl of medium, and maximal release (MR) by including 25 μl of labeled targets, 25 μl of medium, and 50 μl of 1% NP40 (Sigma Chemical Co., St. Louis, MO). The plates are incubated for 4 h at 37°. As a cautionary note, while plating one must ensure the cells are dropped to the bottom and not the sides of the well. If in doubt, the plate is spun for 5 min at 375g before incubating.

5. The release of ^{51}Cr is counted from labeled EL-4 and EG.7 cells. After 4 h, the cells are sedimented by centrifuging the plate at 375g for 5 min, and 70 μl of the supernatant is transferred (without disturbing the cell pellet) to counting tubes. The tip is left in the tube and counted in a gamma counter. The percent specific lysis is calculated using the formula: [(Cpm experimental − cpm spontaneous) / (cpm total − cpm spontaneous)] × 100. Cytotoxic T cell data may be expressed also as lytic units defined as the number of effector cells per 10^6 spleen cells that yield a specific percentage killing (e.g., 20%) of a defined number (e.g., 2.5×10^4) of target cells.

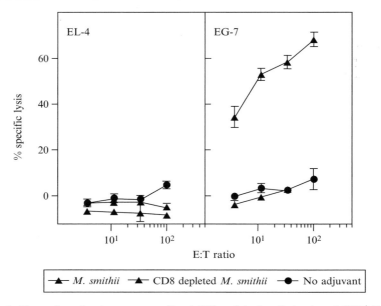

FIG. 6. Abrogation of archaeosome-mediated CTL activity by elimination of CD8$^+$ T-cell effectors. C57BL/6 mice were given intraperitoneal injections on days 0 and 21 with 15 μg OVA in PBS (no adjuvant), or entrapped in *M. smithii* archaeosomes. Spleens were harvested on day 28. CD8$^+$ T cells were eliminated from an aliquot of *M. smithii* effectors (CD8 depleted—*M. smithii*) on day 5 of restimulation with addition of anti-CD8 antibody and rabbit complement. Data are shown from Krishnan *et al.*[7] with permission from *The Journal of Immunology*.

A typical CTL experiment for immunization by either intraperitoneal or subcutaneous routes (Fig. 6) indicates little, to no, killing of OVA-negative EL-4 cells. Furthermore, OVA-specific effectors are essentially absent in splenic cells from mice immunized with nonadjuvanted OVA. However, substantial killing of targets is observed for stimulated splenic cells from mice immunized with *M. smithii* OVA-archaeosomes. Because killing activity may be abolished by removing CD8$^+$ T cells from the splenic population, it is clear that killing is mediated by OVA-specific CD8$^+$ T cells arising *in vivo* from the OVA-archaeosome immunizations.

Acknowledgments

We wish to thank members of the archaeosome team, Mr B.J. Agnew, Ms. Lise Deschatelets, Ms. Chantal Dicaire, and Mr. P. Fleming for excellent technical expertise over the past decade. This is NRC publication 42480.

Section II

Liposomes in Diagnostics

[12]   Magnetoliposomes as Contrast Agents

By JEFF W. M. BULTE and MARCEL DE CUYPER

Introduction

Whereas the original interest in liposomes is founded on their use as a model for biological membranes, these biocolloidal structures have now been used in many biotechnological[1,2] and biomedical engineering[3] applications. In particular, because liposomes are entirely biocompatible, they have been recognized as promising microvehicles for therapeutics and diagnostics *in vivo*. Liposomes can host either polar drugs in their aqueous inner space or lipophilic drugs in the lipid bilayer, protecting the entrapped pharmaceuticals from potential inactivation by external factors and without eliciting unwanted side effects.

Since the inception of targeting liposomes for *in vivo* applications,[4] all major fields of diagnostic imaging, including γ-scintigraphy,[5,6] computed tomography,[7] and magnetic resonance imaging (MRI),[8–13] have been explored for the use of liposome-based reporter molecules or contrast agents. However, two essential prerequisites have to be fulfilled for liposome-based formulations to be useful in virtually any imaging application. First,

[1] D. D. Lasic and Y. Barenholz, "Handbook of Nonmedical Applications of Liposomes—from Design to Microreactors," CRC Press, Boca Raton, FL, 1996.

[2] Q. Yang, *in* "Immobilization of Liposomes in Gel Beads for Chromatography," Dissertation, Acta Universitatis Upsaliensis, Uppsala, 1993.

[3] R. Arshady, "Microspheres, Microcapsules and Liposomes." Citus Books, London, 2001.

[4] I. R. McDougal, J. K. Dunnick, and M. L. Goris, *J. Nucl. Med.* **16**, 488 (1975).

[5] O. C. Boerman, G. Storm, and W. J. G. Oyen, *J. Nucl. Med.* **36**, 1639 (1995).

[6] W. T. Phillips and B. Goins. *in*: "Handbook of Targeted Delivery of Imaging Agents," (V. P. Torchilin, ed.) p. 149. CRC press, Boca Raton, FL, 1995.

[7] A. Sachse, J. U. Leike, T. Schneider, S. E. Wagner, G. L. Rossling, W. Krause, and M. Brandl, *Invest. Radiol.* **32**, 44 (1997).

[8] R. L. Magin, S. M. Wright, M. R. Niesman, H. C. Chan, and H. M. Swartz, *Magn. Reson. Med.* **3**, 440 (1986).

[9] G. Kabalka, E. Buonocore, K. Hubner, T. Moss, N. Norley, and L. Huang, *Radiology* **163**, 255 (1987).

[10] E. C. Unger, D-K Shen, T. A. Fritz, *J. Magn. Reson. Imaging* **3**, 195 (1993).

[11] V. S. Trubetskoy, J. A. Cannillo, A. Milshtein, G. L. Wolf, and V. P. Torchilin, *Magn. Reson. Imaging* **13**, 31 (1995).

[12] R. W. Storrs, F. D. Tropper, H. Y. Li, C. K. Song, J. K. Kuniyosji, D. A. Sipkins, K. C. P. Li, and M. D. Bednarski, *J. Magn. Reson. Imaging* **5**, 719–724 (1995).

[13] D. A. Sipkins, D. A. Cheresh, M. R. Kazemi, L. M. Nevin, M. D. Bednarski, and K. C. P. Li, *Nature Med.* **4**, 623 (1998).

the chosen diagnostic contrast agent should remain stable and bound to the liposome during the relevant imaging interval. For MRI purposes, this can be realized by covalently linking the paramagnetic metal ion–chelate complex to the phospholipids. Dissociation of the complex from the liposome bilayer, however, cannot be completely excluded, especially *in vivo*. A second event that generally hampers the usefulness of liposomes for imaging applications encompasses the fact that liposomes can be seen as "foreign" entities within the body. As a result, they are quickly removed from the blood circulation by macrophage uptake, mainly in the liver and spleen, and to a lesser extent by the bone marrow. This usually rapid uptake, however, can be largely circumvented by including in the lipid bilayer coat a few percent of phospholipids that bear a polymeric poly(ethylene glycol) (PEG) chain.[14–17]

A decade ago, we developed a new type of liposome, so-called magnetoliposomes, in which the aqueous interior is completely occupied by a magnetic iron oxide core.[18] In analogy to the purpose of classical liposomes, magnetoliposomes too were originally developed for studying fundamental membrane processes,[19,20] but over the years, they have found applications as innovative colloids in various biotechnological[21] and biomedical fields,[22,23] including MRI.[24,25] Release of the magnetic imaging

[14] D. Papahadjopoulos, T. M. Allen, A. Gabizon, E. Mayhew, K. Matthay, S. K. Huang, K.-D. Lee, M. C. Woodle, D. D. Lasic, C. Redemann, and F. J. Martin, *Proc. Natl. Acad. Sci. USA* **88,** 11460 (1991).

[15] D. Papahadjopoulos and A. A. Gabizon, *in* "Liposomes as Tools in Basic Research and Industry" (J. R. Philippot and F. Schuber, eds), Chapter 11, p. 177. CRC Press, Boca Raton, FL, 1995.

[16] M. J. Parr, S. M. Ansell, L. S. Choi, and P. R. Cullis, *Biochim. Biophys. Acta* **1195,** 21 (1994).

[17] A. Yoshida, K. Hashizaki, H. Yamauchi, H. Sakai, S. Yokoyama, and M. Abe, *Langmuir* **15,** 2333 (1999).

[18] M. De Cuyper and M. Joniau, *Eur. Biophys. J.* **15,** 311 (1988).

[19] M. De Cuyper and M. Joniau, *Biochim. Biophys. Acta* **814,** 374 (1985).

[20] M. De Cuyper and M. Joniau, *Langmuir* **7,** 647 (1991).

[21] M. De Cuyper, B. De Meulenaer, P. Van der Meeren, and J. Vanderdeelen, *Biotechnol. Bioeng.* **49,** 654 (1996).

[22] D. Müller-Schulte, F. Füssl, H. Lueken, and M. De Cuyper, *in* "Scientific and Clinical Applications of Magnetic Carriers." (U. Häfeli, W. Schütt, J. Teller, and M. Zborowski, eds), p. 517. Plenum Press, New York, 1997.

[23] Y. Masuko, K. Tazawa, E. Viroonchatapan, S. Takemori, T. Shimizu, M. Fujimaki, H. Nagae, H. Sato, and I. Horikoshi, *Biol. Pharm. Bull.* **18,** 1802 (1995).

[24] J. W. M. Bulte, M. De Cuyper, D. Despres, R. A. Brooks, and J. A. Frank, *J. Magn. Reson. Imaging* **9,** 329 (1999).

[25] J. W. M. Bulte, M. De Cuyper, D. Despres, and J. A. Frank, *J. Magn. Magn. Mat.* **194,** 204 (1999).

tag from the phospholipid coat is negligible, and a prolonged blood circulation time can be observed, depending on the type of preparation.

Magnetoliposomes

In classical encapsulation procedures, the material to be trapped into the aqueous interior is present during membrane formation. For relatively small molecules, these procedures generally provide a satisfactory encapsulation. With large structures, such as polymeric molecules or colloidal grains, however, the encapsulation is more problematic, in particular if small unilamellar vesicles are used as a cage. To improve the capture efficiency, different encapsulation recipes have been developed in the past.[26] For instance, in the case in which incorporation of (magnetic) iron oxide cores is envisaged, a clever approach is to first capture Fe^{2+} and/or Fe^{3+} ions within the inner space of vesicles, followed by a controlled permeation of OH^- anions through the membrane.[27,28] As a result, precipitation of the iron ions may result in the formation of magnetizable iron oxides, such as maghemite (γ-Fe_2O_3) or magnetite (Fe_3O_4).[28,29] However, to get a sufficiently large magnetic colloidal structure that is well attracted in a magnetic field, one has to start with highly concentrated iron (II) and (III) chloride or sulfate solutions.[30] Because these solutions demonstrate a very low pH, there is a serious risk for phospholipid degradation such as hydrolysis[31] and/or peroxidation.[32,33] In addition, because the affinity of iron ions for phosphate groups is extremely high,[34] complete removal of the divalent and trivalent metal ions from the phospholipid polar headgroups is difficult to realize, even after a thorough dialysis[28] or by cation exchange chromatography.[29] In addition, with the latter technique, it has been reported that vesicle recovery may be rather poor.[35,36]

[26] J. C. Domingo, M. Mercadal, J. Petriz, J. Garcia, and M. A. de Madariaga, *Cell. Mol. Biol. Lett.* **4,** 583, (1999).

[27] S. Mann, J. P. Hannington, and R. J. P. Williams, *Nature* **324,** 565 (1986).

[28] C. Sangregorio, J. K. Wiemann, C. J. O'Connor, and Z. Rosenzweig, *J. Appl. Physics* **85,** 5699 (1999).

[29] S. Mann and J. P. Hannington, *J. Colloid Interface Sci.* **122,** 230 (1987).

[30] L. Shen, P. E. Leibinis, and T. A. Hatton, *Langmuir* **15,** 447 (1999).

[31] M. Grit, N. J. Zuidam, and D. J. A. Crommelin, *in* "Phospholipids: Characterization, Metabolism, and Novel Biological Applications" (G. Cevc and F. Paltauf, eds), Chapter 2, p. 15, AOCS Press, Champaign, IL, 1995.

[32] Q. T. Li, M. H. Yeo, and B. K. Tan, *Biochem. Biophys. Res. Commun.* **273,** 72 (2000).

[33] O. V. Vasiljeva, O. B. Lyubitsky, G. I. Klebanov, and Y. A. Vladimrov, *Membr. Cell Biol.* **14,** 47 (2000).

[34] B. Tadolini, P. Motta, and C. A. Rossi, *Biochem. Molec. Biol. Internat.* **29,** 299 (1993).

To circumvent these potential problems, we follow a completely different strategy. We first generate a batch of nanometer-sized, superparamagnetic iron oxide grains of uniform size, which, in an intermediate phase, are stabilized by a surfactant coat to produce a so-called water-compatible magnetic fluid. Then, the surfactant molecules are exchanged for phospholipid molecules. In this way, the iron oxide particle functions as a template for membrane formation.

Synthesis of Water-Soluble Magnetic Fluid

To synthesize the starting superparamagnetic iron oxide particles (diameter about 15 nm; Fig. 1), the basic procedure of Khalafalla and Reimers[37] is followed. In practice, to prepare 100 ml of a magnetite grain suspension at a concentration of about 60 mg Fe_3O_4/ml, 6 g $FeCl_2 \cdot 4aq$ (= 30.2 mmol) and 12 g $FeCl_3 \cdot 6aq$ (= 44.4 mmol), each solubilized in 25 ml water, are mixed and coprecipitated with 25 ml of a 28%–30% concentrated ammonia solution. To remove excess base, the black, gel-like slurry is put on a permanent magnet, and the clear liquid on top is decanted. The preciptate is then washed thoroughly with 50 ml water, containing 2.5 ml concentrated ammonia, and the supernatant is again decanted. The latter step is repeated twice. Following this, 3 g of lauric acid (Sigma, St. Louis, Mo) is added to the gumlike precipitate, and, to promote

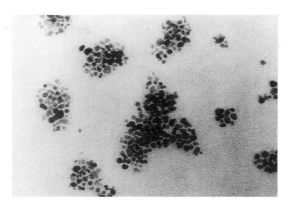

FIG. 1. Transmission electron micrograph of laurate-stabilized Fe_3O_4 cores. The mean diameter of the particles is 15 nm.

[35] L. R. McLean and M. C. Phillips, *Biochemistry* **20,** 2893 (1981).
[36] A. M. H. P. van den Besselaar, G. M., Helmkamp, Jr., and K. W. A. Wirtz, *Biochemistry* **14,** 1852 (1975).
[37] S. E. Khalafalla and G. W. Reimers, *IEEE Trans. Magn.* **MAG-16,** 178 (1980).

solubility of the fatty acid, the whole system is heated for about 5 min in a water bath at 80–90°. Finally, the liquified slurry is dispersed in 60 ml water, adjusted to pH 9.2. Occasionally, it is advisable to include an additional centrifugation step (10 min; 10,000g) to remove a few larger aggregates that may be present. Ultimately, because of Brownian motion, a very finely dispersed solution is formed in which no tendency exists for the formation of precipitates not even after a few years.

In preparing this water-soluble magnetic fluid, the following points have to be kept in mind:

1. Although magnetite can be regarded as being constituted from twice the amount of Fe^{3+} as Fe^{2+}, it is recommended to start from/with a Fe^{2+}/Fe^{3+} ratio of 2:3, because some Fe^{2+} oxidizes spontaneously during synthesis.

2. The architecture and orientation of the individual lauric acid molecules is not known in great detail, but on the basis of some experimental observations, it may be assumed that the laurate detergent molecules adsorb with their carboxyl group facing the iron oxide surface. With the more hydrophobic oleic acid, for instance, an organic solvent–compatible magnetic fluid is created that is no longer dispersible in an aqueous environment.

3. To inhibit a slow oxidation of magnetite, the stock solution is stored in a nonoxidizing atmosphere (e.g., N_2).

Synthesis of Magnetoliposomes

On reacting the laurate-stabilized Fe_3O_4 cores with an excess of sonicated vesicles, constructed with proper types of phospholipids (Avanti Polar Lipids, Alabaster, AL) (see later), and following subsequent dialysis for 3 days (dialysis tubes with a molecular weight cutoff of 12,000–14,000; Medical Industries, Los Angeles, CA), magnetoliposomes are formed. We have found that, vesicles are usually formed by probe-type sonication, but other production methods such as detergent dialysis, reverse-phase evaporation, extrusion, dehydration-rehydration, and freeze-thawing may be equally suited.[38] The rate of dialysis is extremely crucial, because laurate molecules have to be gradually replaced by phospholipids.

Mechanism of Magnetoliposome Formation

A detailed analysis of the mechanism of phospholipid deposition on the solid surface reveals that spontaneous lipid transfer, rather than fusion or aggregation, plays a key role.[20,39,40] Indeed, to explain the lipid movements,

---

[38] M. Paternostre, M. Ollivon, and J. Bolard, in "Manual on Membrane Lipids" (R. Prasad, ed.), Chapter 9, p. 202. Springer-Verlag, Berlin Heidelberg, 1996.

several experimental indications point to a so-called aqueous transfer model. According to this model, the phospholipid molecules under consideration first escape from the donor vesicles, then move in a monomolecular form through the aqueous phase and are adsorbed subsequently onto the solid surface. Within this overall process, the rate-limiting step is the phospholipid exit from the donor structures. In this context, one should keep in mind that, in general, the transferability of a phospholipid molecule increases if (1) the polar headgroup is highly water soluble (e.g., when it is charged;[19]); (2) the fatty acyl side chains are short ($<C_{15}$ for saturated chains) or unsaturated (e.g., dioleoylphosphospholipids are more apt to move than their distearoyl analogues)[41]; (3) the number of apolar chains is low (e.g., lysophospholipids move at a much faster rate than the corresponding diacylphospholipids);[42] (4) charged lipids are embedded in a similarly charged bilayer;[43] (5) the bilayer is in a so-called fluid, disordered state (i.e., the membranes are well above their gel-to-liquid crystal phase transition temperature;[39] (6) the membranes are highly curved;[44] and (7) the ionic strength of the medium is relatively low.[40]

On the basis of these observations, it is possible to predict which phospholipids are qualified to successfully produce magnetoliposomes. For instance, with small unilamellar vesicles made of the anionic dimyristoylphosphatidylglycerol (DMPG) or the zwitterionic phosphatidylcholine analogue (DMPC) (at a phospholipid/magnetite weight ratio of 2.73), the Fe_3O_4 particles become surrounded by a lipid bilayer after dialysis for 3 days at 37° in a 5 mM N-TRIS(hydroxymethyl)methyl-2-aminoethanesulfonic acid (TES) buffer with regular changes of the dialysate. With gas liquid chromatography, it has been shown further that, in the meantime, the laurate molecules are removed completely.[18] In contrast to the observations made with phospholipids having medium-sized fatty acyl side chain residues, under the same experimental conditions, distearoylphospholipids do not demonstrate any transfer tendency, not even after 4 days.[20] With these phospholipid types, a large precipitate of phospholipid-devoid Fe_3O_4 is formed at the bottom of the dialysis bag.

[39] R. E. Brown, *Biochim Biophys. Acta* **1113,** 375 (1992).

[40] M. De Cuyper, M. Joniau, and H. Dangreau, *Biochemistry* **22,** 415 (1983).

[41] J. R. Silvius and R. Leventis, *Biochemistry* **32,** 13318 (1993).

[42] L. R. McLean and M. C. Phillips, *Biochemistry* **23,** 4624 (1884).

[43] M. De Cuyper, M. Joniau, J. F. B. N. Engberts, and E. J. R. Sudhölter, *Colloids Surfaces* **10,** 313 (1984).

[44] M. De Cuyper and M. Joniau, *in* "Liposomes as Drug Carriers" (K. H. Schmidt, ed.), p. 94. G. Thieme Verlag, Stuttgart, 1986.

High-Gradient Magnetophoresis

After formation of the phospholipid bilayer onto the Fe_3O_4 surface, the excess of nonadsorbed phospholipids is separated from the magnetoliposome dispersion by magnetophoresis, operating in a high gradient regimem. Although magnetite is ferrimagnetic in nature, attraction of the extremely small, nanometer-sized phospholipid–Fe_3O_4 complexes by an "ordinary" strong magnetic field (e.g., 4 Tesla) is problematic, because the resultant magnetic moment of the particles is too small. The high-gradient fractionation system that is constructed consists of a set of Silastic medical grade tubings (inner diameter, 0.078 inch; Dow Corning Corp., Midland, MI) that are plugged with loosely packed magnetizable, stainless steel fibers (e.g., type 430, Bekaert Steel Corp., Zwevegem, Belgium) and positioned in the opening of an electromagnet, equipped with 15-cm diameter poles (Bruker, Type BE-15, Karlsruhe, Germany) (Fig. 2). On distinct sites within such a magnetic filter device, the magnetic field lines concentrate, and thus create locally very high magnetic filed gradients, where the small Fe_3O_4–phospholipid complexes accumulate.

Fractionation of excess vesicles and magnetoliposomes is then accomplished by pumping the incubation mixture through the magnetic filter set-up at a uniform rate of 6 ml/h with the magnetic field on. Usually, the electromagnet operates at 30 A and 80 V. Under these conditions, and without the magnetic filter matrix, the magnetic field between the two poles is about 2.3 Tesla. The phospholipid vesicles, which only demonstrate diamagnetic properties,[45] pass through, whereas magnetoliposomes are captured. To remove any nonadsorbed phospholipids remaining in the capillary liquid entrapped within the iron fiber network, the whole filter system is washed further with buffer. After switching off the magnetic field, the magnetoliposomes can be readily released from the filters by a buffer stream at high speed (500 ml/h).

Evidence for Bilayer Formation

The presence of a bilayer can be verified in different ways:

1. Typically, for a 15-nm Fe_3O_4 core (density, 5.1 g/cm^3), a bilayer coating consists of about 3000 phospholipid molecules (of which one third constitutes the inner layer and two thirds the outer one). This results in a mmol phospholipid/g Fe_3O_4 ratio of 0.7–0.9. The values that are experimentally found are in this range.[18]

---

[45] X. Qui, P. A. Mirau, and C. Pidgeon, *Biochim. Biophys. Acta* **1147,** 59, 1993.

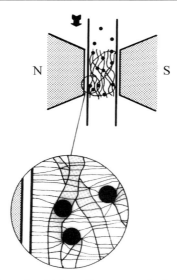

Fɪɢ. 2. Schematic design of a high-gradient magnetic fractionation setup. Magnetic particles pass through the steel wool fibers, which are rendered magnetic by applying an external magnetic field. The particles attach to the grid. When eluting with a buffer in the absence of a magnetic field, the particles are released. Inset, Detail of a high-gradient magnetic separation unit. The field lines (thin lines) between the poles of an electromagnet converge around a magnetizable filter unit (in particular around ridges on the surface irregularities), thereby creating localized regions of high-gradient magnetic fields that capture superparamagnetic colloids. (Reproduced with permission from De Cuyper, "Handbook of Non-Medical Applications of Liposomes," [Y. Barenholz and D. D. Lasic, eds.,] p. 325, CRC Press, Boca Raton, FL, 1996).

2. Adsorption isotherms show two different binding modes: the inner layer, in which the polar headgroups face the solid surface, is adsorbed according to a high-affinity regimen, whereas binding of the outer layer lipids obeys Langmuirian mathemathics (Fig. 3).[18,20]

3. The outer shell of the bilayer can be withdrawn relatively easily by detergents such as Tween 20 at a final concentration of 0.1 weight %[18] or can be extracted by selected organic solvent mixtures (e.g., in $CHCl_3$/CH_3OH/buffer 3.75/1.25/1 or n-hexane/1-propanol/buffer 3/2/1; v/v/v).[46] In contrast, the inner layer lipids are completely resistant to these desorption procedures.

4. After negative staining with 2% sodium phosphotungstate, the presence of a bilayered coating is visible by transmission electron microscopy (Fig. 4).[18]

[46] M. De Cuyper and W. Noppe, *J. Colloid Interface Sci.* **182,** 478 (1996).

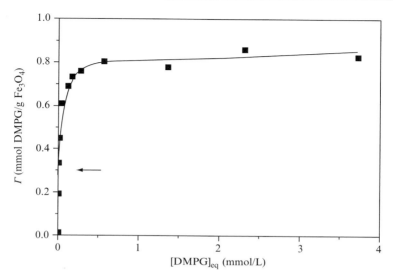

FIG. 3. Adsorption isotherm for DMPG on magnetite. Incubation and dialysis occurred at 37° in 5 mM TES buffer, pH 7.0, for 3 days. The arrow marks the change from the high-affinity to the Langmuir adsorption zone of the isotherm. Γ represents the number of moles of lipid *adsorbed* per gram magnetite; $[DMPG]_{eq}$ is the *free* phospholipid concentration, (i.e., the phospholipid concentration after adsorption has reached a steady state).

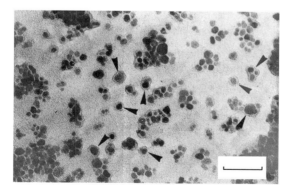

FIG. 4. Transmission electron micrograph of DMPG magnetoliposomes, negatively stained with 2% phosphotungstate. Note the coating around the magnetite particles (indicated by arrows) (Scale bar = 40 nm) (Reproduced with permission from De Cuyper and Joniau, *Eur. Biophys. J.* **15,** 311, 1988).

Stealth® Magnetoliposomes

Classical liposomes are removed from the blood circulation with half-times of only a few minutes. The general consensus, however, is that the blood clearance rate is considerably lowered on incorporation of a few percent (e.g., 5%) of PEGylated phospholipids in the bilayer.[14–17] Ultimately, the biocolloids that are formed in this way are known as Stealth®[47] or sterically stabilized liposomes.

A similar, enhanced blood circulation can be expected for Stealth® magnetoliposomes. Two different routes can be followed to create these magnetizable nanostructures: (1) either, as described previously, by incubating a magnetic fluid sample with vesicles containing the PEGylated lipid or, alternatively (2) by mixing Stealth® vesicles with preformed magnetoliposomes, made with "ordinary," relatively cheap phospholipid species.[48]

Experimentally, the creation of these structures is relatively easy. Independent of the method of preparation, the presence of the PEG residue will clearly improve the phospholipid's water solubility and, thus, in view of the "aqueous transfer" mode (see earlier), its transferability.[49] For instance, in an equimolar mixture of DMPC/DMPG/dipalmitoylphosphatidylethanolamine (DPPE) (8/1/1; molar ratios) donor vesicles and DMPC/DMPG (9/1) acceptor magnetoliposomes, no significant interbilayer transport of DPPE occurs within the time scale of the experiment (4 days). After coupling a 3.35-kDa PEG chain (Shearwater Polymers Europe, Enschede, The Netherlands) onto it, the derivative is clearly able to escape rapidly from the donor membranes. In the latter case, the transfer event can be expressed in a quantitative way by linearizing the initial data points of the kinetic curve according to first-order rate mathematics. The general formula describing this type of kinetics is:

$$\ln \frac{\%PD_t - \%PD_{eq}}{\%PD_o - \%PD_{eq}} = -k_1 \, dt$$

Where $\% PD_t$, $\% PD_o$, and $\% PD_{eq}$ are the phospholipid derivative content in donors or acceptors at any time t during the transport process, at time zero and at equilibrium, respectively. k_1 denotes the first-order rate constant. The half-time ($t_{1/2}$) of the transfer process is related to k_1 by the following equation:

[47] Stealth® is a registered trademark of SEQUUS Pharmaceuticals, Inc.
[48] M. De Cuyper and S. Valtonen, *J. Magn. Magn. Mater.* **225**, 89, 2001.
[49] J. R. Silvius and M. J. Zuckermann, *Biochemistry* **32**, 3153 (1993).

$$t_{1/2} = \frac{0.69}{k_1}$$

For the transfer of DPPE the $t_{1/2}$ value is calculated to be only 96 min.

Because the kinetic mixture just described is made by mixing equimolar amounts (with respect to phospholipid content) of donors and acceptors, at first sight, at equilibrium one should expect an equal distribution of DPPE–PEG over donors and acceptors, (i.e., 5% in each). Experimentally, however, only 30%–35% of the original amount of DPPE–PEG is found in the magnetoliposome acceptors and about 65%–70% remains in the vesicles (Fig. 5). This uneven equilibrium distribution can be understood (Fig. 6) by assuming that in small unilamellar vesicles with a diameter of only 30 nm, about one third of the phospholipids are located in the inner leaflet, whereas two thirds reside in the outer layer.[18,48] Because of the hydrophilic character of the long PEG chain, the inner layer DPPE–PEG molecules are most likely not able to flip-flop through the hydrophobic interior of the donor bilayer, and thus will not participate in the overall transfer process.[48] As a result, only two thirds of the original amount of DPPE–PEG will be able to distribute over donors and acceptors.

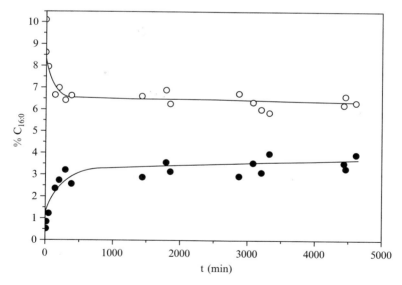

FIG. 5. Transfer kinetics of DPPE-PEG from DMPC/DMPG/DPPE-PEG (molar ratio 8/1/1) donor vesicles to DMPC/DMPG (9/1) acceptor magnetoliposomes. The donor/acceptor molar phospholipid ratio is 1/1. The time-dependent changes in $C_{16:0}$ content in both the vesicles (O) and magnetoliposome population (●) are followed by gas-liquid chromatography.

Kinetic mixture t = 0

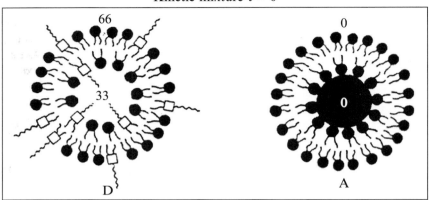

Kinetic mixture t = ∞

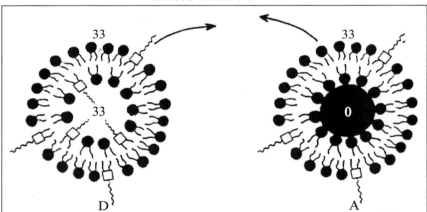

Fɪɢ. 6. Scheme of the distrubution of phospholipids over the outer and inner layer of the phospholipid bilayer of small unilamellar vesicles (Donors, D) and of the magnetoliposome acceptors (A) before transfer occurs (upper part) and after equilibrium was reached (t = ∞) (lower part). The numbers refer to the percentage of the lipid molecules (including DPPE–PEG, represented by open square phospholipds) in both leaflets of the bilayer structures. (Reproduced with permission from De Cuyper and Valtonen, *J. Magn. Magn. Mat.* **225,** 89, 2001).

The final phospholipid–PEG molar content in the Stealth® magneto-liposomes can be easily adjusted at will for each specific application. For instance, if one starts from equimolar donor/acceptor conditions, one can alter the phospholipid derivative content in the starting vesicle donors, or, alternately, at constant lipid–PEG content in the donor vesicle membrane, the donor concentration in the incubation mixture can be varied. In any case, after optimizing the concentration of the PEGylated lipid, the resulting Stealth®magnetoliposomes, in analogy to Stealth® vesicles or liposomes, are able to evade the immediate uptake by cells of the mononuclear phagocyte system.[24,25]

Functionalized Stealth® Magnetoliposomes

For optimal targeting purposes, the lipid colloids should have a great selectivity for the tissues in question.[50,51] This can be achieved by placing a chemically activatable group to the end of the polymer chain, which in turn can be used to anchor target-specific biomolecules such as peptides and monoclonal antibodies. The main drawback of this approach is that a new synthesis has to be initiated for each new target biomolecule. A more efficient approach is to prepare a phospholipid–PEG complex with a functionalized "universal binder" on the end of the polymer residue. To achieve this goal, biotin or (strept)avidin-like molecules are excellent candidates[52,53]; the biotin–streptavidin association constant of about 10^{15} is the highest known in biochemistry.[54]

Synthesis of Phosphatidylethanslamine-Poly (Ethylene Glycol)-Biotin

We have found[55] that a simple route to synthesize the adduct of biotin with the PE–PEG conjugate, with a 100% efficiency, is as follows: In a glass vial, sealed with a Teflon-coated cap, a mixture of 4.7 μmol phosphatidylethanolamine (PE) (3.0 mg of the dimyristoyl or 3.5 mg of the dioleoylanalogue) and a slight excess of N-hydroxysuccinimide-PEG-biotin (NHS-PEG-B) (5.64 μmol, equivalent to 19.2 mg for a PEG polymer with a molecular weight of 3.35 kDa) is stirred at room temperature

[50] A. L. Klibanov, *in* "Long Circulating Liposomes: Old Drugs, New Therapeutics" (M. C. Woodle and G. Storm, eds), p. 269. Springer-Verlag, Berlin – Heidelberg/Landes-Bioscience, Georgetown, TX, 1998.

[51] K. Maruyama, T. Takizawa, T. Yuda, S. J. Kennel, L. Huang, and M. Iwatsuru, *Biochim. Biophys. Acta* **1234**, 74 (1995).

[52] M. De Cuyper, M. Hodenius, Z. G. M. Lacava, R. B. Azevedo, M. F. da Silva, P. C. Morais, and M. H. A. Santana, *J. Colloid Interface Sci.* **245**, 274 (2002).

[53] T. M. Allen, A. K. Agrawal, I. Ahmad, C. B. Hausen, and S. Zalipsky, *J. Liposome Res.* **4**, 1 (1994).

[54] M. Gonzáles, L. A. Bagatolli, I. Echabe, J. L. R. Arrondo, C. E. Argaraña, C. R. Cantor, and G. D. Fidelio, *J. Biol. Chem.* **272**, 11288 (1997).

[55] M. Hodenius, M. De Cuyper, L. Desender, D. Müller-Schulte, A. Steigel, and H. Lueken *Chem. Phys. Lipids* **120**, 75 (2002).

in 0.75 ml of a $CHCl_3/CH_3OH$ mixture (volume ratio, 8/2) in the presence of 1 μl of $N(C_2H_5)_3$ (final concentration 0.15%). The reaction progress is followed by analytical thin-layer chromatography on silica gel plates with $CHCl_3/CH_3OH/25\%$ NH_3 (70/25/4 v/v) as developing solvent. For comparison, the starting PE type is spotted as well. Spraying the plate with molybdenum blue reagent shows that a stain with an R_f of 0.8 appears and that, concomitantly, the original PE with an R_f of 0.3 gradually but completely disappears over 3 h. The complete conversion of the starting PE is verified by the absence of a positive spot after developing the plate with ninhydrin, used as a marker for free $-NH_2$ groups. Further control experiments testify that both the organic medium and the presence of $N(C_2H_5)_3$ are vital to achieve such a high, quantitative conversion. Omitting triethylamine, for instance, drastically reduces PE derivatization; after a 3-h reaction period and even with a-four-fold excess of NHS-PEG-B over PE, only 60% conversion is found. Also, in an aqueous medium (pH 6.0 or 8.0), no reaction occurs at all, probably because of hydrolysis of the PEG–NHS carbonate reagent, releasing NHS and CO_2 and regenerating the underivatized PEG hydroxyl.

Protein Binding to Functionalized Stealth® Magnetoliposomes

The binding potency of the derivatized phospholipid for streptavidinylated biomolecules can be assessed easily with alkaline phosphatase as a model. The latter enzyme is relatively stable, and its catalytic properties can be monitored accurately by spectrophotomety.[56] In practice, 1.34 ml of a dispersion of DMPG magnetoliposomes, doped with 10% of the functionalized dioleoylPE-PEG–spector photometry biotin (0.30 μmol phospholipid/ml, 434.7 μg Fe_3O_4/ml) is prepared and incubated for 100 min with 33 μl of the stock solution of streptavidinylated alkaline phosphatase (SAP) (1000 U/ml; Boehringer, Mannheim, Germany). Then, 1.2 ml of the incubation mixture is subjected to high-gradient magnetophoresis to separate bound from unbound enzyme. The degree of SAP binding is determined by measuring the enzymatic activity present in the retentate, eluate, and—as a control—in the original incubation mixture. In practice, the kinetic mixture consists of 100 μl of the enzyme sample, 1.5 ml 100 mM borate buffer, pH 10.0, and 1 ml of the p-nitrophenylphosphate substrate stock solution (concentration, 0.354 mmol·l^{-1}; Sigma 104 reagent). The kinetics are followed at 410 nm. Linearizing the kinetic data by first-order reaction mathematics (and taking into account the dilution of the eluate, brought about by the washing step; see earlier) yields k_1

[56] G. G. Forstner, S. M. Sabesin, and K. J. Isselbacher, *Biochem. J.* **106,** 381 (1968).

values of 0.182_9 min^{-1}, 0.169_6 min^{-1}, and 0.020_3 min^{-1} for the Stealth®-magnetoliposome-SAP stock solution, for the retentate and the eluate fraction, respectively. These (preliminary) results highlight the suitability of our Stealth® magnetoliposomes to immobilize streptavidinylated proteins.

Magnetic Resonance Imaging

Superparamagnetic Contrast Agents

MR contrast agents generate contrast by shortening the MR relaxation times and can affect T1 and T2 by dipole–dipole interactions. Their net effectiveness is expressed as relaxivity (R), which represents the reciprocal of the relaxation time per unit concentration of metal, with units of mM^{-1}s^{-1}. For simplicity, MR contrast agents are classified generally as "T1 agents" and "T2 agents." This classification is according to their predominant effect on relaxation, but because of dephasing effects, the T2 relaxivity (R2) is always higher than R1 (even for T1 agents). Superparamagnetic iron oxides, in which the magnetically active core is either maghemite (γFe_2O_3) or magnetite (Fe_3O_4), generally act as T2 agents. These superparamagnetic iron oxide nanoparticles have, on a (milli)molar metal basis, a significantly higher relaxivity than paramagnetic T1 contrast agents. On conventional spin-echo MR images, the presence of T2 (superparamagnetic) agents leads to a dark (hypointense) appearance of tissue, whereas T1 agents cause the opposite (hyperintensity).

Superparamagnetic iron oxides have to be stabilized with a surface coating to prevent aggregation. As MR contrast agents, they have been coated with dextran,[57–59] starch,[60] phospholipids,[61–63] polyaminoacids,[64] and dendrimers.[65]

[57] D. D. Stark, R. Weissleder, G. Elizondo, P. F. Hahn, S. Saini, L. E. Todd, J. Wittenberg, and J. T. Ferrucci, *Radiology* **168**, 297 (1988).

[58] T. Shen, R. Weissleder, M. Papisov, A. Bogdanov, and T. J. Brady, *Magn. Reson. Med.* **29**, 599 (1993).

[59] J. W. M. Bulte, and R. A. Brooks, *in* (U. Häfeli, W. Schütt, J. Teller, and M. Zborowksi, eds.), "*Scientific and Clinical Applications of Magnetic Carriers.*" pp. 527–543. Plenum Press, New York, 1997.

[60] A. K. Fahlvik, E. Holtz, U. Schroder, and J. Klaveness, *Invest. Radiol.* **25**, 793 (1990).

[61] A. Bogdanov, Jr., C. Martin, R. Weissleder, and T. J. Brady, *Biochim. Biophys. Acta* **1193**, 212 (1994).

[62] S. Päuser, R. Reszka, S. Wagner, K-J. Wolf, H-J. Buhr, and G. Berger, *Anti-Cancer Drug Design* **12**, 125 (1997).

[63] E. Wittendorp-Rechenmann, I. J. Namer, J. Steibel, D. C. Lam, and D. Pouliquen, *J. Trace Micropr. Techn.* **16**, 523 (1998).

[64] L. X. Tiefenauer, G. Kühne, and R. Y. Andres, *Bioconj. Chem.* **4**, 347 (1993).

Relaxivities of Stealth® Magnetoliposomes

Magnetoliposome (ML) and Stealth®-ML samples containing 0.05–0.5 mM Fe are prepared in 5 mM TES buffer, pH 7.0, as described previously. For these preparations, the T1 and T2 relaxation times are measured on a custom-designed variable-field T1–T2 analyzer (custom built by Southwest Research Institute, San Antonio, TX). The T1 is measured using a saturation recovery pulse sequence with 32 incremental recovery times. The T2 is measured using a Carr-Purcell-Meiboom-Gill sequence of 500 echoes and an interecho time of 2 ms. The field strength is varied from 0.02–1.5 Tesla (T), and measurements are obtained at 3° and 37°. The T1 and T2 results need to be converted to relaxivity in s^{-1}/mM by subtracting the buffer contribution and dividing by the Fe concentration, expressed in mM. Figure 7 shows T1 and T2 relaxivities of ML and Stealth®-ML for the two temperatures. The T1 relaxivites drop exponentially from 85–71 mM^{-1} s^{-1} at 0.02 T to 3–1 mM^{-1} s^{-1} at 1.5 T. There is no T1 relaxivity peak at the lower fields, which is expected for smaller, less magnetic materials according to recently developed theories for (super)paramagnetic nanoparticles.[66,67] The magnitude of the 1/T1 at high field is relatively very low compared with other magnetite-containing nanoparticles,[51] likely resulting from the low surface/volume ratio (i.e., as in the case of larger particles), as well as the limited access of the inner crystal to diffusing water protons, reducing inner sphere dipole–dipole type relaxation. On the other hand, the measured T2 relaxivities (Fig. 7) demonstrate a saturation at relatively low fields, with unusually high values of 210–350 mM^{-1} s^{-1}. Both the saturation and the magnitude of the 1/T2 data are typical of larger iron oxide particle structures. Furthermore, at either temperature, PEG incorporation enhances selectively T2 relaxation by 10%–15%. For Stealth®-ML, the experimentally obtained T1 and T2 relaxivities at 1.5 T and body temperature are 3 and 240 s^{-1}/mM, respectively. This combination of high T2 and low T1 at 37° and clinical field strength (1.5 T) results in an unusually high 1/T2:1/T1 ratio of nearly 80. In the practice of MR imaging, higher 1/T2:1/T1 ratios translate into relatively more T2-dominant contrast effects (i.e., hypointensity using conventional spin-echo sequences).

[65] E. Strable, J. W. M. Bulte, B. Moskowitz, K. Vivekanandan, M. Allen, and T. Douglas, *Chem. Mater.* **13**, 2201 (2001).

[66] A. Roch, R. N. Muller, and P. Gillis, *J. Chem. Phys.* **110**, 5403 (1999).

[67] J. W. M. Bulte, R. A. Brooks, B. M. Moskowitz, L. H. Bryant, Jr., and J. A. Frank, *Magn. Reson. Med.* **42**, 379 (1999).

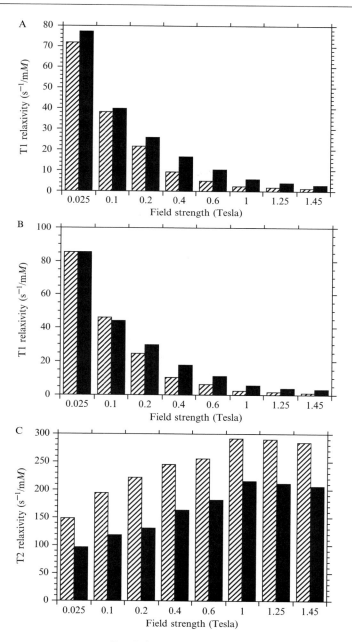

FIG. 7. (*continued*)

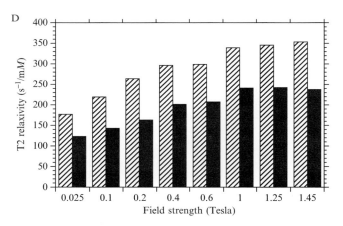

FIG. 7. (A, B) and T2 (C, D) relaxivities of ML (A, C) and Stealth®-ML (B, D). Striped and solid bars represent measurements obtained at 3° and 37°, respectively.

Blood Clearance of Stealth® Magnetoliposomes

Female Harlan Sprague Dawley rats, 6–8 weeks old are initially anesthetized by intraperitoneal (IP) injection of 100 mg/kg ketamine and 10 mg/kg acepromazine. Anesthesia is then maintained by administering 25 mg/kg ketamine and 2.5 mg/kg acepromazine IP by means of a 20-gauge IP catheter (JELCO, Tampa, FL) every 30 min or as deemed necessary. The femoral artery is cannulated with polyethylene tubing and connected to a stopcock. For the experiments described in the following, ML and Stealth®-ML are administered intravenously (IV, tail vein) at 100 μmol Fe/kg.

The blood clearance of the magnetoliposomes can be determined as follows. For each time point following injection of ML ($n = 3$) and Stealth®-ML ($n = 3$), 0.3-ml blood samples are taken from the femoral artery and subsequently mixed with 0.1 ml heparin solution (50 units/ml in 0.9% sodium chloride). To compensate for blood volume loss, 0.3 ml phosphate-buffered saline needs to be injected IV (tail vein) following each collection. For each animal, preinjection baseline blood samples need to be included. The T1 and T2 of these mixtures are then measured at 1.0 T and 37°, according to the measurement procedure described earlier. The blood half-lives can be calculated using the clearance function

$$1/T_t = Ae^{-bt} + Ce^{-dt} + 1/T_0,$$

where

$$t_{1/2} = 0.693/b$$

In these equations, $1/T_t$ is the 1/T1 or 1/T2 value of blood samples taken at time point t following ML and Stealth®-ML injection, $1/T_0$ is the 1/T1 or 1/T2 preinjection baseline value, and $t_{1/2}$ is the calculated blood half-life. For a monoexponential fit, C needs to be set at 0. For each time point, the difference between the measured 1/T1 and 1/T2 blood values of the two groups (ML and Stealth®-ML) need to be analyzed for statistical significance using a two-tailed Student's t test.

Figure 8 shows the experimental 1/T1 and 1/T2 blood values vs time of collection following injection of ML and Stealth®-ML. PEGylation increases the magnetoliposome blood half-life; for t > 30 min, the difference between the measured 1/T1 and 1/T2 blood values can be found to be statistically significant ($p < 0.05$, Student's t test). Both 1/T1 and 1/T2 for ML can be well fitted ($r > 0.99$) using monoexponential clearance kinetics, but for Stealth®-ML, biexponential clearance kinetics are needed. The long (second component, weight = 40%) blood half-life for Stealth®-ML is calculated to be 53.2 ± 13.2 min whereas for ML the blood half-life is 7.4 ± 0.4 min.

Magnetic Resonance Imaging Using Stealth® Magnetoliposomes

For MR imaging studies, 6- to 8-week-old Harlan Sprague Dawley rats are anesthetized as described previously and injected IV (tail vein) with 100 μmol Fe/kg of ML ($n = 2$) and Stealth®-ML ($n = 2$), using a 24-gauge cathether connected to a 100-cm piece of polyethylene tubing and a syringe, so that the contrast agent can be administered without moving the animal out of the magnet. The magnet used in our experiments is a Signa 1.5 T system (General Electric Medical Systems, Milwaukee, WI). Precontrast and postcontrast coronal images are collected using a 5-inch (12.7 cm) circular surface coil with a field of view of 13 cm, a slice thickness of 3 mm, a matrix of 256 × 192, and two excitations. Images are acquired in the following order: T1-weighted spin-echo (SE; repetition time [TR]/echo time time [TE] = 400/15 ms), proton density-weighted SE (TR/TE = 2000/30), T2-weighted SE (TR/TE = 2000/60, 2000/90, and 2000/120), gradient-recalled echo (GRE; TR/TE = 300/20, flip angle = 20°), and fast spin-echo (FSE; TR/TE = 3000/40) images. Animals are imaged before, starting immediately after (0–30 min postcontrast), and starting 30 min after (30–60 min postcontrast) the injection of ML or Stealth®-ML.

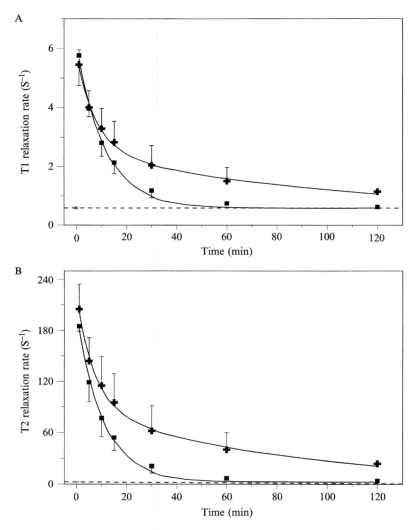

FIG. 8. 1/T1 (A) and 1/T2 (B) blood clearance curves. Solid lines represent best fits of the ML (squares) and Stealth®-ML (crosses) data using monoexponential (ML) and biexponential (Stealth®-ML) clearance functions (Reproduced with permission from J. W. M. Bulte *et al., J. Magn. Reson. Imaging* **9,** 329, 1999).

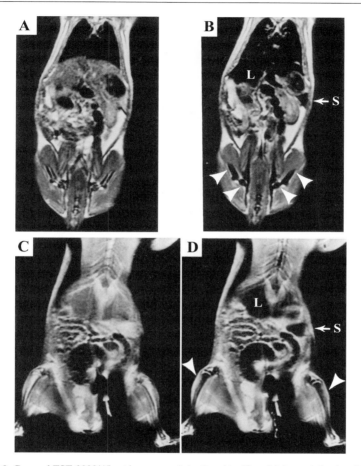

FIG. 9. Coronal FSE 3000/45 rat images preinjection (A, C) and 1 h postinjection (B, D) of ML. Arrowheads outline bone marrow uptake. L, liver; S, spleen. (Partially reproduced with permission from J. W. M. Bulte, *et al.*, *J. Magn. Reson. Imaging* **9,** 329, 1999, and J. W. M. Bulte *et al.*, *J. Magn. Magn. Mat.* **194,** 204, 1999).

We have found FSE images provide the best overall contrast (enhancement) and image quality. On the MR images (Figs. 9 and 10), no difference in organ uptake between the ML and Stealth®-ML preparations is found. MR imaging shows that there is a rapid uptake in the liver and spleen and also a marked uptake in the bone marrow, including the iliac bone, femur, tibia, and lower vertebrae. The bone marrow uptake for both ML

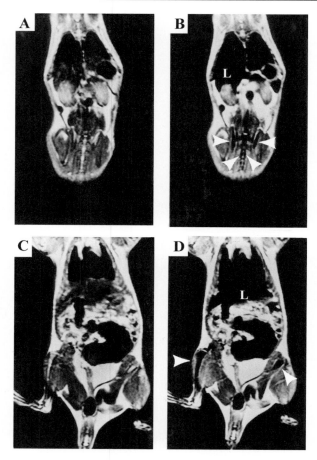

Fɪɢ. 10. Coronal FSE 3000/45 rat images preinjection (A, C) and 1 h postinjection (B, D) of Stealth®-ML. Arrowheads outline bone marrow uptake, L indicates liver (Partially reproduced with permission from J. W. M. Bulte *et al.*, *J. Magn. Reson. Imaging* **9,** 329, 1999, and J. W. M. Bulte *et al.*, *J. Magn. Magn. Mat.* **194,** 204, 1999).

and Stealth®-ML reaches a plateau within the first 30 min following injection, because one can observe no difference in signal intensities between the first and second postcontrast series (data not shown).

Two hours following injection of ML ($n = 3$) and Stealth®-ML ($n = 3$) at 100 μmol Fe/kg, 6- to 8-week-old Harlan Sprague Dawley rats are perfused with a 0.9% sodium chloride solution. Tissue samples of the liver, lung, spleen, cervical lymph nodes, and mesenteric lymph nodes are removed for histochemical processing. Following formalin fixation and embedding in paraffin, deparaffinated tissue sections (10-μm thick) are stained for ferric iron using Perls' reaction (Prussian Blue stain, counterstained with neutral fast red) as follows: A fresh solution of Perls' reagent is prepared by dissolving 1 g of potassium ferro (not ferri) cyanide in 42-ml deionized water, followed by adding 8-ml 37.5 % HCl (these amounts are for one vertical staining jar holding 18 slides). Slides are rinsed three times in PBS and incubated with Perls' reagent for 30 min in the dark. Slides are rinsed three times in PBS and counterstained with neutral fast red (NFR) solution for 3 min. To make an NFR solution, 0.1 g NFR is dissolved in 100-ml deionized water, 5 g aluminium sulfate ($(Al_2SO_4).18H_20$) is added. The solution is heated to boiling, slowly, then cooled, filtered and a grain of thymol is added as a preservative. The NFR solution can be reused several times up to about 2 weeks, or until a precipitate becomes visible. Following NFR counterstaining, slides are rinsed in deionized water and dried overnight in the dark. The next day, the dried slide is embedded in Permount and covered with a coverslip.

Prussian Blue staining of tissue specimens reveals uptake of both ML and Stealth®-ML in the liver. For the ML preparation, iron pigment accumulates primarily within the hepatocytes; involvement of Kupffer cells seems to be secondary. For Stealth®-ML, the iron uptake is more prominent in liver Kupffer cells, with a few positive hepatocytes. In the spleen, there is usually some uptake by macrophages in the red pulp, with no differences between ML and Stealth®-ML. For the lung, no uptake in alveolar macrophages is observed. A remarkable complete lack of uptake can be seen in the lymph nodes for both preparations. Because of the high endogenous iron content of red bone marrow, it is impossible to discriminate between cells containing iron for normal erythropoiesis and cells (macrophages) that have taken up injected contrast material. In our experience, Prussian Blue staining is not useful for assessing uptake in bone marrow smears.

Concluding Remarks

The preceding findings demonstrate that (Stealth®) magnetoliposomes have a strong effect on T2 relaxation, making these particles an ideal platform for further development as MR contrast agents. We have shown that it is possible to increase their blood half-life signficantly by incorporating PEG in the outer lipid bilayer, opening the door to using these Stealth® magnetoliposomes as targeted contrast agents on proper conjugation of selected ligands. In the meantime, untargeted magnetoliposomes seem to have a high affinity for bone marrow. Recently, other studies have shown that IV administered (ultrasmall) superparamagnetic iron oxide ((U)SPIO) colloidal particles are taken up by bone marrow, reducing the signal intensity on MR images.[68-71] This has allowed improved detection of small bone tumors in animal models. Contrary to the use of dextran-coated (U)SPIOs, however, magnetoliposomes may permit incorporation of lipohilic therapeutic drugs or perhaps even gene vectors, with the potential of forming a combined diagnostic and therapeutic drug delivery system.

Acknowledgment

This research was partially supported by a grant from the "Fonds voor Wetenschappelijk Onderzoek – Vlaanderen" (Grant G.0170.96).

[68] E. Senéterre, R. Weissleder, D. Jaramillo, P. Reimer, A. S. Lee, T. J. Brady, and J. Wittenberg, *Radiology* **179**, 529 (1991).
[69] C. H. Bush, C. R. J. Mladinich, and W. J. Montgomery, *J. Magn. Reson. Imaging* **7**, 579 (1997).
[70] H. E. Daldrup, T. M. Link, S. Blasius, A. Strozyk, S. Könemann, H. Jürgens, and E. J. Rummeny, *J. Magn. Reson. Imaging* **9**, 643 (1999).
[71] B. C. Vande Berg, F. E. Lecouvet, J. P. Kanku, J. Jamart, B. E. van Beers, B. Maldegue, and J. Malghem, *J. Magn. Reson. Imaging* **9**, 322 (1999).

[13] Liposomes in Ultrasound and Gamma Scintigraphic Imaging

By SUMEET DAGAR, ISRAEL RUBINSTEIN, and HAYAT ÖNYÜKSEL

Introduction

Liposomes are vesicles in which an aqueous media is enclosed by a bilayer of lipids, usually phospholipids. Liposomes have been considered to be used in numerous disciplines as artificial membranes and for drug-/diagnostic agent delivery since their discovery in 1960s. However, their clinical success was limited because of the short blood circulation time resulting from rapid opsonization and removal from blood by cells of the mononuclear phagocytic system (MPS).[1,2] In addition, because liposomes are particulates, their biodistribution was limited to the circulation and organs of the MPS. In the early 1990s the short half-life of liposomes was

[1] J. H. Senior, *Crit. Rev. Ther. Drug Carrier Syst*, **3**, 123 (1987).
[2] G. Gregoriadis, *Adv. Exp. Med. Biol.* **238**, 151 (1988).

overcome by the surface modification of liposomes with a lipid derivative of polyethylene glycol (PEG) to form long-circulating, sterically stabilized liposomes (SSL).[3,4] Recent advances in the knowledge of pathophysiology and microcirculation of certain diseases such as inflammation and cancer that show leaky vasculature and inadequate lymphatic drainage, have improved significantly the potential use of liposomes in the clinics. Once liposomes were shown to have steric stabilization with long circulation half-life and be prepared at sizes (∼100 nm) to extravasate preferentially at the diseased site,[5] they became attractive for use as targeted drug or imaging agent delivery vehicles. Furthermore, because liposomes are composed of phospholipids, they are relatively safe carriers, and ligands could be attached to their outer surface for active targeting by interaction specifically with biomarkers of certain diseases. These developments resulted in a surge of attempts to develop liposomes as carriers for diagnostic agents for improved imaging and diagnosis of certain diseases.

The use of liposomes in diagnostic medical imaging is a relatively new and rapidly growing field.[6] The role of liposomes in imaging is to provide detectable differences between the pathological and normal tissues and to be retained long enough to permit satisfactory imaging.[7] To date, liposomes have been used to deliver diagnostic agents for gamma-scintigraphy, computed tomography, ultrasonography, and magnetic resonance imaging. This review will focus on the design and development of liposomal imaging agents for ultrasonography and gamma scintigraphy.

Liposomes in Ultrasonography

Ultrasound is one of the most commonly used diagnostic modalities.[8] Since the advent of color Doppler, the practice of ultrasonography for cardiology and the measurement of flow in general has improved greatly. Ultrasound imaging provides structural and functional information, including features of tissue and blood flow characteristics. The action of the ultrasound contrast agents may be due to three different effects: backscatter, sound attenuation, and changes in the sound speed. The most promising agents rely on backscatter. When sound waves from an ultrasound

[3] A. L. Klibanov, K. Maruyama, V. P. Torchilin, and L. Huang, *FEBS Lett.* **268,** 235 (1990).
[4] T. M. Allen, G. A. Austin, A. Chonn, L. Lin, and K. C. Lee, *Biochim. Biophys. Acta,* **1061,** 56 (1991).
[5] D. C. Drummond, O. Meyer, K. Hong, D. B. Kirpotin, and D. Papahadjopoulos, *Pharmacol. Rev.* **51,** 691 (1999).
[6] V. P. Torchilin, *Mol. Med. Today* **2,** 242 (1996).
[7] S. E. Seltzer, *Radiology* **171,** 19 (1989).
[8] C. Tilcock, *Adv. Drug Deliv. Rev.* **37,** 33 (1999).

source pass through a substance, the acoustic properties of that substance will depend on the velocity of the sound and the density of that substance. Acoustic impedance of the substance is most pronounced at interfaces of different substances with greatly different density or acoustic impedance, particularly at the interface between solids, liquids, and gases. When ultrasound waves encounter such interfaces, the changes in acoustic impedance result in a more intense reflection of sound waves (backscatter) and a more intense signal in the ultrasound image. The physical principles and clinical applications of ultrasound image-enhancing agents have been reviewed.[9–11]

Liposomal ultrasound contrast agents generally include gas-filled liposomes and inherently acoustically reflective (echogenic) liposomes.

Gas-Filled Liposomes

Several attempts have been made to make gas-filled liposomes to be used as ultrasound contrast agents. These include liposomes filled with carbon dioxide and nitrogen.[12–14] The most common method to prepare liposomes for ultrasonography is to incorporate gas bubbles into the liposomes or to form a gas bubble directly inside the liposome as a result of a chemical reaction producing a gas such as carbon dioxide.[15] These liposomes are not truly liposomes, because the aqueous core has been replaced by gas. The gas-filled liposomes can be classified broadly into two groups on the basis of whether the gas is water soluble or insoluble.

Liposomes Filled with Water-Soluble Gases such as Carbon Dioxide and Nitrogen. Several methods to produce the gas-filled liposomes involve the presence of gas precursors within the liposome, which are activated by various means.[13] One of these methods involves a gaseous precursor that reacts to form a gas in response to a change in pH within a liposome. For example, bicarbonate salts are entrapped within the interior aqueous space of the vesicle, and an ionophore such as p-trifluoromethoxy-carbonylcyanide

---

[9] B. B. Goldberg, J. B. Liu, and F. Forsberg, *Ultrasound Med. Biol.* **20,** 319 (1994).

[10] J. W. Winkelmann, M. D. Kenner, R. Dave, R. H. Chandwaney, and S. B. Feinstein, *Ultrasound Med. Biol.* **20,** 507 (1994).

[11] P. J. Frinking, A. Bouakaz, J. Kirkhorn, F. J. Ten Cate, and N. de Jong, *Ultrasound Med. Biol.* **26,** 965 (2000).

[12] E. C. Unger, P. J. Lund, D. K. Shen, T. A. Fritz, D. Yellowhair, and T. E. New, *Radiology* **185,** 453 (1992).

[13] E. C. Unger, U. S. Patent Number 5,088,499 (1992).

[14] E. C. Unger, D. Shen, T. Fritz, B. Kulik, P. Lund, G. L. Wu, D. Yellowhair, R. Ramaswami, and T. Matsunaga, *Invest. Radiol.* **29 Suppl 2,** S134 (1994).

[15] V. P. Torchilin, "Handbook of Targeted Delivery of Imaging Agents," CRC Press, Boca Raton, FL, 1995.

phenylhydrazone is present within the liposome membrane matrix to promote hydrogen ion flux across the liposomal membrane. The pH change within the vesicle interior facilitated by the ionophore results in the formation of a highly echogenic carbon dioxide gas within the liposome. In another method, gaseous precursors such as diazonium compounds trapped inside the lipid vesicles react to form echogenic nitrogen gas on exposure to ultraviolet light. Another method involves a gaseous precursor such as methylactate that is transformed from a liquid to a gas in response to an increase in temperature.

Gas-filled liposomes can also be prepared by the basic pressurization and depressurization technique.[13] First, liposomes are added to a vessel, and the vessel is then pressurized with a gas such as carbon dioxide. Under pressure, the gas goes into solution and passes across the liposome membranes. When the pressure is released, gas bubbles form within the liposomes.

The problem with the gas-filled liposomes containing water-soluble gases like carbon dioxide and nitrogen is that they are unstable and have a short half-life. Recently, the focus has been on the encapsulation of gases such as perfluorate gases in liposomes. These gases are more stable and offer a longer lasting effect because of their lower solubility in water.[16]

Liposomes Filled with Water-Insoluble Gases such as Perflourates
There have been a few variations of the perflourate gas-filled liposomes, including perfluorobutane-filled liposomes (MRX-113®),[14] perfluorobutane-filled liposomes targeted to thrombi (MRX-408®),[17] and perfluoropropane-filled liposomes (MRX-115®).[18] All of these are developed by ImaRx Therapeutics, Inc. (now owned by DuPont Pharmaceuticals USA, and Bristol-Myers Squibb Company).

Among the gas-filled liposomal ultrasound contrast agents, Definity™ or MRX 115® (previously also known as DMP 115, and Aerosomes™) is the only one approved recently by the Food and Drug administration to be used in the clinic for cardiac ultrasound imaging (echocardiography). It consists of perfluoropropane gas–filled lipid bilayers composed of dipalmitoylphosphatidylcholine (DPPC).

The preferred method for preparing the perfluorate gas–filled liposomes consists of shaking an aqueous solution composed of a lipid in the presence of a gas at a temperature below the gel state to the liquid

[16] G. Maresca, V. Summaria, C. Colagrande, R. Manfredi, and F. Calliada, *Eur. J. Radiol.* **27** Suppl 2, S171 (1998).

[17] E. C. Unger, T. P. McCreery, R. H. Sweitzer, D. Shen, and G. Wu, *Am. J. Cardiol.* **81,** 58G (1998).

[18] T. A. Fritz, E. C. Unger, G. Sutherland, and D. Sahn, *Invest. Radiol.* **32,** 735 (1997).

crystalline state phase transition temperature of the lipid.[19] A representative method of making perfluorate gas–filled liposomes is given as follows.[17] The phospholipids are suspended at a concentration of 1 mg/ml in aqueous solution in a sterile 2.5-ml vial. The vials are capped and sealed with perfluorobutane headspace. The vials are then placed on a dental amalgamator (Wig-L-Bug, Crescent Dental, Lyons, IL) and agitated at 2800 revolutions/min for 60 s. The microbubles are sized by quasi-elastic light scattering.

There are certain general guidelines for preparing perfluorate gas–filled liposomes by agitation. These include:

1. Any type of motion that agitates the aqueous solution and results in the introduction of gas may be used for the shaking. The shaking must be of sufficient force to allow the formation of foam after a short time. The higher the revolutions per minute, the better the formation of gas-filled liposomes.

2. The formation of gas-filled liposomes on shaking can be detected by the presence of foam on top of the aqueous solution. This is coupled with a decrease in the volume of the aqueous solution on the formation of foam. The method producing the most foam is preferred. The required duration of shaking time may be determined by detection of the formation of foam.

3. The concentration of lipid required to form a preferred foam level will vary depending on the type of lipid used. For example, DPPC at a concentration of 20–30 mg/ml, on shaking, yields a total suspension and entrapped gas volume four times greater than the suspension volume alone.

4. Incorporation of gas should be carried out at a temperature below the gel-to-liquid crystalline phase transition temperature of the lipid used.

Although all the gas containing ultrasound contrast agents have had varying degrees of success, there are still some associated problems. One of the common problems for gas-containing agents is the control of the size and stability of the gas after preparation and injection. Gas-encapsulating ultrasound contrast agents such as liposomes do not have much *in vitro* and *in vivo* stability. Particles greater than 4–5 μm, when injected intravenously, are trapped in the pulmonary capillaries and cannot pass from the right side to the left side of the heart. In addition, as a result of large particle size, circulation half-life is typically limited to a few minutes. These problems can be mostly overcome by using liposomes that do not require a gas to be acoustically reflective, as described in the following.

---

[19] E. C. Unger, T. A. Fritz, T. Matsunaga, V. Ramaswami, D. Yellowhair, G. Wu, U.S. Patent Number 5,935,553 (1999).

Inherently Acoustically Reflective (Echogenic) Liposomes

Önyüksel et al.[20] were the first to develop the inherently echogenic liposomes that are not gas-filled. These liposomes are acoustically reflective because of their multiple rigid bilayer membranes that are formed by controlling the lipid composition and production method. They have an oligolamellar (vesicles in a vesicle) structure because of the dehydration/rehydration process used in their preparation.[20] These liposomes are superior to the gas-entrapping liposomes, because they are (1) inherently echogenic; (2) more stable; (3) smaller than 1 μm, eliminating pulmonary entrapment; and (4) suitable for conjugation of antibodies to be targeted to a specific tissue.[21]

Önyüksel et al. prepared these inherently echogenic liposomes using a mixture of lipids (egg phosphatidylcholine [E-PC], dipalmitoyl phosphatidylethanolamine [PE], dipalmitoyl phosphatidylglycerol [DPPG], and cholesterol [CH]) at the optimum molar ratio of 60:8:2:30 (Fig. 1), by aqueous reconstitution of the dried lipid film followed by sonication until 500-nm liposomes are obtained.[20] These liposomes are then freeze-dried in the presence of a cryoprotectant to form oligolamellar liposomes (<1 μm) on reconstitution of the dry cake.

Demos et al.[21] further developed the inherently echogenic liposomes to prepare targeted echogenic immunoliposomes by conjugating an antibody to the liposome surface to target them specifically to diseased tissues. To prepare these echogenic immunoliposomes, specific to atherosclerotic plaque, anti-fibrinogen antibody is modified with N-succinimidyl pyridyl dithiopropionate (SPDP), and maleimido-4(p-phenylbutyrate)-phosphatidylethanolamine (MPB-PE) is used instead of PE to prepare echogenic liposomes. The details of the method to conjugate antibody to the liposomes has been described.[21] The same conjugation method is also adapted for the conjugation of fragments of the antibody to the liposomes.[22]

Demos et al.[22] tested the *in vitro* fibrin-binding characteristics of the anti-fibrinogen targeted, inherently echogenic liposomes under various flow conditions. They tested the retention of liposomes conjugated to polyclonal antibodies, monoclonal antibodies, and Fab fragments of monoclonal antibodies onto fibrin-coated filter paper placed in a flow circuit under controlled flow conditions. Under a physiological shear stress of

[20] H. Önyüksel, S. M. Demos, G. M. Lanza, M. J. Vonesh, M. E. Klegerman, B. Kane, J. Kuszak, and D. D. McPherson, *J. Pharm. Sci.* **85**, 486 (1996).
[21] S. M. Demos, H. Önyüksel, J. Gilbert, S. I. Roth, B. Kane, P. Jungblut, J. V. Pinto, D. D. McPherson, and M. E. Klegerman, *J. Pharm. Sci.* **86**, 167 (1997).
[22] S. M. Demos, S. Dagar, M. Klegerman, A. Nagaraj, D. D. McPherson, and H. Önyüksel, *J. Drug Target.* **5**, 507 (1998).

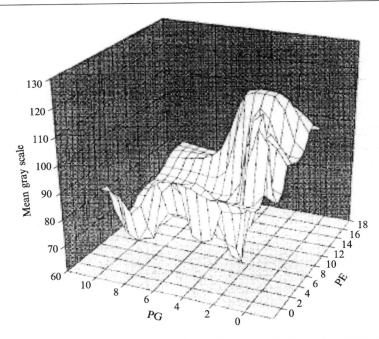

Fig. 1. Effect of Phosphatidylethanomine and phosphatidylglycerol content on the echogenecity of liposomes [Reproduced with permission from (20)].

1.5 N/m^2, more than 70% of the liposomes remained attached to fibrin after 2 hours. Plasma components and temperature had no effect on liposomal retention. In addition, monoclonal antibodies showed a slight trend of reduced retention over time, compared with polyclonal antibodies and Fab fragments of monoclonal antibodies.

The acoustic reflectivity of the immunoliposomes was further demonstrated to detect atherosclerotic (ATH) plaques in a miniswine animal model at different stages *in vivo*.[23] For detection of earlier stage ATH plaques, anti-intracellular adhesion molecule-1 (ICAM-1) antibodies are used and for later stage plaques with thrombi and fibrous portions, anti-fibrinogen antibodies are used. The attachment of the immunoliposomes, conjugated with different kinds of antibodies, to ATH plaques is clearly demonstrated both *in vitro* and *in vivo* using fibrin-coated paper and the miniswine atherosclerosis model, respectively.[22,23] In December 2001, the

[23] S. M. Demos, H. Alkan-Önyüksel, B. J. Kane, K. Ramani, A. Nagaraj, R. Greene M. Klegerman, and D. D. McPherson, *J. Am. Coll. Cardiol.* **33,** 867 (1999).

patents[24,25] describing these nongaseous inherently echogenic immunoliposomes were licensed to Echodynamics Inc. (College Park, MD) for commercial development. The goal of this company is to further develop the echogenic immunoliposomes (now called EGIL) for identification of ATH plaques.

Recently, these inherently echogenic liposomes were further developed by Tiukinhoy et al.[26] to produce ICAM-1–targeted cationic liposomes for gene delivery applications. They demonstrated that these cationic liposomes retained their acoustic reflectivity and specific adherence to fibrin under flow conditions. Significant transfection of luciferase encoding plasmid to human umbilical vein endothelial cells was demonstrated. Higher gene expression using specific antibody-conjugated echogenic liposomes was observed. These novel antibody-conjugated acoustic cationic liposomes present an exciting potential for a vector that allows tissue image enhancement and targeted gene delivery at the same time.[26]

Huang et al.[27] have suggested some modifications to the lipid composition to improve the acoustic reflectivity of the original inherently echogenic liposomes first developed by Önyüksel et al. They varied the CH content in the range from 1–40 mol% and the PG content from 1–16 mol%. They also changed the type and concentration of sugars used as cryoprotectants during lyophilization. Optimal acoustic stability was observed with concentrations of 10–15 mol% CH and a PG concentration greater than 4 mol% compared with 30 mol% CH and 2 mol% PG used in the original preparations of Demos et al. Preparations made with 0.2 M mannitol are more ultrasound-reflective than those made with lactose, trehalose, and sucrose. Relyophilization and freezing temperatures below $-20°$ increased ultrasound reflectivity. The preparations developed seemed to be significantly more acoustically reflective than the previous preparations. However, these liposomes are classical liposomes and will be rapidly taken up by the reticuloendothelial system when injected. Hence, they are limited to local use only. In addition, they are targeted by an antibody, which precludes their repeated use because of possible immune reactions.

[24] G. M. Lanza, H. Alkan-Önyüksel, M. E. Klegerman, D. D. McPherson, B. J. Kane, and S. E. Murer, U.S. Patent Number 5,612,057 (1997).

[25] G. M. Lanza, H. Alkan-Önyüksel, M. E. Klegerman, D. D. McPherson, B. J. Kane, and S. E. Murer, U.S. Patent Number 5,858,399 (1999).

[26] S. D. Tiukinhoy, M. E. Mahowald, V. P. Shively, A. Nagaraj, B. J. Kane, M. E. Klegerman, R. C. MacDonald, D. D. McPherson, and J. S. Matsumura, Invest. Radiol. 35, 732 (2000).

[27] S. L. Huang, A. J. Hamilton, A. Nagaraj, S. D. Tiukinhoy, M. E. Klegerman, D. D. McPherson, and R. C. MacDonald, J. Pharm. Sci. 90, 1917 (2001).

Acoustically Reflective Vasoactive Intestinal Peptide–Sterically Stabilized Liposomes. During our studies on the preparation of vasoactive intestinal peptide (VIP) liposomes, we have by serendipity discovered a novel echogenic composition. These liposomes are prepared by the extrusion/dehydration-rehydration method.[28] Lipids (DSPE-PEG:PC:PG:CH, molar ratio 0.5:5:1:3.5) are mixed in chloroform, and a dry film is formed. The dried lipid film is rehydrated with 0.15 M NaCl. The dispersion is vortexed, sonicated, and then extruded sequentially through polycarbonate filters of pore size 200, 100, and then 50 nm to form liposomes of about 80 nm. VIP and trehalose (cryoprotectant), in powder form, are added to the extruded dispersion. The mixture is then frozen in an acetone–dry ice bath and lyophilized overnight. The freeze-dried cake is resuspended in deionized water and any free VIP is separated by gel permeation chromatography to form inherently echogenic VIP-sterically stabilized liposomes or VIP-SSL.[28] A similar process is used to form inherently echogenic sterically stabilized liposomes without any VIP.

The freeze-fracture electron microscopy picture demonstrates clearly the oligolamellar structure with vesicles within vesicle (Fig. 2). The size of the inherently echogenic SSL and VIP-SSL after freeze-drying and reconstitution is much smaller (~260–290 nm) than the previously discovered inherently echogenic liposomes in our laboratory (~800 nm). More importantly, this second generation of inherently echogenic liposomes is sterically stabilized and therefore expected to have a long circulation half-life. In addition, we have shown previously that VIP incorporates itself within phospholipid bilayers.[29,30] Hence, with VIP on their surface, these liposomes may be an ideal candidate for targeted ultrasound

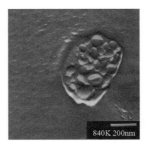

840K 200nm

FIG. 2. Freeze-fracture micrograph showing the structrure of VIP-SSL (X12,000, bar represents 200 nm).

[28] M. Patel, I. Rubinstein, H. Ikezaki, and H. Alkan-Önyüksel, *Proc. Int. Symp. Control. Rel. Bioact. Mat.* **24**, 913 (1997).

TABLE I
GRAY-SCALE VALUES (BRIGHTNESS LEVEL) AND SIZE OF
DIFFERENT INHERENTLY ECHOGENIC LIPOSOMES

Formulation	Average gray scale value	Average size (nm)
Inherently echogenic classical liposomes (Önyüksel et al., 1996)	115	800
Inherently echogenic sterically stabilized liposomes	109	286
Inherently echogenic VIP-sterically stabilized liposomes	119	264

contrast agents for certain diseases such as cancer and arthritis, where the vasculature is leaky and VIP receptors are overexpressed. The acoustic reflectivity of these liposomes (Table I) is consistent with their structure as shown in Fig. 2.

Liposomes in Gamma Scintigraphy

The principal advantage of imaging with radionuclides compared with other imaging modalities is its relatively high sensitivity, which makes it an excellent screening technique as well as valuable for assessing function. Relatively low doses of the radionuclide are required for effective imaging. For the design of liposomes containing radionuclides, the physical state of the bilayer is not so important, because the gamma label is highly penetrating. However, compositions less susceptible to radiolysis (saturated rather than unsaturated) should be favored.

The radionuclides used to label liposomes for gamma-scintigraphy include Gallium-67 (Ga-67), Indium-111 (In-111), and metastable Technetium-99 (Tc-99m).

Labeling Liposomes with Ga-67 and In-111

Gallium-67 can be incorporated into the liposomes during manufacture by making the liposomes in the presence of Ga-67 chelated to nitrilotriacetic acid.[31] However, with this technique low labeling efficiency is obtained. Another method involves loading Ga-67 into preformed liposomes containing desferoxamine.[32] In this method Ga-67 is first complexed to a

[29] I. Rubinstein, M. Patel, H. Ikezaki, S. Dagar, and H. Önyüksel, *Peptides* **20,** 1497 (1999).
[30] H. Önyüksel, B. Bodalia, V. Sethi, S. Dagar, and I. Rubinstein, *Peptides* **21,** 419 (2000).
[31] I. Ogihara-Umeda and S. Kojima, *J. Nucl. Med.* **29,** 516 (1988).

lipophilic chelator oxine, which then carries the Ga-67 into the liposome where it is chelated with desferoxamine. This label is stable *in vivo*. However, there are several reasons why Ga-67 is not preferred for nuclear medicine. These include excessively high-energy gamma emissions (390 KeV and 300 KeV) and a long 78-h half-life.[33] Therefore, other labels such as In-111 have been evaluated.

There are several methods by which liposomes can be labeled with In-111. In one method, preformed liposomes are simply incubated with In-111-oxine. The lipophilic oxine becomes incorporated into the lipid bilayer of the liposomes. However, this label is unstable *in vivo*. There are two methods to label preformed liposomes with In-111. The first method uses lipophilic chelators, oxine to carry In-111 into the liposome where it is entrapped by binding to either nitrilotriacetic acid[34,35] or desferoxamine,[36] as described for Ga-67. The second method uses an ionophore in the lipid membrane to transport externally added In-111 chloride through the lipid bilayer so that once inside, In-111 is entrapped by nitrilotriacetic acid.[37] One disadvantage of the second method is that the temperature at which labeling is carried out should be above the phase transition temperature of the liposomes. These two active-loading techniques have high labeling efficiencies and the label is stable *in vivo*. Although In-111 has somewhat improved characteristics compared with Ga-67 (lower energy gamma emissions at 247 and 172 keV and a relatively shorter half-life of 67 h), it has a disadvantage of being relatively expensive, because it has to be produced by a cyclotron.[33]

The use of liposomal Ga-67 for detection of osteomyelitis,[38] inflammation, and tumor imaging[39,40] has been described. In-111–labeled liposomes (VesCan®, Vestar Inc.) it has been evaluated for detection of various carcinomas, as well as melanoma, sarcoma, and lymphoma.[41,42]

---

[32] A. Gabizon, R. M. Straubinger, D. C. Price, and D. Papahadjopoulos, *J. Liposome Res.* **1**, 123 (1988).

[33] W. T. Phillips, *Adv. Drug Deliv. Rev.* **37**, 13 (1999).

[34] P. L. Beaumier and K. J. Hwang, *J. Nucl. Med.* **23**, 810 (1982).

[35] K. J. Hwang, J. E. Merriam, P. L. Beaumier, and K. F. Luk, *Biochim. Biophys. Acta.* **716**, 101 (1982).

[36] A. Gabizon, D. C. Price, J. Huberty, R. S. Bresalier, and D. Papahadjopoulos, *Cancer Res.* **50**, 6371 (1990).

[37] M. R. Mauk and R. C. Gamble, *Anal. Biochem.* **94**, 302 (1979).

[38] V. Awasthi, B. Goins, R. Klipper, R. Loredo, D. Korvick, and W. T. Phillips, *J. Nucl. Med.* **39**, 1089 (1998).

[39] I. Ogihara, S. Kojima, and M. Jay, *J. Nucl. Med.* **27**, 1300 (1986).

[40] I. Ogihara-Umeda, T. Sasaki, S. Kojima, and H. Nishigori, *J. Nucl. Med.* **37**, 326 (1996).

[41] C. A. Presant, A. F. Turner, and R. T. Proffitt, *J. Liposome Res.* **4**, 985 (1994).

[42] D. J. Hnatowich, B. Friedman, B. Clancy, and M. Novak, *J. Nucl. Med.* **22**, 810 (1981).

Labeling Liposomes with Tc-99m

The most promising research in the area of liposomal imaging agents for gamma-scintigraphy is centered on the development of long circulating SSL with Tc-99m as the label either on their surface or within the interior of the vesicle.[21] Tc-99m is the radionuclide of choice for gamma scintigraphy because of its ready availability, relatively low cost, optimum gamma emission energy, relatively short half-life (6.02 h), and safe decay products.[6,21] There are various ways by which the SSL can be labeled with Tc-99m, which include surface labeling and encapsulation of Tc-99m.

Labeling Liposomes by Surface Chelation of Tc-99m. One of the initial methods of labeling liposomes by surface chelation of Tc-99m involved preparation of liposomes containing lipophilic chelators such as diethylene triamine pentaacetic acid (DTPA) conjugated to the head group region of a long-chain fatty acid or phospholipid molecule.[42,43] These liposomes are then labeled by addition of pertechnetate in the presence of a reducing agent, tin, at a tinDTPA ratio of 0.36:1. This tinDTTA ratio is critical for efficient labeling of the liposomes.[44] Although this method works, it is not as experimentally robust as the following recently developed method.

A method for surface labeling of SSL with Tc-99m has been developed recently using a hydrazine nicotinamide (HYNIC) derivative of distearoyl-phosphatidyl-ethanolamine (DSPE).[45] In this method, first HYNIC is conjugated to DSPE. Distearoylphosphatidyl-ethanolamine is dissolved in warm chloroform followed by addition of S-HYNIC and triethylamine. After incubation at 55° for 1 h under constant stirring, the chloroform is evaporated under a gentle stream of nitrogen. The crude conjugate is stored at −20°. The reaction mixture is analyzed with thin-layer chromatography (TLC) on silica gel plates with chloroform/methanol/water (65:25:4 v/v) as the mobile phase.

Sterically stabilized lysosomes containing DSPE-HYNIC are prepared using a mixture of lipids consisting of partially hydrogenated egg phosphatidylcholine (HE-PC), polyethylene glycol (MW, 2000)-conjugated. DSPE, DSPE-PEG, DSPE, and CH. These lipids are then mixed at the molar ratio of 1.85:0. 15:0.07:1 (HE-PC: DSPE-PEG: DSPE-HYNIC: CH) in methanol: chloroform (10:1 v/v); the solvent is evaporated, and the resulting lipid film is dispersed in phospate-buffered saline (PBS) at room temperature. The size of the liposomes is reduced by multiple extrusions through stacked polycarbonate membranes with an extruder. After sizing by quasi-elastic

[43] R. Goto, H. Kubo, and S. Okada, *Chem. Pharm. Bull. (Tokyo)* **37,** 1351 (1989).
[44] C. Tilcock, Q. F. Ahkong, and D. Land Fisher, *Nucl. Med. Biol.* **21,** 89 (1994).
[45] P. Laverman, E. T. M. Dams, W. J. G. Oyen, G. Storm, E. B. Koenders, R. Prevost, J. W. M. van der Meer, F. H. M. Corstens, and O. C. Boerman, *J. Nucl. Med.* **40,** 192 (1999).

light scattering, the liposomal suspension is dialyzed against PBS overnight at 4° to remove unconjugated S-HYNIC. The final size of these liposomes is approximately 85 nm. The concentration of liposome-associated HYNIC is determined by converting the hydrazine groups to the corresponding hydrazone by a reaction with p-nitrobenzaldehyde and measuring the optical density at 385 nm. To label the HYNIC liposomes, liposomes are added to a mixture of tricine, stannous chloride in saline, and pertechnetate ($^{99m}TcO_4^-$) in saline, and the mixture is incubated for 15 min at room temperature. The radiochemical purity is determined by using instant TLC (ITLC) with 0.15 M sodium citrate (pH 5.0) as the mobile phase. There is no postlabeling purification step, because the labeling efficiency is greater than 95% after a 15-min incubation.

There are two advantages of this method: (1) There is no purification step; and (2) these liposomes can be lyophilized and then reconstituted just before labeling.[46] However, the lyophilization process has to be controlled to reduce the increase in size on rehydration. The least increase in size (from ~85nm to ~100nm) is obtained at lower phospholipid concentrations and 48 h dehydration cycle with 3.6 sucrose/lipid ratio.[46] These liposomes have been developed primarily for the detection of sites of inflammation and infection.[45,47]

Labeling Liposomes by Encapsulation of Tc-99m. Phillips et al.[48] developed an elegant method for labeling liposomes to encapsulate Tc-99m within the internal aqueous compartment of liposomes. This method is based on the mechanism of retention of Tc99m-HMPAO (Tc-99m chelated to d-l–hexamethylpropylene amine oxime) in the brain and the mechanism of labeling of leukocytes using Tc99m-HMPAO.[48–50] In the brain or the leukocytes, the conversion of lipophilic Tc99m-HMPAO to a hydrophilic species is primarily driven by an intracellular reaction with glutathione. Glutathione is present in relatively high concentrations intracellularly, whereas only small amounts exist in plasma. Lipophilic Tc99m-HMPAO complex diffuses through the blood–brain barrier or the lipid bilayer of

[46] P. Laverman, L. van Bloois, O. C. Boerman, W. J. G. Oyen, F. H. M. Corstens, and G. Storm, *J. Liposome Res.* **10,** 117 (2000).

[47] O. C. Boerman, P. Laverman, W. J. G. Oyen, F. H. M. Cortens, and G. Storm, *Prog. Lipid Res.* **39,** 461 (2000).

[48] W. T. Phillips, A. S. Rudolph, B. Goins, J. H. Timmons, R. Klipper, and R. Blumhardt, *Int. J. Rad. Appl. Instrum. B.* **19,** 539 (1992).

[49] R. D. Neirinckx, L. R. Canning, I. M. Piper, D. P. Nowotnik, R. D. Pickett, R. A. Holmes, W. A. Volkert, A. M. Forster, P. S. Weisner, J. A. Marriott, and S. B. Chaplin, *J. Nucl. Med.* **28,** 191 (1987).

[50] R. D. Neirinckx, J. F. Burke, R. C. Harrison, A. M. Forster, A. R. Anderson, and N. A. Lassen, *J. Cereb. Blood Flow Metab.* **65,** 147 (1988).

the leukocytes. Once inside, the complex undergoes reductive decomposition by reaction with the sulfhydryl groups of glutathione. This results in the entrapment of the Tc-99m label within the brain or the leukocyte.[48–50]

Similarly, when incubated with preformed liposomes containing glutathione, the lipophilic Tc99m-HMPAO complex diffuses through the bilayer into the internal aqueous phase. As a result, the enclosed Tc99m-HMPAO complex is first reduced to a hydrophilic form by the glutathione and then trapped irreversibly in the internal aqueous phase.[48] The same approach has been used in the literature for making Tc-99m–labeled liposomes for blood pool imaging and to image sites of infection.[51–53]

Our laboratory has adapted this method to develop targeted Tc-99m-HMPAO–labeled liposomes.[54,55] These liposomes encapsulate Tc-99m-HMPAO in their aqueous volume and are surface-modified withVIP to achieve active targeting. These novel Tc99m-HMPAO–encapsulating targeted liposomes are developed for imaging of cancers that overexpress VIP receptors, including breast cancer, gastrointestinal–endocrine cancer, pancreatic cancer, and prostate cancer.[56,57]

Tc99m-HMPAO–loaded VIP liposomes can be prepared in several steps as described in the following (Fig. 3). First, SSLs are prepared by hydration of a dried lipid film followed by extrusion, as described before[54] with modifications. Egg-phosphatidylcholine, CH, DSPE-PEG, dipalmitoyl phosphatidylglycerol (DPPG) in molar ratios of 0.50:0.10:0.05:0.35 (EPC:DPPG:DSPE-PEG:CH) are dissolved in an organic solvent (chloroform-methanol; 9:1 v/v), and the solvent is evaporated in a rotary evaporator under vacuum to form a film. The film is desiccated under vaccum overnight. The dry lipid film is hydrated with HEPES buffer containing 50 mM glutathione. The solution is vortexed and then bath sonicated. The dispersion is extruded multiple times through a polycarbonate filter (100 nm) using a Liposofast Pneumatic extruder (Avestin Inc., Canada). The size of the liposomes is measured by quasi-elastic light scattering.

[51] B. Goins, W. T. Phillips, and R. Klipper, *J. Nucl. Med.* **37**, 1374 (1996).
[52] W. G. Oyen, O. C. Boerman, G. Storm, L. van Bloois, E. B. Koenders, R. A. Claessens, R. M. Perenboom, D. J. Crommelin, J. W. van der Meer, and F. H. Corstens, *J. Nucl. Med.* **37**, 1392 (1996).
[53] O. C. Boerman, W. J. G. Oyen, L. van Bloois, E. B. Koenders, J. W. M. van der Meer, F. H. M. Corstens, G. Storm, *J. Nucl. Med.* **38**, 489 (1997).
[54] S. Dagar, J. Stastny, M. J. Blend, I. Rubinstein, H. Önyüksel, *AAPS Pharm. Sci. J.* **1**, S, (1998).
[55] S. Dagar, M. Sekosan, B. Lee, I. Rubinstein, and H. Önyüksel, *J. Control. Rel.* **74**, 129 (2001).
[56] J. C. Reubi, *Ann N. Y. Acad. Sci.* **805**, 253 (1996).
[57] I. W. Moody, J. Leyton, I. Gozes, L. Lang, and W. C. Eckelman, *Ann. N Y Acad. Sci.* **865**, 290 (1998).

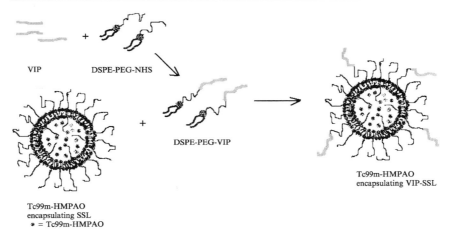

FIG. 3. Schematic representation of the method to prepare Tc99m-HMPAO encapsulating VIP-SSL (Not to scale).

The unentrapped glutathione is removed by gel filtration using a Biogel 10-DG column (Bio-Rad Laboratories, Richmond CA). The glutathione-containing liposomes, visible as turbid fractions, elute in the first few fractions.

Second, sterically stabilized glutathione-containing liposomes are labeled immediately after the free glutathione is removed by gel filtration. Ceretec® (Amersham) is incubated with freshly eluted Tc99m-pertechnate for 10 min at room temperature to form a lipophilic Tc-99m–HMPAO complex. This lipophilic complex is then incubated with preformed glutathione-containing liposomes for 15 min at room temperature, with occasional mixing. The complex, being lipophilic, passes through the bilayer. The Tc-99m–HMPAO complex is then trapped irreversibly in the internal aqueous phase of the liposome because of the reduction of the lipophilic complex by glutathione into a hydrophilic one.[50] The free label is then removed by gel filtration using a Biogel 10DG column. The radioactivity in each fraction is measured using a dose calibrator. The Tc-99m–HMPAO encapsulating liposomes (turbid fractions with high radioactivity) are pooled. The loading efficiency is usually approximately 85%.

Last, the targeting entity (VIP) is conjugated to the surface of the liposomes.[55] An activated DSPE-PEG (DSPE-PEG$_{3400}$-NHS) source is used to conjugate VIP to DSPE-PEG$_{3400}$. This reaction takes place between amines and n-hydroxy succinamide (NHS), which acts as the linking agent. The conjugation is tested using sodium-dodecyl sulfate-polyacrylamide gel electropes. The reaction mixture that is mostly VIP conjugated to DSPE-PEG$_{3400}$ (DSPE-PEG$_{3400}$-VIP) is used for subsequent studies without further purification.

The SSL encapsulating Tc99m-HMPAO prepared as described previously, are then used to form the Tc99m–HMPAO encapsulating VIP-SSL by incubating at elevated temperature ($\sim37^\circ$) for the insertion of DSPE-PEG$_{3400}$-VIP into their outer leaflet.[58] The free label and DSPE-PEG$_{3400}$-VIP are then removed by gel filtration, and the turbid liposome fractions are pooled.

The liposomes prepared by the described technique are novel, because they provide multiple labels per one targeting molecule. Also, the targeting entity is not labeled itself, but just attached to the carrier, which is laden with multiple radionuclide molecules. These properties should improve the sensitivity (caused by multiple labels/targeted entity) and specificity (caused by active targeting) of the imaging technique.

Dagar et al.[55] have tested the ability of the VIP liposomes described previously to bind breast cancer tissues in vitro. For this purpose, they used a carcinogen-induced rat breast cancer model, in which VIP receptors are were press.[59,60] The results demonstrated that VIP liposomes are bound to the breast cancer tissues to a higher extent than liposomes without VIP or liposomes with noncovalently associated VIP. We have also tested the Tc-99m–HMPAO VIP liposomes in vivo in the same carcinogen-induced rat breast cancer model. We found that labeled liposomes with or without VIP accumulate in breast cancer tissue, and in the presence of VIP this accumulation is significantly higher.[61] In addition, the tumor/background ratio was significantly more for VIP liposomes than liposomes without VIP. Work is underway in our laboratory to develop these liposomes further, not only for breast cancer imaging but also to expand the use of the VIP liposomes to image other cancers. We are also in the process of developing VIP liposomes for the targeted delivery of anticancer drugs for the chemotherapy of breast cancer and other cancers that overexpress VIP receptors.

Conclusions

Liposomes are attractive as imaging agents, because they can carry high loads of contrast material, and their surface can be modified to be specific

[58] P. Uster, T. M. Alen, B. E. Daniel, C. J. Mendezm, M. S. Newman, and G. Z. Zhu, FEBS Lett. **386**, 243 (1996).

[59] S. Dagar, M. Sekosan, M. Blend, I. Rubinstein, and H. Önyüksel, Proc. Int. Symp. Control. Rel. Bioact. Mat. **26**, 22 (1999).

[60] S. Dagar, M. Sekosan, I. Rubinstein, and H. Önyüksel, Breast Cancer Res. Treat. **65**, 49 (2001).

[61] S. Dagar, Doctoral thesis (2002).

to a biomarker of a disease state. Development of SSLs and better understanding of the pathophysiology of various disease states have increased the potential use of liposomes for ultrasound and scintigraphic imaging. This suggests that liposomal ultrasound or scintigraphic contrast media can prove to be a better alternative to the existing contrast media for specific applications. Although only one liposomal imaging product is approved by the food and drug administration at this stage, there is reasonable expectation that there would be more clinically and commercially successful liposomal imaging agents in the future. One of the most important developments in liposomal imaging agents has been the development of liposomes that are targeted to specific diseases. This can possibly lead to the early and accurate detection of numerous pathologic conditions such as atherosclerotic plaques and various cancers. On the whole, liposomal imaging agents are rapidly developing for functional imaging of various disease states.

Acknowledgments

This work was aided by Blowitz-Ridgeway Foundation and American Cancer Society, Illinois Division Inc., Grant # BR98-01, Susan G. Komen Breast Cancer Foundation Dissertation Research Grant, UIC Center for Woman and Gender Research, and Department of Defense, BCRP 011 268.

[14] Visualization of the Retinal and Choroidal Microvasculature by Fluorescent Liposomes

By Paul J. Lee and Gholam A. Peyman

History

Age-related macular degeneration and diabetic retinopathy remain leading causes of vision loss. These disease processes involve neovascularization in the choroidal and retinal vasculature in which immature blood vessels are formed in response to various stimuli. These proliferating vessels destroy normal anatomic architecture and cause devastating ocular sequelae with significant vision loss. Because treatment is dependent on the accurate characterization and localization of the neovascular elements, maximal visualization of these elements is crucial. Currently, the most widely used methods of visualizing the ocular circulation include intravenous fluorescein angiography and indocyanine green (ICG) angiography. In this chapter, an alternative method of visualizing the ocular circulation using liposome-encapsulated dye (sodium fluorescein or ICG) and its applications will be discussed.

Fluorescein Angiography

Although Baeyer first synthesized fluorescein dye in 1871, its use in the evaluation of the retinal circulation was not realized until 1959 when MacLean and Maumenee performed fluorescein angioscopy.[1] In this procedure, injected fluorescein was used to visualize choroidal hemangioma during indirect ophthalmoscopy. Also in 1959, two medical students, Novotny and Alvis, described all of the requisites of modern fluorescein angiography.[2] It is interesting that their article titled "A method of photographing fluorescein in circulating blood in human retina" was initially rejected for publication by the *American Journal of Ophthalmology* and not published until a year later.[2]

Fluorescein angiography uses the principle of luminescence in which a fluorescein molecule is stimulated by electromagnetic radiation of a specific wavelength. The excited molecule then emits light of a longer wavelength of less energy as it returns to its ground state. Fluorescein is excited by wavelengths between 465 and 490 nm in the blue spectrum. The excited emission lies between 520 and 530 nm in the green-yellow spectrum.

Fluorescein is an orange-brown crystalline substance belonging to the triphenylmethane group (molecular weight, 376.27 D). It is the product of a reaction between phthalic acid anhydride and resorcinol using hot sulfuric acid as the catalyst. The yellow-green hyperfluorescence appears only when it dissociates in an alkaline sodium-salt solution, because the fluorescence is dependent on the pH of the solvent and is only observed with a pH greater than 6. Injected fluorescein is largely bound to plasma proteins and the surfaces of red blood cells. This interaction between fluorescein and plasma proteins alters its absorption/emission spectrum (Van der Waal reaction). Furthermore, the erythrocytes absorb the exciting and emitted energy, thereby reducing its fluorescence. Fluorescein is excreted by the kidneys and liver.[3]

Once the dye is injected, usually into the antecubital vein, the choroidal and retinal circulations are photographed using a camera equipped with special lens filters. One filter is designed to excite the fluorescein (465–490 nm) and the other to select only the green-yellow light that is

[1] A. L. MacLean and A. E. Maumenee, *Am. J. Ophthalmol.* **50**, 3 (1960).
[2] H. R. Novotny and D. L. Alvis, *Am. J. Ophthalmol.* **50**, 176 (1960).
[3] G. Richard, G. Soubrane, and L. A. Yannuzzi, "Fluorescein and ICG Angiography: Textbook and Atlas." pp 1–9. Georg Thieme Verlag, Stuttgart, 1998.

produced by the excited fluorescein molecules. Although visualization of the choroidal circulation precedes the retinal circulation by 0.5–1 s by virtue of its shorter distance from the ophthalmic artery, both the retinal and choroidal circulations are visualized simultaneously for the most part, overlapping each other on each frame. In fluorescein angiography, hyperfluorescence does not penetrate any hemorrhage or increased pigmentation that blocks the circulation underlying these areas. Furthermore, the juxtaposed choroidal or retinal circulation may obscure the pattern and borders of vascular leakage.

Indocyanine Green

Indocyanine green (ICG) is a water-soluble tricarbocyanine dye with a molecular weight of 775 D. Its empirical formula is $C_{43}H_{47}N_2O_6S_2Na$. It was first used in 1957 to measure cardiac output.[4] In ophthalmology, ICG angiography was first performed by Kogure in animals in 1969[5] and in humans in 1970.[6] Its widespread application was limited by the described technique of intra-arterial injection and the limitations of the infrared film at the time. This method gained more acceptance when Flower[7] described enhanced techniques in 1973 and developed a multispectral camera that allowed simultaneous angiography using both fluorescein and ICG.[8] Further improvements were added with the advent of high-speed digital imaging and scanning laser ophthalmoscopy (SLO), which improved the sensitivity of the images.

Because ICG dye is activated between 790 and 805 nm and excited emission occurs in the infrared range, ICG has several advantages over fluorescein, which emits a shorter wavelength in the yellow-green range. ICG has better penetration through hemorrhage and retinal pigment epithelium. Fifty-nine percent to 75% of the blue-green light is absorbed by the retinal pigment epithelium compared with 21%–38% for near-infrared light. Longer wavelengths are scattered less and potentially are more useful in patients with media opacities, such as cataracts. Also, infrared is less harmful to the retina, thereby allowing continuous exposure for ICG angiography.

Approximately 20 mg of ICG in 1 ml of aqueous solvent is injected rapidly and 10–18 s usually elapse before its arrival in the ocular circulation,

[4] I. J. Fox and E. H. Wood, *Proc. Staff Meet. Mayo Clin* **32,** 541 (1957).
[5] K. Kogure and E. Choromokos, *J. Appl. Physiol.* **26,** 154 (1969).
[6] K. Kogure, N. J. David, U. Yamanouchi, and E. Choromokos, *Arch. Ophthalmol.* **83,** 209 (1970).
[7] R. W. Flower, *Invest. Ophthalmol.* **12,** 881 (1973).
[8] R. W. Flower and B. F. Hochheimer, *Invest. Ophthalmol.* **12,** 248 (1973).

depending on the age of the subject. Choroidal circulation is visible 0.5–1 s before the retinal circulation.

Liposomes

Liposomes were first produced by Bangham et al.[9] while investigating the role of phospholipids in blood clotting. Liposomes are microscopic, spherical vesicles that form when hydrated phospholipids arrange themselves in circular sheets with consistent head–tail orientation. These sheets join others to form a bilayer membrane that encloses some of the water and water-soluble material in a phospholipid sphere. Thus, liposomes enable water-soluble and water-insoluble materials to be used together without the use of surfactants or other emulsifiers. The different types of liposomes that can be formed are unilamellar, multilamellar, and multivesicular.

Liposomes have a wide variety of applications, including the delivery of vaccines, enzymes, and drugs. When used in the delivery of certain cancer drugs, liposomes help to shield healthy cells and vulnerable organs from drug toxicity (site avoidance effect). Liposomes are especially effective in treating diseases that affect the phagocytes of the immune system, because the liposomes tend to accumulate in the phagocytes, which recognize them as foreign invaders. Moreover, liposomes can extravasate selectively into tissues with leaky vasculature (tumors or neovascular membranes) and exhibit target specificity with negligible adverse effects to normal tissues. Encapsulation in liposomes increases both the stability and pharmacokinetics by reducing elimination and increasing circulation half-life.

Liposomes can be custom designed for almost any need by varying the lipid content, size, surface charge, and method of preparation. Factors such as the phase-transition temperature of the lipid and encapsulation coefficient are critical in formulating liposomes. Interestingly, not all phospholipids form a bilayer. For instance, phosphatidylethanolamine forms a hexagonal structure. Altering composition changes intrinsic properties of the liposomes such that inclusion of gangliosides (GM1), sphingomyelin, and polyethylene glycol (PEG)–polymerized lipids results in increased circulation longevity. Cholesterol is usually incorporated in liposomes to confer serum stability; however, its inclusion has effects on vesicular properties.

There are several methods of preparing liposomes, depending on the type of liposomes desired: small unilamellar vesicles (SUV), multilamellar vesicles (MLV), or large unilamellar vesicles (LUV). Some commonly used procedures for the production of liposomes are thin-film hydration,

[9] A. D. Bangham, M. M. Standish, and J. C. Watkins, *J. Mol. Biol.* **13,** 238 (1965).

sonication, French pressure, ethanol injection, detergent dialysis, reverse-phase evaporation,[10] and extrusion.[11]

All phospholipids are capable of existing in either a solid/gel phase or a fluid/liquid crystalline phase.[12] The transition temperature (Tm) is the point at which this transition occurs, thereby allowing Apo-A-1 serum lipo-proteins to insert into the bilayer and destroy the vesicle.[13,14] The end result is the release of liposomal contents. For liposomes composed of dipalmitoylphosphatidylcholine (DPPC) and dipalmitoylphosphatidylgly-cerol (DPPG), the Tm is 41°.

General Methods

Production of Liposomes

This section describes the methods of liposome production used in the experiments described in this chapter. Liposomes are prepared by combin-ing DPPC and DPPG in a 4:1 molar ratio or phosphatidylcholine (PC) and phosphatidylglycerol (PG), also in a 4:1 ratio. The PC–PG mixture is stabilized with cholesterol and the lipid-to-cholesterol molar ratio is 6:4. In the second method, liposomes are prepared from PC, cholesterol, and distearylphosphatidylethanolamine (DSPE) covalently linked to PEG (DSPE-PEG 2000; average molecular weight = 2000 D). DPPC is used in some cases because it increases the stability of the final product, given its nature as a fully saturated phospholipid.[15–23]

[10] F. Szoka Jr. and D. Papahadjopoulos, *Ann. Rev. Biophys. Bioeng.* **9,** 467 (1980).
[11] M. J. Hope, M. B. Bally, G. Webb, and P. R. Cullis, *Biochim. Biophys. Acta* **812,** 55 (1985).
[12] D. L. Melchior and J. L. Steim, *Ann. Rev. Biophys. Bioeng.* **5,** 205 (1976).
[13] L. S. Guo, R. L. Hamilton, J. Goerke, J. N. Weinstein, and R. J. Havel, *J. Lipid Res.* **21,** 993 (1980).
[14] A. Jonas, *Exp. Lung Res.* **6,** 255 (1980).
[15] B. Khoobehi, M. R. Niesman, G. A. Peyman, and M. Oncel, *Retina* **9,** 87 (1989).
[16] B. Khoobehi, G. Peyman, W. G. McTurnan, M. R. Niesman, and R. L. Magin, *Ophthalmology* **95,** 950 (1988).
[17] B. Khoobehi, G. A. Peyman, M. R. Niesman, and M. Oncel, *Ophthalmology* **96,** 905 (1989).
[18] R. C. Zeimer, B. Khoobehi, G. A. Peyman, M. R. Niesman, and R. L. Magin, *Invest. Ophthalmol. Vis. Sci.* **30,** 660 (1989).
[19] B. Khoobehi, C. A. Char, G. A. Peyman, and K. M. Schuele, *Lasers Surg. Med.* **10,** 303 (1990).
[20] B. Khoobehi, G. A. Peyman, and K. Vo, *Ophthalmology* **99,** 72 (1992).
[21] R. C. Zeimer, T. Guran, M. Shahidi, and M. T. Mori, *Invest. Ophthalmol. Vis. Sci.* **31,** 1459 (1990).
[22] B. Khoobehi and G. A. Peyman, *Ophthalmology* **101,** 1716 (1994).
[23] G. Peyman, B. Khoobehi, S. Shaibani, S. Shamsnia, and I. Ribeiro, *Ophthalmic Surg. Lasers* **27,** 459 (1996).

Reverse-Evaporation Method. Large unilamellar vesicles (LUV) (average diameter, 250 nm) with encapsulated carboxyfluorescein are prepared by the reverse-phase evaporation method with occasional substitution of the phospholipid DSPE-PEG 2000 formulation. Lipids are dissolved in the organic phase of 8 ml isopropyl ether and 4 ml chloroform. After washing the ether with 10% sodium bisulfite, the aqueous phase containing the dye is heated to 50° and added to the organic phase, the mixture is flushed with nitrogen, the tube is capped tightly with a Teflon-lined cap, and placed in a cylindrical bath-type sonicator (Laboratory Supplies Co, Hicksville, NY) maintained at 50°. The tube with the resulting homogenous emulsion is placed in a rotary evaporator under reduced pressure (less than 400 mm Hg), where the emulsion begins to bubble extensively as the organic phase is drawn off.[15–23]

As the emulsion foams increasingly with gradually lowered pressure (reaching 150 mm Hg), progressively higher vacuum is used to maintain foaming. Foaming stops at a pressure of approximately 100 mm Hg. Dialysis is used to separate the liposome suspension containing a large amount of encapsulated solute (50% to 75%). Preparations containing liposomes with a maximum diameter of 220 nm are obtained by passing the LUVs through a sterile Millex-GV 0.22-μm filter unit (Millipore, Bedford, MA).[15–23]

Freeze-Thaw Technique. Liposomes can also be prepared with a freezing and thawing technique. After drying the phospholipid solution to remove organic solvents, it is rehydrated with the aqueous solution to be encapsulated. Vacuum is used to dry the organic compound. The resulting mixture of lipid and aqueous is incubated at 50° for 30 min with occasional stirring until a suspension forms. The suspension is frozen in liquid nitrogen and thawed three times to produce MLV 1-2 μm in diameter. Finally, dialysis is used to remove any unencapsulated dye.[14,15,18,21–26]

Extrusion Technique. Small unilamellar vesicles (SUV) (approximately 100 nm average diameter) are made using high-pressure extrusion devices to pass liposomes through two stacked polycarbonate filters of 0.1 μm pore dia diameter and then dialyzed to remove any unencapsulated dye.[11,15–23]

Materials. Lipids and cholesterol are obtained from Sigma Chemical Co. (St. Louis, MO) or Avanti Polar Lipids (Alabaster, AL) and used without further purification. All chemicals are reagent grade or better.

[24] M. B. Yatvin, J. N. Weinstein, W. H. Dennis, and R. Blumenthal, *Science* **202,** 1290 (1978).
[25] E. Ralston, L. M. Hjelmeland, R. D. Klausner, J. N. Weinstein, and R. Blumenthal, *Biochim. Biophys. Acta* **649,** 133 (1981).
[26] H. Diehl and J. L. Ellingboe, *Anal. Chem.* **28,** 882 (1956).

Calcein is obtained from Sigma and is repurified using column chromatography according to a method reported by Ralston et al.[25]

Laser Delivery System

The laser delivery system consists of a computer control system, a modified fundus camera, and an argon laser source. The modifications to the fundus camera consist of:

- A fiberoptic probe carrying the laser
- A 20-nm bandpass filter that filters the 514-nm wavelength
- A polished glass reflector attached to a rotary solenoid that allows visualization of the laser beam
- An exciter field that is activated by another solenoid under computer control.

The camera is activated after a preset delay controlled by the timing circuit that is coupled to the laser. The timing circuit is activated by a phototransistor set to respond to the firing of the laser. The aiming beam of the laser is focused over the retinal vessels.[17]

Animals Used

Albino rabbits, rhesus monkeys, and rats are most frequently used for these experiments. The animals are appropriately anesthetized before all procedures, and the eyes are dilated with cyclopentolate hydrochloride. Liposome-encapsulated calcein, fluorescein, and ICG are used. The injection of liposomes is given over 10 min through a Teflon catheter inserted into the saphenous vein of monkeys; liposomes are injected into the ear veins of rabbits and the tail veins of rats.

Evolution of the Concept

Selective Angiography

The issue of enhancing the image of the neovascular membrane by decreasing the surrounding hyperfluorescence is addressed by use of selective angiography. In this approach, temperature-sensitive liposomes containing a highly concentrated solution of fluorescent dye are injected intravenously. The liposomes with encapsulated dye circulating through the retina and choroid are invisible to the fundus camera, because the fluorescence of the dye is quenched at the high concentrations encapsulated in the vesicles. When the liposomes are exposed to a heat pulse generated by an electromagnetic energy source such as a laser, the vesicles break down, and the dye is released into the bloodstream where it is diluted and becomes intensely fluorescent. This type of targeted system releases dye only in the particular vessel or area of tissue on which the laser beam impinges. This approach offers several potential advantages compared with traditional injection of free sodium fluorescein:

- Enhanced control of timing and location of dye release
- Repeatability of the procedure (numerous times) after a single injection, depending on the availability of circulating liposomes
- Lack of interference from background fluorescence
- Decreased systemic complications
- Absence of the recirculation phenomenon (recirculation starts before first dye passage is completed) that would make measurements impossible if the retinal circulation time is abnormally long.

The specific parameters of this technique (amount and timing of energy delivery; collateral damage) are explained over a series of experiments described in the following.

Validation of the Technique

Selective angiography is validated *in vivo* by showing that a microwave applicator tip held against the limbus of a rhesus monkey is sufficient to release the contents of the liposome. An *in vitro* experiment is designed with liposomes prepared as described previously. A 580-μm (inside diameter) U-shaped capillary tube connected to a syringe pump (Sage Instruments, Orion Research, Cambridge, MA) is used to simulate a blood vessel. The flow is measured at 5 and 10 mm/s and the temperature is set at 37°. Carboxyfluorescein (100 mM) is diluted 1:400 with calf serum to simulate the concentration of fluorescein in fluorescein angiography. The distance between the arms of the "U" is 7 mm. In addition, five U-shaped capillary tubes, 2.2 mm apart, are connected to one another and to a Sage syringe pump. The microwave applicator tip is positioned at the end of the capillary tubing.[16]

Results show a graded response that correlates to the distance of the capillary tubes from the microwave tip and the rate of flow. Partial fluorescence is shown up to 4.4 mm away (distance of two tubes) from the tip at a flow rate of 10 mm/s.[16]

In vivo, this concept is proven in rabbits by holding a microwave tip over the limbus after injecting liposome-encapsulated (100 mM) or free carboxyfluorescein into the ear veins. The right eyes of all rabbits are immediately heated for 2 min to increase the temperature to 40°. Fluorescence measured in the mid-anterior chamber for 2 hours using a fluorophotometer held in the front of the pupil shows that the amount of carboxyfluorescein released in the heated eyes is 8.0 times the amount released in the control eyes ($p = 0.0001$) that receive free carboxyfluorescein.[16]

Parameters

The parameters (total energy, exposure time) and components are further defined in subsequent experiments. The microwave applicator tip is replaced by the laser/photography system described previously. Further experience shows that the encapsulation efficiency of carboxyfluorescein is inconsistent between batches. Hence, calcein ($C_{30}H_{26}N_2O_{13}$), first developed by Diehl and Ellingboe[26] for use in measuring calcium concentration in the presence of magnesium, was considered. Calcein, a combination of ethylenediamine tetraacetic acid and fluorescein with a larger size (622 D compared with 386 D for carboxyfluorescein) and net charge of -3 at pH of 7.4, is thought to cause less leakage during liposome formation and to be easier to handle and store than liposome-encapsulated carboxyfluorescein.[15]

On testing, the encapsulation efficiency of calcein was noted to be superior to that of carboxyfluorescein (28.8% ± 3.1% vs 13.8% ± 3.8%). The release of dye was noted to rise sharply at 40°, enabling a good phase-transition point. These findings proved calcein to be superior to carboxyfluorescein.[15]

Experiments were first performed on rabbits to examine the *in vivo* release and dynamic response of liposome-encapsulated calcein. After systemic injection of the liposome-encapsulated dye, the laser was fired for various intervals. In one case, photographs taken after a total exposure of 500 ms at a power of 3.5 mW (measured at the cornea) and after a delay time of 100 ms showed that the dye had already cleared the artery and early venous filling had started. Thus, the total time between the initiation of the laser exposure and the photograph was 600 ms. Next, the laser exposure time was reduced to 100 ms, and the photograph was taken 1100 ms after the laser was started. As expected, the veins were completely filled. In addition to demonstrating the feasibility of selective angiography *in vivo*, this experiment also demonstrated the unique appearance of the dye release using this system in which the background is almost completely black, highlighting the fluorescence in the vessels. Selective angiography had arrived.[15] Photographs from another set of experiments are shown in Fig. 1.

These results were confirmed and expanded in a preliminary study in a rhesus monkey. Different exposure and delay times were evaluated.[15] The fluorescein findings and time parameters are summarized in Table I.

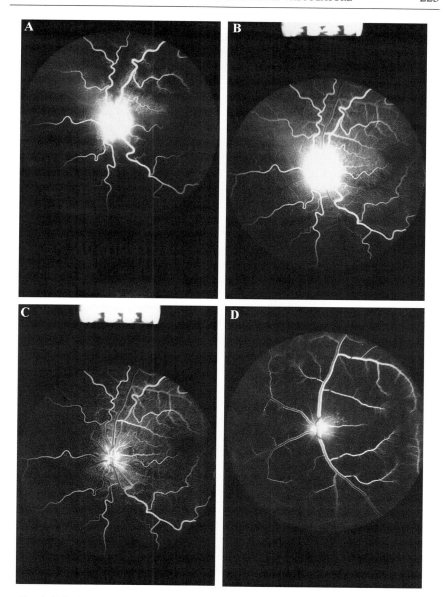

FIG. 1. (A) The fundus is seen 900 ms after the onset of laser application. Note the absence of background fluorescence. (B) Taken 1500 ms after the onset of laser, the angiogram demonstrates the spreading wavefront from dye release. (C and D) Taken 1650 ms and 2650 ms, respectively, after laser exposure, demonstrates the slow advancement of

TABLE I
SELECTIVE ANGIOGRAPHY IN THE RHESUS MONKEY

Exposure (ms)	Delay (ms)	Total time (ms)	Fluorescein angiogram finding
200	0	200	Midperiphery
500	0	500	Periphery
500	40	540	Further periphery

As expected, a longer total delay time resulted in further passage of the dye through the arterial channel. This sequence was repeated in the venous circulation with similar findings. The distance traveled in the veins was shorter, indicating slower velocity in the venous system. Good images could be obtained for up to 2 h after the injection of liposome-encapsulated calcein. However, the best results were obtained within the first hour after the injection.

In some cases, a bright hyperfluorescent dot was observed in the area of the laser delivery. Higher power (8 mW) and longer exposure time (1 s) caused this area to become larger and more intense. Although the small dot at lower energy settings may be explained as an image of the laser-aiming beam, its change to a more diffuse appearance with higher energy implies alteration of the choroidal vasculature. Thus, choroidal dye leakage or slower transit of the dye in the affected choroid may be more plausible explanations. The authors acknowledged that precise focus of the laser was not possible with the equipment used, because reflection occurred from both the front and rear surface of the glass reflector. However, any anatomical alterations were transient, because angiographic and histological follow-up studies of the laser exposure site revealed no damage. The follow-up angiograms were taken after four individual sessions, which encompassed approximately 400 laser applications.[15]

Retinal Blood Flow Velocity

When the dye is activated in a given blood vessel, a definable wavefront results. Because selective angiography provides clear visualization of a single vessel without background hyperfluorescence, these two concepts can be coupled to calculate the flow rate in a given retinal vessel.[17] For instance, the flow velocity can be calculated by measuring the

---

fluorescence and the specificity of this technique as the choroidal vasculature remains unhighlighted. (Reproduced with permission from M. R. Niesman, B. Khoobehi, and G. A. Peyman, *Invest. Ophthalmol. Vis. Sci.* **33,** 2113 [1992]. Copyright, the Association for Research in Vision and Ophthalmology, 1992.)

distance traveled per time frame. Clinically, accurate measurement of retinal blood flow can disclose important information on the state of the retinal circulation in various retinal vascular disorders.

The flow rate in the center of a vessel (centerline velocity) is determined by dividing the distance traveled by dye after release by the total time of dye movement.[17] The total time of dye movement is defined as the sum of the delay time and laser exposure time as previously calculated. Laminar flow is determined as the value of centerline velocity divided by two.

Volumetric flow can be found with the available data as:

$$Q = \pi D^2 V / 4.0, \tag{1}$$

where D is the internal diameter of the vessel of interest.

A sample calculation is provided later in a 100-μm vessel. In this example, the laser exposure time is 100 ms, and the delay time (the time interval after the laser has been turned off until the photograph is taken) is 50 ms. Activation of the liposomes released dye with a sharp wavefront into the bloodstream. The dye traveled 4 mm. Therefore, the maximum centerline velocity is:

$$V_{max} = 4 \text{ mm}/(100 \text{ ms} + 50 \text{ ms}) = 2.6 \text{ mm/s} \tag{2}$$

Inserting this value into the previous equations, a volumetric flow rate of 1.025 μl/min is obtained ($Q = \pi D^2 V/4.0 = \pi$ [100 mm]^2 [1.3 mm/s]/4.0 = 1.025 μl/min. Remember that $V = V_{max}/2$.[17]

The Zeiss image analysis system (Model IBAS 2000, Thornwood, NJ) is used to enhance and enlarge the photographs taken to measure the vessel diameter and distance traveled by the dye in a given time frame. The mean vessel diameters of a single point and of a segment are determined by averaging several measurements taken by different investigators. The distance traveled by the dye per time frame is also determined by averaging several measurements taken by different investigators. After 12 measurements on the same vessel, with each individual measurement representing the mean of two distance measurements by two separate investigators,[17] the V_{max} and volumetric flow rate (Q) are obtained. As expected, larger caliber vessels demonstrate faster V_{max} and larger Q in both the arteries and veins. For instance, an artery with a diameter of 92 μm is associated with V_{max} of 22.6 $\pm$ 1.5 mm/s and Q of 4.5 $\pm$ 0.3 μl/min. In contrast, an artery of 60-μm diameter is associated with 10.0 $\pm$ 0.7 and 0.9 $\pm$ 0.1, respectively. The same relationship holds true in the veins.

Because longer total time yields greater distance traveled by the wavefront, the velocity calculation cannot be performed in those experiments where longer total time yields a wavefront out of the field of view. When the dye is activated at an arterial crossing, allowing dye release in both arterial and venous directions, the distance traveled in both vessels is noted to be equal. Initially, this finding might seem counterintuitive, because the larger vessel (vein) would be expected to display greater velocity. However, equal distance is covered, because the velocity of blood is greater in arteries than veins.[17]

Laser-induced vasoconstriction is considered a confounding factor. However, the facts that the temperature required to release the dye is low $(41°)$[17] and the exposure is short contradict that theory. Perhaps most conclusively, the vessel diameters measured from laser exposures that caused dye release show no vasoconstriction compared with red-free fundus photographs.[17]

These data are consistent with those of Riva et al.[27,28] and Feke et al.[29] For example, the calculated blood velocity in the larger arteries agrees quite well with the data that Riva et al.[27] obtained in humans. However, the difference factor of 2.7 demonstrated by Feke et al.[30] between arterial flow velocity in systole and diastole is not seen in this experiment. The flow velocity measurements obtained using selective angiography likely represent an average of the diastolic and systolic values. The explanation probably lies in the relatively long period of 250 ms required to make a measurement compared with the 50 ms required by Feke.

In a later study, a control system was added to monitor the cardiac pulse and synchronize the laser and camera. This system consisted of an infrared emitter/detector system (UFI, Morro Bay, CA) to measure the pulse and Lab VIEW software (National Instruments, Austin, TX), which recognized the end of the systolic phase and activated the laser and the camera.[31] The results thus obtained showed a difference factor of 3 between the systolic and diastolic values, which correlated with data obtained by Feke.

In practical terms, determination of the leading edge of the dye front is easiest when the distance traveled is short, because the dye is less diluted by the blood. However, given the short distance traveled, an inaccurate measurement would induce a large error. Although it becomes slightly more difficult to visualize the leading edge when the distance traveled in the blood is greater and the dye becomes more dilute, the amount of error

[27] C. E. Riva, J. E. Grunwald, S. H. Sinclair, and B. L. Petrig, Invest. Ophthalmol. Vis. Sci. 26, 1124 (1985).
[28] C. E. Riva, G. T. Feke, B. Ebeni, and V. Behany, Appl. Opt. 18, 2301 (1979).
[29] G. T. Feke and C. E. Riva, J. Opt. Soc. Am. 68, 526 (1978).
[30] G. T. Feke, G. J. Green, D. G. Goger, and J. W. McMeel, Ophthalmology 89, 757 (1982).
[31] B. Khoobehi, O. M. Aly, K. M. Schuele, M. O. Stradtmann, and G. A. Peyman, Lasers Surg. Med. 10, 469 (1990).

induced in the measurement decreases. In Khoobebi *et al.*,[31] the uncertainty of determining the position of the dye front was always less than 5% of the total distance traveled.

Mechanism of Dye Release

Although the mechanism of temperature-induced liposome destabilization has been well defined, the target of the laser beam that causes the rise in temperature has not been established. It is hypothesized that absorption of laser energy by either hemoglobin in the red blood cells or the dye encapsulated in the liposomes is responsible for the elevated temperature. Thus, to differentiate the factors that contribute to dye release, the release of encapsulated dye from the liposomes is measured under various conditions that simulate temperature elevation from either hemoglobin or vesicles.

Specifically, the release is quantified in whole blood (hemoglobin) and a buffer containing high quantities of high-density lipoprotein. A 290-μm diameter capillary tube is used to simulate a retinal vessel and blood flow dynamics. Temperature and flow are maintained at 37° and 40 μl/min, respectively. A laser spot size of 400 ± 20 μm is used to expose all liposomes passing through the capillary to laser energy.[19] A Model 920 argon/dye laser (Coherent Radiation, Palo Alto, CA) is used to generate 488-nm and 577-nm light energy used in the experiment. The fluorescence of all samples is measured with an ocular fluorophotometer (Coherent Radiation) to determine how much calcein has been released from the tip.[19]

This study demonstrates that both the blood and the liposome-encapsulated dye contributes to light absorption, which in turn causes dye release. At 577 nm, release of hemoglobin causes laser energy to be absorbed; release of the encapsulated substance was independent of the dye, making this wavelength ideal for liposome drug delivery without entrapped dye. At 488 nm, both the vesicles and hemoglobin absorb energy; however, the absorption of energy by vesicles exceeds that of hemoglobin. Therefore, a 488-nm laser trigger would be more effective in selective angiography studies where absorption of light energy at the surface of the vesicles might bring about more rapid release. It also might be less disruptive to the red blood cells to rely more on the direct heating of the liposomes that occurs at 488 nm than on the heat transfer from the red blood cells that occurs at 577 nm.[19]

Further Refinement of the Technique

Although selective angiography as described exhibits several characteristics that are superior to traditional fluorescein angiography, two main drawbacks have been noted. The scope of the technique is limited, because dye release is caused by firing the laser at only one artery or vein. The other drawback is that the procedure requires power levels that are unsafe for humans.

These shortcomings were addressed in part by increasing the spot size to the dimensions of the optic disc, reducing the energy density at the retina to a level that might be safe for clinical use in humans.[20] This larger spot also made release more efficient, because the liposomes spent more time in the beam. Furthermore, all four quadrants could be visualized by activating the liposomes as they exited the central retinal artery.

Using this method, laser-triggered repetitive angiography of the fundus became possible at levels that have been suggested as safe for use in humans. In the monkeys used in this study, dye release and visualization could still be elicited for at least 3 h after liposome injection and at energy densities between 0.5 and 3.4 J/cm^2. The dye release could be repeated many times over the course of 2–3 h.[20]

Although some disadvantages of selective angiography are addressed in the studies described previously, other disadvantages remain. These are:

- Technical difficulty of incorporating a high concentration of dye in the liposomes
- The laser beam used to lyse the heat-sensitive liposomes must pass through the cornea, the anterior chamber, the lens, and the vitreous, resulting in some scattering
- A greater amount of energy is required to activate and release the dye.

Visualization of the Fluorescent Vesicle System

An *in vitro* eye model is created to demonstrate the fluorescence of unquenched liposomes and to determine the functional parameters. The system consists of capillary tubing held against a dark background and a double aspheric lens located one focal length from the tubing with a continuously flowing aqueous solution.[22] An *in vivo* system is created in which the fluorescent vesicle suspension (0.03 ml/kg) is injected into the saphenous vein of the monkey or an ear vein of the pigmented rabbit. Large unilameller vesicles of various sizes are filtered to ensure a maximal size of 220 nm in diameter. The fundus is viewed through the SLO (Rodenstock, Danbury, CT), and images of liposomes circulating in all vessels are captured on videotape. The image analysis system (HRX Digital Imaging Processor, Amtronics, New Orleans, LA) is able to digitize and record sequences up to 8 s long. Sampling frequency is 30 frames/s, and every frame is digitized into a 512 × 512-bit mapped image.[22] Software has been developed to compensate for the positional shifting of the eye, which cannot be maintained in a perfectly static position.[22]

Retinal Vessel Velocity

Blood flow velocity can be determined by measuring the distance traveled per unit of time. This is performed by overlaying multiple video frames on a single image to delineate the distance traveled by a particular vesicle in a given time period. The vessel diameter is visualized with red-free video of the fundus using the SLO, and its diameter is derived from the average of five measurements of a single point. Tracing at least 10 liposomes and averaging the results provides distance measurements. For major arteries and veins, the velocities of vesicles are measured at the center of the vessels (maximum centerline velocity).[22]

The *in vitro* study shows that the vesicles fluoresce when scanned with the SLO and that the optimal concentration of encapsulated carboxyfluorescein is 20 mM. Vesicles as small as 100 nm can be visualized, and dilutions as low as 1/25,000 are recorded.[22] Figure 2 shows fluorescent vesicles in the simulated vasculature.

In the retina, as in the *in vitro* model, individual liposomes as small as 100 nm are visible. Measurements of blood flow velocity are possible in vessels of all calibers from the retinal vessels to smaller optic nerve head capillaries. This allowed simultaneous measurement of blood flow

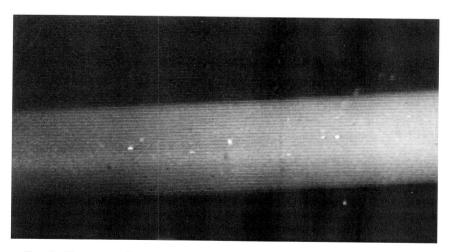

FIG. 2. Fluorescent liposomes are seen traversing the simulated retinal vasculature. (Reprinted with permission from G. A. Peyman, D. M. Moshfeghi, A. A. Moshfeghi, and B. Khoobehi, *Ophthalmic Surg. Lasers* **27,** 279, 1996. Copyright, SLACK Incorporated, 1996.)

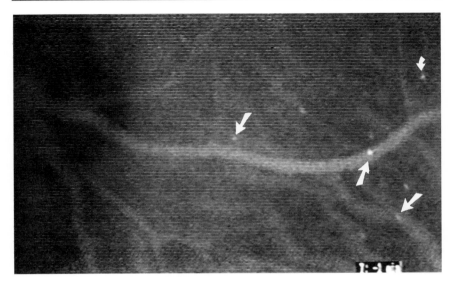

FIG. 3. Fluorescent liposomes are seen in the retinal vessels of different calibers. (Reprinted with permission from G. A. Peyman, D. M. Moshfeghi, A. A. Moshfeghi, and B. Khoobehi, *Ophthalmic Surg. Lasers* **27,** 279 [1996]. Copyright, SLACK Incorporated, 1996.)

velocity in the entire retinal macrocirculation, macular microcirculation, optic nerve head capillaries, and choroidal circulation.[22] Figure 3 shows fluorescent vesicles in retinal vessels of different calibers.

This technique relies on the idea that the dye at a subquenched concentration can be activated within the liposomes and not released; excessive concentrations of encapsulated dye result in quenching and release of the fluorescence. The optimal concentration of carboxyfluorescein in the liposomes is 20 mM. This value is determined from pixel intensity corresponding to the liposome in the digitized frame. The retina flow velocities are summarized in Table II.[22]

Choroidal Vessel Velocity

Choroidal flow velocity is measured by injecting ICG (15 μg/ml with a dose of 0.2 ml/kg) intravenously into a rhesus monkey. The fundus is visualized through the SLO using a diode laser and appropriate barrier filters. Continuous tracking up to 4 h is possible with the SLO, making the location of the dye front less important. Fluorescence of vesicles in the retinal circulation can interfere with the choroidal examination; however, the directions of the flows are quite different and are easily differentiated. Results are shown in Table III.[23]

TABLE II
RETINAL BLOOD FLOW VELOCITY

Location	Flow velocity (mm/s)
Large artery	16.10 ± 5.7
Large vein	9.33 ± 1.67
Macular capillaries	0.76
Optic nerve head capillaries	1.39

TABLE III
AVERAGE CHOROIDAL BLOOD VELOCITIES FOR THREE AREAS OF INTEREST

Region of eye	Average velocity (mm/s)
Macula	5.16 ± 0.81
Vortex	4.04 ± 1.3
Nasal to optic nerve	2.03 ± 0.47

Modified from G. A. Peyman, B. Khoobehi, S. Shaibani, S. Shamsnia, and I. Ribeiro, *Ophthalmic Surg. Lasers* **27**, 459 (1996). Reproduced with permission of SLACK, Inc.

In the routine method of ICG angiography, the dye is visible in the choroidal circulation for only a few seconds before being obscured by fluorescein in the retina. However, the fluorescent vesicle system allows observation of the flow of encapsulated dye over a longer time. The fluorescent vesicles are visualized despite their size being below the resolution limit of the SLO because of their contrast to the surrounding dark area. ICG is also clearly visible in the retinal circulation. In theory, it might be possible to provide a mixture of fluorescent vesicles made of carboxyfluorescein or fluorescein and ICG for distinct analysis of retinal macrocirculation or choroidal circulation, respectively. Figure 4 shows a composite image of the choroidal vasculature as a single liposome is tracked.

Combination of Traditional Angiography with Liposome Technology

One unanticipated drawback to the fluorescent vesicle system is its inability to leak fluorescent vesicles at sites of retinal and choroidal damage caused by inflammation or neovascular tissues. A study was undertaken to determine the feasibility of merging free-dye angiography techniques and the fluorescent vesicle technique to achieve the best characteristics of both methods.[32] In this method, the fluorescent vesicles and free sodium fluorescein were used simultaneously to benefit from both techniques.

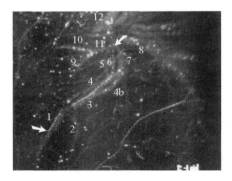

Fig. 4. A composite image of a liposome passing through the choroidal vasculature. White arrows indicate the starting and ending points of the liposome tracked from region 1 through regions 3, 4, and 6. The time frame was derived from the individual frames and distance traveled by the liposome measured on the photograph. (Reprinted with permission from G. A. Peyman, B. Khoobehi, S. Shaibani, S. Shamsnia, and I. Ribeiro, *Ophthalmic Surg. Lasers* **27,** 459 [1996]. Copyright, SLACK Incorporated, 1996.)

The *in vitro* model eye was described previously with the only modification being two different sizes of capillary tubes with inner diameters of 50 μm and 280 μm to mimic ocular flow. The output from the SLO is captured on videotape.[32]

Sodium fluorescein and ICG are tested in conjunction with fluorescent vesicles. Vesicles are encapsulated with carboxyfluorescein or ICG, and the ratios of encapsulated carboxyfluorescein or ICG to free sodium fluorescein are tested at various concentrations. The free and encapsulated dye mixtures are placed in the pump system, scanned by the SLO, and the image recorded to determine which combination provided the optimal viewing of the fluorescent vesicles against the free fluorescein.[32] The optimal viewing of vesicle-encapsulated carboxyfluorescein and free dye occurred at a ratio of 150:1. For ICG, the ratio was 50:1.

One male rhesus monkey and two New Zealand white rabbits were used for an *in vivo* study. The two rabbits had been treated with photodynamic therapy in conjunction with a red neodymium:yttrium–aluminum–garnet frequency-doubled laser within 24 h, resulting in closure of choroidal vessels. The animals were injected intravenously with the combination of free dye and fluorescent vesicles. The ratio used was that determined in the *in vitro* experiment to provide the optimal viewing of vesicles against the free dye. The fundus was scanned with the SLO immediately after injection, and the image was recorded on videotape for 20 min.[32]

---

[32] G. A. Peyman, D. M. Moshfeghi, A. A. Moshfeghi, and B. Khoobehi, *Ophthalmic Surg. Lasers* **27,** 279 (1996).

In vivo, clear delineation of the free dye and liposome-encapsulated dye is possible. Thus, it is possible to visualize leakage with the free dye and potentially evaluate the bloodflow characteristics using liposome-encapsulated dye.[32]

Total Volumetric Blood Flow Measurement

A new technique presented in 2001 measures the total volumetric blood flow in the retina with fluorescent leukocytes and erythrocytes. In this technique, the absolute volumetric flow rate in a retinal vein is measured with leukocytes labeled with SYTO59 nuclear stain and erythrocytes marked with membrane stain that are excited with the SLO by using its helium–neon laser (633 nm). The measured value is compared with its concentration in a blood sample. On the basis of this work, the total volumetric blood flow based on leukocytes is 39.8 nl/s to 63.8 nl/s. Using erythrocytes, this number ranges from 61 nl/s to 78 nl/s.[33]

Concluding Remarks

The technique of liposome-encapsulated, temperature-sensitive dye has undergone many cycles of evolution since its inception. Although initially intended as a vehicle for selectively releasing the encapsulated dye at a given location, this technique evolved to allow visualization of a single dye-enhanced liposome using the SLO. Clearly, this technique has benefited greatly from the advent of technology such as the SLO and digital image processing. The story continues as further refinement and applications are being developed.

---

[33] B. Khoobehi, G. A. Peyman, and R. Hayes, *Invest. Ophthalmol. Vis. Sci* **42**(Suppl), S525 (2001).

[15] Radiolabeling of Liposomes for Scintigraphic Imaging

By Peter Laverman, Otto C. Boerman, and Gert Storm

Introduction

Since their discovery, liposomes have been labeled with radionuclides to trace these particles *in vivo*. A wide variety of radionuclides and labeling techniques has been used, ranging from weak β-radiation emitters such as tritium (^{3}H) and carbon-14 (^{14}C) for tissue distribution and pharmacokinetic studies to γ-radiation emitters such as technetium-99m (^{99m}Tc), indium-111 (^{111}In), and gallium-67 (^{67}Ga), for both biodistribution and scintigraphic studies. For reviews on this subject, see Refs. 1–4.

Labeling of liposomes with ^{3}H or ^{14}C traditionally involves the use of a carefully chosen lipid (e.g., cholesteryloleate, cholesterolhexadecylether) and/or aqueous phase marker (e.g., inulin, sucrose) in which either a hydrogen or a carbon atom is substituted by ^{3}H or ^{14}C, respectively.[5] These markers can be incorporated without changing the actual liposomal formulation and can yield very stable radiolabeled formulations. A disadvantage of this approach is the use of the long-lived radionuclide throughout the whole preparation procedure, thus resulting in long-term contamination of glassware, extruders, etc. Another disadvantage is the laborious method to measure the radioactivity in the dissected tissues, including homogenization and the use of scintillation liquid. Ideally, liposomes should be labeled after preparation and just before the experiments. So-called after-loading or remote-labeling methods are suitable for this purpose. These methods are almost indispensable when using short-lived radionuclides, such as ^{99m}Tc or ^{111}In (physical half-lives of 6.0 and 68 h, respectively). In this situation, the preformed liposomes are labeled before the start of the experiment. The main advantage of the use of γ-radiation–emitting radionuclides is the possibility of using whole-body scintigraphic imaging

[1] P. Laverman, O. C. Boerman, W. J. Oyen, E. T. Dams, G. Storm, and F. H. Corstens, *Adv. Drug Deliv. Rev.* **37**, 225 (1999).

[2] O. C. Boerman, P. Laverman, W. J. Oyen, F. H. Corstens, and G. Storm, *Prog. Lipid Res.* **39**, 461 (2000).

[3] W. T. Phillips, *Adv. Drug Deliv. Rev.* **37**, 13 (1999).

[4] W. T. Phillips and B. Goins, *in* "Handbook of Targeted Delivery of Imaging Agents" (V. P. Torchilin, Ed.), CRC Press, Boca Raton, FL, 1995.

[5] G. L. Pool, M. E. French, R. A. Edwards, L. Huang, and R. H. Lumb, *Lipids* **17**, 448 (1982).

techniques to visualize the biodistribution of the radiotracer. This technique is a valuable tool in addition to so-called *ex vivo* studies, making up the dissection of tissues and subsequent counting for radioactivity. Imaging permits the monitoring of the biodistribution in the same animals for long periods of time, thus eliminating the need of killing animals at several time points after injection to quantify the disposition profiles in time. In addition, dynamic scans can be recorded to monitor the pharmacokinetic behavior on second-based or minute-based intervals. In general, γ-radiation can be measured more easily than the weak β-radiation, providing the opportunity to avoid tissue solubilization with scintillation liquid.

This chapter will focus on several techniques for liposome labeling with γ-radiation–emitting radionuclides and their use in scintigraphic imaging.

Scintigraphic Imaging

Scintigraphic imaging is a basic technique commonly applied in nuclear medicine. Radiolabeled compounds (called radiopharmaceuticals or radiotracers) are administered to patients for diagnostic or, in certain cases, for therapeutic purposes. The *in vivo* distribution can provide noninvasively important physiological information about tissue function. For instance, by administering a radiolabeled compound that is cleared rapidly in the urine by glomular filtration and by recording serial scintigraphic images of the uptake in the kidney and the bladder, it is possible to acquire quantitative information about kidney function. Similarly, when injecting radiolabeled particles with a diameter exceeding that of the lung capillaries, a nuclear medicine physician is able to visualize the extent of the perfusion of the lungs.

For diagnostic applications, the radiopharmaceutical is labeled with a radionuclide emitting photons with energies ranging from 100–500 keV.[6] These photon energies are high enough to allow detection outside the body by a gamma camera. This instrument consists of a lead collimator, an NaI crystal, and photomultiplier tubes (Fig. 1). The photons are collimated by the lead collimator and then strike the NaI crystal, where a scintillation is produced, which is converted into an electronic signal by the photomultiplier tube. This signal is stored digitally in a matrix varying from 64×64 to 512×512.[6] These data are processed to yield the actual image. When radionuclides with a higher gamma energy are used, septa of the lead collimator have to be thicker, which will result in lower resolution. On the other hand, radionuclides with a very low gamma energy will not be

[6] F. A. Mettler and M. J. Guiberteau, "Essentials of Nuclear Medicine," 4th ed., W. B. Saunders, Philadelphia, 1998.

A

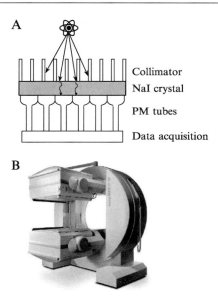

Collimator

NaI crystal

PM tubes

Data acquisition

B

FIG. 1. (A) Schematic representation of a part of gamma camera head. The radiation emitted by the animal will be collimated by the lead collimator in such a way that only gamma rays are detected that are (nearly) perpendicular with the NaI crystal. This will result in a sharp image. When the photon strikes the NaI crystal, scintillation will occur, and the produced "light" flash will be amplified by the photo multiplier (PM) tubes. The PM tube will give an electronic signal, which is processed by the computer system. (B) A dual-headed gamma camera is equipped with two camera heads, enabling the possibility of recording anterior and posterior images at the same time. (Photograph provided by Siemens Medical Solutions, Inc., Hoffman Estates, IL.)

able to excite the NaI crystal. Therefore, radionuclides with an intermediate energy of approximately 120–150 keV, such as ^{99m}Tc, are favorable.

Whereas radiological imaging techniques like computed tomography (CT) and magnetic resonance imaging (MRI) provide images based on anatomical information, scintigraphy provides images based on (patho)physiological processes. Several radiopharmaceuticals are available to image different processes. It has been shown that tumor lesions or inflammatory foci can already be visualized scintigraphically before any disease-induced anatomical changes are apparent. In addition, scintigraphic imaging allows whole-body imaging within a reasonable acquisition time, therefore allowing scanning for abnormalities outside clinically suspected areas.

Because the photons that are emitted by various radionuclides can be quantified simultaneously, but independently from each other, dual labeling approaches can be used. This enables experiments to obtain information regarding both the liposomal carrier (labeled with one radionuclide) and the encapsulated compound (labeled with a different radionuclide) after a single injection in the same animal. When applying this technique, the two radionuclides have to be chosen carefully. Because the gamma camera has to detect the gamma energy peaks of the two nuclides separately, these should not be overlapping.

Animal Experiments

The choice of the animal model and species are important factors to be considered when testing the *in vivo* behavior of a particular liposome. The choice of the animal model obviously depends on the (clinical) application of the liposome formulation. This can range from human tumor xenografts, transgene and knockout mice, to other experimental models. Whichever model is chosen, the biodistribution in a healthy wild-type animal has to be studied first, preventing unnecessary implementation of complicated animal models in case of unexpected biodistribution results.

When using animals for scintigraphic imaging, the size of the animal has to be considered: a mouse is too small to allow clear visualization of various organs. Therefore, somewhat larger animals like rats, hamsters, guinea pigs, and rabbits are more suitable for scintigraphic studies. When using rats, groups of three to five animals, depending on the field of view of the camera, are anaesthetized with a mixture of ethrane/nitrous oxide/oxygen and placed prone on a gamma camera. Rabbits do not need to be anesthetized but can be placed simply in a fixation mold and placed prone on a gamma camera. The animals are then injected with the radiolabeled preparation, and images are recorded at several time points after injection. The amount of radioactivity injected should be sufficient to obtain images of more than 300,000 counts in a reasonable period of time, preferably in less than approximately 4–5 min. For a ^{99m}Tc-label the injected dose is normally 0.1–0.5 mCi for rats and 0.5–1.0 mCi for rabbits. The injected doses of ^{111}In are lower because of its higher gamma energies. After recording and processing of all images, the scintigraphic results can be analyzed by drawing regions of interest over the animals.

The Choice of the Radionuclide

Several factors have to be taken into account when choosing the radiolabel. First, the radiolabeled liposomal formulation should have a high *in vitro* and *in vivo* stability from a radiochemical point of view. Loss of radiolabel from the liposomes during or immediately after injection will result in the injection of free radiolabel and thus lead to underestimation of the injected dose of labeled liposomes, Moreover, release of the radiolabel from the liposomes *in vivo* will complicate interpretation of the scintigraphic images, because in that case the fate of the radiolabel no longer merely reflects the fate of the liposome particles. Depending on its physical characteristics, the unbound radiolabel will be excreted by the hepatobiliary route or in the urine, or taken up by organs such as the thyroid glands and the stomach.

Second, the circulation time of the liposomal formulation is an important factor to consider. When using liposomes with a long circulatory half-life, a radiolabel with a long physical half-life is required to allow imaging at late time points. For a short-circulating liposome, the use of a short-lived radiolabel is sufficient. Using ^{99m}Tc is suitable for imaging up to 24 h after injection, whereas ^{111}In can be used for studies lasting 72 h or more.

The commonly used isotopes for scintigraphic imaging, ^{67}Ga, ^{99m}Tc, ^{111}In, and ^{123}I, are all widely available (Table I). However, generally, ^{99m}Tc is preferred over the other isotopes, because of its optimal imaging characteristics with an ideal photon energy of 140 keV, as discussed previously. In a nuclear medicine department setting, ^{99m}Tc is readily available, because it is the decay product of the reactor-produced ^{99}Mo and can be eluted daily from a ^{99}Mo/^{99m}Tc-generator. Because ^{67}Ga, ^{111}In, and ^{123}I are cyclotron products, they are more expensive and not available in every nuclear medicine department. Moreover, because of their higher gamma energies, the imaging characteristics are less favorable. Although ^{125}I is used widely for labeling proteins and peptides, this radionuclide is unsuitable for scintigraphic imaging because of its low gamma energy.

TABLE I

PHYSICAL CHARACTERISTICS OF SOME COMMONLY USED GAMMA RADIATION EMITTERS

Radionuclide	Half-life	Photons (keV), (abundance [%])
^{67}Ga	3.3 d	93 (38)
		185 (21)
		300 (17)
^{99m}Tc	6.01 h	141 (89)
^{111}In	2.8 d	171 (91)
		245 (94)
^{123}I	13.2 h	159 (83)
^{125}I	59.4 d	36 (7)

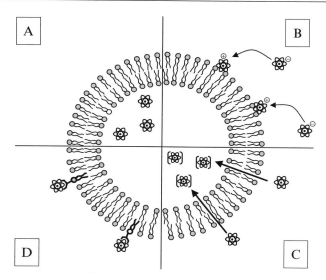

Fig. 2. Overview of the main liposome labeling techniques. (A) Encapsulation during manufacturing. When preparing the liposomes, radioactivity is added to the buffer in which the lipids are hydrated. (B) Tin technetium reduction method. ^{99m}Tc is reduced with SnCl$_2$ in the presence of liposomes, resulting in association of the ^{99m}Tc with the lipid bilayer. (C) After-loading method. A chelator is entrapped in the liposomes during manufacturing. Radioactivity is added and will pass the lipid bilayer by various mechanisms and will be trapped. (D) Chelation method. A chelator is conjugated to a (phospho)lipid and embedded in the lipid bilayer. Adding the radiolabel will lead to binding to the surface exposed chelator.

Labeling Methods

Several liposome labeling methods have been developed in the past decades. These methods can be distinguished on the basis of their labeling mechanism (Fig. 2). Probably the most straightforward method is to encapsulate the radiolabel in the aqueous interior *during* the manufacturing of the liposomes. A second relatively simple method is the reduction of the radiolabel in the presence of the liposomes, resulting in association of the label with the outside of the lipid bilayer. A third approach is to trap the radiolabel in the aqueous phase *after* manufacturing. Finally, the radiolabel can be chelated to a lipid–chelator conjugate incorporated in the lipid bilayer.

Encapsulation during Manufacturing

When preparing liposomes with a radiolabel in the aqueous phase, the radiolabel is added to the aqueous medium in which the liposomes are hydrated (Fig. 2A). This approach has been used to label liposomes with

^{99m}Tc during their manufacture.[7-9] However, only a small fraction of the radiolabel is incorporated in the aqueous phase, and consequently the labeling efficiency (i.e., the percentage of added radiolabel associated with the liposomes) is rather low, generally lower than 10%. The ^{99m}Tc is either encapsulated as pertechnetate[8] or chelated with diethylenetriaminepenta-acetic acid (DTPA).[7] Although the method is basically simple, the approach is laborious, requiring the manufacturing of fresh batches of labeled liposomes for every experiment, thus resulting in contaminated equipment (e.g., glassware, extruder) and relatively high radiation exposure to the technical personnel during the process of liposome preparation.

Tin Technetium Reduction Method

Several groups attempted to label liposomes by reducing the sodium pertechnetate in the presence of preformed liposomes.[10] (Fig. 2B) In this method, stannous chloride ($SnCl_2$) was used as a reducing agent to bring the ^{99m}Tc from its native state (pertechnetate: TcO_4^-, oxidative state, 7+) to a chemically active, lower oxidation state (oxidative state, 5+). Although the first publication on this method by Richardson *et al.*[11] reported a reproducibly high labeling efficiency (less than 2% free pertechnetate), other studies report on significant lower labeling efficiencies (4%–20%).[10,12] *In vivo* studies revealed that neutral or positively charged ^{99m}Tc-liposomes, labeled by this method, were radiochemically unstable *in vivo*,[12] whereas the negatively charged liposomes could be used to visualize experimental bacterial infections in rats.[13] Alafandy and co-workers[14] developed a modified tin reduction method. Tin(II)dioxinate was incorporated in the lipid bilayer, thus preventing the unwanted formation of tin technetium colloids, as observed in other studies.[10] However, to remove free pertechnetate, five time-consuming wash step times had to be introduced.[14] Scintigraphic images of animals injected with these types

[7] V. J. Caride, W. Taylor, J. A. Cramer, and A. Gottschalk, *J. Nucl. Med* **17,** 1067 (1976).

[8] I. R. McDougall, J. K. Dunnick, M. L. Goris, and J. P. Kriss, *J. Nucl. Med* **16,** 488 (1975).

[9] V. J. Caride, *Int. J. Rad. Appl. Instrum. B* **17,** 35 (1990).

[10] G. M. Barrat, N. S. Tüzel, and B. E. Ryman, *in* "Liposome Technology" (G. Gregoriadis, Ed.), CRC Press, Boca Raton, FL, 1984.

[11] V. J. Richardson, K. Jeyasingh, R. F. Jewkes, B. E. Ryman, and M. H. Tattersall, *Biochem. Soc. Trans.* **5,** 290 (1977).

[12] W. G. Love, N. Amos, B. D. Williams, and I. W. Kellaway, *J. Microencapsul.* **6,** 105 (1989).

[13] J. R. Morgan, K. E. Williams, R. L. Davies, K. Leach, M. Thomson, and L. A. Williams, *J. Med. Microbiol.* **14,** 213 (1981).

[14] M. Alafandy, G. Goffinet, V. Umbrain, J. D'Haese, F. Camu, and F. J. Legros, *Nucl. Med. Biol.* **23,** 881 (1996).

of radiolabeled liposomes generally show high uptake of the radiolabel in the bladder and kidneys early after injection, indicating rapid label release.

After-loading Methods

The usefulness of liposome labeling methods strongly improved with the development of the so-called after-loading methods or remote-labeling methods (Fig. 2B). Mauk et al.[15] were among the first who labeled pre-formed liposomes with ^{111}In. They incorporated the ionophore A23187 into the lipid bilayers to facilitate the transport of charged species across the bilayer and enclosed the chelator nitrilotriacetic acid (NTA, 1 mM) in the aqueous interior to keep the ^{111}In trapped inside the liposomes. Radioactivity was added to the liposomes (distearoylphosphatidylcholine:-cholesterol:A23187 or dipalmitoylphosphatidylcholine:cholesterol:A23187 liposomes) in the form of ^{111}InCl$_3$, and liposomes were then heated to 60° or 80°, respectively. The reported labeling efficiency was higher than 90%.[15] In vivo stability was not tested. Disadvantages of this method are the need to incorporate the ionophore into the bilayer, which might change the in vivo behavior of the liposomes, as well as the need to label at high temperatures.

Hwang and colleagues[16] improved the afterloading method by first converting ^{67}Ga and ^{111}In to the lipophilic forms (^{67}Ga-8-hydroxyquinoline (^{67}Ga-oxine) or ^{111}In-8-hydroxyquinoline (^{111}In-oxine) able to cross the lipid bilayers. After the conversion, the lipophilic complex was added to preformed liposomes containing 1 mM NTA encapsulated in the aqueous phase. Once inside the liposome, the radiolabeled complex is chelated by the NTA and consequently trapped inside the liposome. Biodistribution studies with sphingomyelin/cholesterol liposomes labeled with either ^{111}In or ^{67}Ga by this method showed that the circulating liposomes are stable in vivo.[16] An adaptation of this method was described by Beaumier et al.,[17] using acetylacetone as an ionophore and NTA as a chelator. In this case, ^{111}InCl$_3$ was mixed with acetylacetone and incubated subsequently for 1 h at room temperature. With this method, a labeling efficiency of 90% was obtained. Incubation of liposomes in vitro with strong chelating agents, such as ethylenediamine, tetraacetic, acid (EDTA), revealed that only approximately 5% of the entrapped ^{111}In was released over 24 h at room temperature. This, together with the biodistribution data, confirmed that this afterloading method yielded stably radiolabeled liposomes.

[15] M. R. Mauk and R. C. Gamble, Anal. Biochem. **94,** 302 (1979).
[16] K. J. Hwang, J. E. Merriam, P. L. Beaumier, and K. F. S. Luk, Biochim. Biophys. Acta **716,** 101 (1982).
[17] P. L. Beaumier and K. J. Hwang, J. Nucl. Med. **23,** 810 (1982).

A disadvantage of this approach is the laborious method of preparing the ^{111}In-acetylacetone, which involves evaporation of the [111]InCl$_3$ solution to dryness. Because NTA is a relatively weak chelator, metal translocation to serum proteins such as transferrin and other metal-binding proteins may occur. Gabizon et al.[18] studied the use of deferoxamine (DF) as a chelator for [67]Ga. Liposomes containing 25 mM DF in the aqueous interior were incubated with ^{67}Ga-oxine overnight at 4°. Labeling efficiency ranged from 57%–88%, irrespective of their lipid composition. Main differences between the biodistribution of ^{67}Ga-NTA and ^{67}Ga-DF, once released from the liposomes, is their route of clearance; ^{67}Ga-DF is cleared rapidly by the kidneys to the urine, whereas the ^{67}Ga of the ^{67}Ga-NTA complex translocates to transferrin and thus stays in the circulation for extended times.[18,19] The long circulation time of the radionuclide interferes with imaging of the biodistribution of the radiolabeled liposomes.

Similar to these methods, DTPA can be used also as a chelator in the aqueous phase of the liposomes for labeling with [111]In. This results in a stable radiolabeled preparation with a labeling efficiency of approximately 80%. During the preparation of the liposomes, 6 mM of DTPA is encapsulated, and the free DTPA is removed with gel permeation chromatography on a PD-10 column (Sephadex G-25, Amersham Biosciences, Uppsala, Sweden).[20] The liposomes are subsequently labeled with ^{111}In-oxine (Mallinckrodt, Petten, The Netherlands). Briefly, 0.1–1.0 ml of ^{111}In-oxine in 0.2 M TRIS(hydroxymethyl)aminomethane, pH 8.0, is added to 1.0–2.0 ml of liposomes and incubated for 30 min at room temperature. The nonencapsulated ^{111}In-oxine is then removed by gel permeation chromatography as described previously. When using relatively rigid liposomes, the incubation with the ^{111}In-oxine should be carried out at a higher temperature to facilitate transport of ^{111}In-oxine through the bilayer.

Recently, we adapted this method to label two commercially available liposomal formulations: doxorubicin encapsulated in polyethylene glycol (PEG)-coated liposomes (Caelyx®/Doxil®)[21] and daunorubicin encapsulated in small distearoylphosphatidylcholine/cholesterol (DSPC:Chol) liposomes (Daunoxome®).[22] Although no DTPA was enclosed in these liposomes, the labeling efficiency was typically between 70% and 80%,

[18] A. Gabizon, J. Huberty, R. M. Straubinger, D. C. Price, and D. Papahadjopoulos, J. Liposome Res. 1, 123 (1988).

[19] I. Ogihara-Umeda and S. Kojima, J. Nucl. Med. 29, 516 (1988).

[20] M. L. Corvo, O. C. Boerman, W. J. Oyen, L. Van Bloois, M. E. Cruz, D. J. Crommelin, and G. Storm, Biochim. Biophys. Acta 1419, 325 (1999).

[21] P. Laverman, M. G. Carstens, O. C. Boerman, E. T. Dams, W. J. G. Oyen, N. van Rooijen, F. H. M. Corstens, and G. Storm, J. Pharmacol. Exp. Ther. 298, 607 (2001).

[22] P. Laverman and M. G. Carstens, Unpublished results (2000).

and the radiolabeled preparations were stable *in vivo* during the course of the experiment (4 h). Most likely the lipophilic ^{111}In-oxine avidly associates with the lipid bilayer, and encapsulation of DTPA might not be necessary when the experimental observation period does not exceed 4–6 h.

A major step forward in the development of afterloading methods was the introduction of a ^{99m}Tc afterloading method by Phillips and colleagues.[23] This rapid and easy labeling method is adapted from the well-known ^{99m}Tc-leukocyte labeling method,[24] using the lipophilic ^{99m}Tc-hexamethylpropylene-amine-oxime (HMPAO) complex to transport the radionuclide to the inner aqueous phase of preformed liposomes containing reduced glutathione (100 mM). The entrapped glutathione reduces the complex to its hydrophilic form, and as a result the complex is trapped irreversibly in the aqueous interior. This method is generally applicable to most liposome types. This labeling method will be discussed in more detail in the following paragraph.

Preparation of Glutathione–Liposomes. After dissolving the desired lipids (for example, phosphatidylcholine, distearoylphosphatidylethanolamine) in an organic solvent, such as methanol, ethanol, and/or chloroform and subsequent evaporation, the dried lipids are hydrated with an isotonic buffer containing reduced glutathione (100 mM). After the sizing steps, either by using an extruder, sonicator, or other technique, the liposomes are passed over a gel permeation column to remove the unencapsulated glutathione. A PD-10 column is prerinsed with the same buffer (without glutathione) in which the liposomes are prepared, usually phosphate-buffered saline (PBS), pH 7.4, and subsequently the liposomes are applied. The column is eluted with the same buffer, and fractions are collected. Liposomes appear in the fractions immediately after the void volume. Once the liposome-containing fractions are pooled, the liposomes are stored at 4° until radiolabeling. It has been observed that the labeling efficiency of the liposomes is decreased after 3 months of storage at 4°. Experiments to improve the shelf-life of the preformed liposomes show that the presence of 300 mM sucrose in the hydration medium greatly improves the shelf-life; 8 months after preparation, negatively charged liposomes could still be labeled with an efficiency of >80%.[25]

Radiolabeling of Preformed Glutathione–Liposomes with ^{99m}Tc-HMPAO. Preformed glutathione liposomes are labeled by incubation with

[23] W. T. Phillips, A. S. Rudolph, B. Goins, J. H. Timmons, R. Klipper, and R. Blumhardt, *Nucl. Med. Biol.* **19**, 539 (1992).

[24] A. M. Peters, H. J. Danpure, S. Osman, R. J. Hawker, B. L. Henderson, H. J. Hodgson, J. D. Kelly, R. D. Neirinckx, and J. P. Lavender, *Lancet* **2**, 946 (1986).

[25] B. Goins, R. Klipper, and W. T. Phillips, *J. Nucl. Med.* **38**, 179P (1997).

fresh ^{99m}Tc-HMPAO-complex (Ceretec$^®$, Amersham Imaging, Bucking-hamshire, UK) prepared according to the manufacturer's guidelines.[23] One hundred microliters of liposomes (50–100 μmol phospholipid/ml) is incubated with approximately 5 mCi (185 MBq) of ^{99m}Tc-HMPAO for 20 min at room temperature. To remove unbound radiolabel, the lipo-somes are passed over a PD-10 column as described previously. Radio-activity in all fractions is measured in a shielded well-type gamma counter. The (radiolabeled) liposomes elute in the fractions immediately after the void volume, and the unbound radiolabel elute in the later frac-tions. Fractions containing the labeled liposomes can be pooled and are then ready for injection. To verify the radiochemical purity of the final preparation, the gel permeation chromatography procedure can be repeated with a trace amount of the pooled fractions

This labeling method yields a highly stable radiolabeled product. The feasibility of long-circulating PEG–liposomes labeled with ^{99m}Tc-HMPAO for the scintigraphic imaging of infectious or inflammatory foci has been studied in patients suspected of having an infection or inflammation.[26] Foci were detected with high sensitivity and specificity. A typical example of the whole-body distribution of radiolabeled PEG–liposomes at two different phospholipid doses is shown in Fig. 3. Advantages of this labeling method are the rapid labeling procedure and the good radiochemical stability of the radiolabeled product. Although the labeling efficiency is generally 85%–90%, a postlabeling purification is required to remove the unbound radiolabel and to achieve a radiochemical purity exceeding 95%. The high cost of the HMPAO-kits might limit the application of this labeling technique.

External Surface Chelation Methods

The first method that uses the presence of a chelator on the surface of the liposome (Fig. 2D) was based on the conjugation of DTPA to octode-cylamine, a single-chain fatty acid.[27] This conjugate was incorporated in the lipid bilayer. Although the labeling efficiency was high, a major drawback was the relatively poor *in vivo* stability, most likely caused by the lipid ex-change with blood components. An improvement was the conjugation of diethylenetriaminetetraacetic acid (DTTA) to dipalmitoylphosphatidyl-ethanolamine (DPPE).[28] This double-chain fatty acid showed a higher *in vivo* stability regarding the loss of radiolabel.

[26] E. T. Dams, W. J. Oyen, O. C. Boerman, G. Storm, P. Laverman, P. J. Kok, W. C. Buijs, H. Bakker, J. W. van der Meer, and F. H. Corstens, *J. Nucl. Med.* **41,** 622 (2000).

[27] D. J. Hnatowich, B. Friedman, B. Clancy, and M. Novak, *J. Nucl. Med.* **22,** 810 (1981).

[28] Q. F. Ahkong and C. Tilcock, *Int. J. Rad. Appl. Instrum. B* **19,** 831 (1992).

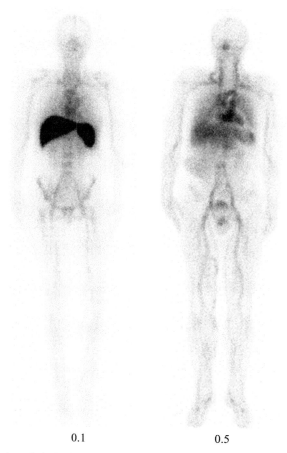

0.1   0.5

Fig. 3. Anterior whole-body scintigram of two patients, 4 h after injection of ^{99m}Tc-PEG-liposomes at different doses. At a dose of 0.1 μmol/kg, the PEG-liposomes are cleared rapidly from the circulation compared with a normal biodistribution (i.e., a long circulation time) at a dose of 0.5 μmol/kg.[29] (Reprinted with permission of the American Society for Pharmacology and Experimental Therapeutics from: P. Laverman *et al. J. of Pharmacol. Exp. Therap.* **293,** 966, [2000].)

More recently, we developed a new chelation method, based on the technetium chelator hydrazino nicotinamide (HYNIC), which is well known for its use in labeling peptides and proteins with high efficiency and excellent stability.[30] HYNIC is conjugated to distearoylphosphatidyl-

[29] P. Laverman, A. H. Brouwers, E. T. Dams, W. J. Oyen, G. Storm, N. van Rooijen, F. H. Corstens, and O. C. Boerman, *J. Pharmacol. Exp. Ther.* **293,** 996 (2000).

ethanolamine (DSPE) and subsequently incorporated in the lipid bilayer. DSPE (28 μmol) is dissolved in 500 μl warm chloroform (55°) followed by the addition of 7 μmol N-hydroxysuccinimidyl hydrazino nicotinate hydrochloride (S-HYNIC) and 36 μmol triethylamine. After incubation at 55° for 1 h under constant stirring, the chloroform is evaporated under a gentle stream of nitrogen. This HYNIC-DSPE conjugate can be then incorporated in the lipid bilayer during the liposome preparation.

Preparation of HYNIC–Liposomes. Polyethylene glycol–liposomes are prepared essentially as follows: a lipid mixture (egg phosphatidylcholine:-PEG-DSPE:HYNIC-DSPE:cholesterol) in methanol/chloroform (10:1) was prepared with a molar ratio of 1.85:0.15:0.07:1. After evaporation of the organic solvents, the resulting lipid film is dispersed in PBS at room temperature. After sizing, the suspension is dialyzed extensively against PBS overnight at 4°, with four buffer changes to remove unconjugated S-HYNIC, using a Slide-a-Lyzer® (Pierce, Rockford, IL) with a cutoff value of 100 kDa. Liposomes are stored in PBS at 4° and can be labeled efficiently for approximately 3–4 months. When the liposomes are freeze-dried, the shelf-life can be extended to more than 1 year.[31]

The number of liposome-associated HYNIC molecules is determined spectrophotometrically by converting the hydrazino group of HYNIC into the corresponding hydrazone by a reaction with o-sulfonic benzaldehyde and measuring the optical density at 343 nm.[32] Samples of the HYNIC-liposomes (10–100 μl) are diluted with 0.1 ml o-sulfonic benzaldehyde (1 mg/ml, 0.1 M NaAc, pH 4.7). Samples are incubated at room temperature in darkness overnight. Liposomes without the HYNIC-DSPE conjugate are used as a negative control. The absorption of the hydrazone is measured at 343 nm. The hydrazine concentration is calculated using a molar extinction coefficient of 17,000 $1 \cdot mol^{-1} \cdot cm^{-1}$. On average, PEG–liposomes with a diameter of 85 nm contain 230 HYNIC groups per particle.

Radiolabeling of HYNIC–Liposomes. Labeling of HYNIC–liposomes is essentially performed as described previously.[32] An optimization study of this labeling technique was published recently.[31] Tricine-$SnSO_4$ kits are prepared containing 20 mg tricine (N-[Tris(hydroxymethyl)-methyl]-glycine) and 10 μg stannous sulfate in 0.2 ml PBS. The $SnSO_4$ dissolved in 2 M HCl is added to a solution of tricine in PBS, and the pH is adjusted

[30] M. J. Abrams, M. Juweid, C. I. tenKate, D. A. Schwartz, M. M. Hauser, F. E. Gaul, A. J. Fuccello, R. H. Rubin, H. W. Strauss, and A. J. Fischman, *J. Nucl. Med.* **31,** 2022 (1990).

[31] P. Laverman, L. Van Bloois, O. C. Boerman, W. J. Oyen, F. H. Corstens, and G. Storm, *J. Liposome Res.* **10,** 117 (2000).

[32] H. J. Rennen, O. C. Boerman, E. B. Koenders, W. J. Oyen, and F. H. Corstens, *Nucl. Med. Biol.* **27,** 599 (2000).

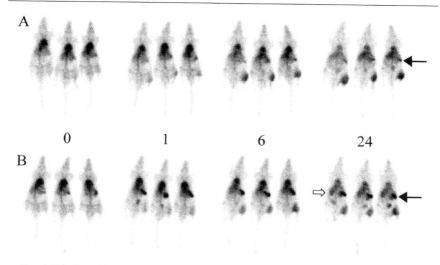

FIG. 4. Scintigraphic images of rats with unilateral *Staphylococcus aureus* abscess in the left calf muscle. Images were recorded at 0 h (= 5 min), 1, 6 and 24 h after injection of ^{99m}Tc-(HYNIC)-PEG-liposomes (A) and ^{99m}Tc-(HMPAO)-PEG-liposomes (B). Open arrow indicates the right kidney, and closed arrows indicate the spleen. (Reprinted by permission of the Society of Nuclear Medicine from: P. Laverman *et al.* A novel method to label liposomes with Tc-99m by the hydrazino nicotinyl derivative. *J. Nucl. Med.* **40**, 192, [1999].)

to 7.0 with 1.0 M NaOH. Kits (200–1000 μl) are stored at $-20°$. One hundred microliters of a tricine-SnSO$_4$-kit is added to 100 μl liposomes ($\sim$85 μmol phospholipid/ml) and 100–300 μl of 20–30 MBq ^{99m}TcO$_4^-$ in saline are added, and the mixture is incubated at room temperature for 20 min. Radiochemical purity is determined using instant thin-layer chromatography (ITLC) on ITLC-SG strips (Gelman Sciences, Ann Arbor, MI) with 0.15 M sodium citrate (pH 5.0) as the mobile phase. The radiolabeled liposomes (1–2 μl) are applied 1 cm from the bottom on a ITLC-SG-strip, and the strip is placed in a developing chamber until the mobile phase runs to 1 cm from the top of the strip. The strip is then divided into two equal parts by cutting the strip in the middle, and the two parts are counted in a shielded well-type gamma counter. The liposomes remain at the origin ($R_f = 0$), whereas the unbound ^{99m}Tc will migrate with the solvent front ($R_f = 1$).

This method is relatively inexpensive and easy to apply. Liposomes are labeled rapidly and with a labeling efficiency generally greater than 95%. ^{99m}Tc-(HYNIC)-PEG-liposomes showed high *in vitro* and *in vivo* stability as described previously.[33] The *in vivo* biodistribution was compared with that of PEG-liposomes labeled with ^{99m}Tc-HMPAO in rats with an intramuscular *Staphylococcus aureus* abscess (Fig. 4). Besides a strongly decreased kidney uptake (compared with HMPAO-liposomes), reflecting an advantage in favor of the HYNIC–liposomes, the *in vivo* behavior was similar.

The feasibility of ^{99m}Tc-(HYNIC)-PEG–liposomes to image colitis in patients with Crohn's disease has been studied.[34] The radiolabeled liposomes clearly visualized inflammatory foci in this group of patients. Examples of the use of this labeling technique for pharmacokinetic studies have been published recently.[21,35] Besides PEG–liposomes, non-PEGylated DPPC:Chol liposomes have been labeled with ^{99m}Tc by HYNIC. Also in this case, satisfactory results were obtained.

Concluding Remarks

Scintigraphic techniques have proven to be useful when investigating the *in vivo* behavior of liposomes. The major advantage of scintigraphy is the ability to obtain images over a longer time, which rapidly gives an impression of the *in vivo* behavior of the liposomes without the need for dissecting a large number of animals.

Several techniques have been developed in the past decades. Probably the most fundamental method is to encapsulate the radiolabel in the aqueous interior during the hydration of the lipids of the liposomes. However, this method is laborious and has a low labeling efficiency.

A second relatively simple method is the reduction of the radiolabel in the presence of the liposomes, resulting in association of the label with the outside of the lipid bilayer. Studies have shown that instability may occur when applying this technique for neutral or positively charged liposomes.

A third approach is to trap the radiolabel in the aqueous phase of preformed liposomes. This method has proven useful in several studies and is widely used. A chelator with high affinity for the radionuclide is entrapped during preparation, and the radionuclide is added to the liposomes just before the experiment. These methods generally yield radiolabeled liposomes with good *in vivo* stability.

Finally, the radiolabel can be chelated to a lipid–chelator conjugate in the lipid bilayer. This method has been used mainly for ^{99m}Tc-labeling and yields liposomes with high radiochemical purity and high stability.

[33] P. Laverman, E. T. Dams, W. J. Oyen, G. Storm, E. B. Koenders, R. Prevost, J. W. van der Meer, F. H. Corstens, and O. C. Boerman, *J. Nucl. Med.* **40**, 192 (1999).

[34] A. H. Brouwers, D. J. de Jong, E. T. Dams, W. J. Oyen, O. C. Boerman, P. Laverman, T. H. J. Naber, G. Storm, and F. H. Corstens, *J. Drug Target.* **8**, 225 (2000).

[35] P. Laverman, M. G. Carstens, G. Storm, and S. M. Moghimi, *Biochim. Biophys. Acta.* **1526**, 227 (2001).

[16] Preparation of a Liposomal Reagent and its Use in an Immunoassay for Albumin

By STEPHEN J. FROST

Introduction

Imunoassays were introduced in the early 1960s,[1] initially using radioisotopes as labels, and have greatly simplified the measurement of analytes at very low concentrations. As examples, clinical laboratories routinely measure estradiol and thyroid hormones in the picomolar range in human blood. Since the early days, various types of immunoassay have been developed, driven by the desire for more stable and convenient labels, for increases in assay performance, and for the ability to perform assays on many samples more quickly using automation. In clinical as well as in other areas, there is also a trend to move from laboratory testing to point of care or field testing, which requires simple and stable reagent kits and equipment.

Liposomal immunoassay is a promising technique that has several potential advantages over other immunoassays. Conventional immunoassays use a reagent in which an antigen or antibody is chemically attached to a single, or a few, molecule of the label (also known as the marker). This is often an enzyme, dye, or fluorophor. In a liposomal immunoassay, on the other hand, each liposome usually contains many molecules of marker. If antigen or antibody is attached to each liposome rather than to individual marker molecules, signal amplification may lead to improved assay sensitivity or a more rapid measurable response.

Liposomes have other potential advantages, which may arise because of the unique properties of the liposomal membrane and the separation of liposomes' contents from the surrounding aqueous milieu. One way these can be exploited is in the use of liposomes as reagents in unique homogeneous immunoassays. Conventional immunoassays are heterogeneous, which means that they require a separation step. This step can be time-consuming and technically complex. The separation may involve, for example, centrifugation of antibody-antigen-label complexes or their binding to a solid support. Homogeneous assays that avoid the need for the separation stage are easier to automate and more applicable to point of care applications.

---

[1] R. S. Yalow and S. A. Berson, *J. Clin. Invest.* **39**, 1157 (1960).

Although liposomes can be used in heterogenous assay formats,[2,3] their particular strength lies in their applicability to homogeneous assays, and this chapter describes the preparation of liposomes and their use in one such assay. The assay uses complement-mediated immune lysis of the liposomes, which is probably the most common approach. In this procedure, the liposomes undergo antigen–antibody reactions at their surfaces and are then lysed by complement. The intensity of signal depends on how much of liposomes' contents are released into the surrounding buffer, which in turn is related directly to the concentration of antigen in the sample measured.

The illustrated assay adopts a simple way to generate a measured signal. This is the measurement of the change in absorbance that occurs on the release of a chromogenic dye, sulforhodamine B, from the liposomes and the dye's dilution in the external buffer.[4] The absorbance change is due to a shift in the wavelength of maximum absorbance of the chromogen. This shift is probably due to the dye existing in a dimerized form at the very high concentrations within the liposomes, but it becomes monomeric when the dye is released and diluted.

The liposomes are first formed and covalently coated in Fab' fragments from antibody raised against human albumin. Fab' fragments are used instead of intact antiserum to avoid spontaneous lysis when complement is added to antibody-coated liposomes.[5] When liposomes are coated with Fab' fragments, complement-mediated lysis occurs only on reaction with albumin and intact anti-albumin antiserum.[6] The degree of lysis and the resultant change in absorbance are dependent on the albumin concentration in the added samples. Albumin in human urine is measured in the assay that is described.

Liposome Formation

In this procedure a small volume (several milliliters) of liposome reagent is prepared. This is suitable for a substantial initial study, because the liposomes are diluted for use, although volumes can be scaled up later if desired. Chemicals and AnalaR grade reagents are used throughout, supplied by BDH, Poole, England, unless otherwise stated. The presence of detergents must be avoided during the described liposome preparation,

[2] J. O'Connell, R. L. Campbell, B. M. Fleming, T. J. Mercolino, M. D. Johnson, and D. A. McLaurin, *Clin. Chem.* **31,** 1424 (1985).
[3] S. J. Frost, G. B. Firth, and J. Chakraborty, *J. Immunol. Methods* **134,** 207 (1990).
[4] S. J. Frost, G. B. Firth, and J. Chakraborty, *J. Liposome Res.* **4,** 159 (1994).
[5] D. W. Bowden, M. Rising, G. Akots, A. Myles, and R. Broeze, *Clin. Chem.* **32,** 275 (1986).
[6] S. J. Frost, J. Chakraborty, and G. B. Firth, *J. Immunol. Methods* **194,** 105 (1996).

or they will not form. Once the liposomes are prepared, subsequent contamination by organic solvents or detergents may mean the liposomes are lost through lysis.

Phosphatidylethanolamine (PE) (Sigma Chemical Company, Poole, England) is first coupled to the heterobifunctional coupling agent, *N*-succinimidyl 3-(2-pyridyldithio) propionate (SPDP) (Pharmacia, Milton Keynes, England).[3,4,6,7] Phosphatidylethanolamine (6.9 mg) (in 300 μl chloroform), 3.75 mg of SPDP (in 300 μl methanol), and 2.7 μl of triethanolamine are mixed, solubilized by brief heating to 30° and then incubated for 2 h at room temperature. The resultant organic phase is washed once in sodium phosphate buffer (0.05 M, pH 7.4) and twice in distilled water, centrifuging at 2000g for 5 min after each addition and discarding each aqueous layer. After the last addition of water, 1 ml of chloroform is added to facilitate complete removal of the aqueous phase without excessive loss of the organic phase. If a white precipitate forms at the interface, this is reduced, if not eliminated, by the addition of 1 ml of methanol, and any residual precipitate should be discarded leaving a clear organic solution.

To confirm formation of dithiopyridyl phosphaditylethanolamine (DTP-PE), its concentration is estimated by the following procedure: 25 μl of the organic phase is evaporated gently to dryness in a glass tube and immediately dissolved in 1 ml of buffer (pH 8.0, containing 10 mM *N*-(2-hydroxyethyl) piperazine -*N'*-(2-ethanosulphonic acid) (HEPES) and 145 mM sodium chloride), using vigorous vortex mixing. The absorbance of the solution at 343 nm is measured before and after adding 50 μl of dithiothreitol (100 mM). The molar concentration of DTP-PE is calculated as the difference in absorbance at 343 nm multiplied by the dilution factor and divided by the molar absorbance coefficient of dithiopyridine at 343 nm (8.08×10^3 M^{-1} cm^{-1}).

It is preferable to immediately incorporate the DTP-PE into liposomes to avoid gradual cross-linkage of thiol groups. A reverse-phase evaporation method[8] is used to form the liposomes. Phosphatidyl choline (40 μmol), cholesterol (20 μmol) and phosphatidic acid (5 μmol) are dissolved in 2 ml of chloroform and 2 ml of petroleum ether. To this solution is added 2 μmol of the synthesised DTP-PE, which typically has a volume of approximately 0.5 ml. Then 1 ml of sulforhodamine B (0.1 M) (Sigma), dissolved in distilled water, is added. The mixture is sonicated in a glass vial and held in a cold water bath for 2 min at an amplitude of 10 μm (Soniprep probe sonicator, MSE, Crawley, England). One milliliter

[7] A. Truneh, P. Machy, and P. K. Horan, *J. Immunol. Methods* **100**, 59 (1987).
[8] F. Szoka and D. Papahadjopoulos, *Proc. Natl. Acad. Sci. U.S.A* **75**, 4194 (1978).

of phosphate buffer is added, and the suspension is transferred to a long-necked evaporation flask warmed to 45° in a water bath. The organic solvent is then removed by evaporation in a rotary evaporator (Corning, Stone, England) under reduced pressure. The suspension rapidly forms a gel, which quickly collapses to create an opalescent aqueous suspension of liposomes.

The entrapment of the dye is estimated by applying 0.2 ml of the mixture to a small gel filtration column (0.7 mm × 20 cm, Bio-Rad, Hemel Hempsted, England) containing Sepharose 4B (Pharmacia, Milton Keynes, England). Half-milliliter fractons are collected on elution with phosphate buffer. The liposomes elute as a dark purple-colored peak in the void volume, followed by a separate purple peak of free dye. Twenty microliters of ethanol is added to each fraction to lyse any liposomes, and each fraction is diluted in phosphate buffer to give a measurable absorbance at 565 nm, which for the peaks is usually a 1/1000 dilution. The absorbance of each fraction is measured in a spectrophotometer, and the percentage incorporation of dye is calculated from the relative size of the two peaks. An incorporation of the added sulforhodamine B close to 20% is typically achieved.

The remaining liposome preparation is passed through a 200-nm polycarbonate filter (Sartorious, Epsom, England) to improve homogeneity. The preparation then is dialyzed overnight using 1.5-cm diameter dialysis tubing (Medicell International, London, England) against 200 times the liposomes' volume of phosphate buffer. This dialysis removes most of the unencapsulated dye. The liposomes so produced are coupled to Fab' fragments within the next 2 days.

Coupling with Fab' Fragments

DTP-Fab' is prepared by mixing 4 ml of an aqueous solution of sheep anti-human Fab' fragments (approximately 10 g/liter, Immunogen International, Llandysul, Wales) with SPDP in methanol at a molar ratio of SPDP to Fab' of 10:1. The mixture is incubated for 30 min at 23° with occasional stirring, and the reaction is then terminated by addition of 250 μl acetic acid, pH 3. Immediately before coupling with the liposomes, 50 μl of dithiothreitol (500 mM) is mixed with the DTP-Fab' and incubated for 20 min at room temperature to thiolate the Fab'. The thiolated Fab' is then separated from the reaction mixture by passage through a short gel-filtration column containing Sephadex G50 (Pharmacia) equilibrated with phosphate buffer. The protein fraction, containing thiolated Fab', elutes in the void volume. This is demonstrated by measuring the protein content of the eluted fractions using a standard Coomassie blue method (Bio-Rad). From each 0.5-ml fraction, 20 μl is mixed with 1 ml of saline and 200 μl of

Coomassie blue reagent, then incubated at room temperature for 15 min. The absorbances of each fraction are measured at 580 nm and compared with albumin standards (Bio-Rad) to estimate the protein concentration.

Equal volumes of the liposome and the thiolated Fab' preparations are mixed and incubated overnight at room temperature to form Fab'-coated liposomes. The mixture is then passed through a gel filtration column containing Sepharose 4B, eluting with phosphate buffer. The liposomes, now coupled to Fab', elute in the void volume, whereas the free dye and uncoupled protein are retarded by the gel. This can be demonstrated by the Coomassie blue procedure (described previously).

The protein and lipid contents of the liposome preparation are measured to characterize it. The protein is measured by the Coomassie blue method, modified by solubilizing the lipid on addition of 20 μl ethanol to 20 μl of the liposome preparation. The standard protein measurement described previously is then followed. A reagent blank is used to compensate for the contribution to the absorbance at 580 nm caused by sulforhodamine B. For this reagent blank, saline replaces the Coomassie blue reagent.

The lipid content is estimated by measuring the cholesterol. This is achieved by a standard cholesterol oxidase method ("CHOD-PAP," measured on an EPOS analyser, BDH). However, first the intense color of the dye is removed from the preparation by overnight dialysis into a thousand-fold volume of distilled water, containing 0.5% detergent (Triton X-100) to lyse the liposomes. The material that is retained by the dialysis tubing is collected, and the cholesterol concentration is measured, from which the total lipid concentration is calculated as the cholesterol concentration $\times$ 3.35, with a further adjustment for the small volume change during dialysis. The measurements of cholesterol and protein enable calculation that the liposomes contain typically about 240 μg protein/micromole lipid.

When a small volume of the final liposome preparation is dialyzed against 40 times its volume of phosphate buffer, there is a rapid but decreasing dialysis of any residual free dye in the preparation, which is virtually complete within 2–4 h. There follows a more linear and slower dialysis of dye as it is released by liposomal leakage or rupture. By use of sequential absorbance measurements of the dialyzed dye after 4 h, an estimate of the rate of dye release from the liposomes is made. The estimation of the absorbance of the dye in the liposome preparation is complicated by the dye within the liposomes being far too concentrated for Beer's law to be obeyed. The total absorbance of the dye in the liposomes must therefore be estimated in terms of its absorbance when diluted to a measurable value, multiplied by the dilution factor. This is achieved by adding 20 μl of ethanol to 20 μl of liposome preparation and diluting in distilled water so that the absorbance can be measured (usually $\times$ 1000). The rate of dialysis of dye at room temperature can then be estimated as a percentage release of the total dye and is found be approximately 0.1% of the liposomal contents of dye per day, which is adequate stability for application in an immunoassay. The liposomes are stored at 4° before use in the assay, at which temperature there is negligible leakage of dye from the liposomes over several months.

Complement-Mediated Immunoassay

The immunoassay is performed on a Cobas Bio analyzer (Roche, Lewes, England). Using an automated analyzer greatly improves the ease and precision of the assay, although a spectrophotometer that can monitor absorbance changes can be used as an alternative. The reaction is performed using 50 mM TRIS buffer, pH 7.5, containing 0.15 M NaCl, 0.5 mM $MgCl_2$, and 0.15 mM $CaCl_2$. This buffer optimizes the rate of complement activity. The liposome reagent is prediluted in the buffer to a dilution of 1/20. Immediately before the assay, guinea pig complement (Sera-lab, Crawley Down, England) is added at a 1:4 dilution. Twenty microliters of the combined liposome-complement reagent is dispensed into each reaction cuvette. Twenty microliters of the standard or sample to be measured is also added to the cuvette and mixed. After a 5-min preincubation, at a temperature of 25°, the reaction is initiated by the addition of 10 μl of human anti-albumin antiserum diluted to 1:4 in TRIS buffer. The resultant absorbance change is monitored at 565 nm at minute intervals over 30 min at 25°. (The analyzer adds 10 μl of distilled water diluent when dispensing the sample. It also takes a first reading at 3 s. Neither the diluent nor such a short time of first reading is critical.) The absorbance change is related directly to the concentration of albumin in the added sample (Fig. 1). A standard curve is plotted from the absorbance change of the standards over the reaction period. This standard curve is used to assess the albumin concentration in the samples to be measured by comparison of the absorbance changes of the samples to those of the albumin standards.

The standards, which are pure human albumin diluted in phosphate buffer, are assayed at concentrations appropriate to the biological or other fluid to be analyzed. In this example, urine albumin is measured requiring standards between 0 and 200 mg/liter. If serum albumin is to be measured from blood, the albumin standards will be in the approximate range of 0–60 g/liter.

Early morning urine samples (approximately 40 ml) from patients attending a hospital diabetic clinic are frozen on collection and thawed immediately before assay, with removal of any precipitated material by centrifugation at 2000g for 10 min. The samples are checked using a

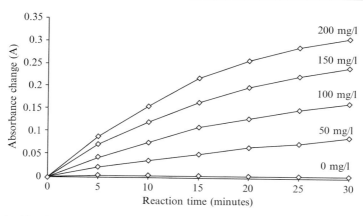

FIG. 1. Absorbance changes obtained in the presence of samples containing various albumin concentrations up to 200 mg/l. The reaction conditions are described in the text. Although the analyzer measures at minute intervals, for clarity only the absorbance readings at 5-min intervals are shown.

chemical strip test (Combur-Test, Roche) to exclude samples with gross proteinuria, which only occurs when the albumin concentration exceeds the measurement range of the assay.

The urine assay is used to differentiate patients with normal urine albumin from patients with slightly increased albumin excretion (typically concentrations between 20 and 200 mg/liter). The latter concentrations are below the limits measurable with any accuracy by chemical methods or strip tests. These low levels of albumin are found in the condition known as microalbuminuria, and this suggests incipient diabetic nephropathy.[9]

When the liposomal results are compared with those of a standard radioimmunoassay (RIA) for urine albumin in this assay range (NETRIA, London, England), there is a correlation of 0.94 and a regression equation (using Deming's correction) of y (liposomal immunoassay) = 1.09 × (RIA) − 1.54 (mg/liter).[6] Imprecision is assessed by measuring samples in replicate, and this shows acceptable coefficients of variation (CVs) of below 10% at concentrations greater than 10 mg/liter.

Characterization of the Immunoassay

The described assay has been optimized for human albumin estimation. If other antigens or antibodies are measured, it will probably be necessary

[9] A. H. Barnett, K. Dallinger, P. Jennings, J. Fletcher, and O. Odugbesan, *Lancet* **1 (8419),** 53 (1985).

to modify the amounts of antiserum and complement used in the assay. This is because the Fab' fragments may have different immunoreactivity with antigen, intact antibody, and complement. These optimization experiments can be achieved on the Cobas Bio analyzer (or another spectrophotometer) by varying the order of addition of reagents. Tables I and II show how the components of the assay are adjusted to optimize the complement and the antiserum concentrations, and the effect of each of these is illustrated in Fig. 2. The antiserum optimization is particularly important, because a decrease in the rate of absorbance change is seen if too much antiserum is added.

The effect of potential interference from other components of the sample also needs assessing. The potential interfering substances are added to the analyte in varying concentrations that encompass the likely ranges of the interferences in the samples. In the illustrated assay for urinary albumin, ranges of pH (5.5–7), phosphate (10–37 mM), sodium (44–166 mM), potassium (3–99 mM), calcium (0.4–4.1 mM), and magnesium (0.9–4.8 mM) are assessed using samples containing either no albumin or an albumin concentration of 75 mg/liter. No effects on the measured absorbances are detected for any of the ions over the stated ranges when a zero albumin standard is used. The only small interference that is found at an albumin concentration of 75 mg/liter is a slight effect of

TABLE I

EXPERIMENTAL CONDITIONS TO OPTIMIZE COMPLEMENT ADDITION

Reagent 1:   TRIS buffer containing liposomes (1/20), 20 μl per cuvette
Sample: Guinea-pig complement dilutions in TRIS buffer plus 200 mg/1 human albumin (pure); 20 μl per cuvette

Preincubate for 5 min at 25°
Reagent 2:   Anti-albumin (1/4); 10 μl per cuvette
Measure absorbance change at minute intervals over 30 min at 25°

TABLE II

EXPERIMENTAL CONDITIONS TO OPTIMIZE ANTISERUM ADDITION

Reagent 1:   TRIS buffer containing liposomes (1/20) plus 200 mg/l albumin (pure); 20 μl per cuvette
Sample: Anti-albumin antiserum diutions in TRIS buffer; 20 μl per cuvette

Preincubate for 5 min at 25°
Reagent 2:   Guinea-pig complement (neat); 10 μl per cuvette
Measure absorbance change at minute intervals over 30 min at 25°

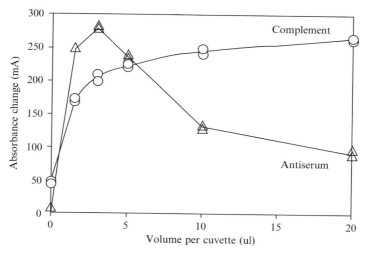

FIG. 2. Relationship between volume of neat complement (O) or antiserum (△) added to each reaction cuvette and the resultant absorbance change. The reaction conditions are described in table I and II. All measurements are duplicates, some points of which are overlaid on the graph.

calcium ions. Over the range of calcium concentration found in urine, however, this affects the measured albumin by less than 2 mg/liter, which is acceptable. The pH range is found to affect the measured albumin by less than 2 mg/liter in a zero albumin standard and less than 4 mg/liter in a 75 g/liter albumin standard. This interference can be minimized, however, by making up the albumin standards in a sodium phosphate buffer (0.05 M) with a pH of 6.5.

In an assay for a blood analyte, the observed ranges of pH and common ions is much narrower. However, other interference, such as from drugs, will need careful assessment before the assay can be used in a clinical setting.

Variations on the Theme

The amounts of individual lipids can be varied experimentally in the described liposome-formation procedure, which has been optimized for albumin measurement and dye incorporation. There is considerable flexibility in the selection of lipids used to form the liposomes. The best choice of lipids depends on the nature of both the analyte and the type of marker that is incorporated and used to produce a signal. The aims are to produce liposomes with high entrapment of marker, stable liposomes with low rates of leakage, and, not least, immune-reactivity of the liposomes to the desired assay performance.

Phosphatidylcholine is a neutral lipid and often forms the bulk of phospholipid used to form liposomes. The addition of cholesterol at up to about 40% of the total lipid stabilizes the membrane and tends to improve incorporation and reduce leakage.[8] If cholesterol is incorporated above that amount, the stability of the liposomes begins to decrease. Phosphatidic acid is added to the lipid mixture in small amounts and introduces a negative charge. This increases both the stability and the amount of incorporation of sulforhodamine B. The beneficial effect of phosphatidic acid on incorporation is probably mainly because the dye is negatively charged at near neutral pH. Higher mol fractions should be avoided, because divalent cations interact strongly with phosphatidic acid. Different lipid combinations may give optimal entrapment if neutral or basic markers are encapsulated. Stearylamine can be added to produce positively charged liposomes but does not form very stable vesicles with sulforhodamine B.

Phosphatidylethanolamine is usually introduced in small amounts, with little effect on stability or incorporation, as a means of introducing amine groups to couple to proteins. Some applications do not require surface-coupling of proteins. In one such example, cardiolipin can be incorporated directly into liposome membranes and used to measure anti-cardiolipin antibodies in blood samples.[10] The assay exploits the destabilization of cardiolipin-containing liposomes in the presence of anti-cardiolipin antibody and divalent cations.

Rather than using a dye, enzymes have been entrapped frequently as the markers in liposomal immunoassays. Examples have included a homogeneous assay for total complement using alkaline phosphatase.[5] Another enzyme that has been used in a complement-mediated assay is glucose-6-phosphate dehydrogenase.[11] Enzyme-based assays make use of compartmentalization of the enzyme from the substrate that surrounds the liposomes. The compartmentalization is broken down on liposomal lysis, which initiates the enzyme reaction. The reagents for an enzyme-based assay have more components than those for a dye-based assay, but the sensitivity may be greater because of amplification afforded by the enzymatic reaction.

Fluorophores are another type of marker. A fluorimetric measurement typically has a greater sensitivity than an absorbance measurement of a chromophore such as a dye. However a fluorometric detector is less readily available than a spectrophotometric one in an automated analyzer, and fluorescence is not a practical measurement for point of care assays. Sulforhodamine B is fluorescent as well as chromogenic, and the sulforhodamine

---

[10] S. J. Frost, J. Chakraborty, and G. B. Firth, *Clin. Chem.* **42,** 874 (1996).
[11] H. R. Six, K. Uremura, and K. S. Kinsky, *Biochemistry* **12,** 4003 (1973).

B–containing liposomes described in this chapter can also be used in a fluorescence-based assay. The dye's fluorescence is quenched within the liposomes because of the dye's high concentration, but on liposomal lysis the dye fluoresces once it is released and forms a more dilute solution. Another fluorescent marker that has been incorporated and used in this way is carboxyfluorescein, for the assay of mycotoxin T-2.[12]

An alternative chromogenic marker, reported in a complement-mediated assay for thyroxine, is 2-(2-thiazolylazo)-4-methyl-5-(sulfomethylamino) benzoic acid (TAMSMB), which becomes colored only when in solution with cobalt ions.[13]

Several homogeneous assays not requiring complement have been described, which make use of alternative ways of destabilizing or lysing liposomal membranes. The assay for anti-cardiolipin antibodies has been mentioned previously.[10] A further example is the use of mellitin, which is a biological cytolysin and can lyse liposomes. When the liposomes are incubated with antibodies to an antigen on the liposomes' surface, the mellitin-mediated lysis is inhibited. The lysis can be restored by the same antigen in an added sample, which competes for the antibody with the liposome-bound antigen. This has formed the basis of an immunoassay for digoxin.[14]

Another example is the use of dioleoylphosphatylethanolamine (DOPE) liposomes stabilized by glycophorin A, a membrane glycoprotein. On the addition of antibodies to glycophorin, the glycophorin becomes localized and unevenly distributed in the liposomal membrane. This causes the membrane to become unstable, and lysis occurs. The effect forms the basis of a competitive assay to any component, such as glycophorin, that can stabilize the liposomes.[15] A variation to this approach is for measurement of Herpes simplex virus. The liposomes that contain DOPE are stabilized in this case by acetylated antibody to the virus. On binding to the virus the acetylated antibodies are localized, and the antibody-depleted areas become unstable, producing lysis,[16] which can be used to detect the virus.

Acknowledgments

Supported by grants from the British Heart Foundation, the Television South Trust, and the South West Thames Regional Health Authority Locally Organised Research Scheme.

[12] F. S. Ligler, R. Bredehorst, A. Talebian, L. C. Shriver, C. F. Hamer, J. P. Sheridan, C. Vogel, and B. P. Gaber, *Anal. Biochem.* **163**, 369 (1987).
[13] Y. Nomura, K. Imai, M. Suzuki, and K. Okuda, *Clin. Chem.* **33**, 1565 (1987).
[14] J. W. Freytag and W. J. Litchfield, *J. Immunol. Methods* **30**, 1441 (1984).
[15] R. J. Ho and L. Huang, *J. Immunol.* **134**, 4035 (1985).
[16] J. Y. Ho, B. T. Rouse, and L. Huang, *Biochem. Biophys. Res. Commun.* **138**, 931 (1986).

[17]　Liposome Immunoassays Using Phospholipase C or Alkaline Phosphatase

By CHONG-KOOK KIM and SOO-JEONG LIM

Introduction

Enzyme immunoassays can be carried out heterogeneously or homogeneously. Homogeneous immunoassays[1,2] rely generally on the modulation of enzymatic activity of enzyme-analyte conjugates by binding of anti-analyte antibody. Homogeneous immunoassays are relatively simple and convenient, because these methods can avoid the tedious step separating bound and unbound labeled immunoreagents as in heterogeneous assays. However, most of these methods have been restricted to low-molecular-weight analytes and suffered from the limitation of low sensitivity.

In recent years, a significantly increasing number of samples in clinical fields and in bioprocesses has increased the demand for rapid and automatic immunoassay methods.[3,4] Homogeneous enzyme immunoassays are more attractive than heterogeneous ones in satisfying the increasing needs. The demands have encouraged the attempts for developing novel homogeneous immunoassay methods with improved sensitivity. One of these attempts, the liposome immunoassay (LIA), has been given considerable attention.[5-7] Liposome immunoassays have been applied to the quantitative determination of drugs,[8] hormones,[9,10] environmental contaminants,[11] blood proteins,[12] and other molecules.[13,14]

[1] A. I. Maclean and L. G. Bachas, *Anal. Biochem.* **195,** 303 (1991).

[2] P. Y. Huang and C. S. Lee, *Biotech. Bioeng.* **40,** 913 (1992).

[3] D. S. Hage, *Clin. Chem.* **65,** 420R (1993).

[4] J. P. Gosling, *Clin. Chem.* **36,** 1408 (1990).

[5] H. A. H. Rongen, A. Bult, and W. P. van Bennekom, *J. Immunol. Methods* **204,** 105 (1997).

[6] A. K. Singh and R. G. Carbonell, *in* "Handbook of Nonmedical Applications of Liposomes" (Lasic, D. D. and Barenholz, Y.), vol. IV, pp. 209–228, CRC Press, Boca Raton, FL, 1996.

[7] D. Martorell, S. T. A. Siebert, and R. A. Durst, *Anal. Biochem.* **271,** 177 (1999).

[8] K. Kubotsu, S. Goto, M. Fujita, H. Tuchiya, M. Kida, S. Takano, S. Matsuura, and I. Sakurabayashi, *Clin. Chem.* **38,** 808 (1992).

[9] H. A. H. Rongen, H. M. van der Horst, G. W. K. Hugenholtz, A. Bult, and W. P. van Bennekom, *Anal. Chim. Acta* **287,** 191 (1994).

[10] G. P. Vonk and D. B. Wagner, *Clin. Chem.* **37,** 1519 (1991).

[11] M. A. Roberts and R. A. Durst, *Anal. Chem.* **67,** 482 (1995).

[12] C. K. Kim and S. J. Lim, *J. Immunol. Methods* **159,** 101 (1993).

[13] S. Bystryak, I. Goldiner, A. Niv, A. M. Nasser and L. Goldstein, *Anal. Biochem.* **225,** 127 (1995).

Liposome immunoassays are based on the lysis of liposomes in response to the attack of activated complement,[15–17] cytolysins,[18] and enzymes.[19,20] Because of the separation of signal-producing molecules (detectors) inside the liposomes, LIAs can be performed in a single assay mixture. Encapsulation of a large number of markers in liposomes allows the amplification of signals, resulting in a lowered detection limit. Moreover, these novel methods can allow the quantitative determination of analytes whose molecular weights are higher than haptens (analytes) that can be assayed by conventional homogeneous enzyme immunoassays.[20]

In homogenous LIAs, enzymes such as alkaline phosphatase and phospholipase C can be used as (1) detectors entrapped in liposomes or (2) liposome lytic agents on antigen-antibody binding.

This chapter describes two different types of LIA methods that use complement and phospholipase C as liposome lytic agents. In the complement-mediated LIA format (1) a method preparing analyte-conjugated liposomes entrapping alkaline phosphatase and (2) an LIA method using these immunoliposomes by complement are described. In the phospholipase C–mediated LIA format (1) the analyte–phospholipase C conjugation method, (2) preparation of phospholipase C–labile liposomes, and (3) the LIA method using analyte–phospholipase C conjugates are described.

Preparation of Analyte-conjugated Liposomes Entrapping Alkaline Phosphatase

Analytes should be coupled to the surface of liposomes to prepare immunoreactive liposomes for performing complement-mediated LIA. The conjugation can be achieved by use of the thioether bond formation method with heterobifunctional reagents.[21] Maleimide derivatives of lipids react with thiols to produce a thioether bond at neutral pH. These reactions can provide high coupling efficiency and bond stability. The overall conjugation procedure is described in Fig. 1.

[14] S. Yamamoto, K. Kubotsu, M. Kida, K. Kondo, S. Matsuura, S. Uchiyama, O. Yonekawa, and T. Kanno, *Clin. Chem.* **41,** 586 (1995).
[15] K. Tomioka, F. Kii, H. Fukuda, and S. Katoh, *J. Immunol. Methods* **176,** 1 (1994).
[16] Y. Sohma, R. Fujita, S. Katoh, and E. Sada, *Appl. Biochem. Biotech.* **38,** 179 (1993).
[17] C. K. Kim and S. J. Lim, *J. Immunol. Methods* **159,** 101 (1993).
[18] M. Haga, S. Hoshino, H. Okada, N. Hazemoto, M. Y. Kato, and Y. Suzuki, *Chem. Pharm. Bull.* **38,** 252 (1990).
[19] C. K. Kim and K. M. Park, *J. Immunol. Methods* **170,** 225 (1994).
[20] S. J. Lim and C. K. Kim, *Anal. Biochem.* **247,** 89 (1997).
[21] L. D. Leserman, J. Barbet, F. M. Kourilsky and J. N. Weinstein, *Nature* **288,** 602 (1980).

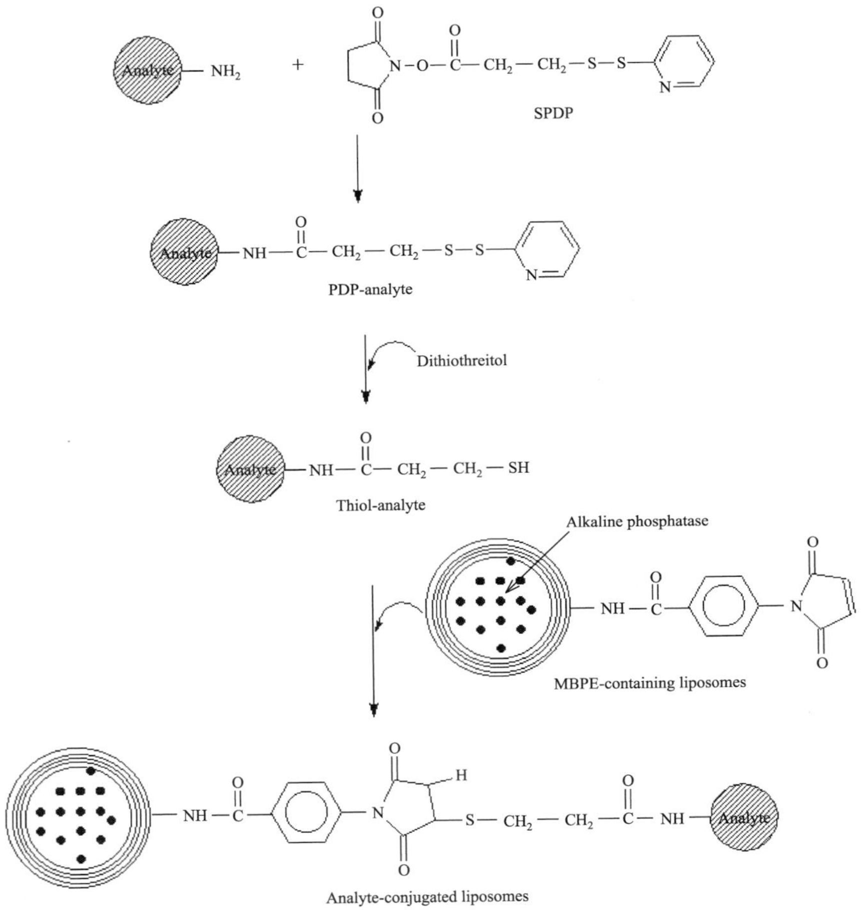

FIG. 1. Overall scheme for conjugation of analytes to the surface of liposomes.

Activation of Phospholipid Derivatives

For the preparation of analyte-conjugated liposomes, phospholipid derivatives are first activated by slight modification of the reported method.[22] Eighty miligrams of dipalmitoylphosphatidylethanolamine (DPPE) (Avanti Polar Lipids, Alabaster, AL) are dried by rotary evaporation in a round-bottom flask and redissolved in 32 ml of dried chloroform (dried over a

---

[22] F. J. Martin and D. Papahadjopoulos, *J. Biol. Chem.* **257**, 286 (1982).

molecular sieve type 4A). Forty-eight milligrams of m-maleimidobenzoyl N-hydroxysuccinimide ester (MBS) (Sigma, St. Louis, MO) dissolved in 4 ml of dried methanol containing triethylamine (200 μmol) are added to the flask. The resulting reaction is allowed to proceed under nitrogen gas at room temperature until completion. The progress of the reaction can be checked by thin-layer chromatography using silica gel (Merck) developed in chloroform/methanol/water (65/25/4). The derivative moves further ($R_f = 0.63$) than free DPPE ($R_f = 0.44$). The organic solvent is then removed under reduced pressure, and the product is redissolved in chloroform. The product is immediately applied to a silica gel column (1.7 × 50 cm) equilibrated with chloroform. The column is washed with 10 ml of chloroform and eluted with 12 ml portions of chloroform/methanol mixtures, 16/1, followed by 16/2, 16/3, and finally by 16/4. The fractions containing the desired product, maleimidobenzoyl-phosphatidylethanolamine (MBPE), are pooled after checking by thin-layer chromatography and concentrated by evaporation at reduced pressure in a rotary evaporator. The product is redissolved in chloroform and stored at 4° under nitrogen.

Activation of Analytes

Efficient conjugation of protein analytes to liposomes requires the thiolation of proteins before being conjugated to the MBPE-containing liposomes. The reagent 3-(2-pyridyldithio)propionic acid N-hydroxysuccinimide ester (SPDP) (Pierce, Rockford, IL) is used to introduce these thiols to analytes, which can be made reactive by subsequent reduction with dithiothreitol (DTT).[23] Thiols are quite unstable, because they are prone to oxidation. It is, therefore, essential to generate thiols immediately before conjugation. Analytes are dissolved in 0.1 M phosphate buffer (Na_2HPO4 81 mM, NaH_2PO_4 19 mM, NaCl 100 mM, pH 7.4) at a concentration of 20 mg/ml. SPDP (20 μmol/ml in dried methanol) is slowly added to the analyte solution with stirring (20:1 molar ratio of SPDP/analyte). The resulting reaction is allowed to proceed for 30 min at room temperature. The product is then separated from the mixture by gel chromatography on a Sephadex G-75 column equilibrated with citrate-phosphate buffer (pH 7.0, $C_6H_5P_7\ Na_3$, 50 mM Na_2HPO4, 50 mM NaCl, pH 7.0). The protein concentration of fractions is checked by reading the absorbance at 280 nm. The protein fractions are pooled, concentrated, and stored at 4°.

Just before the conjugation with liposomes, PDP-analytes are reduced to thiol-analytes. The PDP-analyte solution is adjusted to pH 4.5 by the addition of 0.1 N HCl and then treated with dithiothereitol to reach a final

---

[23] P. N. Shek and T. D. Heath, *Immunology* **50,** 101 (1983).

concentration of 25 mM. The mixture is incubated with stirring under nitrogen gas for 30 min at room temperature. The reduced protein is separated by gel chromatography on a Sephadex G-50 column equilibrated with TES buffer.

Preparation of Liposomes

Liposomes are prepared by reverse-phase evaporation[24,25] with minor modifications. Lipid mixtures composed of dipalmitoylphosphatidylcholine (DPPC), MBPE, and cholesterol (molar ratio of 12:11:2) (Avanti Polar Lipids) are dissolved in 3 ml of chloroform prewarmed to 50°. The organic solvent is removed using a rotary evaporator at reduced pressure. The dried lipid film is then redissolved in 4.5 ml of a 1:2 mixture of chloroform/isopropyl ether. One mililiter of TES buffer (10 mM TES, 135 mM sodium chloride, pH 6.7) containing 5 mg of alkaline phosphatase is then added, and the two phases are emulsified by vortex mixing. The emulsion is left for 30 min at room temperature. The organic phase is then removed under reduced pressure at room temperature for 25 min. The resulting lipid vesicles are centrifuged at 30,000g for 20 min, and the pelleted vesicles are resuspended in 1 ml of TES buffer.[26]

The redispersed liposomes are then incubated with 10 mg/ml of trypsin in deionized water for 1 h at 37° to digest the alkaline phosphatase associated with the outer surface of the vesicles or remaining free. These enzyme molecules should be removed or inactivated to decrease the noise when LIAs are carried out using these liposomes. After digestion, vesicles are centrifuged at 30,000 g to remove the digested enzymes. The final liposomal pellets are resuspended to the original volume in the same TES buffer.

Conjugation of Analytes to Liposomes

Immunoliposomes, analyte-conjugated liposomes, are prepared by incubation of liposomes (3.15 μmol/ml) with thiol-analytes (780 μg/ml) as described in Fig. 1. After overnight incubation at 4°, immunoliposomes are separated from unbound analytes by centrifugation at 30,000 g, and the pellets, analyte-conjugated liposomes, are resuspended to the original volume. The amount of protein analytes coupled to the liposomes is determined according to the Lowry method[27] in the presence of sodium deoxycholate. The amount of conjugated analytes can be controlled by

[24] F. Szoka and D. Papahadjopoulos, *Proc. Natl. Acad. Sci. USA* **75**, 4194 (1978).
[25] C. K. Kim and B. J. Lee, *Arch. Pharm. Res.* **10**, 110 (1987).
[26] C. K. Kim, H. S. Kim, B. J. Lee, and J. H. Han, *Arch. Pharm. Res.* **14**, 336 (1991).
[27] O. H. Lowry, N. J. Rosebrough, A. L. Farr, and R. J. Randall, *J. Biol. Chem.* **193**, 265 (1951).

changing the analyte/liposome ratio in the initial reaction mixture. In the case of bovine serum albumin, approximately 55 μg of albumin is coupled to 1 μmol of liposomes after 12 h of incubation.[17]

Conjugation of macromolecular analytes to liposomes by means of thioether linkage as described previously does not cause the loss of any immunoreactivity of immunoreactants. The linkage is also quite stable during storage.

Complement-Mediated Lysis of Liposomes

Before performing LIA, the dilution of antibody required for the maximal release of alkaline phosphatase by complement-mediated lysis of analytes-conjugate liposomes should be determined to optimize the assay condition. Aliquots of antiserum (containing anti-analyte antibody) are added to 0.1 M phosphate buffer (pH 7.5) to get appropriate dilutions. Guinea pig complement (Sigma) is inactivated by preincubation at 56° for 30 min. Fifteen microliters of the liposomal suspension is added to the test tubes containing antiserum and guinea pig complement. The mixtures are further diluted to 500 μl with TRIS buffer (TRIS 50 mM, NaCl 150 mM, CaCl$_2$ 0.15 mM, MgCl$_2$ 0.5 mM, pH 7.4). All the mixtures are incubated at 25° for 30 min and then added to 2 ml of a solution containing 1 mg/ml of alkaline phosphatase substrate (Sigma) in borate buffer (pH 9.5). These mixtures are incubated for another 10 min at 25°. The enzyme reaction is terminated by adding 2 ml of 2 N sodium hydroxide, and the absorbance is measured at 410 nm. The absorbance indicates the amount of substrate hydrolyzed by alkaline phosphatase released from inside of liposomes. On antibody binding to analytes conjugated on the surface of liposomes, complement is activated, resulting in the immune-specific lysis of liposomes. An example is shown in Fig. 2.

LIA by Complement-Mediated Lysis of Analyte-Conjugated Liposomes

Homogeneous LIA by complement-mediated lysis of analyte-conjugated liposomes is carried out in a mixture of (1) alkaline-entrapped liposomes labeled with analytes, (2) substrate for alkaline phosphatase, (3) complement, (4) antibody solution, and (5) test sample containing an unknown concentration of analytes.

A standard curve for the analyte is constructed with varying concentrations of standard solution instead of the test sample. In performing the LIA, the components listed previously (1–5) are mixed in the order described in Fig. 3. As the analyte concentration in the test sample increases,

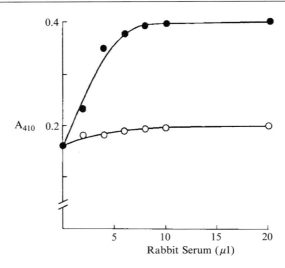

FIG. 2. Effect of antiserum concentration on the immune lysis of analyte-conjugated liposomes in the presence of complement. As an analyte, bovine serum albumin is used. Albumin-conjugated, alkaline phosphatase–entrapped liposomes are mixed with varying dilutions of antiserum containing anti-analyte antibody (●) or normal serum without anti-analyte antibody (○). The mixtures are incubated as described in the text, and the absorbance is read at 410 nm, after adding alkaline phosphatase substrate (Reproduced from the *Journal of Immunological Methods* **159,** 101–106, 1993. Reprinted with permission from Elsevier Science).

the amount of antibody available for analytes on the liposomal surface decreases, leading to a lower release of alkaline phosphatase from liposomes. Therefore, the absorbance decreases linearly as the analyte concentration in the test sample increases, enabling quantitative determination of analytes in the samples.

This method does not require any separation procedure and involves no radiation hazard. Because the test results can be determined easily in a single step, the assay can be carried out by relatively untrained personnel. It should be noted, however that nonspecific lysis of liposomes can sometimes occur for many reasons. The incubation time with substrates should also be controlled carefully to avoid the diffusion of substrates into the liposomes.

Preparation of Analyte-Phospholipase C Conjugates

Phospholipase C from *Clostridium perfringens* binds and hydrolyzes phospholipids such as phosphatidylcholine in the presence of calcium.[28,29] Because the hydrolysis of phospholipids can result in the disruption of

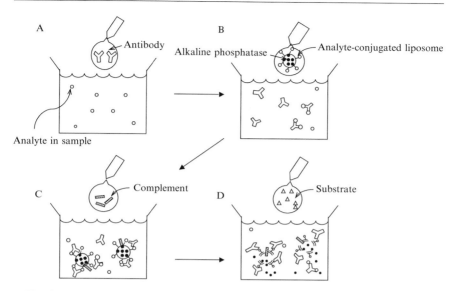

FIG. 3. Homogeneous complement-mediated LIA proccedure. (A) Antibody is added to the test sample containing unknown concentration of analyte. (B) Analyte-conjugated, alkaline phosphatase-entrapping liposomes are added. (C) Complement is added. (D) After adding alkaline phosphatase substrate, lysis of liposomes (enzyme release) is detected by measuring the absorbance at 410 nm.

liposomal structure, phospholipase C activity could be quantified by using marker-entrapped small unilamellar vesicles composed of egg phosphatidylcholine.[30] Using the liposome lytic activity of phospholipase C, a homogeous LIA method was designed.[19,20]

The homogeneous LIA method using phospholipase C, instead of complement, can be performed without the tedious step of preparing analyte-conjugated liposomes. It is also the advantage that analyte-conjugated phospholipase C is more stable during storage compared with analyte-conjugated liposomes. Phospholipase C also has some advantages over melitin, another liposome lytic agent that has been used frequently in LIA. These include better solubility, easier separation after conjugation, safety when handling, and lowered cost of assay.

[28] P. R. Young, W. R. Snyder, and R. F. Mcmahon, *Biochem. J.* **280,** 407 (1991).

[29] R. A. Demel, W. S. M. Kessel, Geurts Van Kessel, R. F. A., Zwaal, B. Roelofen, and L. L. M. van Deenen, *Biochim. Biophys. Acta* **406,** 97 (1975).

[30] T. B. Buxton, B. Catto, J. Horner, R. Yannis, and J. P. Rissing, *Microchem. J.* **34,** 349 (1986).

Homogeneous LIA using phospholipase C allowed us to lower the detection limit of analytes, which was comparable with those of conventional heterogeneous enzyme immunoassays (EIA). This new assay approach may offer a simple, sensitive, and inexpensive testing procedure for determining ultra-trace amounts of substances in clinical fields and bioprocesses.

Conjugation of Phospholipase C to Analytes by Homobifunctional Cross-Linking Reagents

Homobifunctional cross-linking reagents such as carbodiimide[31] and glutaraldehyde[32] have been used frequently for coupling analytes to enzymes. Twenty miligrams of phospholipase C dissolved in 1.0 ml of PBS (0.1 M, pH 6.4) is mixed with 7 mg of 1-ethyl-3-(3-dimethylaminopropyl)-carbodiimide and 14 mg of N-hydroxysuccinimide (Sigma). The mixtures are incubated for 30 min at 4° with a magnetic stirrer. Fifty miligrams of analytes dissolved previously in 0.5 ml of PBS (0.1 M, pH 6.4) are added dropwise to the activated phospholipase C solution. The reaction mixtures are left overnight at 4° and then separated by gel chromatography on a Sephadex G-25 column (Pharmacia LKB Biotechnology, Uppsala, Sweden) equilibrated with PBS (pH 7.4). Fractions containing the desired products are pooled and concentrated by rotary evaporation at reduced pressure. The conjugates obtained are dialyzed against distilled water and then kept at 4°.

Conjugation of Phospholipase C to Analytes by Heterobifunctional Cross-Linking Reagents

Homobifunctional reagents tend to cause homopolymerization of enzymes. On the other hand, heterobifunctional cross-linking reagents, such as succinimidyl-4-(N-maleimidomethyl)cyclohexane-1-carboxylate (SMCC) and SPDP, can conjugate phospholipase C to analytes without homopolymerization or significant loss of enzymatic activity. This conjugation method can be applied for many kinds of analytes containing functional amine groups (Fig. 4).

Introduction of Maleimide Groups to Analytes. Maleimide labeling of analytes is performed by a method reported by Yoshitake *et al.*[33] with slight modification. Here we describe the conjugation procedure for insulin as a

[31] V. P. Torchilin and A. L. Klibanov, *Enzyme Microbial. Technol.* **3**, 27 (1981).
[32] V. P. Torchilin, B. A. Khaw, V. N. Smirnov, and E. Haber, *Biochem. Biophys. Res. Commun.* **89**, 1114 (1979).
[33] S. Yoshitake, M. Imagawa, E. Ishikawa, Y. Niitsu, I. Urushizaki, M. Nishiura, R. Kanazawa, H. Kurosaki, S. Tachibana, N. Nakazawa, and H. Ogawa, *J. Biochem.* **92**, 1413 (1982).

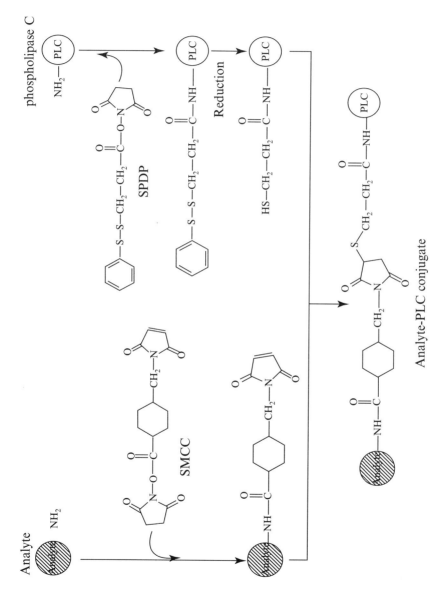

Fig. 4. Conjugation chemistry of phospholipase C to the analyte by use of the heterobifunctional reagents SMCC and SPDP.

model analyte. Fifty miligrams of bovine insulin (Sigma) (6 mg/ml) is dissolved in PBS (14 mM KH$_2$PO4, 57 mM Na$_2$HPO4, 70 mM NaCl, pH 7.4). SMCC (Pierce) in 2–10-fold molar excess of insulin is dissolved in freshly distilled dimethylformamide with slight heating (33 mg/ml) immediately before use and then added slowly to the insulin solution during a period of 15 min. The reaction mixture is left stirring for an additional 30 min. The unreacted cross-linking reagents are removed by Sephadex G-50 chromatography using the same buffer. The fractions corresponding to the first peak are pooled and stored at $-80°$ after concentration. No significant degradation is observed during storage at $-80°$.

Determination of the Number of Maleimide Groups Introduced to Analytes. The analyte concentration is determined by the Lowry method,[27] and the number of maleimide groups introduced per molecule is determined as follows.[34] One hundred fifty microliters of a known amount of derivatized analytes in 100 mM sodium phosphate buffer (pH 6.0) and 50 μl of 2.5 mM 2-mercaptoethylamine in the same buffer are mixed in a final volume of 300 μl. The mixture is left for 15 min at $30°$, and the fraction of thiol groups remaining free is determined by the method of Grassetti and Murray.[34] Two hundred microliters of 0.88 mM 4,4'-dithiodipyridine is added to the mixture and incubated in 2.5 ml of final volume of 100 mM PBS, pH 7.0, for 20 min at room temperature. Absorbance at 324 nm for both treated and control samples is read, and the decrease in the number of sulfhydryl groups caused by maleimide derivatization is calculated (molar extinction coefficient of 19,800 for the modified, pyridine-4-thione, thiol groups).[35]

The numbers of coupled maleimidyl groups increase linearly with increasing SMCC/analyte ratio, as shown in Table I. Therefore, it can be controlled by modulating the SMCC/analyte ratio in the initial reaction mixture.

Thiolation of Phospholipase C. Phospholipase C obtained from Sigma is dissolved in a minimum volume of PBS buffer and passed through a Sephadex G-75 column (2 × 60 cm) equilibrated with the same buffer. The peak of protein fractions indicating liposome-lytic activity are pooled and concentrated to approximately 5 ml. A 10-fold molar excess of SPDP in absolute ethanol (15 mg/ml) is added rapidly to this solution, but dropwise, with stirring, and allowed to stand for 30 min at $25°$ with occasional stirring. The excess reagents and low-molecular-weight by-products are removed by

[34] P. Tijssen, *in* "Practice and Theory of Enzyme Immunoassays" (R. H. Burdon and P. H. Knippenberg), pp. 221–278, Elsevier Co., the Netherlands, 1985.
[35] D. R. Grassetti and J. F. Murray, *Arch. Biochem. Biophys.* **119,** 41 (1967).

TABLE I
Effect of Molar Ratio of SMCC to Insulin in the Reaction
Mixture on the Introduction of Maleimidyl Group to Insulin.

SMCC:insulin in reaction mixture (molar ratio)	Maleimidyl/insulin (number)
1.98:1	0.90 ± 0.35
3.00:1	1.37 ± 0.04
4.03:1	2.46 ± 0.54
10.00:1	5.22 ± 0.00

(Reproduced from the *Analytical Biochemistry* **247**, 89–95, 1997. Reprinted with permission from Academic Press).

gel filtration on a Sephadex G-50 column preequilibrated with the same buffer. The protein peaks are collected by reading the absorbance of fractions at 280 nm. The pooled fractions are concentrated to approximately 2 ml. The amount of 2-pyridyl disulfide groups is determined by measuring the release of thiopyridone, following reduction of the disulfide bond with 50 mM dithiothreitol.[23]

To obtain a thiolated product of phospholipase C, 2-pyridyl disulfide groups are reduced with dithiothreitol at a final concentration of 25 mM. The mixture is left for 30 min at room temperature under nitrogen. The excess reducing agent and by-products are removed by gel filtration on Sephadex G-25. The first peak is collected and used for the subsequent conjugation with analytes as soon as possible.

Under suggested experimental conditions, two dithiopyridyl groups are coupled to one phospholipase C molecule, assuming the molecular weight of phospholipase C as 43,000.[36]

Coupling of Derivatized Phospholipase C and Analytes

The thiolated phospholipase C solution is added to the derivatized analyte solution (in a molar ratio of approximately 1:10). The mixture is concentrated to 3 ml in an Amicon® ultrafiltration cell (Amicon division, Beverly, MA) under nitrogen at room temperature and incubated overnight at 4°. The conjugated products are applied on a Sephadex G-100 (superfine) column (1.5 × 88 cm) preequilibrated and eluted with PBS. The final products collected from the column are concentrated and stored in aliquots at −80° after adding 0.1% bovine serum albumin as a stabilizer.

[36] J. Carlsson, H. Drevin, and R. Axen, *Biochem. J.* **173,** 723 (1978).

Preparation of Phospholipase C-labile Liposomes

Liposomes are prepared by the reverse-phase evaporation method[24] to enable the encapsulation of a large amount of markers. Dipalmitoylphosphatidylcholine (DPPC) (Avanti Polar Lipids) and cholesterol in a molar ratio of 9:1 (total 7.5 μmol) are dried to a thin film using a rotary evaporator and then dissolved in 1.5 ml of 2:1 mixture of isopropyl ether and chloroform. Then, 0.5 ml of TRIS buffer (10 mM TRIS, pH 7.4) containing 100 mM calcein is added dropwise, and the two phases are vortexed briefly to suspend the lipid. The suspension is sonicated in a bath-type sonicator for 10 min at 37°. The organic solvent is removed under slightly reduced pressure until a clear suspension is obtained. Half a milliliter of TBS buffer (10 mM TRIS, 150 mM sodium chloride, pH 7.4) is added again, and the residual organic solvent is eliminated under greatly reduced pressure. The resulting liposomal dispersions stand for 30 min at room temperature and are sized by extrusion through polycarbonate filters (0.1-μm pore size, Nuclepore, Costar, Cambridge, MA) assembled in an ultrafiltration cell under nitrogen pressure at a slightly raised temperature. The sized liposomes are eluted on a Sepharose CL-4B column equilibrated with TBS buffer to remove the unentrapped calcein. Liposome fractions are collected and stored in aliquots at 4° until use.

Liposome Immunoassay Method Using Analyte-Phospholipase C Conjugates

Lysis of Liposomes by Phosholipase C and Analyte Conjugate

To assess the enzymatic activity of phospholipase C conjugates, the liposome lytic activity of phospholipase C before and after conjugation should be determined. The concentration of conjugates is determined by the Lowry method.[27] Aliquots of phospholipase C before and after conjugation are mixed with the liposome solution in PBS buffer (14 mM KH$_2$PO4, 57 mM Na$_2$HPO4, 70 mM NaCl, pH 7.4). The resulting mixtures are kept at room temperature for 1 h. The fluorescence intensity of calcein released from liposomes is determined by fluorometry at 494 nm excitation wavelength and 520 nm emission wavelength.

The percentage of lysis is calculated from Equation 1.

$$\% \, \text{lysis} = (F_S - F_R)/(F_T - F_R) \times 100 \tag{1}$$

where F_S is the fluorescence intensity of the sample solution after lysis and F_R is the fluorescence intensity of the reference solution that contains 5 μl of liposome solution and 240 μl of buffer. F_T is the total fluorescence

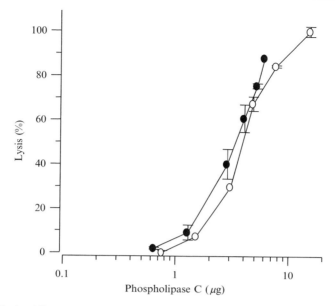

FIG. 5. Lysis of liposomes by phospholipase C before (○) and after (●) conjugation with insulin. Insulin was used as an analyte and thus conjugated to phospholipase C as described in the text. (Reproduced from *Analytical Biochemistry* **247,** 89–95, 1997. Reprinted with permission from Academic Press).

intensity when liposomes are ruptured completely by excess phospholipase C or Triton X-100 (final concentration, 0.2%, w/v). As shown in Fig. 5, the lytic activity of phospholipase C before and after conjugation to insulin is not significantly different. The slight difference of lytic activity of conjugate might be due to the further purification of phospholipase C on a Sephadex column during the conjugation procedure.

Determination of Antiserum Dilution for Performing Liposome Immunoassay

The basic principle of the homogeneous enzyme immunoassay relies on the modulation of enzymatic activity of enzyme-hapten conjugates on antibody binding. Although the mechanism by which the antibody modulates the activity of the analyte-phospholipase C conjugate is not known clearly, perhaps the binding of antibody to phospholipase C conjugate inhibits sterically the exposure of active sites of phospholipase C.

To optimize the homogeneous LIA condition, the antibody dilution appropriate for constructing standard curves should be predetermined. Fifty

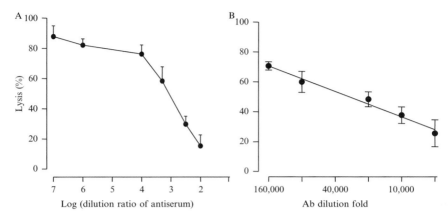

FIG. 6. Effect of antiserum dilution on the inhibition of analyte-phospholipase C conjugate activity against liposomes. (A) Gentamicin and (B) insulin was used as analytes. (Reproduced from the *Journal of Immunological Methods* **170,** 225–231, 1994 (A) and reprinted with permission from Elsevier Science; and from *Analytical Biochemistry* **247,** 89–95, 1997 (B) and reprinted with permission from Academic Press).

to 100 μl antiserum solution prediluted with PBS and 4.9 μg of the analyte-phospholipase C conjugate is mixed in a final volume of 240 μl. The mixture is incubated for 3 h at 37°. Five microliters of liposomes is added to each reaction tube and then incubated for an additional 1 h. The fluorescence intensity of calcein released from liposomes is measured as described in the previous section. For example, the appropriate antibody dilutions for performing LIA with phospholipase C conjugates to gentamicin and insulin are shown in Fig. 6.

Homogeneous Liposome Immunoassay Procedure

Analyte standards are prepared by diluting the analyte solution (known concentration) with PBS (pH 7.4). Gentamicin standards are prepared in the concentration range from 0.1 ng/ml to 10 μg/ml. In case of insulin as an analyte, insulin standards are prepared in the range of 1.7~331 μIU/ml. Five microliters of the standard solution are diluted 20-fold with PBS and mixed with 100 ml of a predetermined dilution of antiserum, as described in the previous section (100-fold-diluted antiserum in case of gentamicin, and 5000-fold-diluted antiserum in case of insulin). The mixture is incubated for 30 min at 37°. Then, 4.9 μg of conjugates are added, and the resulting mixtures are incubated in a final volume of 240 μl PBS for 3 h at 37°. Next, 5 μl of liposomes are added to the mixture and incubated for an additional

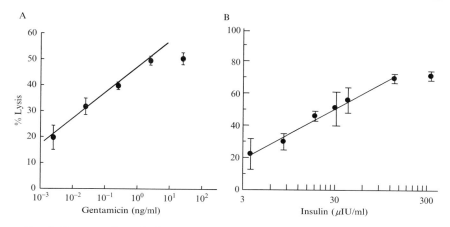

FIG. 7. Constructed standard curve for (A) gentamicin and (B) insulin by phospholipase C–mediated LIA as described in the text. (Reproduced from the *Journal of Immunological Methods* **170,** 225–231, 1994 (A) and reprinted with permission from Elsevier Science; and *Analytical Biochemistry* **247,** 89–95, 1997 (B) and reprinted with permission from Academic Press).

1 h. Increasing concentration of free analytes competes with analytes conjugated to phospholipase C for antibody binding. Therefore, increasing the concentration of free analytes can inhibit the binding of antibody on the phospholipase C–analyte conjugates, enabling the phospholipase C conjugates to attack the liposomes. As a result, the fluorescence intensity increases with increasing concentration of the analytes. The percentage of lysis at varying concentrations of analyte standards can be calculated and plotted against the concentration of standards. Examples are shown in Fig. 7. The standard curve could be obtained in the range of 0.0025–2.5 ng/ml of gentamicin ($r = 0.997$), and in the range of 4–130 μIU/ml of insulin ($r = 0.994$). The detection limits of gentamicin are known to be 1 μg/ml, 0.2 μg/ml, and 5 ng/ml by conventional enzyme immunoassay,[37] bioassay,[38] and high-performance liquid chromatography.[39] Therefore, the detection limit obtained by this LIA method (approximately 10 pg/ml) is much lower than that with other assay methods. In the case of insulin, the detection limit (8 μIU/ml) was comparable with those of

[37] J. E. Gorsky, P. R. Bach, and W. Wong, *Clin. Chem.* **29,** 1994 (1983).
[38] J. Strassburger, *Z. Klin. Med.* **41,** 634 (1986).
[39] M. Nakano, *Hokkaidoritsu Eisei Kenkyushoho* **36,** 78 (1986).

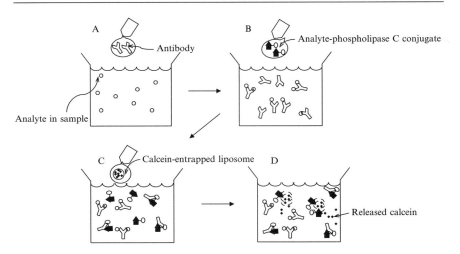

FIG. 8. Phospholipase C–mediated homogeneous LIA procedure. (A) Antibody is added to the test sample containing unknown concentration of analytes. (B) Analyte–phospholipase C conjugates are added. (C) Calcein-containing liposomes are added. (D) Calcein is released from lysed liposomes, and the fluorescence intensity is measured.

the conventional heterogeneous enzyme immunoassays.[40,41] The coefficients of variation in the intra-assay and interassay with this LIA system were within 8% and 10%.

For performing homogeneous LIA using phospholipase C–analyte conjugates, the following reagents are required as final components of the LIA kit:

1. Analyte-containing test samples
2. Antibody with appropriate dilution
3. Phospholipase C–analyte conjugates
4. Liposomal aliquots

The overall assay procedure is carried out as described in Fig. 8.

Conclusion

This chapter summarizes the homogeneous LIA methods using alkaline phosphatase as a marker or using phospholipase C as a liposome lytic agent. Although the assessment studies such as cross-reactivities, recovery

[40] E. L. Krug and C. Kent, *Arch. Biochem. Biophys.* **231,** 400 (1984).
[41] H. Shinkai, M. Sohma, Y. Takahashi, R. Kojima, M. Hashimoto, and N. Ogawa, *Mol. Immunol.* **17,** 377 (1980).

test, and application for many serum samples are still required and the improvement of stability of liposomes in human serum are required for the practical application of this LIA system, the sensitivity of homogeneous LIA is comparable or somewhat better than those obtained by conventional heterogeneous techniques, suggesting a potential as a novel homoegeneous immunoassay technique.

Section III

Liposomes in Gene Delivery and Gene Therapy

[18] Lipoplex Assembly Visualized by Atomic Force Microscopy

By VOLKER OBERLE, UDO BAKOWSKY, and DICK HOEKSTRA

Introduction

The era of genomics has led to an increased interest in experimental approaches to deliver DNA into eukaryotic cells, a process also known as "transfection." The purpose of such applications may range from substituting malfunctioning genes in cells ("gene-therapy") to expressing novel genes for fundamental cell biological investigations. Over the years, the development of an appropriate vehicle for the effective (intra-)cellular delivery of the gene(s) of interest has represented a major challenge. In fact, researchers embarking in this particular field were aware of the existence of a highly effective *natural* gene delivery system (i.e., viruses).[1] Unfortunately, such virus-based delivery systems have a number of serious disadvantages, such as their ability to effectively trigger an immune response,[2] which will make repeated administration ineffective. In addition, although their reproduction machinery is inactivated and only the delivery properties of the virus are exploited, the risk of recombinant viral particle formation cannot be excluded. Also, the size of the genes that can be delivered by such viral particles is limited, whereas large-scale production of the viral carriers creates a considerable economic burden. Therefore, alternatives have been developed such as DNA-containing particles that are assembled on mixing DNA with polymers ("polyplexes") or cationic lipids ("lipoplexes"). Although both these complexes do not suffer from most of the disadvantages noted before, their transfection efficiencies are very modest relative to a viral vectors, and further improvement is highly desirable. An essential part of the strategy to accomplish this will involve the need for a better understanding of the physicochemical properties of the particles used. This knowledge is crucial in the molecular design of particles that should be refractory to modulation by the physiological environment (serum, lymph) and simultaneously display an improved efficiency of gene-complex processing by the cells.

Here, we will focus on the application of atomic force microscopy as a useful tool in the characterization of gene complexes composed of plasmid

---

[1] W. H. Gunzburg and B. Salmons, *Mol. Med. Today* **1,** 410 (1995).
[2] N. Somia and I. M. Verma, *Nat. Rev. Genet.* **1,** 91 (2000).

DNA and cationic lipids.[3,4] Because of their positive charge, cationic lipids interact readily with the negatively charged DNA, resulting in the assembly of the so-called lipoplexes. When added to cells in culture, these complexes can be internalized, release their DNA, and, eventually, the genes encoded by the DNA are expressed. Hundreds of different cationic amphiphiles have been synthesized; the vastness of this activity is rationalized by the notion that a clear structure–function relationship is as yet not defined clearly. Although some important biophysical parameters of the cationic lipids that may drive distinct steps in the overall transfection process are gradually revealed,[5,6] many others still remain to be determined. For characterization of parameters such as the effect of heterogeneity in size and morphology,[7] microscopic techniques of sufficient resolution will be highly useful, because they may provide further insight into the mechanism of lipoplex assembly. Atomic force microscopy (AFM) offers such possibilities. In this chapter we will describe the application of this versatile technique in studies involving the characterization of the lipoplex assembly.

Some Principles of the Use of Atomic Force Microscopy

In atomic force microscopy a small needle of atomic size scans over the surface of a sample. The interaction between the needle and the surface is measured, and an image is reconstructed from the data collected in this manner. With AFM, it is possible to reach an extremely high resolution and, because it can be applied under standard conditions in an aqueous environment, any significant perturbation of the sample can be avoided. In contrast to light microscopy and (scanning) electron microscopy, AFM provides the most optimal means to investigate the topography of surface structures in three dimensions, with a resolution as high as 0.1–0.2 nm. Because the geometry of the needle or tip represents a major parameter that determines the resolution of the measurement, a key element of the microscope is the tip–cantilever system. This system is integrated into a stylus profilometer, which allows the tip to be moved over the sample surface. As a result of such a movement, the attractive and repulsive forces between

[3] J. H. Felgner, R. Kumar, C. N. Sridhar, C. J. Wheeler, Y. J. Tsai, R. Border, P. Ramsey, M. Martin, and P. L. Felgner, *J. Biol. Chem.* **269,** 2550 (1994).

[4] S. Audouy and D. Hoekstra, *Mol. Mem. Biol.* **18,** 129 (2001).

[5] I. Koltover, T. Salditt, J. O. Radler, and C. R. Safinya, *Science* **281,** 78 (1998).

[6] J. Smisterova, A. Wagenaar, M. C. Stuart, E. Polushkin, G. ten Brinke, R. Hulst, J. B. F. N. Engberts, and D. Hoekstra, *J. Biol. Chem.* **276,** 47615 (2001).

[7] D. Simberg, D. Danino, Y. Talmon, A. Minsky, M. E. Ferrari, C. J. Wheeler, and Y. Barenholz, *J. Biol. Chem.* **276,** 47453 (2001).

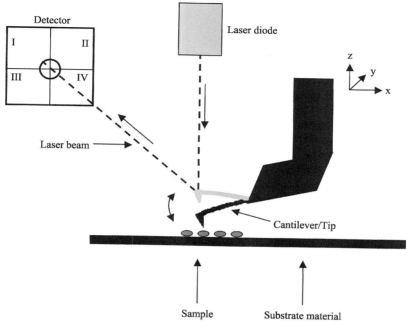

Fig. 1. Schematic presentation of an AFM system with a movable cantilever. When scanning a sample, the vertical movement of the tip is detected by the change in reflection of a laser beam, which is focussed on the backside of the cantilever. In the tapping mode, the tip makes in addition an independent periodic vertical movement at defined frequencies. From both the vertical displacement of the tip and changes in frequency, digital images are generated, which provide information about height or friction of the sample, allowing a three-dimensional reconstruction of the surface architecture of the material examined.

the tip and surface will change, and the ensuing forces are measured by sensing the deflections of the cantilever (Fig. 1).

In principle, a number of different scanning modes can be distinguished that differ primarily in the type of interaction between the tip and the sample being examined. The most relevant techniques for the visualization of surface morphologies are contact AFM, TappingMode™ AFM, and lateral force microscopy. TappingMode™ AFM (Digital Instruments, Santa Barbara, CA) imaging is a key advance in AFM of soft, adhesive, or fragile samples, as in the case of biological materials. This scanning mode overcomes many of the problems encountered in the contact mode and allows in particular the imaging of soft samples with high resolution, because the tip does not come into direct contact with the material, thereby precluding damage of the sample.

The approach can be applied for examination of surfaces at air interfaces but also of liquid-covered surfaces, without the need for any specific sample preparation. The latter is clearly of additional advantage, compared with the usually laborious and expert procedures necessary for electron microscopic studies.

Specifics of the Microscope and Tips/Cantilever System

We have used the scanning mode technique to analyze plasmid DNA, liposomes prepared from cationic lipids, and the resulting lipoplexes obtained when vesicles and plasmids are mixed. All experiments are performed with a Digital Nanoscope IIIa Dimension 5000 (Digital Instruments, Santa Barbara, CA), as described elsewhere.[8] The microscope is vibration-damped. AFM tips, commercially available pyramidal Si_3N_4 tips (NCH-W, Digital Instruments, Santa Barbara, CA) are used on a cantilever with a length of 125 μm. The resonance frequency is about 220 kHz, and the nominal force constant of the cantilever is 36 N/m. The scan speed is proportional to the scan size, and the scan frequency is between 0.5 and 1.5 Hz. Images are obtained by displaying the amplitude signal of the cantilever in the trace direction and the height signal in the retrace direction; both signals are simultaneously recorded. The results are visualized either in the height or in the amplitude mode.

Requirements for Sample Substrates

A proper sample preparation and the selection of a suitable substrate for deposition of the sample of interest are crucial for high-resolution and artifact-free scanning force microscopy. The substrate must be flat and as smooth as possible, so that the surface roughness is substantially less than that of the size of objects under investigation. Substrate materials that have been successfully used in AFM include freshly cleaned surfaces of highly oriented pyrolytic graphite, mica, evaporated or single crystal gold, glass slides, and silicon wafers.[9–12] It is particularly relevant that the interaction between the sample and the substrate surface is not too tight to avoid artifacts of the morphology of the sample because of tight sample–substrate

[8] V. Oberle, U. Bakowsky, I. S. Zuhorn, and D. Hoekstra, *Biophys. J.* **79**, 1447 (2000).

[9] T. Thundat, D. P. Allison, R. J. Warmack, and T. L. Ferrell, *Ultramicrosc.* **42**, 1101 (1992).

[10] V. J. Morris, *Prog. Biophys. Molec. Biol.* **61**, 131 (1994).

[11] O. Marti, V. Elings, M. Haugan, C. E. Bracker, J. Schneir, B. Drake, S. A. Gould, J. Gurley, L. Hellemans, and K. Shaw, *J. Microsc.* **152**, 803 (1988).

[12] M. Radmacher, J. P. Cleveland, M. Fritz, H. G. Hansma, and P. K. Hansma, *Biophys. J.* **66**, 2159 (1994).

interactions. On the other hand, the interaction has to be strong enough to prevent detachment of the sample during the scanning process. A hydrophilic surface should be chosen for examination of biological samples, because hydrophilicity will reduce the contact angle of the wetting fluid that covers the sample and the surface. Materials of choice include mica and a chemically modified mica, which have been applied for the characterization of both DNA,[9,13–16] and lipids.[17,18] In addition, the use of silicon wafers as substrate for DNA, DNA gene transfer complexes, and lipids has also been described.[8,19,20]

For sample recruitment and visualization of the plasmids and lipoplexes prepared in bulk phase, or in monolayer systems like Langmuir-Blodgett troughs, as will be described later, unmodified hydrophilic silicon wafers can be used with a surface roughness of 0.3 nm (1 μm^2) and a contact angle of 45° (water). For the Langmuir-Schäfer transfer, we used silanized (C18 phase) silicon wafers with a contact angle of 112°.

The silicon wafers (Wacker Chemie, Burghausen, Germany), are silanized by placing them for 1 h at room temperature in a solution of 1 mM dimethoxy-hexadecyl-chlorosilane in chloroform. Subsequently the wafers are washed extensively with chloroform and heated at 105° for 2 h in a stove. The thickness of the resulting silane layer is 2.4 nm, and the contact angle in relation to water is greater than 100°.

Characterization of Plasmids and Cationic Lipid Vesicles

To obtain insight into the properties of lipoplexes by a microscopic approach, it is essential to first define the exact morphology of the pure compounds, in this case plasmid DNA and the liposomes. In 1991,

[13] H. G. Hansma, J. Vesenka, C. Siegerist, G. Keldermann, H. Morrett, R. L. Sinsheimer, V. Elings, C. Bustamante, and P. K. Hansma, *Science* **256**, 1180 (1992).
[14] H. G. Hansma, R. L. Sinsheimer, M. Q. Li, and P. K. Hansma, *Nucleic Acids Res.* **20**, 3585 (1992).
[15] H. G. Hansma, M. Bezanilla, F. Zenhausern, M. Adrian, and R. L. Sinsheimer, *Nucleic Acids Res.* **21**, 505 (1993).
[16] W. M. Heckl, D. P. E. Smith, G. Binnig, H. Klagges, T. W. Hansch, and J. Maddocks, *Proc. Nat. Acad. Sci. USA* **88**, 8003 (1991).
[17] R. W. Tillmann, U. G. Hofmann, and H. E. Gaub, *Chem. Phys. Lipids* **73**, 81 (1994).
[18] P. A. Ohlsson, T. Tjärnhage, E. Herbai, S. Löfas, and G. Puu, *Bioelectrochem. Bioenerg.* **38**, 137 (1995).
[19] C. Kneuer, M. Sameti, U. Bakowsky, T. Schiestel, H. Schirra, H. Schmidt, and C. M. Lehr, *Bioconjug. Chem.* **11**, 926 (2000).
[20] U. Bakowsky, W. Rettig, G. Bendas, J. Vogel, H. Bakowsky, and U. Rothe, *Phys. Chem. Chem. Phys.* **2**, 4609 (2000).

Hansma et al.[21] published the first high-resolution AFM image of nucleic acids. The most common method for the preparation of DNA samples for AFM involves the evaporation of a droplet of a DNA solution[9,22,23] placed on a proper substrate. Mica or modified mica has been especially popular as a substrate for DNA.[9,14,22–26] However, special care should be taken in preparing the sample, because it has been reported[9] that the final organization of the visualized DNA may very much depend on sample history. Humidity and the stress applied during the evaporation procedure are particularly relevant in this regard. It has further been reported that the use of pure mica can lead to imaging artifacts, and occasionally the DNA might even be removed from the substrate surface by the cantilever's motion.[9] Hence, lack of sample stability and substrate-induced changes to the DNA conformation are of general concern in DNA imaging. Consequently, different methods have been evaluated to overcome these problems.[27] In our studies, we applied a method in which the DNA self-assembles on the silicon substrate. This is accomplished by immersing a silicon wafer for 10 min into a suspension of sample DNA (0.8 μg/ml). At these steady-state conditions, the DNA can adhere to the carrier substrate without any direct mechanical stress. Subsequently, the wafer is removed from the sample solution and examined immediately to avoid dehydration. The most common DNA species used for transfection are plasmids, which may range in size from 3–7 kBp. The structure adopted by plasmids can be open circular, nicked circular, and/or supercoiled. These different structural features can be revealed readily by agarose gel electrophoresis, which showed that the plasmids used here display the supercoiled form. The morphological appearance of such DNA as revealed by AFM is shown in Fig. 2. The dimension of the plasmid indicates a width of approximately 40–50 nm, whereas its average length is approximately 200 nm, as measured from one end of the curved plasmid to the other (Fig. 2). Compared with the width of relaxed double-stranded DNA, which is 2 nm, it may be assumed that the plasmid consists of a

[21] H. G. Hansma, A. L. Weisenhorn, A. B. Edmundson, H. E. Gaub, and P. K. Hansma, *Clin. Chem.* **37**, 1497 (1991).
[22] T. Thundat, D. P. Allison, R. J. Warmack, G. M. Brown, K. B. Jacobson, J. J. Schrick, and T. L. Ferrell, *Scanning Microsc.* **6**, 911 (1992).
[23] E. Henderson, *J. Microsc.* **167**, 77 (1992).
[24] C. Bustamante, J. Vesenka, C. L. Tang, W. Rees, M. Guthold, and R. Keller, *Biochemistry* **31**, 22 (1992).
[25] E. Henderson, *Nucleic Acids Res.* **20**, 445 (1992).
[26] T. Thundat, X. Y. Zheng, S. L. Sharp, D. P. Allison, R. J. Warmack, D. C. Joy, and T. L. Ferrell, *Scanning Microscopy* **6**, 903 (1992).
[27] H. G. Hansma, *Proc. Natl. Acad. Sci. USA* **96**, 14678 (1999).

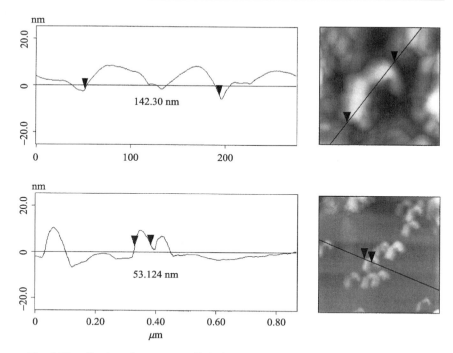

Fɪɢ. 2. Visualization of pure supercoiled plasmids on a silicone wafer. A silicone wafer was dipped in a plasmid solution and examined immediated to avoid sample dehydration. After scanning the height profile of the sample in Tapping mode, several plasmids, showing a curved appearance, can be discerned. When measured from tip to tip, the average length of a plasmid is about 200 nm, whereas the width is approximately 50 nm (lower panel).

three-fold spiralized double strand (threefold superspiralization $2^3 = 8$), which means eight double helices, which are wound around each other.

Because the imaging of soft biological samples is potentially associated with risks of artifacts, it is useful to verify such findings with results obtained by other techniques, such as x-ray crystallography, when available. Thus, such measurements have demonstrated that the periodic length of one turn of the B-form α-helix corresponds to 3.4 nm[28] and that the thickness of hydrated DNA is 2.2–2.5 nm for a single double strand. In Fig. 3, a very high-resolution image of a small part of the supercoiled plasmid as obtained by AFM is shown. The measurements indicate a distance of 3.388 nm from turn to turn and a thickness of 2.228 nm, which are

[28] J. O. Radler, I. Koltover, T. Salditt, and C. R. Safinya, *Science* **275,** 810 (1997).

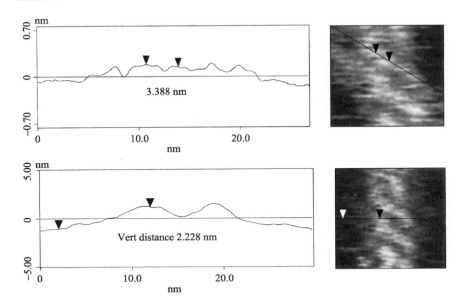

FIG. 3. High-resolution AFM image of the plasmid DNA structure A zoom into the structure of a plasmid reveals two double helices (as reflected by the scan in the lower left panel, obtained from the image in the right lower panel) twisted around each other. In the upper panel, the distance from one helix turn to the other was measured. In the lower panel, the thickness of one DNA double helix is indicated.

entirely consistent with the dimensions obtained by x-ray diffraction. This indicates that the images represent a reliable reflection of the plasmids in their native state.

The second partner of the lipoplex is the cationic lipid, or synthetic amphiphile, which is not a natural compound, but rather is obtained by chemical synthesis. Cationic amphiphiles consist of a cationic hydrophilic headgroup and a hydrophobic tail. Typical headgroups are amines,[29] polyamines,[30] or a pyridinium group.[6] Diverse hydrophobic chains have been used, ranging from long aliphatic chains[31] to complex cholesterol-like structures.[32] For preparing the lipoplexes, liposomes consisting of the cationic lipids and, if necessary, a helper lipid are prepared: they are then

[29] M. E. Ferrari, D. Rusalov, J. Enas, and C. J. Wheeler, *Nucleic Acids Res.* **29,** 1539 (2001).
[30] A. Kichler, W. Zauner, M. Ogris, and E. Wagner, *Gene Ther.* **5,** 855 (1998).
[31] P. R. Clark and E. M. Hersh, *Curr. Opin. Mol. Ther.* **1,** 158 (1999).
[32] K. Yang, X. S. Mu, R. L. Hayes, Y. H. Qiu, F. L. Sorgi, L. Huang, and G. L. Clifton, *Neuroreport* **8,** 2355 (1997).

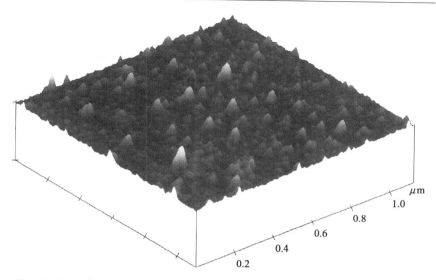

Fig. 4. Three-dimensional image of liposomes used for lipoplex formation. Liposomes, consisting of SAINT-2, were prepared by sonication and transferred to a silicone wafer by the dipping technique. The height profile of the liposomes shows a homogeneous population of vesicles varying in size between 50 and 100 nm.

mixed with the plasmid. Here we have used the cationic lipid N-methyl-4(dioleyl)methylpyridiniumchloride, abbreviated as SAINT-2. A typical protocol for the preparation of SAINT-2 lipid vesicles is as follows. One milliliter of ultrapure water is added to a dried film of 1 mmol SAINT-2 (stored in chloroform at $-20°$), followed by vigorous vortexing at room temperature. The resulting multilamellar vesicles are subjected to sonication (usually approximately 20 min) in a bath sonicator at room temperature until a clear solution of small unilamellar vesicles is obtained.[33] The liposomes thus obtained are characterized by AFM. To prepare the sample of the vesicles, a silicon wafer is dipped into the vesicle solution and, to avoid dehydration, immediately placed at the AFM. As shown in Fig. 4, the lipid vesicles show a fairly homogenous distribution of sizes; the particles display an average diameter between 50 and 100 nm.

[33] I. van der Woude, A. Wagenaar, A. A. Meekel, M. B. ter Beest, M. H. Ruiters, J. B. Engberts, and D. Hoekstra, *Proc. Natl. Acad. Sci. USA* **94,** 1160 (1997).

Lipoplex Formation in Bulk Phase

Lipoplexes used for transfection studies are conveniently prepared in bulk phase. However, unlike at a phase border, the reaction between plasmid DNA and cationic amphiphile in bulk phase is a very fast process, making it difficult to discern clearly early events. Nevertheless, AFM offers the possibility to visualize lipoplexes after the initial phase and monitor long-term morphological changes. A typical protocol to prepare and visualize such lipoplexes is as follows. Plasmid DNA is solubilized in a phosphate-free buffer or in pure water. Note that the inertness of the solvent is essential, because there should be no substance other than the DNA that is interacting with the cationic lipid. Subsequently, cationic lipids in the form of sonicated vesicles (as described earlier) are added at a concentration such that the ratio of the cationic lipid charge versus the charge of the phosphate of the single nucleotides is 2:1. Thus typically, 1.4 μmol of SAINT-2 lipid is mixed with 0.22 μg plasmid DNA in a final incubation volume of 2 ml. The surplus of charges guarantees that the entire plasmid is covered with lipid, which is important for its protection against DNAse attack and also for obtaining optimal transfection efficiencies. The resulting lipoplexes are spread on a silicon wafer by immersing the wafer in the lipoplex solution for 30 s.

The lipoplexes are examined by AFM immediately after formation and after 24 h. Under the former conditions, only spherical complexes with a diameter of approximately 200–300 nm are observed (Fig. 5A). The surface morphology of the complexes, composed of only the cationic lipid (i.e., without helper lipid), is fairly irregular. Because DNA strands are only occasionally seen, most of the DNA must have been contained inside the lipoplexes. Because the size of the pure plasmid used corresponds to a length of about 200 nm and a width of 50 nm (Fig. 2; see also Fig. 7), the organization of the lipoplex would be such that the supercoiled plasmids are positioned parallel to each other (see also later) surrounded by cationic lipids and that one layer may contain three to four plasmids. This architecture is consistent with a proposal based on small-angle x-ray diffraction.[28]

During storage at room temperature for 24 h, the complexes rearrange and aggregate into clusters of sizes of 2.5 μm and larger. Apparently, the initial complexes formed are unstable in the sense that, as a function of time, they form aggregates, because the contours of these early complexes are still clearly distinguishable, implying that massive fusion of the intially assembled lipoplexes into a single complex does not occur (see later). By agarose gel electrophoresis, it can also be shown readily that, within these complexes, the DNA is still present in a condensed supercoiled structure, well shielded from the outside medium.[8]

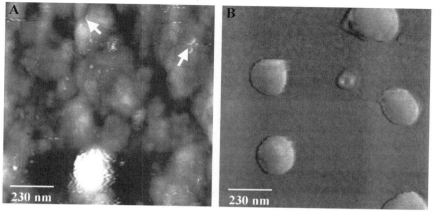

Fɪɢ. 5. Lipoplexes formed under bulk phase conditions. (A) Lipoplexes were formed from the cationic amphiphile SAINT-2 and pCAT plasmid at a charge ratio of amphiphile DNA of 2:1. The resulting lipoplexes were transferred by the dipping technique to a silicone wafer and subsequently visualized by AFM. Single lipoplexes reveal a size between 200 and 300 nm. Some uncovered DNA (arrows) is occasionally visible on the surface of the lipoplexes. (B) Lipoplexes were prepared in an identical manner as in (A) except that the liposomes were composed of SAINT-2/DOPE (1:1).

Optimal transfection with SAINT-2 lipoplexes is obtained when 50 mol% of the helper lipid dioleoylphosphatidylethanolamine (DOPE) is included in the complex. Mixed SAINT–2/DOPE vesicles are prepared as follows: SAINT-2 and DOPE are dried from stock solutions, dissolved in methanol, and mixed at equimolar concentrations at a final lipid concentration of 1 mM. The solvent is then removed under vacuum, and the lipid film is suspended in Millipore water, followed by sonication in a water-bath until a clear solution is obtained. The size of the resulting vesicles is approximately 170 nm in diameter, with low polydispersity. For preparation of the lipoplexes, the lipid vesicles (15 nmol lipid) and 1 μg pDNA (pEGFP-N1, Clontech,) are suspended separately in equal volumes (0.5 ml) of ultrapure water, after which the DNA is added immediately to the liposome suspension. The relative amounts of lipid and DNA correspond to a molar charge ratio of 2.5:1.

As observed for complex formation with the pure SAINT-2 in bulk phase, complex formation in case of the mixed system is very fast. The morphology is investigated following sample preparation (silicon wafer) as described previously. As shown in Fig. 5B, uniform and round lipoplexes are obtained with typical diameters between 300 and 400 nm. In contrast to the images obtained for the lipoplexes prepared from pure SAINT-2, the mixed lipoplexes show a very smooth surface. As in the case of the pure SAINT-2 complexes, relatively few SAINT-2/DOPE complexes show DNA structures on their surface, and essentially all the plasmid become complexed (i.e., free DNA is not visibly detectable). The striking differences in surface morphology of both type of complexes presumably reflect differences in the structural organization of the amphiphiles and has prompted further investigations described elsewhere.[6]

Finally, it should be reemphasized that the AFM technique does not allow conclusions to be drawn concerning the architecture of the inner core of the lipoplex (i.e., only morphological insight of the surface structure is obtained reliably).

Examination of Lipoplex Formation under Equilibrium Conditions

Because of the fast kinetics of lipoplex assembly in bulk phase, little insight on the mechanism of the assembly process is obtained with AFM under such conditions. However, when carried out at equilibrium conditions, a fairly detailed view of individual steps in the overall assembly process can be obtained. For that purpose, a rectangular (140 cm^2 × 1 cm depth) Wilhelmy-type film balance (Rieger & Kirstein GmbH, Golm, Germany) is combined with an epifluorescence microscope (Olympus Optical Co. GmbH, Germany) to characterize the lateral morphology of the lipid film and the resulting lipid–DNA aggregates.

The cationic lipid SAINT-2 is dissolved in chloroform at a final concentration of 1 mM. Thirteen microliters of this solution is spread onto the surface of the ultrapure water subphase and equilibrated for 20 min to allow for evaporation of the organic solvent. In all experiments, the lipid film is compressed dynamically at a speed of 3 × 10^{-2} Å^2/s·molecule. The resulting SAINT-2 monofilm shows typical fluid film behavior with a high mobility and a high compressibility. The film has a lift-off value of 150 Å^2 and collapses at a final molecular area of 51 Å^2 at a surface pressure of 55 mN/m. When 0.5 mol% of the fluorescent lipid marker N-NBD-PE (Avanti Polar Lipids, Alabaster, AL) is included in the film, a homogeneous bright epifluorescense image is apparent, reflecting the complete miscibility of the dye in a homogenous fluid lipid phase and the lack of phase-separated regions in the film.

When monitoring lipoplex formation in this system, the SAINT-2 film is compressed to a surface pressure of 25 mN/m. This pressure is chosen to mimic as closely as possible the lateral surface pressure in lipid vesicles, which is typically between between 20 and 30 mN/m.[34–36] After a further equilibration for 20 min, the film tension remains constant over time. Under these conditions, at which it is assumed all lipid molecules are organized in a monomolecular interface film without the occurrence of molecular desorption, it can be calculated that 7.8×10^{15} SAINT-2 molecules are adsorbed at the air–water interface. This value corresponds to a charge density of $2.16 \times 10^{-10}/cm^2$ at 25 mN/m. To start the experiment (Fig. 6),

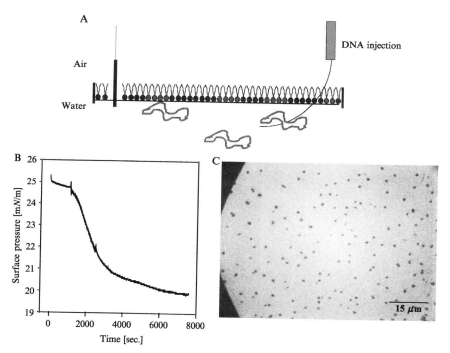

FIG. 6. Lipoplex assembly in a monolayer system. Plasmid DNA was injected with a Hamilton syringe into the subphase under a monolayer of the cationic amphiphile SAINT-2, at a surface pressure of approximately 25 mN/m (A). A sudden decrease in surface pressure (B) indicates that on plasmid–monolayer interaction, an ordering of the cationic amphiphiles occurs at the phase border. After approximately 30 min, the lateral surface pressure drops rapidly, which may reflect an active recruitment of lipids from the monolayer onto the plasmid, followed by a dissociation of the assembled complexes from the interphase into the subphase. (C) The ordering of the lipids, occurring on plasmid–cationic lipid interaction is evidenced by a phase separation of the fluorescent lipid analogue N-NBD-PE from crystalline DNA-lipid domains, as reflected by black spots. The fluid regions still show a bright and homogenous fluorescence distribution as observed before addition of the plasmid.

[34] H. Fischer, R. Gottschlich, and A. Seelig, *J. Membr. Biol.* **165,** 201 (1998).

[35] M. R. Wenk, T. Alt, A. Seelig, and J. Seelig, *Biophys. J.* **72,** 1719 (1997).

[36] R. B. Weinberg, V. R. Cook, J. B. Jones, P. Kussie, and A. R. Tall, *J. Biol. Chem.* **269,** 29588 (1994).

20 μg of the plasmid pCAT is distributed homogeneously within the sub-phase using a mycrosyringe. A gradual decrease in surface pressure can be observed immediately after DNA injection. After approximately 30 min, the surface pressure decreases rapidly by 4–5 mN/m, and after another 30 min, the system reaches a quasi-equilibrium state. The driving force for complex formation is the electrostatic interaction between the positively charged surface film and the negatively charged DNA. The initial step of this process is controlled presumably by diffusion, and its duration is determined by the time it takes for the DNA to transfer from the subphase to the air–water interphase. This assumption is supported by the zero-order kinetics of the pressure-time diagram during this initial period, as Fick's law of diffusion would predict. A decrease of the lateral pressure is usually related to a decrease in molecular density (number of molecules per surface area), which results from the diffusion of molecules from the surface into the subphase. Alternatly, a decrease in lateral pressure may be the consequence of a smaller molecular area after condensation of the lipid phase. Such changes can be followed by monitoring changes in the distribution of the fluorescent lipid marker N-NBD-PE by epifluorescence microscopy (Fig. 6B). When phase separation occurs, small black domainlike areas devoid of the fluorescent marker become apparent because of self-quenching of fluorophores in close proximal. From the kinetics of the pressure–time curve, it can be inferred that complex formation or re-arrangements that result in the appearance of the black domains must be a fast process. The sequence of events reflected by the surface pressure changes suggests that the decrease is related to an initial ordering of the lipids, giving rise to phase separation of the fluid phase marker N-NBD-PE, followed by a dissociation of lipoplexes, assembled at the interphase, into the subphase.

Lipoplex assembly occurring in this manner is then visualized by AFM. For sample preparation, the monolayer has to be transferred to a solid surface, which can be carried out by either the Langmuir-Schäfer[37] or the Langmuir-Blodgett technique.[38] The Langmuir-Schäfer technique uses a hydrophobic silanized silicon wafer that is taken between tweezers and horizontally layered over the lipid film. When the substrate is lifted subsequently in a vertical direction from the air–water interface, the lipid film and all structures localized near the surface will remain attached to the wafer as a result of van der Waals interactions. In the case of the Langmuir-Blodgett technique, a hydrophilic silicon wafer is dipped

---

[37] G. L. Gaines, in "Insoluble Monolayers at Liquid-gas Interfaces" (I. Prigogine, Ed.), Wiley & Sons Interscience, New York, 1966.
[38] C. W. Hollars and R. C. Dunn, *Biophys. J.* **75**, 342 (1998).

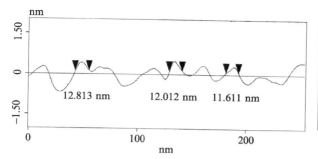
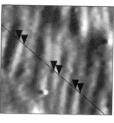

FIG. 7. Visualization of monolayer–plasmid interaction in samples transferred by the Langmuir–Schäfer technique to a silanized silicon wafer. Immediately after injecting the plasmids underneath the monolayer (see Fig. 6A), their ordered parallel organization under the layer of the cationic amphiphile can be seen. The parallel ordering is due to long-range orientational interactions between the single plasmids. The width of one plasmid is approximately 10–20 nm, whereas the length is stretched to approximately a 300 nm.

vertically into the subphase. Then, the silicon wafer attached to a lift is pulled out of the subphase at constant speed while the lateral pressure of the lipid film is kept constant by the movement of the film balance barrier. During this process, the surface film will be transferred directly onto the substrate. In our studies, both techniques are used to exclude the risk of artifacts. Those films transferred by the Langmuir-Blodgett technique are more homogenous, but the overall appearance and morphology of the DNA complexes are comparable.

Only minutes after the start of the experiment (i.e., following injection of the plasmid into the subphase) different structures can be visualized. Most of the plasmid DNA is adhered onto the cationic lipid headgroups in its supercoiled form.[39] Interestingly, the attached single plasmids do not display a random orientation of attachment but rather are organized on the monolayer in a highly ordered manner either parallel to each other or in a 90-degree angle (Fig. 7). This organization likely results from long-range interactions between the single plasmids.[40] Because a DNA double strand in its B-form has a width of 2–2.5 nm, the structures shown in Fig. 7 can be interpreted as multiple rather than extended plasmids that are aligned in parallel. From the distances between periodic structures in the height profile along the molecules, it can be inferred that the plasmids are still supercoiled, although slightly stretched out, because the width is smaller (12 nm width vs 50 nm, respectively) and the length

[39] O. Zelphati, X. Liang, P. Hobart, and P. L. Felgner, *Hum. Gene Ther.* **10**, 15 (1999).
[40] A. A. Kornyshev and S. Leikin, *Proc. Natl. Acad. Sci. USA* **95**, 13579 (1998).

slightly larger than that of a plasmid without cationic lipid interactions (200–250 nm vs 250–300 nm, respectively). Within the same sample, typical "beanlike" complexes of lipid and DNA can be discerned also (Fig. 8). These structures accumulate over time, while concomitantly, uncomplexed DNA disappears. The conversion into beanlike complexes is virtually completed after 30 min under the given conditions. Interestingly, the dimensions of these structures are 50–70 nm in width and approximately 200 nm in length. Given the dimension of the pure plasmid, it may thus be inferred that the supercoiled plasmid determines the size of the bean structure, and that it is surrounded by three to five bilayers of the amphiphile. Subsequently, single complexes start to aggregate, leading to the formation of larger aggregates, the smoothness of the edges suggesting that membrane fusion was the underlying mechanism of the assembly event. Surprisingly, this aggregation process is not chaotic but occurs mainly in parallel or under a 90-degree interaction angle between two smaller lipoplexes (Fig. 8).

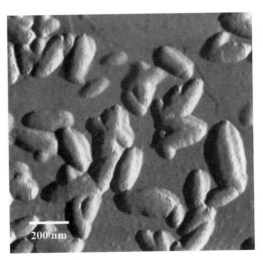

FIG. 8. Lipoplexes prepared with the film balance under equilibrium conditions. The monolayer (see Fig. 6A) was transferred by the Langmuir–Schäfer technique to a silanized silicon wafer and examined subsequently by AFM. Beanlike structures are seen; the smallest ones show a length of 250–300 nm and a width of 50–60 nm, representing a single supercoiled plasmid surrounded by several layers of lipid (see text). This lipoplex "unit" seems to represent the core building block for assembly into larger aggregates. Aggregated lipoplexes reveal a smooth surface and connect preferentially in parallel or in a 90° angle.

Limitations and Prospects of Atomic Force Microscopy in Lipoplex Research

Atomic force microscopy provides the ability to scan at very high resolution (on the order of a few tenths of a nanometer) the surface of objects. It is therefore crucial to inspect samples as a function of time to reconstruct the overall process that led to the structure examined at equilibrium conditions. Specifically, AFM does not allow an investigation of the inner structures of the system under study. For characterization of the inner core of lipoplexes, additional techniques are available, including electron microscopy.[41] However, with this technology, the resolution will be less, whereas, more importantly, lipidic samples are often not inert toward complex fixation and embedding procedures. A potential drawback of the AFM technique is the inability to determine dynamic processes (e.g., the changes the lipid undergoes during lipoplex formation or afterwards during DNA release), unless the experimental system can be adapted for such purposes, as in the case of monolayer studies. Despite these limitations, AFM nevertheless seems to be a powerful tool for the characterization of distinct steps of the lipoplex assembly process. On the basis of observations like those made for monolayer–plasmid interactions, the next challenge will be to carry out similar studies in the case of interactions between lipoplexes and the cellular plasma membrane, which should provide insights into the mechanism by which lipoplexes are internalized by cells.

Acknowledgments

Part of this work was supported by the Stiftung Deutscher Naturforscher Leopoldina BMBF 9901/8–6 and BMBF 03C0301D1.

---

[41] A. A. P. Meekel, A. Wagenaar, J. Smisterova, J. E. Kroeze, P. Haadsma, B. Bosgraaf, M. C. A. Stuart, A. Brisson, M. H. J. Ruiters, D. Hoekstra, and J. B. F. N. Engberts, *Eur. J. Org. Chem.* **4,** 665 (2000).

[19] Biophysical Characterization of Cationic Liposome–DNA Complexes and their Interaction with Cells

By Maria C. Pedroso de Lima, Henrique Faneca, Miguel Mano, Nuno Penacho, Nejat Düzgüneş, and Sérgio Simões

Introduction

The use of cationic liposomes as gene-delivery agents has been based on the following assumptions: (1) a spontaneous electrostatic interaction between the positively charged liposomes and the negatively charged DNA, which results in condensation of the macromolecule; (2) the fact that the resulting cationic liposome–DNA complexes could exhibit a net positive charge that promotes their association with the negatively charged cell surface; and (3) the fusogenic properties exhibited by the cationic liposomes that can induce fusion and/or destabilization of cell membranes, thus facilitating the intracellular release of complexed DNA. These assumptions paved the way for studies aiming at improving the biological activity of lipoplexes and facilitated the identification of several critical parameters that affect their efficacy. These parameters underlie the various stages followed by lipoplexes, from their formation to their interaction with cells. Among others, the physicochemical characteristics including size, charge density, and colloidal stability are relevant properties of lipoplexes that are determined by the mode of lipoplex formation. Moreover, these properties play a major role in lipoplex–cell interactions, including binding to the cell surface, internalization, and intracellular delivery of the carried DNA. In this chapter, we describe some experimental procedures whose application can provide information on the mechanisms involved in the aforementioned processes.

Cationic Liposomes

Cationic liposomes can be purchased from several companies or easily prepared in the laboratory. Liposomes of different lipid compositions are prepared by first drying a lipid film under argon and then in a vacuum oven at room temperature (RT), followed by hydrating with 1 ml deionized water at a final concentration of 5 mg/ml. The multilamellar vesicles (MLVs) obtained are then sonicated briefly under argon, extruded 21 times through polycarbonate filters of 50-nm pore diameter using a Liposofast device (Avestin, Toronto, Canada), diluted five times with deionized water, and filter-sterilized with Millex 0.22-μm pore-diameter filters. Because, 1,2-dioleoyl-3-trimethylammonium-propane (DOTAP) and other cationic lipids do not contain phosphate groups, total lipid concentration may be estimated by measuring the concentration of the colipid by the Bartlett method[1] (phospholipids) or other appropriate method. For some experiments, liposomes are labeled by incorporating a fluorescent lipidic probe (e.g., Rh-PE) into the lipid mixture (1–5 mol% of total lipid). Liposomes should be stored at 4° under nitrogen and used within 3 weeks after being prepared.

A large number of cationic lipids (monovalent and multivalent) are available[2] and can be used alone or in combination with other lipids that contribute not only to the formation of the liposomes but also to improve their ability to transfect cells. Among others, 1,2-dioleoyl-*sn*-glycero-3-phosphoethanolamine (DOPE), or cholesterol (ChoL) are those that have been often included with the cationic lipids.

Cationic Liposome–DNA Complexes

Preparation of Complexes

Complexes are usually prepared by sequentially mixing 100 μl of a solution of 100 mM NaCl, 20 mM HEPES, pH 7.4 (HBS), containing an appropriate amount of liposomes (depending on the +/− charge ratio required) with 100 μl of buffer containing 1 μg of plasmid DNA and gently mixed. The mixture is incubated for 15 min at RT. Complexes are prepared under aseptic conditions (flow cabinet) using sterile solutions and should be used immediately after their preparation to avoid any aggregation.

It is relevant to emphasize that the mode of lipoplex preparation strongly determines the final properties. Therefore, critical parameters should be considered and adapted to the relevant applications. These include structure and size of cationic liposomes, the concentrations of cationic lipid and DNA, the ionic strength and temperature of the suspending medium, the order of addition and the mixing rate of the components, as well as the extent of complex formation.

A novel method to optimize the properties of lipid–DNA complexes was proposed, consisting of both extrusion of the liposomes (to obtain large unilamellar liposomes, LUVs) and controlled mixing of lipid and DNA.

[1] G. R. Bartlett, *J. Biol. Chem.* **254,** 466 (1959).
[2] M. C. Pedroso de Lima, S. Simões, P. Pires, H. Faneca, and N. Düzgüneş, *Adv. Drug. Deliv. Rev.* **47,** 277 (2001).

This procedure has been shown to result in lipoplexes exhibiting small sizes with a narrow distribution that present a high colloidal stability and adequate features for their *in vivo* use.[3,4] Essentially, this procedure avoids the concerns associated with the order of addition of cationic liposomes and DNA by mixing them simultaneously at a constant ratio and at a precisely controlled rate. The mixing occurs at a T-connector joining the two syringes into which DNA and the liposomes are separately loaded. The semi-automated control of the mixing rate (through the use of a pump) should be adjusted, depending on the structure and size of the cationic liposomes used for lipoplex preparation. Monodisperse and stable lipoplexes are obtained when using extruded cationic liposomes (LUVs) over MLVs and when prepared in sorbitol/sodium acetate buffer (5% sorbitol, 20 mM sodium acetate, pH 6) rather than in water or saline buffer.

Physicochemical Characterization of the Complexes

The relative proportion of cationic lipid and DNA determines the properties of the lipoplexes, namely surface charge (zeta potential), size, efficacy of complexation, colloidal stability, and biological activity. Zeta potential measurements of the different lipid–DNA complexes can be performed using a Coulter® DELSA 440 instrument (Coulter Electronics, Miami, FL). The DELSA 440 is a laser-based multiple-angle particle electrophoresis analyzer that measures the electrophoretic mobility and zeta potential distribution simultaneously with the hydrodynamic size of particles in suspension. The principle of the electrophoretic light scattering is based on the measurement of the velocity of particles moving under the influence of an electric field. Electrophoretic mobility is detected by the Doppler shift in frequency, $\Delta\nu$, using a heterodyne method,[5] and calculated by the equation:

$$\Delta\nu = U(Ek) \tag{1}$$

where U is the electrophoretic mobility, E the applied electric field, and k is the scattering angle determined by:

$$k = (4\pi n/\lambda)\,sin(\theta/2) \tag{2}$$

where λ is the wavelength of the helium–neon laser light in vacuum, n the refractive index of the medium, and θ is the angle between the incident

[3] O. Zelphati, C. Nguyen, M. Ferrari, J. Felgner, Y. Tsai, and P. L. Felgner, *Gene Ther.* **5**, 1272 (1998).
[4] S. Hirota, C. T. de Ilarduya, L. G. Barron, and F. C. Szoka, *Biotechniques* **27**, 286 (1999).
[5] M. C. Woodle, L. R. Collins, E. Sponsler, N. Kossovsky, D. Papahadjopoulos, and F. J. Martin, *Biophys. J.* **61**, 902 (1992).

laser beam and the detector. From both equations, electrophoretic mobility can be calculated by the equation:

$$U = \frac{(\Delta\nu\lambda)}{2nE\,sin(\theta/2)} \tag{3}$$

Electrophoretic mobility is related to the zeta potential by the assumption derived from the Smoluchowski approximation, which assumes that the double-layer thickness is small compared with the colloidal particle diameter[5]:

$$U = \frac{\varepsilon\zeta}{\eta} \tag{4}$$

where ζ is the zeta potential and ε is defined by $\varepsilon = \varepsilon_0 D$, ε_0 being the permitivity of free space and D the dielectric constant of water, η is the viscosity, this being a valid assumption for aqueous systems of defined electrical conductivity.

Samples of the prepared complexes are placed in the measuring cell, whose position is adjusted to cover a previously determined stationary layer, and an electric current of 3.0 mA is applied. Measurements are recorded, and the zeta potential (ζ) is calculated for each scattering angle (8.6°, 17.1°, 25.6°, and 34.2°).

For the evaluation of the hydrodynamic size of the complexes two techniques can be used:

1. Electrophoretic light scattering in the absence of any electric field and considering only the Doppler shift in light scattering at a detection angle of 34.2°[5–7]
2. Photon correlation spectroscopy (PCS)

The PCS technique uses autocorrelation spectroscopy of scattered laser light to determine its time-dependent fluctuations resulting from the Brownian motion of particles in suspension. The light intensity scattered at a given angle is detected by a photomultiplier whose output current is passed to an autocorrelator, which analyzes time dependence, determining the rate of diffusion or Brownian motion of the particles and hence their size. With this technique, the detection angle is fixed at 90°. This assay can be performed using a Coulter N4 Plus instrument.

Results obtained in our laboratory demonstrate that the relative proportion of cationic lipid and DNA determines the size, surface charge (zeta potential), and biological activity of the resulting complexes. Figure 1 shows

[6] E. Tomlinson and A. P. Rolland, *J. Control. Rel.* **39,** 357 (1996).
[7] C. Lourenço, M. Teixeira, S. Simões, and R. Gaspar, *Int. J. Pharm.* **138,** 1 (1996).

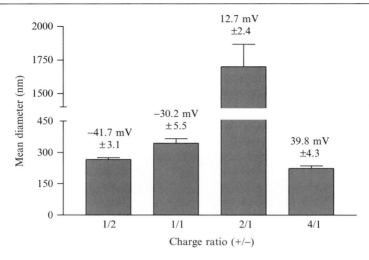

FIG. 1. Dependence of zeta potential and mean diameter of DOTAP:DOPE/DNA complexes on their lipid/DNA charge ratio. DOTAP:DOPE/DNA complexes are prepared at the lipid/DNA charge ratios (+/−) ((1/2), (1/1), (2/1) and (4/1)) as described. Zeta potential measurements of the complexes are performed using a Coulter® DELSA 440 Instrument. Data represent the mean ± standard deviation of at least three independent experiments.

a typical result obtained on the relationship between the mean diameter and surface charge of the complexes and their lipid/DNA charge ratio. As can be observed, highly positively charged complexes, in which DNA is completely sequestered and condensed, exhibit homogeneous size distribution (mean diameter between 200 and 250 nm). A similar size distribution is also observed when complexes are prepared with an excess of DNA over cationic lipid (i.e., negatively charged complexes), although in this case, the presence of free DNA is generally observed.[6,8–10] On the other hand, complexes prepared from a lipid/DNA charge ratio of approximately 3/2 exhibit a neutral zeta potential, suggesting that all the cationic lipid molecules are neutralized by DNA.[10–13] Surprisingly, the use of different batches of cationic lipids for the preparation of the complexes

[8] J. O. Radler, I. Koltover, T. Salditt, and C. R. Safinya, *Science* **275,** 810 (1997).

[9] S. J. Eastman, C. Siegel, J. Tousignant, A. E. Smith, S. H. Cheng, and R. K. Scheule, *Biochim. Biophys. Acta* **1325,** 41 (1997).

[10] P. Pires, S. Simões, S. Nir, R. Gaspar, N. Düzgüneş, and M. C. Pedroso de Lima, *Biochim. Biophys. Acta - Biomembranes* **1418,** 71 (1999).

[11] S. Simões, V. Slepushkin, R. Gaspar, M. C. Pedroso de Lima, and N. Düzgüneş, *Gene Ther.* **5,** 955 (1998).

[12] G. S. Koe, H. L. Way, G. M. Quetingco, J. G. Smith, T. Wedeking, and R. W. Niven, *Pharm. Res.* **14,** S57 (1997).

[13] F. D. Ledley, *Pharm. Res.* **13,** 1595 (1996).

results sometimes in changes of the lipid/DNA charge ratio corresponding to neutrality, which reinforces the importance of controlling the zeta potential. Such neutral complexes are characterized by a heterogeneous size distribution (mean diameter from 350 to 1200 nm) and usually present a much lower colloidal stability than those exhibiting an excess of net positive or negative charge. This can be attributed to a lack of electrostatic repulsive forces among the complexes that would prevent their aggregation.[3,10,11,14,15] As mentioned previously, the influence of lipid–DNA stoichiometry on the physicochemical properties of the complexes becomes even more difficult to evaluate considering that, for a fixed lipid/DNA charge ratio, the increase in concentration of lipid and DNA results in a significant change in their size and colloidal stability, which can be attributed to enhanced precipitation at higher concentrations caused by smaller interparticle separation.[6]

Assessment of DNA Protection

Two different assays are currently used to evaluate the degree of DNA protection conferred by its association with cationic liposomes.

Ethidium Bromide Intercalation Assay

The principle of this assay is based on the properties of ethidium bromide (EtBr), a monovalent DNA intercalating agent whose fluorescence is dramatically enhanced on binding to DNA and quenched when displaced by higher affinity compounds or by condensation of the DNA structure. Therefore, this probe has been used to examine the ability of cationic lipids to protect DNA.

The accessibility of EtBr to the DNA associated with the complexes is monitored at 37° using a fluorometer. The fluorescence is read at excitation and emission wavelengths of 518 and 605 nm, respectively. The sample chamber should be equipped with a magnetic stirring device, and the temperature should be controlled with a thermostated circulating water bath. The fluorescence scale is calibrated such that the initial fluorescence of EtBr (20 μl of a 2.5 mM aqueous solution added to a cuvette containing 2 ml HBS solution) is set as residual fluorescence. The value of fluorescence obtained on addition of 1 μg DNA (control) is set as 100%. Cationic liposome–DNA complexes (1 μg DNA) are added to the cuvette containing 2 ml HBS solution followed by addition of 20 μl of

[14] J. H. Felgner, R. Kumar, C. N. Sridhar, C. J. Wheeler, Y. J. Tsai, R. Border, P. Ramsey, M. Martin, and P. L. Felgner, *J. Biol. Chem.* **269,** 2550 (1994).

[15] R. I. Mahato, A. Rolland, and E. Tomlinson, *Pharm. Res.* **14,** 853 (1997).

EtBr. The amount of DNA available to interact with the probe is calculated by subtracting the values of residual fluorescence from those obtained for the samples and expressed as percentage of the control.[16]

The EtBr intercalation assay, although being simple, rapid, and sensitive, may lead to misinterpretation regarding the access of nucleases to the complexed DNA. In fact, EtBr, being a very small molecule, can intercalate into DNA fragments that, while being accessible to the probe, cannot be degraded by nucleases.

A more biologically relevant assay to evaluate the protection conferred by cationic liposomes to DNA degradation can be performed by following DNase I resistance mediated by the complexes.

Resistance to DNase I Action

Resistance of cationic liposome–DNA complexes to DNase I action (Sigma, St. Louis, MO), can be determined either by electrophoresis or spectrofluorimetry.

DNase I is maintained in a buffer solution at a stock concentration of 50,000 units/ml (50 mM TRIS-HCl (pH 7.5), 10 mM MnCl$_2$, 50 mg/ml bovine serum albumin [BSA]). Complexes are submitted to DNase I action (5 units DNase I/μg of DNA), for 30 min at 37°, followed by inactivation of the enzyme by addition of an aliquot of 0.5 M ethylenediamine tetraacetic-acid (EDTA) (final concentration: 1 μl/unit of DNase I). Parallel experiments are performed by incubating samples under the same experimental conditions, except that DNase is previously inactivated. Following treatment of the complexes with DNase I (active and inactive), 5 μl of "loading buffer" (15% [v/v], Ficol 400, 0.05% [w/v], bromophenol-blue, 1% [w/v], sodiumidodecy sulfate (SDS), 0.1 M EDTA, pH 7.8) are added to aliquots corresponding to 200 ng of DNA. These samples are loaded on a 1% agarose gel prepared in triborate EDTA (TBE) solution (89 mM TRIS-base (pH 8.6), 89 mM boric acid, 2.5 mM EDTA) containing 1 μg/ml EtBr.

The sensitivity and accuracy of this assay may be limited by the cationic lipid associated to DNA, which not only can prevent EtBr intercalation but can also decrease the DNA electrophoretic mobility. Therefore, in some experiments, lipid extraction is required following incubation of the complexes with DNase I. A mixture of phenol, chloroform, and isoamilic alcohol (25:24:1) is added to the complexes at a volume ratio of 1:1 and subsequently centrifuged at 10,000g for 10 s. Isopropanol 1:2 (v/v) is added to the aqueous phase containing DNA, and the mixture is maintained at −20° for 30 min to facilitate DNA precipitation. After centrifugation (14,000 g, 15 min, 4°), the supernatant is aspirated and the pellet washed

[16] H. Faneca, S. Simões, and M. C. Pedroso de Lima, *Biochim. Biophys. Acta.* **1567,** 23 (2002).

again with 1 ml isopropanol. Finally, the pellet is allowed to dry off, dissolved in HBS, and electrophoresed.

Even with the lipid extraction, the results obtained using this assay can be inconclusive, because the agarose gel frequently does not have enough sensitivity to allow a quantitative evaluation of such small amounts of DNA. Alternatively, in these cases, nondegraded DNA can be quantified by measuring EtBr fluorescence. For these measurements, the same experimental procedure previously described for the EtBr access to the complexes should be carried out. The extent of DNA degradation can then be determined according to the following equation:

$$\text{DNA degradation } (\%) = \frac{F_i - F_a}{F_{100} - F_0} \times 100 \qquad (5)$$

where F_i and F_a are the fluorescence values emitted by EtBr in the presence of the complexes treated with the inactive or active enzyme, respectively, F_{100} is the fluorescence value emitted by EtBr in the presence of 1 μg of DNA, and F_0 is the residual fluorescence of EtBr.

Dissociation of Cationic Liposome–DNA Complexes

Following lipoplex internalization, DNA release into the cytoplasm is crucial to avoid DNA degradation at the lysosomal level. The mode by which lipoplexes induce the disruption of the endosome to gain access into the cytoplasm and the degree of DNA condensation/compaction when it reaches the nucleus are questions that still need to be resolved. Among others, a possible way to study the dissociation of the lipid–DNA complexes involves the application of a fluorometric assay based on resonance energy transfer between two fluorophores, a donor, N-(7-nitrobenz-2-oxa-1,3-diazol-4-y1 (NBD) and an acceptor, Rh (Lissamine™ rhodamine B). These two probes can be used for this purpose because of the overlapping of the NBD and Rh fluorescence emission and excitation spectra, respectively.[17] Cationic liposomes are prepared as described previously, except that 1 mol% NBD-PE should be added to the formulation.

Labeling of DNA with Rh can be achieved easily by using commercially available kits based on nonenzymatic protocols (e.g., Label IT® Rhodamine Labeling Kit, Panvera Corporation, WI). The DNA solution is mixed with buffer A to obtain a final concentration of 0.1 μg of DNA/μl. One microliter of Label IT reagent is added per 1 μg of DNA in a reaction volume ranging between 5 and 50 μl for maximal labeling efficiency. The mixture is incubated for 1 h at 37° and the unreacted label removed by gel filtration using a G50 Microspin purification column.

[17] D. K. Struck, D. Hoekstra, and R. E. Pagano, *Biochemistry* **20**, 4093 (1981).

Once complexed, Rh-labeled DNA quenches the fluorescence of NBD-labeled cationic liposomes. Dissociation of the complex can be monitored by following either the dequenching of NBD fluorescence, using $\lambda_{exc} = 460$ nm and $\lambda_{em} = 534$ nm or the decrease in Rh fluorescence at 590 nm.

It should be noted that an excess of any of the probes may affect the results obtained: (1) If NBD is in excess, complexation may not result in the quenching of NBD fluorescence; (2) if Rh is in excess, complex dissociation may not result in dequenching of NBD fluorescence. Therefore, the amount of NBD should be carefully tested and optimized for the amount of labeled DNA to be complexed with the cationic liposomes.

Assessment of Lipoplex–Cell Interactions

Different experiments can be performed to examine the mechanisms by which cationic liposomes deliver genetic material into cells. Standard protocols to assess the relative contribution of membrane fusion (involving the cytoplasmic and/or endosomal membranes) or of endocytosis in the process of intracellular gene delivery mediated by lipoplexes are carried out as in the following.[18]

Fluorimetric Measurements (Quantitative Assays)

Cell Association and Binding. Cell association experiments are performed at 37°. After growth for 24 h in six-well microplates (5×10^5 cell/well), cells are washed twice with 1 ml serum-free Dulbecco's modified Eagle's medium-high glucose (DMEM)-HG and incubated with 5 μl of fluorescently labeled lipoplexes (approximately 7 μM cationic lipid concentration) in a final volume of 1 ml (serum-free DMEM-HG). For these experiments, cationic liposomes containing 1 mol% Rh-PE are currently used.

After different incubation times, the medium containing the non-associated lipoplexes is collected and diluted to a final volume of 2 ml of serum-free DMEM-HG. Fluorescence is measured at 37° following addition of Triton X-100 (E. Merck, Darmstadt, Germany) at a final concentration of 0.5% (v/v).

To assess the fluorescence associated with the cells, cells are rinsed and detached from the culture dishes with disposable scrapers (Corning Costar Corporation, Cambridge, MA) and then suspended in 2 ml of medium. The fluorescence of the cell suspension is measured in the presence of Triton X-100 as described previously.

---

[18] M. T. Girão da Cruz, S. Simões, P. P. Pires, S. Nir, and M. C. Pedroso de Lima, *Biochim. Biophys. Acta* **1510**, 136 (2001).

The extent of cell association is determined according to the following equation:

$$\text{Cell association } (\%) = \frac{F_{cells}}{F_{non\text{-}associated} + F_{cells}} \times 100 \tag{6}$$

where F_{cells} is the value of fluorescence associated with the cells and $F_{non\text{-}associated}$ is the value of fluorescence of nonassociated lipoplexes.

Binding studies are performed at $4°$. Cells (two times freshly washed with 1 ml serum-free DMEM-HG) are incubated with the lipoplexes in a final volume of 1 ml (serum-free DMEM-HG). After different incubation times, the medium with the nonassociated lipoplexes is collected and diluted to a final volume of 2 ml of serum-free DMEM-HG. Cells are rinsed and detached from the culture dishes with disposable scrapers and then suspended in 2 ml of medium. The experimental procedure for binding quantification is the same as that described for the cell association measurements. The extent of binding is determined according to Equation 6.

Fusion (Lipid Mixing). Lipid mixing between cells and lipoplexes can be evaluated by monitoring the increase of fluorescence of Rh-PE incorporated in the liposomal membrane at a self-quenching concentration (5 mol%). Cells (washed twice with 1 ml serum-free DMEM-HG) are incubated with the lipoplexes in a final volume of 1 ml (serum-free DMEM-HG). After 1 h incubation on ice (to allow binding, but not internalization of the complexes), the medium containing the nonassociated lipoplexes is removed. Cells are rinsed with 1 ml of culture medium and then incubated in 2 ml of the same medium at $37°$ to promote fusion.

After different incubation times, culture dishes are transferred to ice to stop fusion. Cells are detached as described previously, and the extent of fusion is determined by measuring the fluorescence of the cell suspension at $20°$. Values are given as a percentage of the maximal fluorescence, which is obtained on addition of Triton X-100 at a final concentration of 0.5% (v/v). The initial fluorescence of the bound Rh-PE–labeled lipoplexes (and any residual light scattering caused by cells) is set as 0% fluorescence. This zero value is determined, for each experimental condition, after 60 min incubation of the lipoplexes with the cells at $4°$; the unbound lipoplexes are removed, and the fluorescence of the cell suspension is measured at $20°$.

The extent of fusion is determined according to the following equation:

$$\text{Fusion } (\%) = \frac{F_t - F_0}{F_{max} - F_0} \times 100 \tag{7}$$

where F_t is value of fluorescence after t min, F_{max} is the value of fluorescence after addition of Triton X-100, and F_o is the time zero value of fluorescence.

The results of lipid mixing are expressed as a percentage of the maximal fluorescence (F_{cells} + $F_{bound\ lipoplexes}$, after addition of Triton X-100). Because different extents of binding are obtained for each charge ratio tested, fusion results are normalized, taking into account the respective extent of binding.

Fusion (Lipid Mixing) Versus Endocytosis. Parallel experiments of cell association, binding, and fusion may be performed in the presence of drugs that interfere with the endocytotic pathway. Cells should be pretreated with the metabolic inhibitors of endocytosis, antimycin A (1 μg/ml), sodium fluoride (10 mM), and sodium azide (0.1%, w/v) for 30 min at 37°, before their incubation with the lipoplexes in the presence of the drugs.

All fluorescence measurements are performed in a fluorometer. The fluorescence is read at excitation and emission wavelengths of 568 nm and 590 nm, respectively. For these experiments, 4-ml volume and 10-mm path length disposable fluorimetric cuvettes (Hughes & Hughes Limited, Tonedale, Wellington, UK) are used. The sample chamber should be equipped with a magnetic stirrer, and the temperature should be controlled with a thermostatic circulator.

Microscopy Studies (Qualitative Evaluation)

Fluorescence Microscopy Studies. To confirm the results provided by the assays described previously, both fluorescence and confocal microscopy can be used. These methods allow the observation of the cellular localization of the lipoplexes and each of their components under different experimental conditions.[19]

For that purpose, cells (0.19 × 10^6 cells/well) are seeded in an eight-chambered coverslip (Nalge Nunc International, Naperville, IL) in a final volume of 0.4 ml. When approximately 90% confluence is reached, cells are incubated with the labeled complexes.

DNA is labeled either with fluorescein or Rh using the Mirus Label IT^{TM} fluorescein or Rh nucleic acid kit (Mirus Corporation Madison, WI), as previously described for complex dissociation studies.

If DNA is labeled with fluorescein, liposomes are labeled by incorporation of 1 mol% (Texas Red 1,2-dihexadecanoyl-*sn*-glycerol-3-phosphoethanolamine, triethylammonium salt) (DHPE) from Molecular Probes (Eugene, OR). If the DNA is unlabeled or labeled with Rh, 1 mol% fluorescein-DHPE   (*N*-(fluorescein-5-thiocarbamoyl)-1,2-dihexadecanoyl-*sn*-glycero-3-phosphoethanolamine, triethylammonium salt) should be used

---

[19] Pires, S. Simões, N. Düzgüneş, and M. C. Pedroso de Lima, unpublished data (2003).

to label the liposomes. Labeled DNA and cationic liposomes are then used to prepare lipoplexes following the protocol described previously.

To allow the detection of lower amounts of DNA, as may be necessary to visualize DNA inside the nucleus, parallel experiments can be performed using pGene Grip™ vectors (Gene Therapy Systems, San Diego, CA). These plasmids are labeled with Rh and encode the green fluorescent protein (GFP). This approach allows the identification of transfected cells (following GFP expression), as well as the detection of intracellular DNA localization by following Rh-labeled DNA.

The lipoplexes are incubated with cells for 4 or 24 h at 37°. After incubation, cells are washed with CellScrub™ buffer (Gene Therapy Systems, San Diego, CA) (10–15 minutes incubation at 37°) to remove lipoplexes bound to cell surface. Following a further wash with phosphate-buffered saline (PBS) (without $CaCl_2$ and $MgCl_2$), cells are fixed with a 2% paraformaldehyde solution in PBS (30-min incubation at RT). The excess of paraformaldehyde is removed by washing cells with PBS. Cells can then be observed either by fluorescence microscopy or laser scan confocal microscopy.

Negative Staining Electron Microscopy. To characterize lipoplex–cell interactions at the ultrastructural level, electron microscopy (EM) studies can be performed. Electron microscopy offers the possibility of visualizing DNA inside the cell on its labeling with gold particles. For use in EM, cells are seeded in a six-well plate (1.7×10^6 cells/well) in a final volume of 2 ml 3 days before the experiments to allow growth to the adequate confluence.

DNA is labeled with biotin using the Mirus Label IT™ biotin nucleic acid kit (Mirus Corporation Madison, WI), following the protocol described previously for complex dissociation studies. Labeled DNA is then used to prepare lipoplexes, as described previously. After incubating the cells with the complexes for 4 or 24 h at 37°, cells are fixed with a mixture of 2% paraformaldehyde and 0.25% glutaraldehyde (incubation for 30 min), followed by sequential dehydration in ethanol at 70% (2×5 min), 80%, 90%, and 100% (3×15 min). The dehydrated cells are released from the wells with propylene oxide, transferred to polypropylene centrifuge tubes, and washed with series of 1:1 propylene oxide: lactated Ringers (LR) white resin (30 min) and $3 \times 100\%$ LR white resin (30 min each). Embedding of the cells is achieved after overnight incubation with LR white resin. Sections of 100 nm are mounted on a nickel grid for EM investigation. For postembedding labeling, section thickness is unimportant, because the antibodies do not penetrate the resin sections.

Grids are placed on a droplet of heat-inactivated horse serum for 10 min. After the incubation time, serum excess is gently blotted off, and the grids are transferred to the surface of 50-μl droplets of buffer (1% goat serum, 0.1% Tween 20, 1% BSA, and 0.1% sodium azide in PBS, pH 8.2) to reduce the background labeling. PBS provides the tonicity for antibody reaction, goat serum and BSA are used to reduce nonspecific binding of labeled antibodies, and Tween 20 reduces the possibility of hydrophobic attraction between tissue components and the gold particles. DNA labeling is achieved after grid incubation for 1 to 4 h with two successive 50-μl droplets of 10-nm gold-avidin conjugate (diluted [1:100] in appropriate buffer). The first drop removes excess buffer and the second drop provides the incubation medium. To remove the unbound gold conjugate, the grids are transferred to a series of 50-μl droplets of distilled water (5 $\times$ 2 min or more), the water excess being blotted with filter paper. The grids are then observed under the EM.

Interaction of the Complexes with Serum

Transfection of certain cell types by some cationic liposome compositions is sensitive to the presence of serum.[20–22] The inhibition of gene delivery by serum is considered to be one of the limitations to their application *in vivo*.[22] Therefore, the development of novel cationic liposome formulations for gene delivery should be accompanied by studies on the effect of serum on the interaction of lipoplexes with cells and on their transfection activity. These studies are performed using the protocols described previously and in the next chapter, except that cells are incubated with the lipoplexes in the presence of various amounts of serum.

According to results obtained in our laboratory using HeLa cells, and summarized in Table I, transfection activity mediated by highly positively charged complexes is significantly inhibited by serum, whereas neutral complexes are shown to be resistant to the presence of serum. It is interesting to note that the extent of lipoplex–cell interaction, including both cell association and fusion, is not affected by the presence of serum.

Interaction of Cationic Liposome–DNA Complexes with Plasma Proteins

The assay for determining the interaction of cationic liposome–DNA complexes with plasma proteins is based on the retention of the complexes by the 200-nm Anopore membrane (Whatman, Kent, UK), as described by Ogris *et al.*[23] The complexes are incubated with 2% human plasma for

[20] P. L. Felgner, T. R. Gadek, M. Holm, R. Roman, H. W. Chan, M. Wenz, J. P. Northrop, G. M. Ringold, and M. Danielsen, *Proc. Natl. Acad. Sci. USA* **84,** 7413 (1987).

[21] V. Ciccarone, P. Hawley-Nelsen, and J. Jessee, *Focus* **15,** 80 (1993).

[22] S. Li and L. Huang, *J. Liposome Res.* **6,** 589 (1996).

[23] M. Ogris, S. Brunner, S. Schuller, R. Kircheis, and E. Wagner, *Gene Ther.* **6,** 595 (1999).

TABLE I
EFFECT OF SERUM (10%) ON LIPOPLEX–CELL INTERACTIONS AND
TRANSFECTION ACTIVITY (HeLa CELLS)

Lipid/DNA charge ratio (+/−)	Cell association[18]	Fusion[18]	Transfection activity[16]
1/2	No effect	Slight enhancement	No effect
1/1	Strong inhibition	No effect	No effect
2/1	No effect	No effect	Slight inhibition
4/1	No effect	Slight enhancement	Strong inhibition

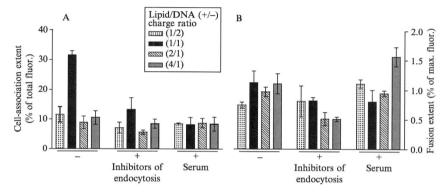

FIG. 2. Effect of inhibitors of endocytosis on association and fusion (lipid mixing) of DOTAP:DOPE lipoplexes with HeLa cells. (A) The extent of cell association is measured at 37°, after 60 min incubation at 37°, as described in the text. Values are expressed as a percentage of the total fluorescence. (B) Fusion experiments are performed as described in the text. The effects of inhibitors of endocytosis are evaluated as described previously. Values are expressed as a percentage of the maximal fluorescence. The effect of inhibitors of endocytosis was investigated by treating HeLa cells for 30 min with 1 μg/ml of antimycin A, 10 mM sodium fluoride, and 0.1% (w/v) sodium azide, at 37°. Lipoplexes are prepared at 1/2, 1/1, 2/1, and 4/1 lipid/DNA (+/−) charge ratios. For each culture dish, the cationic lipid concentration is approximately 7 μM, and the cell density is approximately 1.6×10^6 cells/ml. The data represent the mean ± standard deviation obtained from experiments carried out in triplicate (Adapted from Girão da Cruz et al.[18])

30 min at 37°. The membrane of microcentrifuge tube filters (Whatman, Kent, UK; 200-nm Anopore membrane) is saturated with 500 μl bovine serum albumin (1 mg/ml) to reduce nonspecific interaction of plasma proteins with the membrane and washed three times with 500 μl of HBS. The mixture of complexes and plasma is applied onto the membrane, filtered by centrifugation at 10,000g, and washed three times with 500 μl HBS. The residue (complex-bound plasma proteins) is eluted with 100 μl

HBS containing 5% (w/v) SDS (eluate). Samples (eluate) and control (HBS solution treated in the same way) are diluted 10 times with HBS and 2 times with TRIS-HCl buffer (100 mM TRIS-HCl (pH 6.8), 4% (w/v) SDS, 20% (v/v) glycerol, 0.2% (w/v) bromophenol-blue), and the associated proteins are denatured in boiling water for 3 min. Aliquots of the treated and control samples are then applied onto a 7.5% SDS polyacrylamide gel and separated in a Mini-PROTEAN II electrophoresis cell (Bio-Rad, Hercules, CA). For immunological identification of the proteins, the gel is blotted on to a Hybond P (PVDF) membrane by means of semi-dry blotting (Bio-Rad) at 15 V for 10 min. Unspecific binding is blocked by incubating the blot in 5% milk powder in TBST (2.5 mM TRIS-HCl (pH 7.6), 15 mM NaCl, 0.1% Tween 20) for 2 h at RT. The blots are incubated for 1 h at RT in the antibody solutions (antibodies are diluted in TBST): rabbit anti-human complement C_3 (Serotec, Oxford, England) 1:4000; rabbit anti-human albumin (Rockland, Gilbertsville, PA) 1:4000. As second antibody, an alkaline phosphatase conjugated with goat–anti–rabbit IgG, is used (Amersham Pharmacia Biotech) 1:20000. After 1 h incubation at RT, the blots are washed several times with TBST and incubated with the alkaline phosphatase substrate ECF^{TM} substrate (Amersham Pharmacia Biotech, Freiburg, Germany) for 5 min at RT. Fluorescence detection is performed at 570 nm in a Storm-860 instrument (Molecular Dynamics, World Headquarters, Sunnyvale, CA), with the excitation at 450 nm.

Concluding Remarks

Although much progress has been made in gene delivery by use of cationic liposomes, we are still far from obtaining "ideal" complexes that would mimic viral vectors. It would be desirable to produce complexes of small size with a narrow distribution, while presenting a neutral or negatively charged surface (to prevent nonspecific interactions with blood components) that ensure complete protection of DNA, exhibit specific targeting, have the ability to promote efficient intracellular delivery of carried material and to facilitate its translocation into the nucleus, thus leading to high and sustained levels of transgene expression without causing cytotoxicity. Thus, before embarking on extensive and expensive animal experiments and clinical trials, it is advisable to pursue fundamental research focused on the mechanisms by which lipoplexes are formed and deliver their DNA, as well as on strategies through which the different biological barriers they face can be overcome. The techniques outlined in this chapter will help in the identification of lipoplex parameters that enhance gene delivery and expression.

[20] Calorimetry of Cationic Liposome–DNA Complex and Intracellular Visualization of the Complexes

By Abdelatif Elouahabi, Marc Thiry, Véronique Pector, Jean-Marie Ruysschaert, and Michel Vandenbranden

Introduction

Cationic liposome–mediated gene transfer has become a widely used tool for *in vitro* transfection of eukaryotic cells and a promising and safe alternative for *in vivo* gene therapy applications. It involves the formation of a cationic lipid–DNA complex that interacts efficiently with the cell surface, leading to the entry and expression of the exogenous DNA. However, the use of cationic lipids as DNA delivery systems for gene therapy is still limited by the relatively low efficiency of the gene transfer process compared with viral vectors (e.g., adenoviral vectors).[1] Improvement of the *in vitro* and *in vivo* cationic lipid–mediated gene transfection efficiency can be attempted by use of trial and error approaches consisting of synthesizing and testing a large number of cationic lipid derivatives.[2–4] Alternately, a deeper understanding of the intermediate steps and mechanisms involved in the gene transfer process mediated by cationic lipids could allow the design of new rational strategies to improve the transfection efficiency.

Two key steps are involved in the cationic lipid–mediated gene transfer process:

1. Interaction between the cationic liposomes and DNA leading to formation of a complex.

2. Interaction between the complex and the target cells leading to DNA entry and expression of the transgene. We have studied the former step using calorimetric approaches, (i.e., isothermal titration calorimetry (ITC) and differential scanning calorimetry (DSC), whereas the later was addressed using cell-imaging techniques (i.e., immunofluorescence and

[1] R. G. Crystal, *Science* **270,** 404 (1995).

[2] J. H. Felgner, R. Kumar, C. N. Sridhar, C. J. Wheeler, Y. J. Tsai, R. Border, P. Ramsey, M. Martin, and P. L. Felgner, *J. Biol. Chem.* **269,** 2550 (1994).

[3] E. R. Lee, J. Marshall, C. S. Siegel, C. Jiang, N. S. Yew, M. R. Nichols, J. B. Nietupski, R. J. Ziegler, M. B. Lane, K. X. Wang, N. C. Wan, R. K. Scheule, D. J. Harris, A. E. Smith, and S. H. Cheng, *Hum. Gene Ther.* **7,** 1701 (1996).

[4] R. P. Balasubramaniam, M. J. Bennett, A. M. Aberle, J. G. Malone, M. H. Nantz, and R. W. Malone, *Gene Ther.* **3,** 163 (1996).

immunogold staining combined with confocal and electron microscopy, respectively).

Calorimetry of Cationic Liposome–DNA Complex

The first limiting step in gene transfer mediated by cationic liposomes is the formation of an appropriate lipid–DNA complex. Parameters such as DNA/lipid ratio, concentrations of complexants, ionic strength, temperature, the number and degree of ionization of the cationic functions, and, finally, the chemical structure of the cationic lipid all determine the size, shape, global charge, and structure of the complexes and have a profound effect on the success of transfection.

Such a sensitivity to initial conditions could be explained by the fact that cationic liposomes, rather than being inert spheres, can undergo spectacular rearrangements in the presence of DNA. This is perhaps best illustrated by the large diversity of size, shape, and molecular structure that has been described for cationic lipid–DNA complexes. From stacks of lipid bilayers sandwiching DNA rolls to the "spaghetti-meatballs" imagery, there are many different descriptions for the structure of lipid–DNA complexes that can be attributed to the different cationic molecules being used, although it also depends on the conditions of complex formation.[5–10] In fact, the complexation process seems to proceed through a series of steps, the last ones involving reorganization of the liposome structure. The very last steps are fully achieved when reaching a critical DNA/lipid ratio.[11] In such a perspective, it would be interesting to monitor the complexation process while "titrating" the cationic lipids with DNA (or the opposite way).

Interaction between two macromolecules can be followed by ITC.[12] It relies on the measurement of the heat of the complexation reaction. This approach has been used for DNA and several polycationic ligands.[13,14]

[5] J. O. Radler, I. Koltover, T. Salditt, and C. R. Safinya, *Science* **275,** 810 (1997).
[6] I. Koltover, T. Salditt, J. O. Radler, and C. R. Safinya, *Science* **281,** 78 (1998).
[7] B. Sternberg, F. L. Sorgi, and L. Huang, *FEBS Lett.* **356,** 361 (1994).
[8] D. Simberg, D. Danino, Y. Talmon, A. Minsky, M. E. Ferrari, C. J. Wheeler, Y. Barenholz, *J. Biol. Chem.* **276,** 47453 (2001).
[9] C. R. Safinya, *Curr. Opin. Struct. Biol.* **11,** 440 (2001).
[10] V. Cherezov, H. Qiu, V. Pector, M. Vandenbranden, J.-M. Ruysschaert, and M. Caffrey. *Biophys J.* **82,** 3105 (2002).
[11] V. Pector, J. Backmann, D. Maes, M. Vandenbranden, and J. M. Ruysschaert, *J. Biol. Chem.* **275,** 29533 (2000).
[12] L. Indyk and H. F. Fisher, *Methods Enzymol.* **295,** 350 (1998).
[13] I. Haq and J. Ladbury, *J. Mol. Recognition* **13,** 188 (2000).
[14] D. Matulis, I. Rouzina, and V. A. Bloomfield, *J. Mol. Biol.* **296,** 1053 (2000).

However, there are currently very few ITC studies available describing cationic liposome–DNA interactions.[11,15,16] In this chapter, we describe the experimental conditions to perform such a study. In addition, we show how another calorimetric technique, DSC, can be used with certain classes of cationic lipids to characterize further the complexation process. Suggestions are made about other methods that can provide information on processes occurring at the molecular level.

Choice of a Cationic Lipid

Although the situation is improving with the advent of very sensitive calorimetric systems, most calorimetric techniques require large amounts of lipids (of the order of 5–10 mg/measurement for ITC and 0.5–1 mg/ measurement for DSC). Many commercial lipids used for transfection are expensive and are already formulated as a liposome suspension, making them unsuitable for physicochemical studies in which concentration and buffer choice are essential. Nevertheless, a few cationic lipids are either available in bulk quantities (such as N-[1-(2,3-Dioleoyloxy)]-N,N,N-trimethylammonium propane, dimyristoyltrimethylammonium propene, didodecy-dimethyl-diamonium bromide) at an affordable price or can be easily synthesized.[17,18] If DSC experiments are to be planned, lipids with a well-defined gel-to-liquid-crystal transition temperature in the 0–70° range (at higher temperatures DNA melting starts) must be selected, which, in most cases, restricts the choice to lipids bearing two saturated alkyl chains with 12–16 carbon atoms or long (C18) monounsaturated chains. Here we describe the use of a purely synthetic cationic lipid with two saturated C14 chains: N-t-butyl-N'- tetradecyl-3-tetradecylaminopro-pionamidine, which is synthesized from tetradecylamine, tertiobutylchloride, and acrylonitrile by the general method of amidine synthesis described by Gordon et al.[19] and further adapted by Fuks.[20,21] The reaction yield is close to 60%. The resulting cationic lipid is purified by chromatography

[15] M. T. Kennedy, E. V. Pozharski, V. A. Rakhmanova, and R. C. MacDonald, *Biophys. J.* **78,** 1620 (2000).

[16] B. A. Lobo, A. Davis, G. Koe, J. G. Smith, and C. R. Middaugh, *Arch. Biochem. Biophys.* **386,** 95 (2001).

[17] X. Gao and L. Huang, *Biochem. Biophys. Res. Commun.* **179,** 280 (1991).

[18] R. C. MacDonald, V. A. Rakhmanova, K. L. Choi, H. S. Rosenzweig, and M. K. Lahiri, *J. Pharm. Sci.* **88,** 896 (1999).

[19] J. E. Gordon and G. C. Turell, *J. Org. Chem.* **24,** 269 (1956).

[20] R. Fuks, *Tetrahedron.* **29,** 2147 (1973).

[21] F. Defrise-Quertain, P. Duquenoy, R. Brasseur, P. Brak, B. Caillaux, R. Fuks, and J.-M. Ruysschaert, *J. Chem. Soc. Chem. Commun.* **13,** 1060 (1986).

on alumina (alumina 90, acidic, Merck, Darmstadt, Germany) and crystallization ($3\times$) at $4°$ in hexane and stored at $-20°$.

Choice of Nucleic Acid

The large quantities of cationic lipids necessary for calorimetric measurements require equivalent quantities of nucleic acids. Many applications of cationic liposomes concern transfer of bacterial plasmid DNA into mammalian cells. Transfer of synthetic oligonucleotides has also been performed, but less frequently. Synthetic oligonucleotides can be produced on a large scale and can be purified satisfactorily. Crude genomic DNA from mammalian tissues, although available in very large quantities at reasonable price, is heterogeneous, and its purification is less improved than plasmid DNA. When plasmid DNA is available, it is preferred over genomic DNA, because subsequent comparisons with transfection of reporter genes are feasible. Many types of common plasmid DNA can be purified easily in quantities of 1–10 mg at once. In the present contribution, we use a classical pCDNA3.1 plasmid amplified in *Escherichia coli* DH5 alpha and purified with a Qiafilter plasmid kit (Qiagen Inc., Qiagen, Valencia, CA, USA), following exactly the recommendations of the supplier. The resulting DNA always has an absorbance ratio (A_{260nm}/A_{280nm}) of 1.8–1.9 and gives reproducible transfection results using the luciferase reporter gene inserted in its multiple cloning site.

Liposome Preparation

The cationic lipid (diC14-amidine) is dissolved in chloroform, dried under a nitrogen stream, and stored overnight in a desiccator under vacuum. Liposomes are formed by addition of 10 mM HEPES buffer, pH 7.3, to the lipid film and vortex mixing at $60°$ for 10 min, which is above the transition temperature (Tc = $23°$)[21] of the diC14-amidine liposomes. A stock suspension of liposomes at 5 mg/ml is made. In the case of cationic lipids available in the base form, which is the case for diC14-amidine, care is taken to adjust the pH slightly with microliter portions of 1 M HCl when necessary. For critical micellar concentration (cmc) determination using ITC measurements, liposomes are dialyzed overnight against 50 volumes of the same batch of 10 mM HEPES buffer, pH 7.3, as the one used for filling the calorimeter cell. The liposomes can be stored at $4°$ provided they are heated up again before each experiment (this is important if the transition temperature is higher than $4°$). Especially for ITC measurements, it is important before each experiment to degas the liposomal suspension under vacuum while stirring for 10 min.

Isothermal Titration Calorimetry

With the recent development of new calorimeters that can measure heat change of 1 microcal/s corresponding to a temperature change of 10–6° in a volume of approximately 1 ml, measurements of biological macromolecule interaction have become more feasible than previously. At this order of magnitude, however, very subtle effects such as heat of dilution of the reactant that is injected in the reaction chamber, speed of injection, stirring rate, and thermal equilibration of the reactants have to be considered.

Experiments described here are made with an Omega titration microcalorimeter (MicroCal Inc., Northhampton, MA). The small volume of the reaction chamber (1.3 ml) and its shape allow rapid mixing of the reactants. The reactant to be injected is transferred into the injection syringe, which is preequilibrated at room temperature. After insertion of the syringe into the reaction chamber, thermal equilibrium is reached after 10 min. The stirrer is then turned on, and the calorimeter is calibrated. The baseline is then recorded with the stirrer on at constant speed (400 rpm). Stepwise injection by portions of 10 μl is computer controlled. After each injection, mixing occurs within a few seconds, and an endothermic or exothermic peak is recorded that lasts for approximately 1 min, followed by a slow return to the baseline (Fig. 1A,B, upper part). Usually 6–7 min is necessary until equilibrium is reached. The reaction enthalpy is evaluated by integrating the area under each peak using the Origin (Microcal) software (Fig. 1A,B, lower part). A correction is made to take into account the heat of dilution of the injected sample at each injection step. This is performed by injecting the reagent in the reaction chamber containing buffer alone.

Isothermal titration colorimetry measurements give the enthalpy of the global reaction taking place and include contributions that can be either exothermic, such as electrostatic interaction between cationic lipid headgroups and negatively charged phosphate groups of DNA, or endothermic, such as partial dehydration of anionic or cationic groups on complexation, and more importantly, possible rearrangement of liposomes into new structures. In the latter case, the phenomenon is slower than the initial complexation process and can be discriminated partially on the thermogram as a slower endothermal component (positive lobe beside the peak), whereas the first step of the reaction (complexation) occurs at a fast rate because of a high concentration of the reactants. In this case, the second (endothermic) step is the limiting step. This is typically the case with diC14-amidine when working at 10 mM cationic lipid (Fig. 1B). At lower reactant concentrations, however, both endothermal and exothermal contributions are superimposed and give a single peak whose result depends

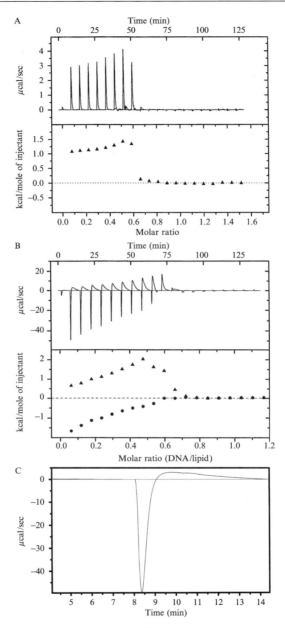

Fig. 1. Titration calorimetry (ITC) of diC14-amidine liposomes with plasmid DNA (pcDNA 3.1) in a 10 mM HEPES buffer (pH 7.3) at 28°. (A) Low lipid concentration (0.93 mM diC14-amidine). Upper part, heat capacity tracings as a function of time. Each peak corresponds to the injection of 10 μl of a DNA solution (8.8 mM in nucleotides) in a 1.33-ml

on the relative magnitudes of the exothermic and endothermic contributions (Fig. 1A). In the case of diC14-amidine, the endothermal component seems to originate from rupture and rearrangement of vesicles as suggested by a series of vesicle stability assays.[11] Numerical simulation of a simplified kinetic model taking into account experimental concentrations of free liposomes and DNA can predict the conditions for observing a separate exothermal component at the beginning of the reaction.[11] This model uses the reaction scheme:

$$D + L \rightleftharpoons C \rightarrow F$$

where D represents the DNA population, L the lipid, C the electrostatic complex between DNA and lipid, and F the rearranged complex after destabilization. The model predicts a fast exothermal component at the beginning of the reaction at high-lipid concentrations. This component is absent at low-lipid concentrations. The exothermal component can be interpreted as resulting from a temporary excess of electrostatic complex waiting to proceed to the next (rearrangement) step. In this model, the rearrangement step is the limiting step.

Another useful application of ITC can be the determination of the (liposomal) cmc of cationic lipids. It corresponds to the limiting monomer concentration above which a monomer–liposome or monomer–micelle (depending on the type of molecule) equilibrium exists. The cmc value can be used to compare cationic lipids with well-known membrane lipids. Perhaps a more crucial point concerning transfection is the possibility to check whether low-lipid concentrations (10^{-5}–10^{-6} M) used in many experiments on cultured cells are compatible with the existence of liposomes or monomers only, and this parameter may vary greatly depending on the cationic molecule (or mixture) being used. The cmc can be evaluated by measuring the heat of demicellization when liposomes are diluted into buffer. At the (liposomal) cmc, there is an increase in the endothermic effect of dilution (Fig. 2). The difference between the apparent heat of dilution before and after the cmc limit corresponds to the heat of micellization per mole of lipid. The average heat of micellization is 8 ± 3 kcal/mole ($n = 5$) and the average cmc, $3.9 \times 10^{-7} \pm 0.2$ M.

cell filled with the liposomal suspension. Lower part, binding isotherm resulting from integration over time; the reaction enthalpy (kcal/mol of injectant) is plotted as a function of the DNA/lipid molar ratio. (B) High lipid concentration (10 mM diC14-amidine). Upper part, calorimetric recording as a function of time. Each peak corresponds to the injection of 10 μl of the DNA solution (76 mM nucleotides) in the liposomal suspension. Lower part, binding isotherms for endothermal (triangles) and exothermal (dots) components. (C) Enlargement of the first peak of titration observed in A showing the negative (exothermal) and positive (endothermal) components of the peak observed at high lipid concentration. (Adapted from *J. Biol. Chem.* **275**, 29533 (2000), with permission.)

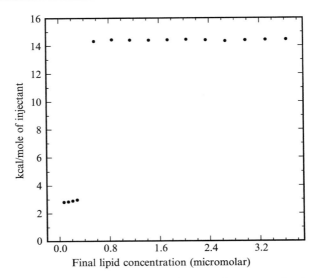

Fig. 2. Determination of the critical micellar (liposomal) concentration (cmc) of diC14-amidine by ITC. The dots correspond to the apparent heat of dilution per mole of injected lipids as a function of the final lipid concentration in the ITC cell. Concentration of liposomes injected in the calorimeter, 1.87×10^{-5} M.

The ITC experiments at various DNA/lipid ratios we have shown in the preceding are based on titration of cationic liposomes with DNA. It is conceivable to perform the titration of DNA (contained in the reaction chamber) with liposomes. Although there is a recent example of such an experiment,[15] injecting the liposome in a DNA solution makes interpretation of the data rather puzzling. A possible reason could be the strong heat contribution caused by a liposome-to-monomer equilibrium shift when liposomes are diluted in the reaction chamber. In principle, this process should affect the first injection only. Considering a final concentration of 1×10^{-5} M cationic lipids in the reaction chamber, the cmc is largely exceeded, and there is a saturating concentration of monomers. It seems clear that other factors are responsible for the irregular aspect of thermograms we have observed. In particular, we do not know what could be the effect of the binding of a monomer to DNA.

Differential Scanning Calorimetry

Whereas ITC is sensitive to all the interactions taking place during the DNA–liposome complexation process, except those for which the enthalpic contribution is weak, it gives no precise information about the possible physical changes occurring specifically at the level of the lipid matrix. Under certain circumstances, DSC can be used to characterize the physical state of the lipid phase after interaction with DNA.

During a DSC experiment, a sample is heated over a range of temperature. At some point, the material starts to undergo a chemical or physical change that releases or absorbs heat. If one plots the heat released or absorbed as a function of temperature, the area under the curve represents the total heat or enthalpy change (ΔH) for the entire process.

The DSC technique has been used extensively for the determination of the transition temperatures between various lipid phases, mainly the gel-to-liquid crystal phase transition in bilayers made of phospholipids. There are far fewer data for cationic lipids.[22–26] Figure 3 shows an experiment with diC14-amidine liposomes, using a MC-2 ultrasensitive differential scanning calorimeter (Microcal, Inc.) with twin-1.2 ml cells. The samples are scanned at a 1°/min rate in the 10–40° range. At very low DNA/lipid (D/L) molar ratios, the diC14-amidine lipid displays a gel-to-liquid crystal transition at 23°, as revealed by the strong enthalpic peak at this temperature. The area under the peak corresponds to the enthalpy change (ΔH) associated with the phase transition and is integrated using the Microcal[TM] software. When DNA is added, the ΔH decreases as a function of the D/L ratio, and there is a slight shift in the temperature of the maximum that rises until D/L = 0.8 and then decreases to a temperature slightly lower than the transition temperature of the pure lipid. Loss of peak sharpness and decrease of ΔH are generally associated with a loss of cooperative interactions between the hydrocarbon chains of lipids. In this example, interaction of DNA with cationic lipids clearly modifies their thermotropic phase properties. Another application of DSC consists in revealing a phase separation in a binary mixture made of a phospholipid and a cationic lipid.[22,23] This will not be addressed here.

Complementary Methods

Although ITC and DSC are convenient methods for monitoring complex formation as a function of the D/L ratio and allow determination of

[22] M. Subramanian, J. M. Holopainen, T. Paukku, O. Eriksson, I. Huhtaniemi, and P. K. Kinnunen, *Biochim. Biophys. Acta.* **1466,** 289 (2000).

[23] R. Zantl, L. Baicu, F. Artzner, I. Sprenger, G. Rapp, and J. O. Rädler, *J. Phys. Chem.* **103,** 10300 (1999).

[24] R. P. Balasubramaniam, M. J. Bennett, A. M. Aberle, J. G. Malone, M. H. Nantz, and R. W. Malone, *Gene Ther.* **3,** 163 (1996).

[25] A. Koiv, P. Mustonen, and P. K. Kinnunen, *Chem. Phys. Lipids.* **70,** 1 (1994).

[26] J. Wang, X. Guo, Y. Xu, L. Barron, and F. C. Szoka, Jr., *J. Med. Chem.* **41,** 2207 (1998).

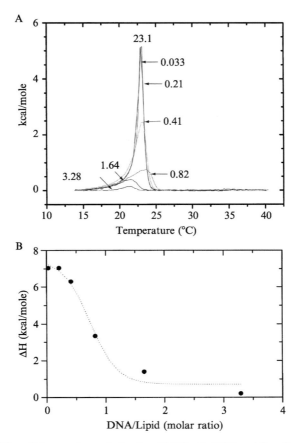

Fig. 3. Differential scanning calorimetry (DSC) of diC14-amidine liposomes in the presence of increasing amounts of plasmid DNA (pcDNA 3.1). Upper part, DSC trace as a function of temperature. Lower part, excess heat capacity peak area (kcal/mol) versus the DNA/lipid ratio. Concentration of the lipid in the calorimetric cell: 0.93 mM. Buffer, HEPES 10 mM, pH 7.3, scan rate: 1°K/min. (Adapted from *J. Biol. Chem.* **275,** 29533 (2000), with permission.)

energies associated with macroscopic processes, they do not give access to features at the structural or molecular level. We and others[10,11,15,16] have used various methods such as liposome leakage assays, lipid bilayer intermixing, particle size analysis, and finally small-angle x-ray diffraction analysis to get a more detailed view of structural changes occurring at different D/L ratios.

Intracellular Visualization of the Complex

Monitoring the intracellular trafficking of the cationic liposome–DNA complex requires detection of the nucleic acid and the lipidic parts of the complex. Ideally, both components of the complex should be visualized simultaneously using the same detection technique. We describe later a double-fluorescence labeling technique combined with confocal laser scanning analysis to follow the cellular entry and fate of the two components (DNA and liposomes) of the complex. Immunogold staining combined with electron microscopy is used to study the subcellular localization of the complex.

Labeling of the Plasmid DNA with Bromodeoxyuridine

Specific immunodetection of exogenous plasmid DNA in the cell often requires chemical modification of the plasmid before its transfection. Bromodeoxyuridine (BrdU) is a nucleotide analogue that, once incorporated into a DNA strand, can be detected specifically using specific anti-BrdU antibodies. The plasmid DNA pCMV-CAT (containing the chloramphenicol acetyl transferase, CAT, bacterial gene under the control of the CMV immediate early promoter/enhancer) is extracted using the alkaline lysis method[27] and further purified using a Qiagen Kit according to manufacturer's instructions. To label the pCMV-CAT plasmid DNA with BrdU, a modified nick-translation technique is used.[28] A 100-μl reaction mixture is prepared to contain 50 mM TRIS/HCl, pH 7.5, 10 mM MgSO$_4$, 0.1 mM Dithiothreitol, 50 μg/ml (BSA), 0.1 mM each nucleotide (except deoxythymidine [dTTP]), 10 μg DNA, 2 μl (10 units/μl) DNA polymerase I (Boehringer), 2 μl of the DNA polymeraseI/DNase I mixture containing 0.4 pg of DNase I (cat no: N5000, Amersham, U.K.), and 0.1 mM BrdU triphosphate (Sigma). The mixture is incubated for 2 h at 16°; then the reaction is stopped by adding 5 μl of 0.5 M ethylenediamine tetranectic acid (EDTA). Unincorporated nucleotides are eliminated on a Sephadex G50 column (gravity flow or spin column) (Amersham Biosciences, uppsala, Sweden). Under the conditions used here for nick translation, the low DNase I/DNA ratio and the increased DNA polymerase I/DNase I ratio allowed the incorporation of BrdU into the plasmid DNA without significant alteration of plasmid integrity. Agarose gel electrophoresis analysis revealed that the labeled plasmid DNA still migrated as two main bands (i.e., relaxed

[27] J. Sambrook, E. F. Fritsch, and T. Maniatis, *in* "Molecular Cloning: A Laboratory Manual," Cold Spring Harbor Laboratory Press, New York, 1989.
[28] A. El Ouahabi, M. Thiry, V. Pector, R. Fuks, J. M. Ruysschaert, and M. Vandebranden, *FEBS Lett.* **414,** 187 (1997).

and supercoiled) even though the supercoiled form is present at lower amounts than untreated plasmid DNA control. The nick translation procedure described previously is highly reproducible by us and has also been used successfully by others.[29] As an alternative approach to nick translation, BrdU may be metabolically incorporated into plasmid by adding free BrdU into the bacteria growth medium during plasmid-containing bacteria amplification. We have Found that this method generates poorly BrdU-labeled plasmid, making its detection in the cells, after transfection, less sensitive than the nick-translated plasmid. One possible explanation is that the bacteria incorporates endogenous dTTP more efficiently into newly synthesized DNA during replication than the exogenous analogue BrdU.

Liposome Preparation and Labeling

The cationic lipid diC14-amidine is synthesized as previously described.[30–32] DiC14-amidine liposomes are prepared by the ethanol injection method as described.[33] N-(lissamine rhodamine B sulfonyl) phosphatidylethanolamine (Rh-PE) (Avanti Polar Lipid Inc., Alabaster, AL, USA) is incorporated into the liposomes as follows: 1 mg of diC14-amidine (in chloroform) is mixed with 1 mol% Rh-PE (in chloroform) in an Eppendorf tube, and the solvent is evaporated under N_2. The lipid film is dissolved in 20 μl of ethanol, pipetted up into an automatic pipette, and injected into 1 ml of prewarmed (55–60°) HEPES-buffered saline (HBS) under vigorous vortexing (avoiding contact between the pipette tip and the buffer). Lipofectin[TM] (a commercially available cationic liposome suspension consisting of an equimolar mixture of the cationic lipid DOTMA and the neutral helper lipid phosphatidylethanolamine) is purchased from Invitrogen, Carlsbad, CA, USA. One milligram of Lipofectin suspension is lyophilized in a glass tube. Lipids are dissolved in chloroform, and 1 mol% Rh-PE is added. The chloroform is evaporated in a rotary evaporator under a stream of N_2, and trace chloroform is removed by vacuum dessication overnight. The lipid film is resuspended in bidistilled water by vigorous vortexing and briefly sonicated in a Branson-1200 bath sonicator at 30 W (Branson Europa, The Netherlands).

[29] R. Wattiaux, M. Jadot, N. Laurent, F. Dubois, and S. Wattiaux-De Coninck. *Biochem. Biophys. Res. Commun.* **227,** 448 (1996).
[30] F. Defrise-Quertain, P. Duquenoy, R. Brasseur, P. Brak, B. Caillaux, R. Fuks, and J. M. Ruysschaert, *J. Chem. Soc, Chem. Commun.* **13,** 1060 (1986).
[31] D. G. Neilson, *in* "The Chemistry of Amidines and Imidates," Wiley, New York, 1975.
[32] M. Van Den Bril and R. Fuks, *Bull. Soc. Chim. Belg.* **89,** 433 (1980).
[33] A. El Ouahabi, V. Pector, R. Fuks, M. Vandebranden, and J. M. Ruysschaert, *FEBS Lett.* **380,** 108 (1996).

Transfection, Immunofluorescence Staining, and Confocal Microscopy Analysis

BHK21 cells (obtained from the ATCC Manassas, VA, USA) are grown in Glasgow modified Eagle medium supplemented with 5% Fetal Bovine Serum, 10% tryptose phosphate broth, 2 mM glutamine, and antibiotics. Cells are kept at 37° in humidified, 5% CO_2 atmosphere and are passaged every 2 days. For immunofluorescence experiments, 2×10^4 BHK21 cells are plated per well in light-well Lab-Tek chamber slides (Nunc, Denmark). BrdU-labeled DNA (0.2 μg) and 0.4 μg of Rh-PE–labeled diC14-amidine or 1 μg of Rh-PE–labeled Lipofectin are diluted each in 10 μl of HBS mixed and incubated at room temperature for 10–15 min. The complexes are further diluted with 80 μl of serum-free medium and applied to cells for 5, 15, 30, 60, 120, and 360 min. At the end of the incubation, the cells are washed three times with serum-free medium, once with phosphate-buffered saline (PBS), and then fixed with freshly prepared 2% paraformaldehyde in PBS at room temperature for 30 min. Fixed cells are incubated in 2 N HCl for 1 h at 37°. The HCl is washed out, and acidity is neutralized by incubation in 0.1 M borate buffer, pH 8.5, for 2×5 min. Cell are washed with PBS for 3×5 min and incubated with the primary anti-BrdU antibody (Roche, Switzerland) at 3 μg/ml in PBS containing 0.5% bovine serum albumin (BSA) for 1 h at room temperature. After 2×5 min washes with PBS and a 1×5 min wash with PBS containing 0.5% BSA, the cells are incubated with goat anti-mouse IgG coupled to Fluorescein isothiocyanate (FITC) (Amersham, Uppsala, Sweden) (diluted 1/100 in PBS containing 0.5% BSA) for 1 h at room temperature. Cells are washed again 3×5 min with PBS and 1×5 min with water. The chamber-forming plastic structures are removed from the slides, and the latter are allowed to air dry for 5–10 min. A small drop (± 5 μl) of Vectashield mounting medium (Vector Laboratories, Inc. Burlingame, CA) is added to each well, and a 2×2 cm coverslip is put on the slide, allowing the mounting medium to spread over the cell surface (two coverslips per slide). It is strongly recommended for confocal observations to make the mounting medium layer as thin as possible. This can be achieved by applying slight pressure onto the coverslip and removing excess mounting medium using a piece of blotting paper. Finally, the coverslip is sealed with nail varnish. Cells are observed under a Zeiss Axiovert fluorescence microscope, and images are obtained using a laser-scanning confocal microscope (MRC 1000, Bio-Rad, Hercules, CA, USA) equipped with an argon-krypton laser and Comos software (Bio-Rad Hercules, CA, USA). Serial images of rhodamine and FITC fluorescence at 0.72-μm Z intervals are recorded separately and then combined. Images are further analyzed using Imagespace software (Molecular Dynamics, Inc., Uppsala, Sweden).

Intracellular Distribution of Doubly Labeled Cationic Liposome–DNA Complex

Figure 4 shows fluorescence confocal micrographs of BHK21 cells treated with doubly labeled Lipofectin–DNA complex (Lipofectin is labeled with Rh-PE = red and DNA with FITC = green). After a 30-min incubation of cells with the labeled Lipofection–DNA complex, yellow aggregates (corresponding to the colocalization of red-associated liposome and green-associated DNA colors) are detectable at the surface of some cells (Fig. 4A, arrows). Yellow aggregate-associated cells are observed more frequently with increasing incubation times. Some yellow aggregates appear inside the cells, whereas others were adsorbed to the cell surface (Fig. 4B, arrowheads). The existence of intracellular yellow aggreagates supports the idea that the Lipofectin–DNA complex penetrates as a whole inside the cell before it dissociates. The cytosol becomes progressively red-stained, and after 6 h of incubation with the complex, much the cells show a strong cytosolic red staining (Fig. 4B). The observation that the cell cytosol stains red suggests that after entry of complexes into the cytoplasm, Lipofectin vesicles fuse with intracellular membranes, leading to redistribution through the intracellular membrane network. The DNA is released in the cytoplasm and diluted to such an extent that its fluorescence becomes too weak to be detected. This could explain why no green color is visible inside

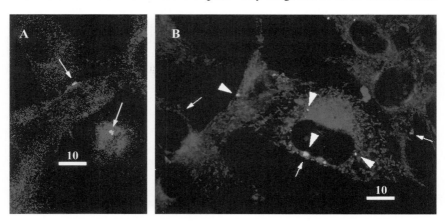

FIG. 4. Simultaneous visualization of DNA and Lipofectin in cells by confocal microscopy; BHK21 cells were treated for 30 min (A) or 6 h (B) with complexes made of BrdU-labeled DNA (detected by FITC-stained anti-BrdU antibodies) and Rh-PE–labeled Lipofectin liposomes. Yellow aggregates result from the superimposition of green (DNA) and red (liposomes). (A) Arrows indicate yellow aggregates located at the surface of the cells. (B) Arrows indicate green staining at the cell surface. Arrowheads indicate cell surface–bound or intracellular-located yellow aggregates. Note that the cytosol is red stained. Bars = 10 μm.

the cells. Interestingly, green staining is often observed at the cell surface (Fig. 4B, arrows), suggesting that some complexes could have fused with the cell membrane. This fusion process leads to accumulation of the DNA at the cell surface rather than its entry deep inside the cells. On the other hand, we failed to detect any significant green fluorescence inside the nucleus even 12 h after adding the complexes to cells (data not shown). Similar results are obtained when using doubly labeled diC14-amidine–DNA complex, except that no green staining is observed at the cell surface. This suggests that the diCl4-amidine–DNA complex, in contrast to the lipofectin–DNA complex, does not fuse with the cell surface.

Several controls are necessary to rule out possible artefacts caused by cell autofluorescence or nonspecific binding of the antibodies. Nontransfected cells or cells transfected with unlabeled plasmid DNA complexed to unlabeled liposomes are devoid of any red or green colors. Likewise, when cells are transfected with BrdU-DNA complexed to unlabeled liposomes and stained for BrdU without the acid treatment or the primary antibody, no green fluorescence can be detected in the cell. On the other hand, the use of this double-fluorescence detection technique requires additional controls to rule out the possibility that energy transfer could occur from one probe to another during laser scanning. For instance, FITC emission spectra partially overlap rhodamine excitation spectra, so that FITC emission could excite rhodamine, resulting in a fluorescence emission from the latter that can be mistakenly attributed to FITC. This can be ruled out by including controls labeled with FITC only. This inconvenience can be minimized by using consecutive laser scanning for each probe instead of simultaneous scanning both probes.

Detection of BrdU-labeled plasmid DNA after transfection into the cells offers several advantages. First, the same transfected cells can be used for immunofluorescence staining followed by confocal microscopy or immunogold staining followed by electron microscopy detection system (see later). BrdU offers also a flexibility in the choice of the primary and secondary antibodies and the possibility of using enhancement methods (e.g., biotin/streptavidin). A disadvantage of the method comes from the requirement of an acidic treatment of the fixed cells for BrdU recognition by the anti-BrdU antibodies. As an alternative method, the plasmid DNA can be labeled directly with a fluorescent nucleotide such as 2′-deoxyuridine-5′-triphosphate (dUTP) FITC using the nick translation technique described previously. Another recently reported possibility to fluorescently label the plasmid DNA is to use a peptide nucleic acid (PNA) covalently attached to a fluorophore.[34] However, these methods

[34] O. Zelphati, X. Liang, P. Hobart, P. L. Felgner, *Hum. Gene Ther.* **10,** 15 (1999).

introduce significant chemical modifications of the plasmid, which may alter its cellular behavior.

Immunogold Staining and Electron Microscopy Observations

To follow the cellular entry and fate of the BrdU-labeled plasmid DNA by electron microscopy, cells are seeded onto 24-well plates at 50,000 cells/well and transfected 24 h later. One microgram of BrdU-labeled plasmid DNA and 2 μg of diC14-amidine liposomes are separately diluted into 25 μl of HBS before being mixed and incubated for 10–15 min at room temperature. The complexes are further diluted into 1 ml of serum-free medium before being applied to cells for 5, 15, 30, 60, and 120 min. At the end of the incubation, cells are washed with PBS (0.14 M NaCl; 0.006 M Na$_2$HPO$_4$; 0.004 M KH$_2$PO$_4$; pH 7.2) and fixed in $situ$ for 120 min min at 4° in 4% formaldehyde in 0.1 M Sorensen's buffer (Na$_2$HPO4 and NaH$_2$PO$_4$, 0.1 M in phosphate, pH 7.4). After three washes with PBS, cells are dehydrated in graded ethanol solutions (30%, 70%, 95%, 100%) and embedded in Epon. The resin is polymerized over 3 or 4 days at 40°. Detection of BrdU incorporated into plasmid DNA is performed as follows. Ultrathin sections (60–90 nm) are prepared and mounted on gold grids, then incubated at room temperature in 5 N HCl for 30 min. The latter are incubated by floating them down on a drop of PBS containing normal goat serum blocking reagent (Sigma-Aldrich, Bornen, Belgium) diluted 1/30 and 1% BSA, then rinsed with PBS containing 1% BSA. The next step of the treatment is incubation with monoclonal anti-BrdU antibody (Boehringer Mannheim, Germany) diluted 1/300 in PBS containing 0.2% BSA and normal goat serum diluted 1/50 for 4 h at room temperature. After five washes with PBS containing 1% BSA, the sections are incubated for 1 h with goat anti-mouse IgG coupled to colloidal gold/10-nm diameter (Amersham) diluted 1/50 with PBS (pH 8.2) containing 0.2% BSA at room temperature. After four washes with PBS containing 1% BSA, the sections are rinsed in double-distilled water and dried before being stained with uranyl acetate and lead citrate. It should be emphasized that these experimental conditions are compatible with the in $situ$ immunodetection of BrdU incorporated into the DNA but do not make possible a clear recognition of cellular membranes.

Labeling is associated with aggregates composed of electron-dense rings of varying size, reminiscent of liposomes (Fig. 5A, arrows). Gold particles are specifically located over filaments that extend from the rings, suggesting that they may be plasmid DNA molecules (Fig. 5A, arrowheads). These filaments are not observed when liposomes alone are incubated with cells, supporting the conclusion that they represent plasmid DNA. After a

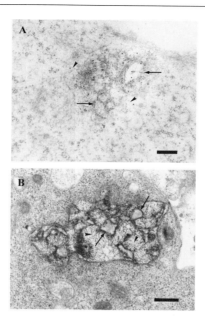

FIG. 5. Visualization of diC14-amidine/DNA complex entry into cells by electron microscopy; BHK21 cells were treated with diC14-amidine/DNA complexes for 3 h, fixed, and processed for immunogold labeling of BrdU-DNA (A) or nonspecific staining of the complexes (B). Arrows denote examples of vesicle structures, whereas arrowheads indicate examples of filament networks. Note that gold particles are located preferentially over the filaments. Bar in "A" = 0.3 μm. Bar in "B" = 1 μm.

15-to 30-min incubation with cells, these diC14-amidine–DNA aggregates are detected on the surface of a few cells (data not shown), but for longer incubation times (2 h), the aggregates can be detected in the cytosol (Fig. 5A). These results support the hypothesis that the diC14amidine–DNA complex penetrates into the cytosol as a whole and does not dissociate at the cell surface. No gold label is observed in the nucleus up to 24 h after the initial incubation of cells with the complexes. Similar results are obtained using the Lipofectin–DNA complex. When untransfected cells or cells transfected with liposomes alone or with BrdU-unlabeled DNA complexed with liposomes are used, the sections are devoid of label. Likewise, no label is observed when the primary antibody is omitted. Finally, gold lacking the antibody tag did not bind to the sections.

For monitoring entry of the transfecting plasmid DNA and its fate within cells, other strategies have been used. The first strategy, applied to ultrathin sections obtained after freeze-substitution and embedding of rapidly cryoimmobilized cells, involves detection of the unmodified DNA by *in situ* hybridization with digoxygenin-tagged oligonucleotide probes.[35] The hybrids are revealed with boronated anti-digoxygenin antibodies and imaged by energy-filtering transmission EM. This strategy offers the advantage of using unmodified DNA. Drawbacks come from the requirement of the denaturing steps, which may alter some cell compartments. Furthermore, the technique is not always sensitive enough to detect small quantities of unmodified transfecting DNA, because only DNA present at the surface of ultrathin sections of plastic-embedded cells can hybridize with the probe. Another strategy consists of linking gold particles (10 nm in diameter) through streptavidin to the biotinylated DNA before transfection.[35–37] After transfection, the gold-labeled DNA is detected by EM. The gold particles can be replaced with nanogold[38] and the ultrastructure imaged by energy filtering transmission EM.[35] Both approaches require labeling of the DNA before transfection. This raises the question of whether the modifications introduced into the DNA might alter the transfection process. This is a definite possibility when the DNA is tagged with large gold particles (10 nm in diameter). The nucleotide analogue BrdU is not expected to alter significantly the properties of the DNA. BrdU-labeled DNA migrates like unlabeled DNA in agarose gels,[39] and when the analog is incorporated into the genomic DNA of cells by metabolic labeling, DNA synthesis/replication is not impaired.

Nonspecific Staining of the Complexes and Electron Microscopy Observations

The immunogold strategy described previously has been optimized to facilitate specific detection of the DNA transfected into the cells, but it does not make possible the clear visualization of the cellular structures and especially cell membranes. During sample preparation for immunogold staining, the osmium tetroxide fixation step, which is required for good cell membrane visualization, is omitted, because it interferes with BrdU detection. In this section, we describe an EM approach that was optimized for cell membrane visualization.

BHK21 cells are seeded onto six-well plates at 500,000 cells/well and incubated overnight. Cells are treated with diC14-amidine–DNA complexes

[35] M. Malecki, *Scanning Microsc.* **10**, 1 (1996).

[36] J. Zabner, A. J. Fasbender, T. Moninger, K. A. Poellinger, and M. J. Welsh, *J. Biol. Chem.* **270**, 18997 (1995).

[37] D. S. Friend, D. Papahadjopoulos, and R. J. Debs, *Biochim. Biophys. Acta* **1278**, 41 (1996).

[38] J. F. Hainfeld and F. R. Furuya, *J. Histochem. Cytochem.* **40**, 177 (1992).

[39] A. El Ouahabi, M. Thiry, S. Schiffmann, R. Fuks, H. Nguyen-Tran, J. M. Ruysschaert, and M. Vandenbranden, *J. Histochem. Cytochem.* **47**, 1159 (1999).

for 2–3 h then washed and scraped off the plates. Cells are centrifuged at 350 g for 3 min to form pellets. Small fragments of the various pellets are fixed for 30 min at room temperature in 2.5% glutaraldehyde in 0.1 M Sorensen's buffer (pH 7.4). After washing in buffer, cells are postfixed in 2% osmium tetroxide for 30 min at room temperature, dehydrated in ethanol (30%, 1 × 10 min; 70%, 3 × 10 min), and embedded in Epon 812 resin (70% ethanol/Epon, 1/1, 2 × 10 min; 70% ethanol/Epon, 1/2, 2 × 10 min; pure Epon, overnight). The polymerization is achieved at 60°. Ultrathin sections (90 nm) are placed on 200-mesh nickel grids and stained with uranyl acetate and Reynold's lead citrate.[40] The samples are examined in a Jeol CX100 electron microscope at 60 kV.

DiC14-amidine–DNA complexes are identified on the ultrathin sections by their high electron density and appear as aggregations of ovoid structures (Fig. 5B, arrows) and filament network (Fig. 5B, arrowheads). After a 2- to 3-h incubation, these aggregates appear to be contained in large endocytotic vesicles that are quasi-filled with the aggregates (Fig. 5B), supporting the fact that endocytosis is involved in complex internalization into the cells. This approach was applied to transfected COS cells and permitted to reveal disruptions/destabilization of the endosomal membrane at the complex attachment sites.[28] The frequency of endosomal membrane destabilization depended on the cationic lipid used and could be directly correlated to the transfection efficiency.[28]

Concluding Remarks

Uses of cationic lipids for gene transfer have seen a dramatic development during the last decade. A few years after the first report by Felgner et al.[41] describing cationic lipid-mediated transfection of cultured cells in 1987, a gene therapy clinical trial using cationic lipids was initiated in 1993 by Nabel et al.[42] Despite this rapid progress in cationic lipid–mediated gene delivery applications, little was known about their mechanisms of action. Here we have described novel methods to evaluate the dynamics of complex formation and to follow the cellular pathways involved in the gene transfer process mediated by cationic lipids.

Incremental titration calorimetry and differential scanning calorimetry may serve as tools to better characterize the complexation process of

[40] E. Reynolds, *J. Cell. Biol.* **17**, 208 (1963).

[41] P. L. Felgner, T. R. Gadek, M. Holm, R. Roman, H. W. Chan, M. Wenz, J. P. Northrop, G. M. Ringold, and M. Danielsen, *Proc. Natl. Acad. Sci. USA* **84**, 7413 (1987).

[42] G. J. Nabel, E. G. Nabel, Z. Y. Yang, B. A. Fox, G. E. Plautz, X. Gao, L. Huang, S. Shu, D. Gordon, and A. E. Chang, *Proc. Natl. Acad. Sci. USA* **90**, 11307 (1993).

cationic liposomes with DNA. It may give an idea of the destabilization state of liposomes as a function of DNA added. This parameter may be correlated with transfection activity, because D/L ratios that give best transfection activity show nearly complete complexation in ITC measurements and complete bilayer destabilization in DSC measurements.[11] When generalized to different cationic lipids, it could help to establish and perhaps to predict at what stage of complexation and degree of destabilization cationic liposome–DNA complexes display their highest transfection activity.

The BrdU-based plasmid DNA detection system developed so far, combined with immunofluorescence and immunogold staining, permits us to follow unambiguously the cellular entry process of cationic lipid–DNA complexes. Furthermore, we have shown that the method can be applied to different cationic lipids, indicating that it may be used as a tool for studying cationic-lipid–mediated DNA entry into cells. BrdU is a nucleotide analogue that is not expected to alter significantly DNA properties and, therefore, is not likely to interfere with DNA trafficking in the cells. However, the BrdU-based detection system, under the used conditions, seems not to be sensitive enough to detect intranuclear plasmid DNA either by confocal microscopy or EM. One possibility to improve the sensitivity of BrdU-labeled DNA detection is to use cytohistochemical staining systems (e.g., using alkaline phosphatase–coupled secondary antibodies instead of fluorescence or gold labels), but this possibility still needs further investigation.

[21] Cationic Liposome–Protamine–DNA Complexes for Gene Delivery

By Jing-Shi Zhang, Song Li, and Leaf Huang

Introduction

Liposome–protamine–DNA complex (LPD), as a novel nonviral vector, was developed on the basis of our understanding about the cellular and molecular barriers to cationic lipid–based gene delivery systems.[1,2] Although cationic liposome–DNA complexes ("lipoplexes") have shown promise in gene transfection both *in vitro* and *in vivo*, their efficiency of

[1] X. Gao and L. Huang, *Biochemistry* **35**, 1027 (1996).
[2] F. L. Sorgi, S. Bhattacharya, and L. Huang, *Gene Ther.* **4**, 961 (1997).

gene transfer is not as good as viral vectors. Freeze-fracture electron microscopic studies have shown that cationic liposomes form "spaghetti and meatball"-like structures with DNA, indicating that DNA in the lipoplex is not well condensed.[3] Moreover, several studies have indicated that endocytosis-mediated cellular uptake of particles, including lipoplex, is limited by particle size.[4,5] On the basis of these studies, Gao and Huang[1] proposed to incorporate polycations into lipoplexes to condense the particles. The idea is based on a report that some polycations, such as polylysine, histone, and protamine, are effective to condense DNA to a highly compact structure about 30–100 nm in diameter.[6] The condensed structure may interact with cationic liposomes to form a viruslike vector, which could facilitate effective delivery of DNA into cells.

Protamines, small peptides (MW, 4000–4250), play a unique role in condensing DNA to form a compact structure in the sperm and delivering the sperm DNA into the nucleus of the egg after fertilization. Protamines are often obtained from mature testes of fish. They are highly positively charged because of a high arginine content. One mole of protamine sulfate contributes 21 moles of positive charge. In addition, protamine sulfate is a United States Pharmacopoeia compound used clinically as an antidote to heparin-induced anticoagulation. These facts suggest that protamine sulfate should be a safe and efficient condensing agent in nonviral gene delivery systems.

Formulations

Currently, there are two LPD formulations in use, which are different in cationic lipid composition. One is composed of dioleoyl-trimethylammonium propane (DOTAP) and cholesterol, and the other is composed of $3\beta(N-(N',N'-dimethylaminoethane)$ carbamoyl) cholesterol (DC-Chol) and dioleoylphosphatidylethanolamine (DOPE). The former has been developed for systemic gene delivery, whereas the latter has been mainly used for local gene transfer.

DOTAP/Chol LPD

DOTAP	1200 n
Cholesterol	1200 n
Protamine sulfate	60 μg
Plasmid DNA	100 μg

[3] B. Sternberg, F. L. Sorgi, and L. Huang, *FEBS Lett.* **356,** 361 (1994).
[4] P. Machy and L. D. Leserman, *Biochim. Biophys. Acta* **730,** 313 (1983).
[5] X. Zhou and L. Huang, *Biochim. Biophys. Acta* **1189,** 195 (1994).
[6] E. Wagner, M. Cotten, R. Foisner, and M. L. Birnsteil, *Proc. Natl. Acad. Sci. USA* **88,** 4255 (1991).

DC-Chol/DOPE LPD
    DC-Chol   36 n
    DOPE   24 n
    Protamine sulfate   80 μg
    Plasmid DNA   100 μg

Preparation of LPD Complex

Materials and Equipment

  A. Chemical reagents

    1. DOTAP (25.0 mg/ml) stock solution in chloroform (Avanti Polar Lipids, Alabaster, AL).
    2. DC-Chol (2.0 mg/ml) stock solution in chloroform (Avanti Polar Lipids).
    3. Cholesterol (20.0 mg/ml) (Sigma, St. Louis, MO) stock solution in chloroform.
    4. DOPE (20.0 mg/ml) stock solution in chloroform (Avanti Polar Lipids).
    5. Protamine sulfate-USP (10 mg/ml) (Elkins-Sinn, Cherry Hill, NJ).
    6. Plasmid DNA (2 mg/ml) in Tris-EDTA (ph 8.0) (TE) buffer.
    7. Sterile water (autoclaved or filtered by 0.2-μm-pore sterile filter).
    8. Dextrose solution (5 $\times$) (26.0% [w/v]) in sterile water.

  B. Equipment

    1. LiposoFast^{TM} extruder, with 1.0-ml syringes (Avestin, Ottawa, ON, Canada).
    2. Nuclepore$^{®}$ (1.0, 0.4, 0.1 μm) polycarbonate membrane filters (Corning$^{®}$, available from VWR International, West Chest, Pennsylvania).
    3. Cortex glass tube (30-ml).
    4. Polypropylene conical tube (50-ml).
    5. Eppendorf tube (1.5-ml).
    6. N_2 gas tank.
    7. Vacuum desiccator.
    8. Bath sonicator.

Procedure

The preparation procedure is illustrated in Fig. 1.

  A. Preparation of liposome

Liposomes are prepared by hydration of a thin film, followed by extrusion. This method is a rapid and simple protocol for the preparation of a

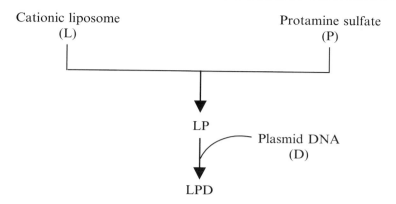

FIG. 1. Preparation of cationic liposome–protamine–DNA (LPD).

concentrated and homogenous suspension of small unilamellar liposomes for laboratory-scale experiment. Two steps are involved in the preparation procedure, the hydration of a thin lipid film formed by desiccation and the extrusion of the lipid suspension through a polycarbonate filter of desired pore size. The shear force produced by passage of the lipid suspension through the pores results in the formation of liposomes with unilamellar liposomes.

The preparation of DOTAP–Chol (10 mg/ml DOTAP) and DC-Chol–DOPE (2 mM DC-Chol) liposome stock solutions is described in the following. Both cationic liposome stock solutions are stable for several months when stored at 4°.

1. A 30-ml Corex tube is rinsed three times with chloroform. There is no need to dry the residual chloroform.
2. According to the proceeding formulations, the desired amount of lipids is mixed in the glass tube.
   a. For 2 ml of 10 mg/ml DOTAP–Chol liposomes, 0.8 ml DOTAP stock solution and 0.55 ml cholesterol solution are mixed.
   b. For 2 ml of 2 mM DC-Chol–DOPE liposomes, 1.07 ml DC-Chol stock solution and 0.1 ml DOPE stock solution are mixed.
3. In a chemical hood, chloroform is evaporated to form a thin lipid film in a glass test tube by blowing a stream of N_2 gas down the side of the tube while rotating the tube. To speed up the process, the tube can be dipped in a warm water bath.
4. The film is dried to completion by vacuum desiccation for 2–3 h.
5. Two milliliters of sterile water is added to the thin film.
6. The lipids are suspended by vortex mixing for 15 s on maximum speed several times. Several quick (e.g., 10–15 s) bursts in a bath

sonicator are needed in case lipids still remain attached to the tube side even after maximum speed vortexing.

7. The suspension is incubated either at room temperature for 2–3 h or at 4° overnight to allow complete hydration of lipids. The longer the hydration, the easier the extrusion. Overnight hydration is recommended as a common practice.

8. The lipid suspension is vortexed and heated for 5–10 min in a 65° water bath. If lipid aggregates are observed, the suspension is sonicated in a bath sonicator until all aggregates disappear and then returned to the water bath.

9. Two 1.0-μm polycarbonate filters are placed in the extruder.

10. The lipid suspension and the extruder are heated to about 65° for 5 min. This will allow easier extrusion and improved lipid mixing.

11. The lipid dispersion is extruded by passing the suspension through the extruder five times. The extruder is returned to the water bath.

12. Steps 9 through 11 are repeated using 0.4- and 0.1-μm polycarbonate filters sequentially to obtain small unilamellar liposomes with a mean diameter of about 100 nm.

13. The liposomes are stored at 4°.

B. Preparation of LPD

LPD used for *in vitro* transfection can be prepared by two different protocols, with comparable efficiency. The first involves premixing protamine with DNA followed by the addition of cationic liposomes, whereas the second involves premixing of cationic liposomes with protamine followed by the addition of DNA. To prepare concentrated LPD for *in vivo* applications, the latter protocol is preferred because preformed protamine–DNA complexes tend to form aggregates at high concentrations (see Fig. 2). The LPD complex is usually prepared immediately before use. The required amounts of liposome and protamine sulfate USP are mixed in a half volume of a master solution, and then an equal volume of diluted DNA solution is added dropwise to the liposome–protamine solution. For *in vitro* transfection, liposome–protamine and DNA are diluted with Hank's balanced salt solution (HBSS) or serum-free medium; in the case of intravenous injection, dextrose is added to the solutions of liposome–protamine and DNA to a final concentration of 5.2%. The following volumes and quantities are for 100 μg of plasmid DNA in a final volume of 600 μl (DOTAP–Chol liposomes) or 400 μl (DC-Chol–DOPE liposomes).

1. The liposomes and protamine are mixed in a diluted solution.

    a. For DOTAP–Chol LPD, the liposome-protamine (LP) solution is prepared in a 50-ml conical tube by adding 60 μl 5× dextrose

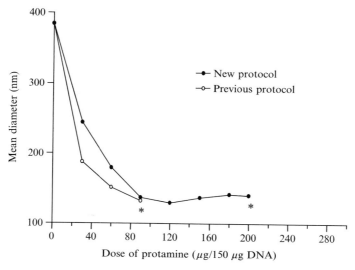

FIG. 2. Formation of aggregates and size of LPD as a function of the amount of protamine. Previous protocol: LPD is prepared by mixing protamine with plasmid DNA followed by the addition of DOTAP–Chol liposomes. New protocol: LPD is prepared by mixing DOTAP–Chol lipoosmes with protamine sulfate followed by the addition of diluted DNA. * Indicates the upper limit of protamine concentration above which aggregation will occur. (Reprinted with permission from Li *et al.*[7])

   solution, 148 μl sterile water, 86 μl DOTAP–Chol liposomes, and 6 μl protamine sulfate solution. The mixture vortexed gently for 10 s.

b. For DC-Chol–DOPE LPD, the LP solution is prepared in a 50-ml conical tube by adding 40 μl 5× dextrose solution, 122 μl sterile water, 30 μl DC-Chol–DOPE liposomes, and 8 μl protamine sulfate solution.

2. Dilute plasmid DNA

a. For DOTAP–Chol LPD, a DNA solution is prepared in an Eppendorf tube by adding 60 μl 5× dextrose solution, 190 μl sterile water, and 50 μl plasmid DNA. The tube is tapped gently to mix. Vortexing should be avoided.

b. For DC-Chol–DOPE LPD, a DNA solution is made up in an Eppendorf tube by adding 40 μl 5× dextrose solution, 110 μl

[7] S. Li, M. A. Rizzo, S. Bhattacharya, and L. Huang, *Gene Ther.* **5,** 930 (1998).

    sterile water, and 50 μl plasmid DNA. The tube is tapped gently to mix, but not vortexed.

3. While gently swirling the LP solution, the DNA solution is added dropwise. It takes about 5–10 s to add 200–300 μl of the DNA solution.

4. The complex is incubated at room temperature for 10–15 min before injection to allow complex maturation.

Notes

1. The use of protamine generally increases the transfection efficiency for most of the lipids tested. A variety of cationic liposome formulations are commercially available for lipofection of cells in tissue culture. However, the optimal formulations depend on the cell line studied and typically do not correlate with optimal formulations used for gene transfer of corresponding tissues *in vivo*. Thus, for *in vitro* transfection, we recommend testing commercially available liposomes following the manufacturer's instructions. Meanwhile, the lipid dose has to be reoptimized when the liposomes are used for cell transfection.

2. Transfection efficiency of LPD is affected by the overall charge ratio. Excess positive charge is of benefit to transfection efficiency of LPD in the lung. A 1:1 charge ratio of protamine to DNA is an optimal choice for DOTAP–Chol LPD.[8] In practice, one often needs to change the +/− charge ratio according to requirement. In this case, it is recommended to keep the ratio of protamine to DNA (1:1) as a constant and vary the amount of lipids. The charge ratio can be calculated according to how much charge each composition carries. One nanomole of DOTAP and protamine sulfate contribute 1 and 21 nanomoles of positive charge, respectively. Cholesterol is neutral. One microgram DNA contains approximately 3.1 nm of negative charge.

3. Plasmid DNA should be highly purified and endotoxin free. In addition, it is recommended that plasmid DNA (or other forms of DNA) be diluted in water or 5.2% dextrose solution. This is because the salt interferes with the charge–charge interaction between DNA and cationic lipid or polymer and causes aggregate formation.

4. White, stringlike precipitates may form during mixing. This is commonly caused by salts in the solution (TE, NaCl, etc.). The precipitates appear to be toxic to mice following intravenous

---

[8] S. Li and L. Huang, *Gene Ther.* **4,** 891 (1997).

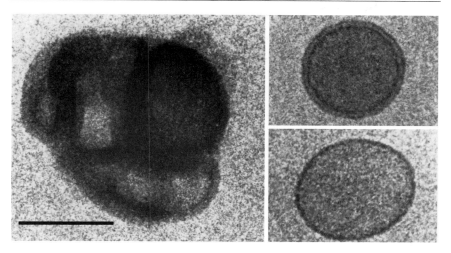

FIG. 3. Cryomicrograghs of liposomes (right lower panel), liposome–DNA complexes (left panel), and LPD particles (right upper panel). Bar = 100 nm. (Reprinted with permission from Li *et al.*[9] and Tan *et al.*[10])

injection. A proper mixing vessel, slow swirling of the LP solution, and dropwise addition of the DNA solution help to prevent the formation of the precipitate. If the precipitate forms, it is advisable to try preparing the complex again instead of proceeding with injection. With practice, precipitate formation can be avoided routinely. Moreover, it should be kept in mind that the complex should never be vortexed.

Evaluation of Size and Stability of LPD

A.   Particle size

The size of the LPD complex can be examined by negative stain electron microscopy and cryoelectron microscopy, and both methods have shown similar micrographs of LPD particles with an average size of less than 100 nm.[1,9] Cryoelectron microscopy shows clearly the highly condensed particles coated with a lightly stained, ring-shaped lipid structure (Fig. 3), suggesting the formation of viruslike particles with a core of protamine-condensed DNA inside and a lipid bilayer outside.[9,10]

[9] S. Li and Z. Ma, *Curr. Gene Ther.* **1,** 201 (2001).
[10] Y. Tan, M. Whitmore, S. Li, P. Frederik, and L. Huang, *in* "Gene Therapy Protocols" (J. R. Morgan, ed.), Vol. 69, p. 73, Humana Press Inc., Totowa, NJ, 2001.

B.   Stability on storage

The formation of lipoplexes is due to a polyelectrostatic interaction, so it is difficult to obtain a homogenous size distribution in a controlled manner.[11] The heterogeneous population of the complex usually further forms aggregates during storage, resulting in a decrease in transfection efficiency.[12] The condensation by protamine improves effectively the stability of the system. It can be stored at 4° for at least 4 weeks without any changes in size and activity.[10] In the presence of 10% sucrose, LPD is even more stable and can be left at 4° for more than 8 weeks.[13] However, LPD dispersion is highly sensitive to storage temperature, so the LPD complex should not be stored at room temperature.[13] Lyophilization is a common method to stabilize the pharmaceutical agents. The lyophilized LPD is of the same transfection efficiency as nonlyophilized LPD and can be stored at room temperature for a period of 8 weeks.[13]

In Vivo Gene Delivery of Liposome–Protamine–DNA

A.   DOTAP–Chol vector

The vector containing cholesterol has a slow and low disintegration in serum compared with DOPE-containing vector.[14] DOTAP–Chol LPD, with a net charge ratio of 5:1 (+/−), is used routinely for systemic injection. Following systemic administration to mice, the LPD complex transfects effectively all major organs (Fig. 4). The highest level of gene expression is found in the lung, predominantly in the pulmonary endothelial cells.[8] By using the LPD formulation, retinoblastoma (Rb), a tumor suppression gene, is expressed effectively in the lung of Rb (+/−) mice, which have primary tumors develop in the pituitary and later metastases develop in the lung. With repeated injections, metastatic tumor cells undergo spontaneous apoptosis, and the frequency of lung tumor metastases is decreased.[15]

[11] L. Huang, *in* "Nonviral Vectors for Gene Therapy" (L. Huang, ed.), p. 3, Academic Press, San Diego, CA, 1999.
[12] T. J. Anchordoquy, L. G. Girouard, J. F. Carpenter, and D. J. Kroll, *J. Pharm. Sci.* **87**, 1046 (1998).
[13] B. Li, S. Li, Y. Tan, D. B. Stolz, S. C. Watkins, L. H. Block, and L. Huang, *J. Pharm. Sci.* **89**, 355 (2000).
[14] S. Li, W. C. Tseng, D. B. Stolz, S. P. Wu, S. C. Watkins, and L. Huang, *Gene Ther.* **6**, 585 (1999).
[15] A. Y. Nikitin, M. I. Juarez-Perez, S. Li, L. Huang, and W. H. Lee, *Proc. Natl. Acad. Sci. USA* **96**, 3916 (1999).

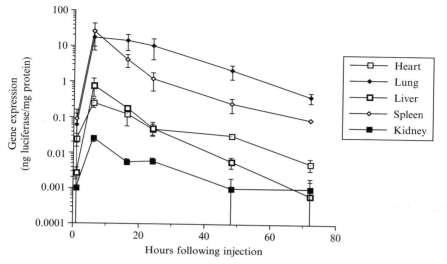

Fig. 4. Time-course of transgene expression following intravenous injection of LPD to mice. (Reprinted with permission from Li *et al.*[14])

B. DC-Chol–DOPE vector

DC-Chol–DOPE LPD, with a net charge ratio of 1:1 (+/−), is efficient for brain gene transfer by intracranial injection. It was used for intraventricular delivery of a gene-encoding aspartoacylase (ASPA) to treat Canavan disease.[16] Following intracranial injection, the LPD complex diffused in the rat brain in the order of millimeters, and gene expression of a reporter gene extended beyond the location of delivery. After extensive preclinical studies, a gene therapy study was carried out on two children with Canavan disease. Significant clinical improvement was obtained with LPD delivery.

Safety

Intracranial injection of the LPD complex is safe in animals and human patients without any noticeable side effects.[16] Systemic administration of LPD can trigger the response of proinflammatory cytokines, such as interferon-γ (IFN-γ), tumor necrosis factor (TNF-α), interleukin-12 (IL-12),

[16] P. Leone, C. G. Janson, L. Bilianuk, Z. Wang, F. Sorgi, L. Huang, R. Matalon, R. Kaul, Z. Zeng, A. Freese, S. W. Mcphee, E. Mee, and M. J. During, *Ann. Neurol.* **48,** 27 (2000).

and IL-6.[17] This dose-dependent response not only directly inhibits transgene expression but also induces apoptosis of lung cells at a high dose.[17] Induction of cytokines by LPD is mainly due to the unmethylated CpG sequences in the plasmid DNA.[17] Several strategies are used to overcome this problem, including (1) modification of plasmid DNA to reduce the number of CpG motifs[18]; (2) use of polymerase chain reaction (PCR)–amplified fragments instead of the intact plasmid[19]; (3) use of dexamethasone, a common immunosuppressant[20]; and (4) use of NF-κB decoy oligonucletide.[21]

Acknowledgments

Supported by grants to L. H. from NIH (AI 48851, DK 54225, CA 74918, DK 44935, and AR 45925) and a grant to S. L. from NIH (HL 63080).

[17] S. Li, S-P. Wu, M. Whitmore, E. J. Loeffert, L. Wang, S. C. Watkins, B. R. Pitt, and L. Huang, *Am. J. Physiol. Lung Cell. Mol. Physiol.* **276,** L796 (1999).
[18] N. S. Yew, H. Zhao, I. H. Wu, and A. Song, *Mol. Ther.* **1,** 255 (2000).
[19] C. R. Hofman, J. P. Dileo, Z. Li, S. Li, and L. Huang, *Gene Ther.* **8,** 71 (2001).
[20] Y. Tan, S. Li, B. R. Pitt, and L. Huang, *Human Gene Ther.* **10,** 2153 (1999).
[21] Y. Tan, J. S. Zhang, and L. Huang, *Mol. Ther.* **6,** 804 (2002).

[22] Transferrin-Lipoplexes with Protamine-Condensed DNA for Serum-Resistant Gene Delivery

By C. Tros de Ilarduya, M. A. Arangoa, and N. Düzgüneş

Introduction

Gene therapy has emerged as a promising approach for the treatment or prevention of acquired and genetic diseases. The field of nonviral vector–mediated gene therapy, particularly by use of cationic liposomes, has made great strides since the initial report by Felgner *et al.* in 1987[1] to their use in the world's first *in vivo* human gene therapy clinical trial by Nabel *et al.* in 1993.[2] Although the efficiency of transfection by cationic lipids has been substantially lower than that of viral delivery vehicles, cationic lipids do not have the same clinical risks.

[1] P. L. Felgner, T. R. Gadek, M. Holm, R. Roman, H. W. Chan, M. Wenz, J. P. Northrop, G. M. Ringold, and M. Danielsen, *Proc. Natl. Acad. Sci. USA* **84,** 7413 (1987).
[2] G. J. Nabel, E. G. Nabel, Z. Y. Yang, B. A. Fox, G. E. Plautz, X. Gao, L. Huang, S. Shu, D. Gordon, and A. E. Chang, *Proc. Natl. Acad. Sci. USA* **90,** 11307 (1993).

For most nonviral vectors, high-efficiency *in vitro* transfection correlates with a global excess of cationic charges.[3] However, this excess of cationic charges can facilitate nonspecific interactions *in vivo* with many undesired elements such as extracellular matrix and negatively charged serum components. Because of that, when transfecting in the presence of high serum concentration, a low +/− charge ratio might be favorable. Different authors have studied the effect of serum on lipofection,[4–9] achieving efficient *in vitro* gene transfer by precondensation of plasmid DNA with polylysine before formation of the lipoplexes[10] or by gradual addition of the lipid to plasmid DNA.[11]

To improve the efficiency of transfection mediated by cationic liposomes, several investigators have attempted to devise strategies to specifically target, transcribe, and translate DNA. As in this study, one approach has been to introduce DNA into cells by attaching it to a ligand and exploiting the natural specificity of receptor-mediated endocytosis. Another direction of research is based on the effort to condense DNA in a manner similar to natural vectors such as virus and sperm. Protamine is present in sperm, and its role is to bind with DNA, assist in forming a compact structure and delivering the DNA to the nucleus of the egg after fertilization. This unique role overcomes a major obstacle in gene therapy by nonviral vectors, the efficient delivery of DNA from the cytoplasm into the nucleus.[12] This pathway has been used in gene transfer studies with polycation–ligand conjugates.[13–15]

[3] K. Takeuchi, M. Ishihara, C. Kawaura, M. Noji, T. Furuno, and M. Nakanishi, *FEBS Lett.* **397,** 207 (1996).

[4] V. Ciccarone, P. Hawley-Nelson, and J. Jessee, *Focus* **15,** 80 (1993).

[5] J.-P. Yang and L. Huang, *Gene Ther.* **5,** 380 (1998).

[6] V. Escriou, C. Ciolina, F. Lacroix, G. Byk, D. Scherman, and P. Wils, *Biochim. Biophys. Acta* **1368,** 276 (1998).

[7] S. Li, W. C. Tseng, D. B. Stolz, S. P. Wu, S. C. Watkins, and L. Huang, *Gene Ther.* **6,** 585 (1999).

[8] S. Audouy, G. Molema, L. de Leij, and D. Hoekstra, *J. Gene Med.* **2,** 465 (2000).

[9] P. V. Dileep, A. Antony, and S. Bhattacharya, *FEBS Lett.* **509,** 327 (2001).

[10] L. Vitiello, A. Chonn, J. D. Wasserman, C. Duff, and R. G. Worton, *Gene Ther.* **3,** 396 (1996).

[11] O. Boussif, M. A. Zanta, and J. P. Behr, *Gene Ther.* **3,** 1074 (1996).

[12] J. Zabner, A. J. Fasbender, T. Moninger, K. A. Poellinger, and M. J. Welsh, *J. Biol. Chem.* **270,** 18997 (1995).

[13] G. Y. Wu and C. H. Wu, *J. Biol. Chem.* **262,** 4429 (1987).

[14] M. Cotten, E. Wagner, and M. L. Birnstiel, *Meth. Enzymol.* **217,** 618 (1993).

[15] T. Ferkol, J. C. Perales, E. Eckman, C. S. Kaetzel, R. W. Hanson, and P. B. Davis, *J. Clin. Invest.* **95,** 493 (1995).

For efficient delivery, receptors that mediate endocytosis of their ligand(s) and recycle to the plasma membrane are particularly attractive. In this way, ligands such as transferrin (Mw 80,000), a well-characterized plasma glycoprotein, is recognized by specific cell membrane receptors that are gate keepers responsible for physiological iron acquisition by most cell types in the organism. Transferrin receptor levels are found to be elevated in various types of cancer cells; therefore, they are considered to be useful as a prognostic tumor marker and as a potential target for drug delivery in the therapy of malignant cell growth.[16,17] Transferrin has been complexed to different polymers such as polylysine,[18–21] polyethylenimine,[22–24] or lipids[17,25–30] to get formulations for gene delivery by means of the receptor-mediated pathway. However, the long poly-L-lysines commonly used as DNA- binding or condensing moieties are described as toxic[31] and activate the complement system *in vitro.*[32]

[16] T. Miyamoto, N. Tanaka, Y. Eishi, and T. Amagasa, *Int. J. Oral Maxillofac. Surg.* **23**, 430 (1994).

[17] L. Xu, K. F. Pirollo, W. H. Tang, A. Rait, and E. H. Chang, *Hum. Gene Ther.* **10**, 2941 (1999).

[18] M. Cotten, F. Langle-Rouault, H. Kirlappos, E. Wagner, K. Mechtler, M. Zenke, H. Beug, and M. L. Birnstiel, *Proc. Natl. Acad. Sci. USA* **87**, 4033 (1990).

[19] D. T. Curiel, S. Agarwal, M. U. Romer, E. Wagner, M. Cotten, M. L. Birnstiel, and R. C. Boucher, *Am. J. Respir. Cell Mol. Biol.* **6**, 247 (1992).

[20] E. Wagner, C. Plank, K. Zatloukal, M. Cotton, and M. L. Birnstiel, *Proc. Natl. Acad. Sci. USA* **89**, 7934 (1992).

[21] C. E. Harris, S. Agarwal, P.-C. Hu, E. Wagner, and D. T. Curiel, *Am. J. Respir. Cell Mol. Biol.* **9**, 441 (1993).

[22] R. Kircheis, A. Kichler, G. Wallner, M. Kursa, M. Ogris, T. Felzmann, M. Buchberger, and E. Wagner, *Gene Ther.* **4**, 409 (1997).

[23] M. Ogris, S. Brunner, S. Schuller, R. Kircheis, and E. Wagner, *Gene Ther.* **6**, 595 (1999).

[24] R. Kircheis, L. Wightman, A. Schreiber, B. Robitza, V. Rossler, M. Kursa, and E. Wagner, *Gene Ther.* **8**, 28 (2001).

[25] P.-W. Cheng, *Human Gene Ther.* **7**, 275 (1996).

[26] S. Simões, V. Slepushkin, R. Gaspar, M. C. Pedroso de Lima, and N. Düzgüneş, *Gene Ther.* **5**, 955 (1998).

[27] S. Simões, P. Pires, N. Düzgüneş, and M. C. Pedroso de Lima, *Curr. Opin. Mol. Therapeut.* **1**, 147 (1999).

[28] C. Tros de Ilarduya and N. Düzgüneş, *Biochim. Biophys. Acta* **1463**, 333 (2000).

[29] C. Tros de Ilarduya, M. A. Arangoa, M. J. Moreno-Aliaga, and N. Düzgüneş, *Biochim. Biophys. Acta* **1561**, 209 (2002).

[30] K. Kono, Y. Torikoshi, M. Mitsutomi, T. Itoh, N. Emi, H. Yanagie, and T. Takagishi, *Gene Ther.* **8**, 5 (2001).

[31] S. Gottschalk, J. T. Sparrow, J. Hauer, M. P. Mims, F. E. Leland, S. L. Woo, and L. C. Smith, *Gene Ther.* **3**, 48 (1996).

[32] C. Plank, K. Mechtler, F. C. Szoka Jr., and E. Wagner, *Hum. Gene Ther.* **7**, 1437 (1996).

Protamine sulfate USP has been shown to be able to condense plasmid DNA efficiently for delivery into different types of cells *in vitro* by various cationic liposomes.[33–35] Liposome–protamine–DNA (LPD) complexes gave a high level of gene expression also *in vivo*.[36] A protocol in which liposomes were mixed with protamine before the addition of plasmid DNA was shown to produce small condensed particles with a diameter of about 135 nm. These particles produced a high level of gene expression in all tissues examined including lung, heart, spleen, liver, and kidney.[37] Moreover, intravenous administration of LPD elicits a systemic proinflammatory cytokine response that mediates the antitumor activity of the lipopolyplex.[38] Nuclear transport of oligonucleotides in HepG2 cells mediated by protamine sulfate and negatively charged liposomes has also been reported.[39]

Principle of Method

"Transferrinfection" is the process by which DNA is transferred into cells by transferrin–polycation conjugates based on transferrin-dependent, receptor-mediated endocytosis. DNA–lipid complex (lipoplex) formation is a very complex process that depends on thermodynamic, as well as kinetic factors. Lipid–DNA complexes generally are prepared by the simple mixing together of the two components; however, it is also important to consider that the applied protocols for complex formation and subsequent modifications strongly influence the properties of the transfection particle. Also, the order of addition of components to form the lipoplex affects considerably lipofection activity.[40] In this respect, Cheng[25] reported both efficient and inefficient lipoplexes in the presence of transferrin for *in vitro* gene delivery, depending on the protocol of mixing of the components.

Traditional protocols for preparing protamine lipoplexes involve precomplexation of plasmid DNA with protamine, followed by the addition of liposomes. Under these conditions, however, large amounts of

[33] F. L. Sorgi, S. Bhattacharya, and L. Huang, *Gene Ther.* **4,** 961 (1997).
[34] J. You, M. Kamihira, and S. Iijima, *J. Biochem.* **125,** 1160 (1999).
[35] S. Mizuarai, K. Ono, J. You, M. Kamihira, and S. Iijima, *J. Biochem.* **129,** 125 (2001).
[36] S. Li and L. Huang, *Gene Ther.* **4,** 891 (1997).
[37] S. Li, M. A. Rizzo, S. Bhattacharya, and L. Huang, *Gene Ther.* **5,** 930 (1998).
[38] M. Whitmore, S. Li, and L. Huang, *Gene Ther.* **6,** 1867 (1999).
[39] C. Welz, W. Neuhuber, H. Schreier, M. Metzler, R. Repp, W. Rascher, and A. Fahr, *Pharm. Res.* **17,** 1206 (2000).
[40] X. Gao, and L. Huang, *Biochemistry* **35,** 1027 (1996).

excess cationic lipids are required to achieve maximal gene expression. Moreover, the charge-neutralizing effect of protamine on DNA makes preparation of complexes difficult at high concentrations, especially when a high protamine/DNA ratio is required. Taking this into consideration, we hypothesized that by precomplexing protamine with transferrin (Tf) (slightly negatively charged) before addition of the plasmid, more stable complexes could be prepared for *in vivo* application. By use of this protocol, the competition between cationic liposomes and protamine for interaction with plasmid DNA would be greatly reduced. At the same time, protamine would maintain its ability to condense the DNA without reducing further interactions with the cationic lipids. Also, lipid and DNA could interact with each other inside our Tf–lipoplexes in the presence of serum to form a new structure that would be much more efficient in transfection *in vivo*. It is also interesting to note that although most of the current protocols imply covalent binding of the ligand, our complexes have been formulated by simple mixing of the four components in a studied and established order of addition.

Experimental Procedures

Materials

The cationic lipid 1,2-dioleoyl-3-(trimethylammonium) propane (DOTAP), dioleoylphosphatidylethanolamine (DOPE), and cholesterol are purchased from Avanti Polar Lipids (Alabaster, AL). Iron-saturated, heat-inactivated human transferrin is obtained from Collaborative Biomedical Products, Becton Dickinson (Bedford, MA). The plasmids, pCMVLuc (VR-1216, Vical Inc., San Diego, CA) encoding luciferase and pCMVLacZ (Clontech, Palo Alto, CA) encoding β-galactosidase are used for carrying out the transfection experiments. NaCl and N-(2-hydroxyethyl(piperazine-N'-[2-ethanesulfonic acid] (HEPES) and protamine sulfate (derived from salmon) are obtained from Sigma (Madrid, Spain). DNase I, ethidium bromide, and fetal bovine serum (FBS) are purchased from GibcoBRL Life Technologies (Barcelona, Spain).

Cell Culture

HeLa and HepG2 cells (American Type Culture Collection, MD) are maintained at 37° under 5% CO_2 in Dulbecco modified Eagle's medium-high glucose (DME-HG) (Irvine Scientific, Santa Ana, CA) supplemented with 10% (v/v) heat-inactivated fetal bovine serum (FBS) (Sigma, St. Louis,

MO), penicillin (100 units/ml), streptomycin (100 μg/ml), and L-glutamine (4 mM) (Irvine Scientific Santa Ana, CA). Cells are passaged by trypsinization 1:10 twice a week.

Liposome Preparation

DOTAP–DOPE (1:1 molar ratio) liposomes are prepared by drying a chloroform solution of the lipids by rotary evaporation under reduced pressure and then by hydrating the film with 1 ml of deionized water to a concentration of 5 mg total lipid/ml. The tube is vortexed for 1 min, and the suspension is diluted to a final concentration of 1 mg/ml of total lipid. The multilamellar vesicles are extruded through polycarbonate membranes with 100-nm pore diameter using a Liposofast device (Avestin, Toronto, Canada) to obtain a uniform size distribution and filter-sterilized (Millex 0.22 μm, Millipore, Bedford, MA). The average diameter of the liposomes is 102 ± 2 nm, as determined by dynamic light scattering. Liposomes are stored at 4° under nitrogen and are used within 1 month after preparation. DOTAP–CHOL (1:0.9 molar ratio) liposomes used for *in vivo* studies are hydrated with a 10 mM HEPES, 10% (w/v) glucose buffer (pH 7.4) to give a final concentration of 10 mM DOTAP/9 mM CHOL.

Preparation of Lipoplexes

Complexes for transfection of HeLa cells are prepared by sequentially mixing 100 μl of a solution of 100 mM NaCl plus 20 mM HEPES, pH 7.4 (HEPES buffer), with (Tf-lipoplexes) or without (plain lipoplexes) 32 μg of transferrin, with 2, 4, 8, or 16 μl of the liposome suspension and incubated for 15 min at room temperature. One hundred microliters of HEPES buffer containing 1 μg of pCMVLacZ plasmid encoding β-galactosidase is then added and gently mixed.

Complexes for *in vitro* experiments in HepG2 cells are prepared by first mixing 400 μl of a solution of HEPES buffer, with (Tf-lipoplexes) or without (plain lipoplexes) 128 μg of transferrin, with 8, 16, 32, or 64 μl of the DOTAP–DOPE liposome suspension and incubating for 15 min at room temperature. Four hundred microliters of HEPES buffer containing 3.5 μg of pCMVLuc is then added and gently mixed. Complexes for *in vivo* experiments, prepared at a 5:1 (+/−) charge ratio, contain 19 mM of total lipid (DOTAP/CHOL) and 32 μg of transferrin per μg of DNA. The final concentration of plasmid in the complexes is 0.3 mg/ml.

Protamine–Tf–lipoplexes are prepared by two different protocols: Protocol 1 involves the precomplexation of plasmid DNA with protamine followed by the addition of preformed cationic liposome–transferrin complexes, prepared as described previously. Protocol 2 involves mixing transferrin with protamine, followed by their addition to the liposomes, incubation for 15 min, addition to the plasmid, and further incubation for 15 min. The lipid/DNA charge ratio is always calculated as the mole ratio of DOTAP (one charge per molecule) to nucleotide residue (average MW, 330).

Transfection Protocol

For transfection, 2×10^5 HeLa cells are seeded in 1 ml of medium in 48-well culture plates (11.3-mm well diameter, Costar, Acton, MA) 24 h before the addition of the complexes and used at approximately 80% confluency. In the case of HepG2, 10^6 cells are seeded in 2 ml of medium in six-well culture plates (35-mm well diameter, Costar, Acton, MA). Cells are washed twice with DME-HG without antibiotics, and then 300 μl of DME-HG and 200 μl of complex are added to HeLa cells. If the lipofection is performed in the presence of serum, nonheat-inactivated FBS is added to a final concentration as indicated in the experiments. Transfection in HepG2 cells is always performed in 60% FBS; 1.2 ml of fetal bovine serum and 0.8 ml of complexes are added gently to each well. After a 4-h incubation (at $37°$ in 5% CO_2), the medium is replaced, and the cells are incubated further for 48 h in medium containing 10% FBS. In all experiments, the number of viable cells plated and the cell density at the moment of transfection should be constant.

In Vitro Transfection Activity Measurement

β-Galactosidase Assay. Cells are washed 48 h after transfection with phosphate-buffered saline (PBS) and lysed with 100 μl of lysis buffer (0.25 M TRIS, pH 8) at room temperature for 10 min, followed by alternating freeze-thaw cycles in the culture vessel. The cell lysate is centrifuged for 5 min at 10,000 g to pellet debris. Seventy microliters of the supernatant is assayed for total β-gal activity as described by Invitrogen (Carlsbad, CA) using an Emax microplate reader (Molecular Devices, Sunnyvale, CA). The protein content of the lysates is measured by the DC Protein Assay reagent (Bio-Rad, Hercules, CA) using BSA as the standard. The data are expressed as nanograms of β-gal (based on a standard curve for β-galactosidase activity) per milligram of protein.

Luciferase Assay. After 48 hs, cells are washed with PBS and lysed with 250 μl of reporter lysis buffer (Promega, Madison, WI) at room temperature for 10 min, followed by two alternating freeze-thaw cycles. The cell lysate is centrifuged for 2 min at 12,000g to pellet debris. Twenty microliters of the supernatant is assayed for total luciferase activity using the luciferase assay reagent (Promega), according to the manufacturer's protocol. A luminometer (Sirius-2, Berthold Detection Systems, Innogenetics, Diagnóstica y Terapéutica, S.A., Barcelona, Spain) is used to measure luciferase activity. The protein content of the lysates is measured by the DC Protein Assay reagent (Bio-Rad, Hercules, CA) using BSA as the standard. The data are expressed as picograms or nanograms of luciferase (on the basis of a standard curve for luciferase activity, using luciferase obtained from Promega) per milligram of protein.

DNase I Protection Assay. Tf-lipoplexes at different charge ratios are prepared as described previously. DNase I (1 unit per μg of DNA) is added to 2.5 μg (DNA) of each sample, and the mixtures are incubated at 37° for 30 min. Sodium dodecyl sulfate (SDS) is then added to a final concentration of 1% to release DNA from the complexes. Samples are analyzed by agarose gel electrophoresis and the integrity of the plasmid in each formulation is compared with untreated DNA as a control. Ultraviolet (UV) spectroscopic studies are performed with a reaction mixture containing 5 μg/ml of plasmid DNA. After a 15 min incubation of Tf-lipoplexes at room temperature, 1 unit of DNase I per microgram of DNA is added, and the absorbance is monitored at 20-s intervals for 30 min in a UV spectrophotometer (Hewlett Packard 8452 A, Waldbronn, Germany).

Dye Displacement Assay. The binding of protamine to DNA inside Tf-lipoplexes is examined using a quenching method based on ethidium bromide, a monovalent DNA-intercalating agent that occupies an effective binding site of two base pairs; its fluorescence is dramatically enhanced on binding to DNA and quenched when displaced by higher affinity compounds or by condensation of the DNA structure. Twelve micrograms of DNA in 2 ml of 10 mM HEPES, 10% (w/v) glucose buffer (pH 7.4), are mixed with 1.4 μg of ethidium bromide and a baseline fluorescence (F_o) is determined. Various amounts of protamine are added to the preceding mixture, followed by cationic lipids and transferrin to prepare 4:1 $(+/-)$ charge ratio Tf-lipoplexes. The fluorescence is measured ($n = 3$) after each addition in an LS 50 spectrofluorometer (Perkin Elmer, Mountain View, CA) at an excitation and emission wavelength of 520 and 600 nm, respectively, with a slit width of 6 mm. The relative fluorescence values are determined as follows: $F_r = (F_{obs} - F_e) \times 100/(F_o - F_e)$, where F_r is the relative fluorescence, F_{obs} is the measured fluorescence, F_e is the fluorescence of ethidium bromide in the absence of DNA under the given buffer conditions, and F_o is the initial fluorescence of ethidium bromide-DNA in the absence of the lipid and the polycation.

In Vivo *Gene Expression Measurement*

Female Balb-c mice (8–10 weeks of age), purchased from Harlan Ibérica Laboratories (Barcelona, Spain) and housed in accordance with institutional guidelines, are used for the experiments. Individual mice in groups of eight are injected in the tail vein with 60 μg of DNA formulated in plain, Tf- or protamine–Tf–lipoplexes in a total volume of 200 μl 5% w/v glucose. Twelve hours following intravenous injection, the mice are killed with a sodium pentobarbital overdose, and approximately 1 ml of blood is removed by intracardiac puncture. One milliliter of PBS is perfused through the right cardiac ventricle. The heart, lungs, liver, and spleen are collected and washed with cold saline twice. The organs are homogenized with lysis buffer (Promega) using an homogeneizer (Mini-Beadbeater^{TM}, BioSpec Products, Inc., Bartlesville, OK) and centrifuged at 12,000g for 3 min at 4°. Twenty microliters of the supernatant is analyzed for luciferase activity as described previously.

Application of Complexes

Increased Transfection Activity by Tf-Lipoplexes in the Presence of High Concentration of Serum

Transfection activity mediated by either plain or Tf-lipoplexes, at different lipid/DNA (+/−) charge ratios, was examined as a function of serum concentration. The association of transferrin with cationic liposomes composed of DOTAP and DOPE increases β-galactosidase expression both in the absence or presence of serum compared with lipoplexes without transferrin (Fig. 1).

DOTAP–DOPE liposomes bind transferrin through the negatively charged groups present in the ligand at physiological pH, and the resulting structure forms a complex with DNA through a charge–charge interaction. Association of transferrin with DOTAP–DOPE–DNA complexes seems to facilitate the internalization of the complex because of the ability of the ligand to stimulate endocytosis.

Expression of β-gal mediated by 1/2, 1/1, and 2/1 (+/−) Tf-lipoplexes is reduced as the serum concentration is increased. The inhibitory effect of serum is diminished with increasing charge ratio (+/−). For 4/1 (+/−) Tf-complexes, the inhibitory effect of serum is overcome, and the transfection activity in 60% FBS increases by five-fold compared with that in the absence of serum. Gene delivery and expression by plain lipoplexes even at the lowest serum concentration used (20% FBS) is less than 0.2 ng β-galactosidase per milligram of total cell protein. In Tf-lipoplexes,

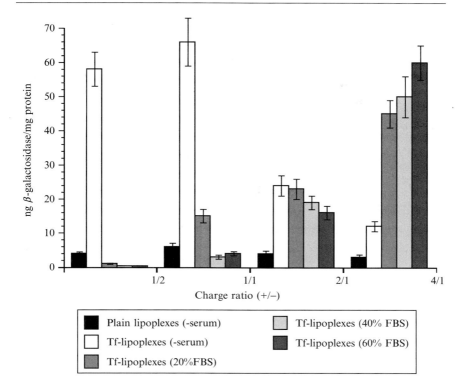

FIG. 1. Effect of serum concentration on β-galactosidase gene expression by HeLa cells transfected with plain DOTAP–DOPE or transferrin lipoplexes. The data represent the mean $\pm$ SD of three wells and are representative of three independent experiments. (Adapted from *Biochim. Biophys. Acta* **1463,** 333 (2000), with permission.)

the transferrin molecule, in addition to acting as a ligand also neutralizes part of the positive charge on the liposomes, leading to diminished interactions with polyanionic molecules in serum. Consequently, complexes resistant to higher concentrations of serum can be prepared. It is also interesting to note that increasing amounts of serum increases transfection levels. The reasons for this increase are not clear at present.[28]

The enhancement of Tf-lipoplex–mediated gene transfer by high concentrations of serum was also observed for DOTAP–cholesterol and DOTAP–DOPE–cholesterol liposomes, indicating that this effect is not specific for DOTAP–DOPE. At the same time, conditions that enhance transfection activity also enhance transfection efficiency, although the percentage of enhancement cannot be directly correlated. The difference in transfection activity between these sets of complexes cannot be explained by different toxicities of the formulations or of the serum, because cell viability was similar in both cases. It is also important to note that increased gene expression by use of the prepared complexes in the presence of transferrin was also observed in other cell lines such as HepG2 and 3T3-L1, as well as in cells known to be hard to transfect, including primary hepatocytes and adipocytes.[29]

Nuclease Resistance of Tf-Lipoplexes

Another possible mechanism for the enhancement in transfection by Tf-lipoplexes is improved protection of DNA inside the complexes. Gel electrophoresis data indicate a high level of nuclease resistance for DNA formulated in Tf-lipoplexes, which confirms its stable nature. At the 4:1 (+/−) charge ratio, the plasmid is almost completely protected from digestion by DNase I (Fig. 2a, lane 6). Ultraviolet spectroscopy data confirmed the preceding results. When DNA itself is treated with DNase I under physiological conditions, complete degradation occurs within 4 min. In contrast, the percentage of degraded plasmid in Tf-lipoplexes at this time, for the maximal charge ratio, is only 7% of that for native DNA (Fig. 2b). This feature of high resistance toward nuclease attack is surely an advantage of using the Tf-lipid complexes as a reservoir for DNA under physiological circumstances, when DNA degradation through nuclease attack takes place readily.

Improved Gene Delivery In Vitro *and* In Vivo *by Tf-Lipoplexes with Protamine-Condensed DNA*

Figure 3 shows a dose-dependent gene expression by protamine included in Tf-lipoplexes compared with plain and Tf-complexes in the absence of the polycation. The maximum level of transfection is observed at a protamine concentration of 0.5 $\mu g/\mu g$ of DNA and is nine-fold over Tf-lipoplexes and 25-fold over the levels seen with plain complexes. The polycation itself in combination with DNA had no significant effect on gene expression. The same figure shows also the influence of preparing the complexes by protocol 2, by which the highest level of transfection is obtained. An 18-fold increase over Tf-lipoplexes without protamine and a two-fold increase over protamine-Tf-lipoplexes prepared by the conventional method (protocol 1) are observed.

The displacement of ethidium bromide from DNA in Tf-lipoplexes is monitored by measuring the fluorescence as protamine sulfate was added to the complexes. The addition of the polycation results in a rapid decrease in fluorescence. At the optimal amount of protamine obtained

A

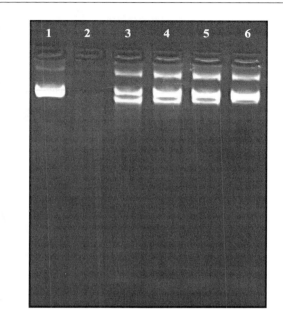

B

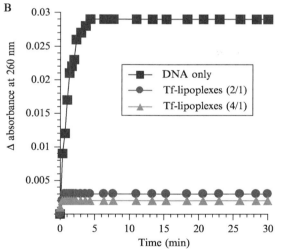

FIG. 2. (A) Susceptibility of plasmid DNA or Tf-lipid–DNA complexes to degradation by DNase I treatment. Untreated DNA (lane 1), DNA treated with DNase: naked plasmid (lane 2), DNA inside Tf-lipoplexes at 1/2 (lane 3), 1/1 (lane 4), 2/1 (lane 5), and 4/1 (lane 6) (+/−) charge ratios. (B) Degradation profile of DNA alone or Tf-lipid–DNA complexes at 2/1 and 4/1 (+/−) charge ratios with DNase I. (Adapted from *Biochim. Biophys. Acta* **1561,** 209 (2002), with permission.)

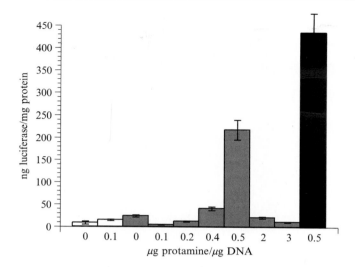

FIG. 3. Gene expression in HepG2 cells by plain (□) and Tf-(▨ protocol 1; ■ protocol 2) lipoplexes in the absence or presence of protamine sulfate. Complexes were prepared at 4/1 (+/−) charge ratio. The data represent the mean ± SD of three wells and are representative of three independent experiments. (Adapted from *Biochim. Biophys. Acta* **1561,** 209 (2002), with permission.)

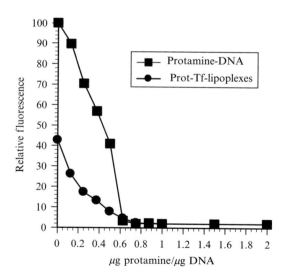

FIG. 4. Binding behavior of protamine to DNA, studied by ethidium bromide assay. Complexes were prepared at 4:1 (+/−) lipid/DNA charge ratio. Each point represents the mean of three replicate measurements. (Reproduced from *Biochim. Biophys. Acta* **1561,** 209 (2002), with permission.)

for *in vitro* studies (0.5 μg), DNA is almost completely condensed inside the Tf-lipoplexes at the 4:1 (+/−) charge ratio (Fig. 4). The residual fluorescence (Fres), at which the polycation cannot further exclude ethidium bromide from the DNA, has a low value of 1.9, indicative of the stability of the complex.

To examine the *in vivo* activity of Tf-lipoplexes and protamine-Tf-lipoplexes, complexes are injected systemically into mice. Figure 5 shows that complexes in the presence of the ligand lead to a higher level of gene expression compared with plain lipoplexes in the four studied organs. Protamine-Tf-lipoplexes, prepared by the newly developed protocol, resulted in an increase of 5, 2, 1.3, and 5–fold in the liver, lung, heart, and spleen, respectively, over Tf-lipoplexes. Although charge ratios (+/−) close to 10 are used generally for *in vivo* transfections, protamine-Tf-lipoplexes at 5:1 (+/−) are able to mediate efficient gene expression. Consequently, a reduction in lipid-induced cellular toxicity compared with plain lipoplexes is observed (as indicated by the death of two of the eight mice from toxicity when using plain lipoplexes and by the lack of any signs of toxicity when protamine-Tf-lipoplexes were injected). The

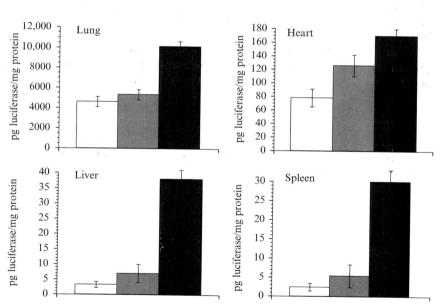

FIG. 5. *In vivo* transfection efficiency of plain (□), Tf-(▨), and prot-Tf-lipoplexes (■). Complexes in the presence of the polycation were prepared with 0.5 μg of protamine per microgram of DNA following protocol 2, as described in "Materials and Methods." (Reproduced from *Biochim. Biophys. Acta* **1561,** 209 (2002), with permission.)

capacity of transferrin to promote cellular entry and the ability of protamine to facilitate nuclear localization of DNA, could be additional factors that cause enhanced gene expression. Complexation of Tf with the cationic liposomes could result in charge shielding, as well as provide steric hindrance to other serum components, thereby minimizing the inhibitory effects of serum. Also, the smaller particle size of the protamine-Tf-lipoplexes (~200 nm) compared with Tf-lipoplexes (~388 nm) and plain lipoplexes (~230 nm) may also contribute to the higher gene expression observed.

Concluding Remarks

The novel protamine-Tf-lipoplex described in this chapter generates nanometer-sized particles with plasmid DNA and protects the DNA to be delivered into cells. The gene carrier has a remarkably low toxicity and good transfection performance *in vitro* and *in vivo*. The described method for the preparation of such lipoplexes offers several advantages over other conventional transfection protocols: (1) They are easy to prepare and economical compared with commercial transfection reagents; (2) they are simple to use; (3) unlike viral methods, this technique does not require the generation of special viral constructs; (4) they can be used with many DNA molecules and may be applicable to other ligands that can be internalized on binding to their receptors. Consequently, protamine-Tf-lipoplexes, or those containing different receptor ligands, could be used widely for gene transfection *in vitro* and *in vivo*.

Acknowledgments

Supported by a grant from the Government of Navarra (Department of Education), and funds from the Echebano Foundation, the University of Navarra, and the University of the Pacific School of Dentistry.

[23] DNA Complexes with Reducible Cationic Lipid for Gene Transfer

By Nathalie Mignet, Gerardo Byk, Barbara Wetzer, and
Daniel Scherman

Introduction

Rationale

Cationic lipids have been developed and investigated widely in the past 10 years as an alternative to viral vectors for gene therapy. They are able to associate with DNA by ionic attraction and protect it to some extent from nucleases in biological media.[1] Lipoplexes (i.e., cationic lipid–(DNA complexes) are readily taken up *in vitro* in numerous cell types.[2,3] However, even if nearly all cell types have been shown to take up lipoplexes, this only leads to transgene expression in a low proportion of these cells. Indeed, several limiting steps take place after lipoplexes entry into the cells.[4] For efficient transfection to occur, DNA should be released from the complexes. Complexes have been localized by electron microscopy in intracellular vesicular structures, the plasmid thus not being released from all the complexes.[5] Moreover, plasmids have been shown to present a very limited mobility in the cytoplasm, rendering it susceptible to nuclease degradation before reaching the nucleus.[6,7] Cell division was also shown to play an essential role to reach high transfection efficacy.[8] To improve DNA release from its cationic vector, lipids susceptible to esterase hydrolysis have been synthesized.[9] This approach showed increased *in vitro* transfection efficiency, and reduced cytotoxicity but could not be applied to polyamines because of intramolecular side reactions. It may also suffer from decreased stability when delivered *in vivo*.

[1] K. Crook, G. McLachlan, B. Stevenson, and D. Porteous, *Gene Ther.* **3,** 834 (1996).
[2] P. Felgner and G. Ringold, *Nature* **337,** 387 (1989).
[3] X. Gao and L. Huang, *Gene Ther.* **2,** 710 (1995).
[4] A. Schatzlein, *Anti-cancer Drugs Rev.* **12,** 275 (2001).
[5] J. Zabner, A. J. Fasbender, T. Moninger, K. A. Poellinger, and M. J. Welsh, *J. Biol. Chem.* **270,** 18997 (1995).
[6] G. L. Lukacs, P. Haggie, O. Seksek, D. Lechardeur, N. Freedman, and A. Verkman, *J. Biol. Chem.* **275,** 1625 (2000).
[7] M. Capecchi, *Cell* **22,** 479 (1980).
[8] V. Escriou, M. Carriere, F. Bussone, P. Wils, and D. Scherman, *J. Gene Med.* **3,** 179 (2001).
[9] R. Leventis and J. Silvius, *Biochim. Biophys. Acta* **1023,** 124 (1990).

We[10,11] and others[12–14] developed an alternative strategy taking advantage of the intracellular reductive environment. The choice of a reducible cationic lipid is based on the property of cationic lipids to strongly interact with the cell membrane and promote cellular uptake of DNA, together with the expected advantage of an increased intracellular DNA release. So far, no general intracellular reductive mechanism has been reported. Nevertheless, several biological mechanisms, such as diphtheria toxin and ricin toxicity or Sindbis virus membrane fusion, have been described as depending on reduction by cellular components.[15,16] High cytosolic glutathione concentrations have also been reported to affect intracellular reduction.[17]

Design

Cationic lipids are composed of a hydrophilic moiety linked to apolar chains. According to the disulfide group position on the lipid, the DNA release process might be different (Fig. 1). Indeed, a disulfide linkage between the hydrophilic and the lipophilic parts will induce a complete lipoplex disruption on reduction and complete DNA release. When the disulfide link is located in the lipophilic part, reduction will not necessarily release DNA, and a partial disruption may occur where DNA is still associated to the remaining cationic lipid part. Our group described several reducible cationic lipids.[10,11] We have selected RPR128522 and RPR202059 (Fig. 2) as two reducible lipids representing the models described in Fig. 1. The described methods can be applied to other similar compounds. Products RPR120535 and RPR203769 are control parent compounds devoid of disulfide bonds.

Description of Methods

Synthesis

The synthesis of RPR128522 is given as an example (Fig. 3). This synthesis includes the generation of the key intermediate, unsymmetrical, disulfide-bridged molecule **1**. Peptide coupling of intermediate **1** to

[10] G. Byk, B. Wetzer, M. Frederic, C. Dubertret, B. Pitard, G. Jaslin, and D. Scherman, *J. Med. Chem.* **43**, 4377 (2000).
[11] B. Wetzer, G. Byk, M. Frederic, M. Airiau, F. Blanche, B. Pitard, and D. Scherman, *Biochem. J.* **356**, 747 (2001).
[12] F. Tang, W. Wang, and J. A. Hughes, *J. Liposome Res.* **9**, 331 (1999).
[13] F. Tang and J. A. Hughes, *Biochem. Biophys. Res. Commun.* **242**, 141 (1998).
[14] F. Tang and J. A. Hughes, *Bioconj. Chem.* **10**, 791 (1999).
[15] B. Abell and D. Brown, *J. Virol.* **67**, 5496 (1993).
[16] E. Feener, W. Shen, and H. Ryser, *J. Biol. Chem.* **265**, 18439 (1990).
[17] A. Meister and M. E. Anderson, *Annu. Rev. Biochem.* **52**, 711 (1983).

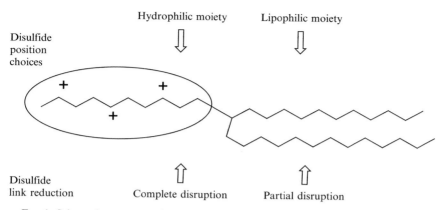

FIG. 1. Schematic representation of a reducible cationic lipid (RCL) with the potential disulfide group position and putative effect of reduction on RCL/DNA complexes.

dioctadecylamine, deprotection, and coupling with the corresponding monofunctionalized protected polyamine, followed by a final deprotection, result in the desired product.

BocNH-CH₂CH₂-S-S-CH₂COOH   1

Triethylamine (TEA) (2.7 ml, 20 mmol) and $HSCH_2COOH$ (0.27 ml, 3.18 mmol) in portions are added to a stirred solution of N,N'-diBoc cystamine (1.2 g, 3.18 mmol) in $CHCl_3$ (20 ml). The mixture is stirred 2 h at room temperature, then TEA is removed from the solution with $KHSO_4$ 0.5 M (3 × 20 ml). The organic layer is dried ($MgSO_4$), filtered, and evaporated under reduced pressure. The crude product is dissolved in Et_2O (100 ml) and extracted with saturated $NaHCO_3$ solution (3 × 50 ml). The combined aqueous phases are washed with Et_2O and acidified with $KHSO_4$ 0.5 M to pH 3 (200 ml). The precipitate is extracted with $CHCl_3$ (3 × 80 ml) and washed further with NaCl, aq (2 × 50 ml), dried over $MgSO_4$, and evaporated to dryness 0.32 g (yield 31%) of the expected product **1** is obtained. TLC Rf = 0.25 ($CHCl_3$/MeOH; 9/1), $MH^+ = 268$.

RPR128522

Product **1** (0.29 g, 1.1 mmol), dioctadecylamine (0.522 g, 1.1 mmol), and TEA (0.7 ml, 6 mmol) are dissolved in $CHCl_3$ (10 ml) in the benzotrice zolyloxy-tris(dimethylancino) presence of phosphonium hexafluoro phosphate (BOP reagent). Then, $KHSO_4$ (0.5 M, 3 × 30 ml) is added with a $NaHCO_3$ aqueous solution (3 × 30 ml) and the organic layer dried with $MgSO_4$ is filtered and evaporated under reduced pressure. The intermediate

(A) Example of lipid bearing a disulfide bond within the hydrophilic moiety of the molecule: RPR128522

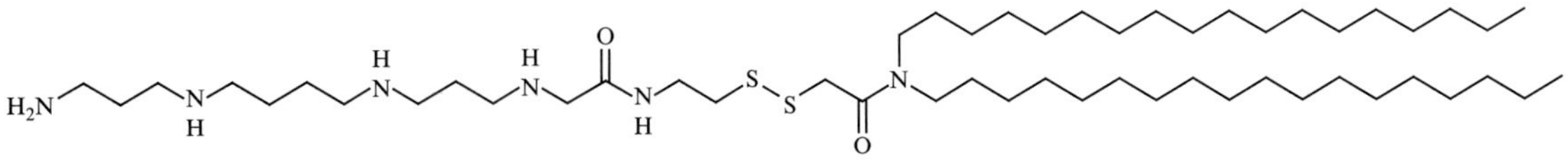

Parent lipid bearing no disulfide bond: RPR120535

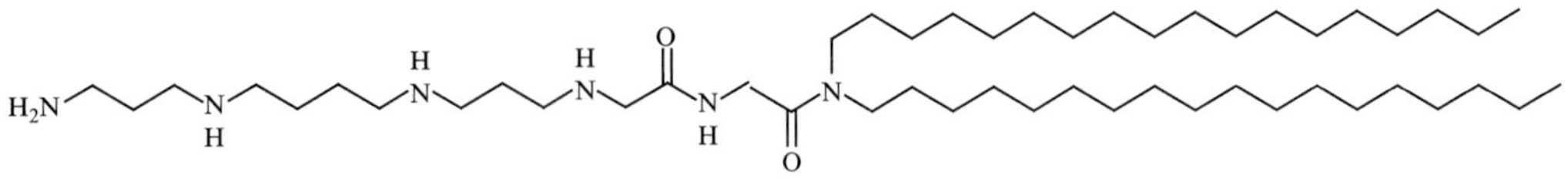

(B) Example of lipid bearing a disulfide bond within the hydrophilic moiety of the molecule: RPR202059

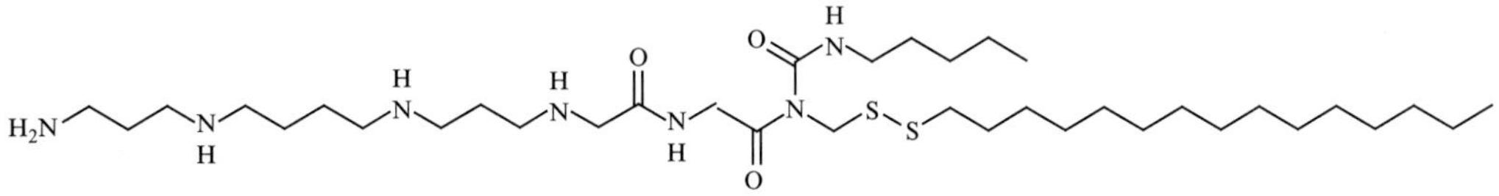

Parent lipid bearing no disulfide bond: RPR203769

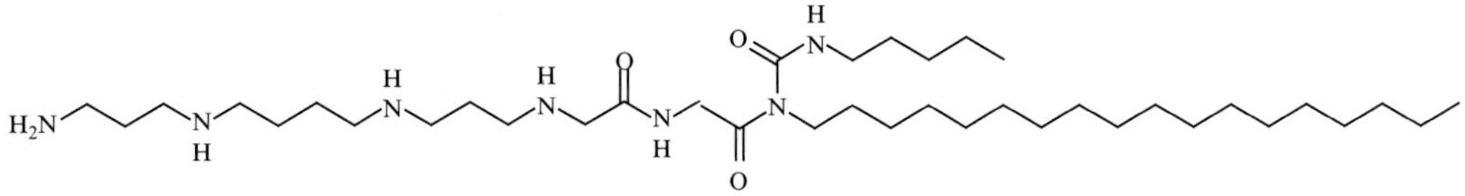

Fig. 2. Examples of cationic lipids bearing a disulfide linkage and their parent compound devoid of a disulfide bond.

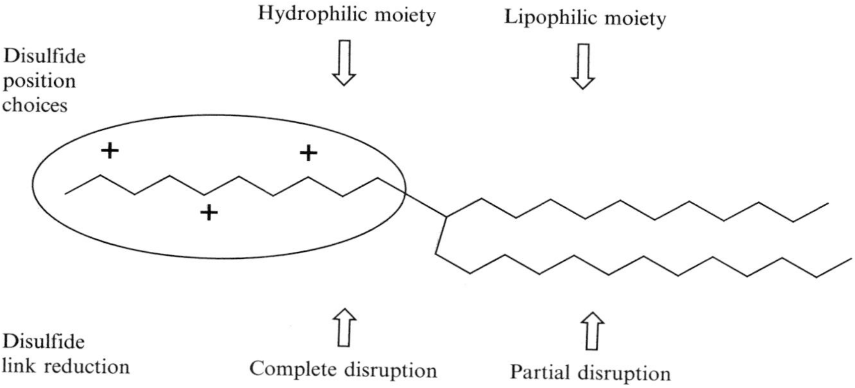

FIG. 1. Schematic representation of a reducible cationic lipid (RCL) with the potential disulfide group position and putative effect of reduction on RCL/DNA complexes.

dioctadecylamine, deprotection, and coupling with the corresponding monofunctionalized protected polyamine, followed by a final deprotection, result in the desired product.

BocNH-CH_2CH_2-S-S-CH_2COOH **1**

Triethylamine (TEA) (2.7 ml, 20 mmol) and $HSCH_2COOH$ (0.27 ml, 3.18 mmol) in portions are added to a stirred solution of *N,N′*-diBoc cystamine (1.2 g, 3.18 mmol) in $CHCl_3$ (20 ml). The mixture is stirred 2 h at room temperature, then TEA is removed from the solution with $KHSO_4$ 0.5 M (3 × 20 ml). The organic layer is dried ($MgSO_4$), filtered, and evaporated under reduced pressure. The crude product is dissolved in Et_2O (100 ml) and extracted with saturated $NaHCO_3$ solution (3 × 50 ml). The combined aqueous phases are washed with Et_2O and acidified with $KHSO_4$ 0.5 M to pH 3 (200 ml). The precipitate is extracted with $CHCl_3$ (3 × 80 ml) and washed further with NaCl, aq (2 × 50 ml), dried over $MgSO_4$, and evaporated to dryness 0.32 g (yield 31%) of the expected product **1** is obtained. TLC Rf = 0.25 ($CHCl_3$/MeOH; 9/1), MH^+ = 268.

RPR128522

Product **1** (0.29 g, 1.1 mmol), dioctadecylamine (0.522 g, 1.1 mmol), and TEA (0.7 ml, 6 mmol) are dissolved in $CHCl_3$ (10 ml) in the benzotrice zolyloxy-tris(dimethylancino) presence of phosphonium hexafluoro phosphate (BOP reagent). Then, $KHSO_4$ (0.5 M, 3 × 30 ml) is added with a $NaHCO_3$ aqueous solution (3 × 30 ml) and the organic layer dried with $MgSO_4$ is filtered and evaporated under reduced pressure. The intermediate

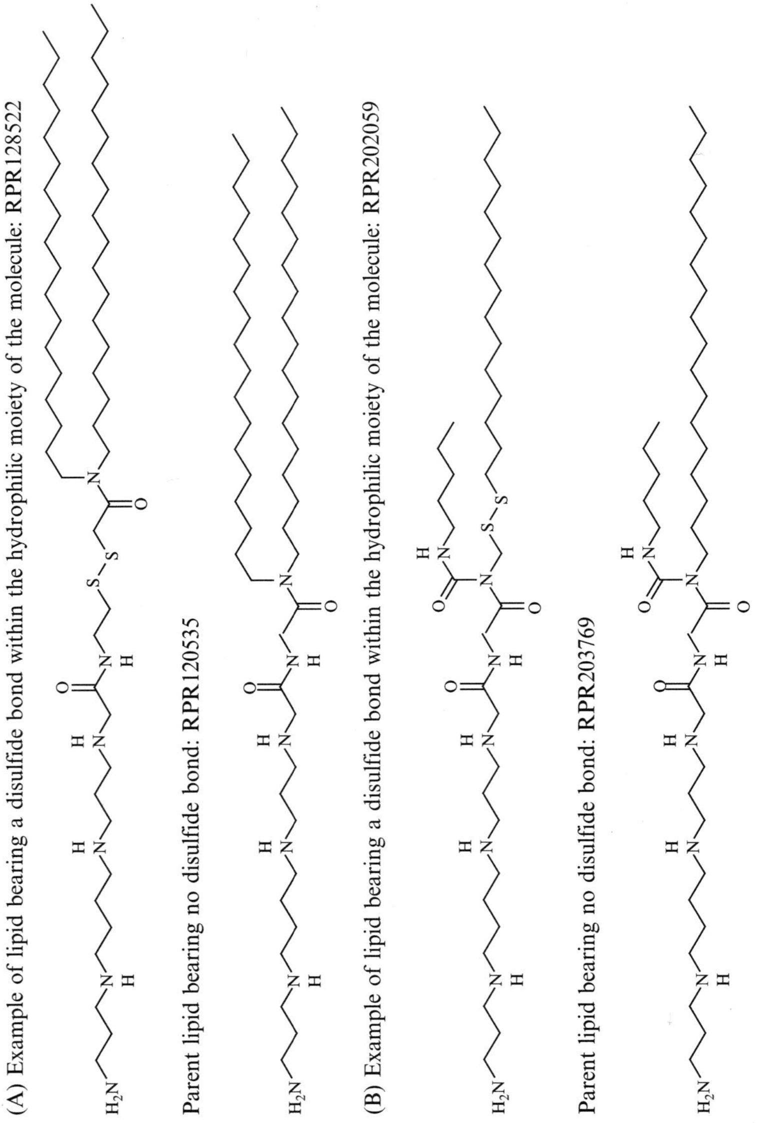

FIG. 2. Examples of cationic lipids bearing a disulfide linkage and their parent compound devoid of a disulfide bond.

(A) Example of lipid bearing a disulfide bond within the hydrophilic moiety of the molecule: RPR128522

Parent lipid bearing no disulfide bond: RPR120535

(B) Example of lipid bearing a disulfide bond within the hydrophilic moiety of the molecule: RPR202059

Parent lipid bearing no disulfide bond: RPR203769

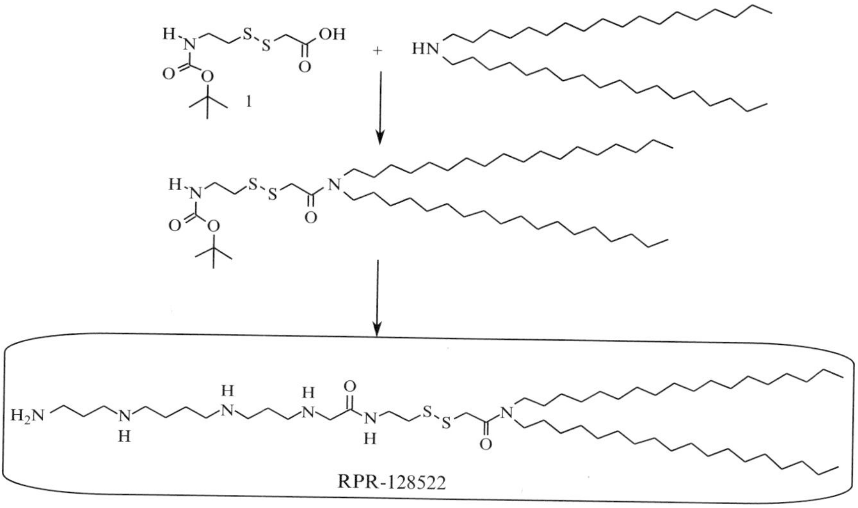

FIG. 3. Synthetic scheme of RPR128522.

lipid (0.7 g, yield 90%) is used without further purification. MH$^+$ = 771. The intermediate (0.35 g, 0.435 mmol) is deprotected using 10 ml trifluoroacetic acid (TFA) For 1 h at RT. Trifluoroacetic acid is evaporated under vacuum, and the product is dissolved in dimethylformamide (DMF) (10 ml). Triethylamine is added until the pH of the solution is basic (about 0.36 ml, 2 mmol). Then BocNH-$(CH_2)_3$-NBoc-$(CH_2)_4$-NBoc-$(CH_2)_3$-NBoc-CH_2CO_2H (0.209 g, 0.44 mmol) and BOP reagent (0.221 g, 0.5 mmol) are added. The solution is left for 2 h until the fluorescamine test is negative. KHSO$_4$ (0.5 M, 100 ml) is then added, and the product is extracted with ethylacetate (3 × 50 ml). The combined organic layer is washed with KHSO$_4$ (0.5 M, 3 × 30 ml), then with saturated NaHCO$_3$ solution (3 × 30 ml), dried over MgSO$_4$, filtered, and evaporated under reduced pressure to give 0.465 g of the protected intermediate (yield 81%). The protective group is removed with 10 ml TFA for 1 h. Trifluoroaceticacid is evaporated under vacuum, and the final product is purified by high-performance liquid chromatography (HPLC). *Analytical HPLC* was performed on a Merck-HITACHI gradient pump equipped with a AS-2000A autosampler, a L-6200A Intelligent pump, an ultraviolet-vis detector L-4000 with tuneable wavelength set at 220 nmoles, and a column BU-300 aquapore Butyl 7 μ, 300A, 300 × 4.6 mm from Applied Biosystems, CA, USA. Analysis of products was carried out using a gradient H$_2$O (0.1% TFA)/acetonitrile (0.08% TFA): 3 min (40/60), 3–20 min (0/100), 20–35 min (0/100) and a flow rate of 1.0 ml/min. *Preparative HPLC* was performed on a Gilson gradient system equipped with two Gilson 305 intelligent pumps, an ultraviolet-vis Gilson 119 detector with double channel set at 220 and 254 nmoles, a fraction collector Gilson 202, and a column C4 214TP1022 from Vydac, Roissy (France). Purification of products was carried out using a gradient H_2O (0.1% TFA)/MeCN (0.08% TFA): 10 min (70/30), 10–80 min (0/100), 80–120 min (0/100) and a flow rate of 18 ml/min to afford, after freeze-drying, 0.25 g of pure RPR128522 (yield 50%). HPLC Rt = 15.73 min. ^1H NMR (400 MHz, $(CD_3)_2SO$ d6 with few drops of CD_3COOD d4, at a temperature of 343 K, δ in ppm): 0.88 (t, J = 6.5 Hz, 6H: CH_3); from 1.20–1.40 (mt, 60H: CH_2); 1.52 (broad band, 4H: CH_2) ; 1.70 (mt, 4H: $(CH_2)_2$ of butyl); from 1.90–2.10 (mt, 4H: CH_2 of propyls); from 2.85–3.10 (mt, 12H: NCH_2 of butyl and NCH_2 of propyls); 2.90 (t, J = 7 Hz, 2H: CH_2S); 3.28 (broad band, 4H: NCH_2); 3.50 (t, J = 7 Hz, 2H: CH_2NCO); 3.68 and 3.76 (2 broad s, 2H each: NCH_2CON and $CONCH_2S$). $MH^+ = 913$.

Complex Formation and Characterization

The ability of the newly synthesized compounds to compact DNA has to be studied. These cationic lipids are expected to compact DNA with the same efficiency as their nondisulfide analogues.[18] For this purpose, usual tests consisting in ethidium bromide (EtBr) exclusion and particle size measurement by dynamic light scattering are performed.

Ethidium Bromide Exclusion

Ethidium bromide fluoresces on insertion between DNA bases. Because compacted DNA is not accessible to EtBr, the formation of lipoplexes induces a decrease in EtBr fluorescence. DNA–lipid complexes are prepared by mixing volume-to-volume 0.01 g/l final DNA concentration in 150 mM NaCl and an excess of lipid, for instance 40 nmol in 400 μl to obtain a charge ratio (+/−) equal to 5. Charge ratios are expressed as mole of positive charges on the lipid per mole of DNA negative charges, assuming that the spermine lipid bears three positive charges at neutral pH. Because 1 μg of DNA corresponds to 3 nM of negatively charged phosphate, the number of nanomoles of cationic lipid per microgram DNA is equivalent to the charge ratio (+/−). Ethidium bromide (3 μl, 1 g/l) is added to the complexes, and the fluorescence intensity is measured after excitation at 260 nmoles. This signal reflects the amount of noncompacted DNA, compared with free DNA taken as 100% fluorescence. Maximal fluorescence is detected at charge ratios (+/−) <1, when negative charges

---

[18] B. Pitard, O. Aguerre, M. Airiau, A. Lachages, T. Boukhnikachvili, G. Byk, C. Dubertret, C. Herviou, D. Scherman, J. Mayaux, and J. Crouzet, *Proc. Natl. Acad. Sci. USA* **94,** 14412 (1997).

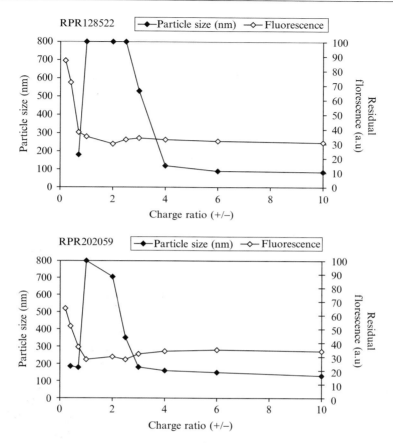

FIG. 4. Size of RPR128522/DNA and RPR202059/DNA complexes (closed symbol) and level of DNA compaction according to ethidium bromide fluorescence intensity measurements (open symbol).

of DNA are in excess compared with the cationic charges of the lipid (Fig. 4). Increasing the amount of cationic lipid at a constant DNA concentration induces complete DNA compaction, as shown by a residual EtBr fluorescence of approximately 10%; that is to say, 90% of DNA is not accessible to the intercalator. As shown in Fig. 4, DNA compaction was not affected by the addition of a disulfide bond on any part of the lipid.

Particle Size

To determine the size of the particle structures formed by DNA compaction with a cationic lipid, dynamic light scattering experiments are performed. Eight hundred microliters of complexes is prepared by mixing equal volumes of 0.01 g/l final DNA concentration in 150 mM NaCl and an excess of lipid, for instance 40 nM in 400 μl to obtain a charge ratio $(+/-)$ equal to 5. The formation of complexes leads to particles 100–200 nmoles in size. Particular care should be taken on the dynamic light scattering measurement. The size indicated by the device is highly dependent on the range of particle counts, on the defined angle, and also on the equilibration and run times. On a Coulter N4 Plus analyzer (Coulter, Amherst, MA), a count range of 5.10^4–5.10^5, a 5-min equilibration time, and a 15-min running time are used to make reliable measurements. Figure 4 shows the unimodal analysis results obtained with the two reducible lipids RPR128522 and RPR202059. Their interaction with DNA at a charge ratio $(+/-)$ of 5 gave small particles, corresponding to a zone of colloidal stability. When the ionic repulsion between the particles is too weak (i.e., in case of a balance between the positive and negative charges), the colloidal suspension collapses, and larger particles are detected. This zone of colloidal instability corresponds with a charge ratio $(+/-)$ lipid/DNA close to 1 (i.e., from 0.5–3 for the two examples given in Fig. 2). This value is also highly dependent on the lipid studied, the method of preparing the lipid–DNA complexes,[19] the DNA size,[20] the pH, and the salt conditions used.[21]

DNA Release–Complex Dissociation

Residual EtBr fluorescence can also be used to monitor DNA release from the complexes, in particular after reduction of a disulfide bond on the cationic lipid. Procedures are performed as described previously, and diothiothreitol (DTT) is added to the complexes to reach a final concentration of 5 mM. To guarantee reduction, some samples are incubated at 37° for up to 24 h. As alternative reducing agents, cysteine (Sigma, France) and glutathione (Sigma, France) are used at the same concentration. The pH of solutions of glutathione is adjusted to 6.0–6.5. DNA release is monitored by an increase in EtBr fluorescence. Figure 5 represents two examples of this experiment.

The release of DNA from lipoplexes, after reducing agent addition, is observed with the lipids RPR128522 and RPR202059, which bear disulfide bonds, but not with their parent lipids, which do not contain any reducible bond.

[19] M. Ferrari, D. Rusalov, J. Enas, and C. Wheeler, *Nucl. Acids Res.* **29,** 1539 (2001).

[20] P. Kreiss, B. Cameron, R. Rangara, P. Mailhe, O. Aguerre-Charriol, M. Airiau, D. Scherman, J. Crouzet, and B. Pitard, *Nucleic Acids Res.* **27,** 3792 (1999).

[21] J. Turek, C. Dubertret, G. Jaslin, K. Antonakis, D. Scherman, and B. Pitard, *J. Gene Med.* **2,** 32 (2000).

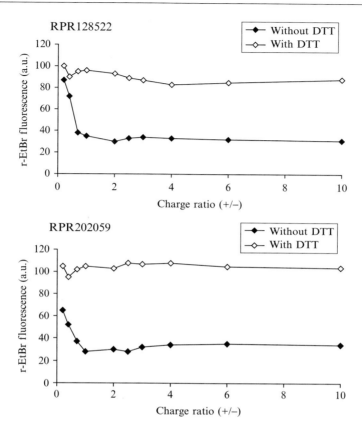

FIG. 5. Induction of DNA release from the complexes measured by ethidium bromide experiments. Effect of an overnight treatment with DTT is shown for RPR128522/DNA and RPR202059/DNA complexes.

The pH value and salt concentrations influence lipid reduction, and thus DNA release. At pH values less than 6.0, DNA release was not observed, whereas a NaCl concentration higher than 50 mM facilitated DNA release.

Small-Angle X-Ray Scattering

Small-angle x-ray scattering (SAXS) experiments can be undertaken to characterize lipid–DNA complexes and to define their structure. Complexes are formed as described previously and are highly concentrated by centrifugation (for colloidally unstable complexes) and ultrafiltration (for colloidally stable and reduced complexes).[18] Pellets or very concentrated complexes are then injected into 1.5-mm diameter capillaries. The results of synchroton X-ray diffraction showed a scattering peak at 83Å for RPR128522 complexes. The half-width of the scattering peak corresponds to an effective domain size of 200 Å, which indicates the repetition of three to four periodicities per particle. The same structural parameters have been found for DNA complexes with the similar parent lipid RPR120535. To reduce the cationic lipid, DTT (final concentration 100 mM) is added to DNA–lipid complexes. After incubation with DTT, complexes of DNA with RPR128522 did not show any organized structure, which indicates lipoplex dissociation. In contrast, with the control RPR120535 lipid, the structure of the complexes after DTT addition did not differ from the structure of the untreated complexes.[11]

Fluorescence Resonance Energy Transfer

Fluorescence resonance energy transfer (FRET) has already been used to evaluate the association/dissociation of nonreducible complexes by the addition of sodium dodecyl sulfate (SDS)[22] or the mechanism of oligonucleotide release from cationic liposomes.[23] By use of a fluorescently labeled plasmid and a mixture of the cationic lipid and a labeled colipid (1% of total lipid), the proximity of the two fluorescent entities is close enough (< 80 Å) in complexes to induce an energy transfer between the two fluorophores.

We describe the energy transfer between cationic liposomes composed of RPR202059/dioleoyphosphatidylethanolamine (DOPE)–DOPE-rhodamine and a 15-mer-phosphodiester-Fluorescein isothiocyanate (FITC) (Fig. 6). When the labeled oligonucleotide (8 μg) is incorporated in a nonlabeled liposome (5 nM cationic lipid/μg oligonucleotide, no DOPE-rhodamine), one emission peak is observed at 516 nmoles, with the excitation wavelength set at 465 nmoles (black triangles). Incorporation of the same oligonucleotide in the labeled liposome analogue induces a reduction of the donor emission peak intensity (524 nmoles) in favor of the acceptor emission signal observed at 586 nmoles, corresponding to the rhodamine emission on the DOPE analogue (circles). Controls show that under a 465-nmoles excitation wavelength, this second peak cannot be attributed to the liposome itself (line). The fluorescence energy transfer is monitored by the decrease in donor emission intensity and by the increase in acceptor emission intensity.

The addition of the reducing agent DTT to the lipoplexes of RPR202059–DOPE–DOPE-rhodamine and 15-mer phosphodiester FITC

[22] S. Li, W. Tseng, D. Stolz, S. Wu, S. Watkins, and L. Huang, *Gene Ther.* **6**, 585 (1999).
[23] O. Zelphati and F. Szoka, *Proc. Natl. Acad. Sci. USA* **93**, 11493 (1996).

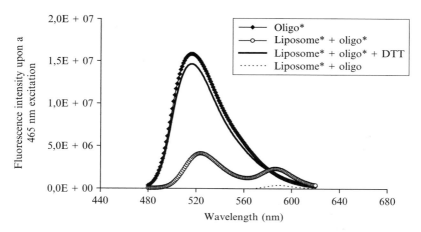

FIG. 6. FRET experiments. Triangles represent the oligonucleotide 15-mer FITC in the same buffer conditions as the other experiments. Circles represent the same oligonucleotide in a RPR128522/DOPE/DOPE-rhodamine liposomes. Crosses indicate the effect of DTT addition on the former liposome (circles). The line shows the control unlabeled oligonucleotide in a labeled liposome.

reverses the FRET effect (Fig. 6). This loss of energy transfer is not immediate, although, and an overnight treatment with 5 mM reducing agent concentration is required. This suggests a low accessibility of the hydrophilic DTT to the disulfide bridge within the hydrophobic moiety of lipoplexes.

Gene Transfer and Cellular Studies

Cellular uptake and gene transfer efficiency have been studied for the reducible cationic lipids and their DNA complexes and compared with their nonreducible analogues. Presumably, a difference between reducible cationic lipids and nonreducible cationic lipids is expected in terms of transfection efficiency. Conversely, cellular uptake should be quite similar, because the cationic entity is the same for both type of lipids.

Transfection efficiency is evaluated with HeLa cells grown at 37° in a 5%CO_2–95% air incubator in Eagle minimum essential medium (MEM) containing 2 mM L-glutamine, 1% MEM nonessential amino acid solution ×100, 50 units/ml penicillin, 50 μg/ml streptomycin, and 10% (v/v) fetal bovine serum (all from Gibco/BRL Life Technologies, Gaithersburg, MD). At 24 h before transfection, 1×10^5 cells/ml are seeded into a 24-well culture plates to obtain approximately 80% confluency after 36 h. The DNA–lipid complex solution (50 μl) containing 0.5 μg of plasmid (4.5 kB containing the firefly luciferase luc gene driven by the CMV-promoter) in

TABLE I
LUCIFERASE ACTIVITY IN HELA CELLS AND UPTAKE EFFICIENCY OF THE LIPOPLEXES ARE
REPRESENTED FOR BOTH REDUCIBLE CATIONIC LIPIDS

Cationic lipid	RPR202059	RPR203769
Luciferase activity (RLU/μg protein)	510^4–10^5	$3 \ 10^3$
Internalization (% ^{3}H DNA in cells)	4%	8%

75 mM NaCl at different charge ratios is added to each well. Two hours after transfection, the serum-free medium is supplemented with 10% (v/v) FBS, and the cells are grown for 36 h at 37° in 5% CO_2.

The luciferase activity is then measured using the Promega Luciferase Assay System (Promega, Madison, WI). Cell layers are rinsed twice with 500 μl PBS and lysed with 250 μl Promega cell culture lysis reagent for 30 min at 20°. After 5 min of centrifugation at 12,000 g and 4°, aliquots of 10 μl supernatant are evaluated for luciferase activity using the Promega luciferase substrate. In our case, luminometric measurements were performed with a Victor^2 1420 Multilabel Counter (Wallac, Turku, Finland). Luciferase activity is expressed as relative light units (RLU) per microgram of protein. The protein concentration of the lysate is obtained with a Pierce BCA assay kit (Rockford, IL).

Results are shown in Table I for the reducible lipid RPR202059 and its unreducible analogue RPR203769. Lipoplexes formulated with the reducible lipid showed a higher transfection than with the unreducible analogues.[11]

To evaluate the cellular uptake of DNA associated to reducible cationic lipids toward nonreducible ones, fluorescent and radiolabeled DNA can be used.[24] For radiolabeling, plasmid DNA is incubated with CpG-methylase and S-adenosyl-L-[methyl-^{3}H]methionine (New England Biolabs, Beverly, MA).[21] The DNA–lipid complexes are formed with [^{3}H]DNA, and HeLa cells are transfected as described previously. Aliquots (150 μl) from the lysate supernatant are mixed with 12.5 ml Ready Safe scintillation liquid (Beckman Instruments, Fullerton, CA), and radioactivity is counted by liquid scintillation on a 1414 WinSpectral (Wallac, Turku, Finland). Radioactivity is measured as counts per minute. The amount of internalized DNA is expressed as a percent of the total radioactivity that is added to cells in the course of transfection. As shown in Table 1, a comparable

[24] C. Miller, B. Bondurant, S. D. McLean, K. A. McGovern, and D. F. O'Brien, *Biochemistry* **37**, 12875 (1998).

amount of DNA is internalized when complexed with the reducible RPR202059 or with the nonreducible analogue RPR203769.

In summary, disulfide lipid–DNA complexes can be destabilized by reduction, which is followed by DNA release. Compared with their nondisulfide analogues, such lipids can provide an improved reporter gene activity in transfected cells.

Acknowledgment

We thank Cécile Bellier for the establishment of the FRET analyses in the laboratory.

[24] Gene Delivery by Cationic Liposome–DNA Complexes Containing Transferrin or Serum Albumin

By SÉRGIO SIMÕES, PEDRO PIRES, M. TERESA GIRÃO DA CRUZ, NEJAT DÜZGÜNEŞ, and MARIA C. PEDROSO DE LIMA

Introduction

Among nonviral vectors, cationic liposome–DNA complexes (lipoplexes) have been used for numerous *in vitro* and *in vivo* gene delivery applications. However, some disadvantages, including limited efficiency of delivery and gene expression, toxicity at higher concentrations, potentially adverse interactions with negatively charged macromolecules in serum and on cell surfaces, and impaired ability to reach tissues beyond the vasculature unless directly injected into the tissue,[1–4] represent restrictions to their *in vivo* use. In view of these limitations we and others considered the possibility that coating cationic liposomes with proteins, such as transferrin, would promote internalization of the complexes presumably by receptor-mediated endocytosis, thus enhancing their transfection activity. Cheng's[5] and our results[6,7] indicated that transferrin complexed to lipoplexes

[1] R. J. Lee and L. Huang, *J. Biol. Chem.* **271,** 8481 (1996).

[2] J. Zabner, A. J. Fasbender, T. Moninger, K. A. Poellinger, and M. J. Welsh, *J. Biol. Chem.* **270,** 18997 (1995).

[3] K. Konopka, E. Pretzer, P. L. Felgner, and N. Düzgüneş, *Biochim. Biophys. Acta* **1312,** 186 (1996).

[4] K. Konopka, G. S. Harrison, P. L. Felgner, and N. Düzgüneş, *Biochim. Biophys. Acta* **1356,** 185 (1997).

[5] P. W. Cheng, *Hum. Gene Ther.* **7,** 275 (1996).

[6] S. Simoes, V. Slepushkin, R. Gaspar, M. C. P. de Lima, and N. Düzgüneş, *Gene Ther.* **5,** 955 (1998).

enhances gene delivery to various cell types, including primary cells. Studies on the mechanisms of gene delivery by such transferrin–lipoplexes suggested, however, that specific ligand–receptor interactions are not involved in this process.[8] On the basis of these findings, we explored whether associating the most abundant plasma protein, albumin, would result in a similar enhancing effect and would also alleviate some of the undesired interactions between liposome–DNA complexes and serum components. Although albumin would not be expected to function as a receptor ligand, our results indicate that it facilitates transfection by lipoplexes, possibly by mediating nonspecific endocytosis.[9]

It has also been demonstrated that besides its role in triggering internalization of the lipoplexes, albumin and transferrin function as fusogenic proteins that destabilize endosomes under acidic conditions, thus enhancing intracellular gene delivery.

In this chapter, various methods and techniques used to prepare protein-associated lipoplexes and to evaluate their biological activity and mechanisms of interaction with cells are described. Overall, these approaches may constitute important tools for the development and application of protein-associated or peptide-associated lipoplexes, especially those designed for targeting purposes.

Preparation of Protein-Associated Complexes

Liposome Preparation

Cationic liposomes composed of DOTAP or DOTAP:DOPE (1:1 weight ratio) are prepared by first drying a film of lipid under argon and then in a vacuum oven at room temperature and hydrating the lipid film with 1 ml deionized water at a final concentration of 5 mg/ml. The multilamellar vesicles obtained are then sonicated briefly under argon, extruded 21 times through polycarbonate filters of 50-nm pore diameter using a Liposofast device (Avestin, Toronto, Canada), diluted 5 times with deionized water, and filter-sterilized using (Millipore, Keene, NH, USA) 0.22-μm pore-diameter filters. Liposomes should be stored at 4° under nitrogen atmosphere and used within 3 weeks after being prepared.

---

[7] S. Simoes, V. Slepushkin, E. Pretzer, P. Dazin, R. Gaspar, M. C. P. de Lima, and N. Düzgünes, *J. Leukocyte Biol.* **65,** 270 (1999).

[8] S. Simoes, V. Slepushkin, P. Pires, R. Gaspar, M. C. P. de Lima, and N. Düzgünes, *Gene Ther.* **6,** 1798 (1999).

[9] S. Simões, V. Slepushkin, R. Gasper, M. C. Pedroso de Lima, and N. Düzgünes, *Biochim. Biophys. Acta* **1463,** 459 (2000).

Preparation of the Ternary Complexes

Complexes are prepared by sequentially mixing 100 μl of a solution of 100 mM NaCl, 20 mM HEPES, pH 7.4, containing 32 μg iron-saturated human transferrin (Collaborative Biomedical Products, via Becton Dickinson, Bedford, MA) or 32 μg human serum albumin (HSA, Sigma, St. Louis, MO) with an appropriate amount of liposomes (depending on the $+/-$ charge ratio required) and incubated at room temperature for 15 min. One hundred microliters of buffer containing 1 μg of pCMVluc (provided by Dr. P. Felgner, Vical, Inc., San Diego, CA) plasmid (encoding luciferase), or 1 μg pCMV.SPORT β-gal (Gibco, Paisley, Scotland) plasmid (β-galactosidase), or 1 μg of modified pHooK™-2 (provided by Dr. John Rossi, City of Hope, Duarte, CA) plasmid encoding the green fluorescent protein (GFP) is then added and mixed gently, and the mixture is further incubated for 15 min at room temperature. Complexes are prepared under aseptic conditions (bioguard hood or flow cabinet) using sterile solutions, because their size frequently hampers sterilization through 0.22-μm filters. Complexes should be used immediately after their preparation to avoid any aggregation. Alternately, complexes can be frozen in liquid nitrogen and stored at $-80°$ without losing their biological activity.[10] A detailed physicochemical characterization of the ternary complexes is described in Chapter 23.

Protocols of Cell Transfection

It has been recognized that transfection activity mediated by lipoplexes is strongly dependent on the type, confluence, and age of the cell. Therefore, for each liposomal formulation different experimental conditions for cell transfection should be tested to define the parameters leading to optimal levels of gene expression. Transfection protocols are described in the following for a variety of cell types, including adherent (epithelial) and suspension (lymphocytic) cell lines, as well as primary culture cells (human macrophages and neuronal cells).

Cultured Cell Lines

Adherent Cells. HeLa and COS-7 cells (American Type Culture Collection, Manassas, VA) are maintained at $37°$, under 5% CO_2, in Dulbecco's modified Eagles medium-high glucose (DME-HG) (Irvine Scientific, Santa Ana, CA) supplemented with 10% (v/v) heat-inactivated fetal bovine serum (FBS) (Sigma, St. Louis, MO), penicillin (100 units/ml), streptomycin

[10] H. Faneca, S. Simões, and M. C. Pedroso de Lima, *Biochim. Biophys. Acta* **1567,** 23 (2002).

(100 μg/ml), and L-glutamine (4 mM). For transfection, 0.2×10^6 HeLa cells are seeded in 1 ml of medium in 48-well culture plates and used at 80%–90% confluence. COS-7 cells are seeded in 1 ml of medium in 48-well culture plates and used at 40%–60% confluence. Both types of cells are rinsed twice with serum-free medium and then covered with 0.3 ml of DME-HG before lipid–DNA complexes are added.

Suspension Cells. H9 cells, a CD4$^+$ clonal derivative of the Hut-78 T-cell line readily infectable by HIV, are grown in RPMI 1640 medium (Irvine Scientific, Santa Ana, CA) supplemented with 10% (v/v) FBS, L-glutamine (2 mM), and antibiotics as described previously. B-lymphocytic TF228.1.16 cells that stably express functional human immunodeficiency virus (HIV) envelope proteins on the cell surface are grown in DME-HG supplemented with 16% (v/v) FBS, L-glutamine (4 mM), and antibiotics. For transfection, cells are rinsed twice with serum-free medium (centrifuged at 180 g for 5 min), and 10^6 cells/0.3 ml of medium are aliquoted into polypropylene culture tubes (Corning Costar, Cambridge, MA) before lipid–DNA complexes are added.

Primary Cell Cultures

Human Macrophages. Monocytes are obtained from buffy coats by centrifugation on a Ficoll-Hypaque (Histopaque-1077; Sigma, St. Louis, MO) gradient and plastic adherence. Mononuclear cells separated by centrifugation are counted and plated in DME-HG without serum at a density of 1.4×10^6 cells/ml per well in 48-well plates. It is assumed that approximately 5%–10% of the cells plated are recovered as macrophages. The cells are allowed to adhere overnight, then washed, and the medium is replaced with DME-HG supplemented with 20% (v/v) heat-inactivated FBS (Sigma, St. Louis, MO), 10% (v/v) human AB serum (Advanced Biotechnologies, Columbia, MD), penicillin (100 U/ml), streptomycin (100 μg/ml), and L-glutamine (2 mM), as described previously.[11] The cells are left undisturbed in this medium for 6–7 days for differentiation to occur. In some experiments, human granulocyte-macrophage colony stimulating factor (hGM-CSF) (Boehringer Mannheim Biochemica, Indianopolis, IN) is added to the wells (final concentration of 100 UI/well) on the second day of differentiation. In other experiments, cells are cultured for 8 more days in medium containing 20% FBS, antibiotics, and L-glutamine but in the absence of hGM-CSF. Cells are rinsed twice with serum-free medium and then covered with 0.3 ml of medium before lipid–DNA complexes are added. It is relevant to note that the differentiation time and the

---

[11] E. Pretzer, D. Flasher, and N. Düzgüneş, *Antiviral Res.* **34,** 1 (1997).

presence or absence of differentiation stimulating factors are determining parameters of the levels and duration of transgene expression.[7]

Hippocampal Neurons. Primary cultures of hippocampal neurons are obtained from 18- to 19-day gestation rat embryos. The animal is killed by cervical dislocation, and the embryos are transferred into a large Petri dish containing Hank's balanced salt solution (HBSS; 137 mM NaCl, 5.36 mM KCl, 0.44 mM KH$_2$PO$_4$, 4.16 mM NaHCO$_3$, 0.34 mM Na$_2$HPO$_4$. H$_2$O, 5 mM glucose, 0.001% phenol red, 1 mM sodium pyruvate, 10 mM HEPES, pH 7.2) (Sigma Chemical, St. Louis, MO). Brains are removed from the skull, the two hemispheres separated, and the *hippocampi* dissected under the microscope using thin, straight tweezers. The tissue is placed into a 35-mm Petri dish containing HBSS and transferred to a centrifuge tube containing 0.1% trypsin and 0.03% DNase (Sigma) in HBSS (final concentration). Following an incubation of 15 min at 37° (while slowly shaking the tube), the sample is centrifuged for 1 min at 88 g. The supernatant is removed, and 10 ml of HBSS enriched with 10% FBS (BioWhittaker, Verviers) is added. The sample is slowly shaken and again centrifuged for 1 min at 88 g. The supernatant is removed, and 10 ml of HBSS is added. The tissue is then dissociated with a pipette, and the resulting suspension is centrifuged for 3 min at 405 g. The supernatant is then removed and the cells resuspended in neurobasal medium (Gibco, Paisley, Scotland), enriched with 2% (v/v) B27 supplement to prevent glial proliferation, 0.005% gentamicin (Gibco), 0.05% glutamine, and 25 μM glutamate (Sigma).

Cell viability is determined by scoring the cells by Trypan blue exclusion,[12] and the suspension is diluted to the appropriate cell density. Cells are plated at 0.1 × 10^6 cells/well on 48-well plates previously coated with poly-L-lysine (prepared from a 0.01% solution in 150 mM boric acid, pH 8.4) and incubated at 37° in a humidified atmosphere containing 5% CO$_2$. Cells are rinsed twice with serum-free medium and then covered with 0.3 ml of medium before lipid–DNA complexes are added.

Cortical Neurons. Primary cultures of cortical neurons are obtained from 15-16-day gestation rat embryos. The animal is killed by cervical dislocation, and the embryos are transferred into a large Petri dish containing Krebs solution (120.9 mM NaCl, 4.83 mM KCl, 1.22 mM KH$_2$PO$_4$, 25.5 mM NaHCO$_3$, 13 mM glucose, 0.0015% phenol red, 10 mM HEPES, pH 7.4) (Sigma). Following dissection, the tissue is placed on a 35-mm Petri dish containing 0.3% bovine serum albumin (BSA) (Sigma) in Krebs solution (solution 1) and cut into small pieces. The tissue is transferred to a tube containing solution 1 and centrifuged briefly at 146 g. The supernatant is removed, the tissue resuspended in solution 2 (0.02% trypsin, 0.004%

---

[12] B. Detrick-Hooks, T. Borsos, and H. J. Rapp, *J. Immunol.* **114**, 287 (1975).

DNase, in solution 1) (Sigma), and incubated for 10 min at 37° (slowly shaking the tube). Solution 3, containing 0.052% trypsin inhibitor, 0.004% DNase (Sigma Chemical, St. Louis, MO) in solution 1, is then added, and the sample is centrifuged for 5 min at 146 g. The supernatant is removed, and the pellet is resuspended in solution 4 (resulting from the mixture of 16.8 ml of solution 1 and 3.2 ml of solution 3). The tissue is then dissociated with a pipette, and the resulting suspension is centrifuged for 3 min at 405 g. The supernatant is removed, and the cells are resuspended in neurobasal medium (Gibco), enriched with 2% (v/v) B27 supplement (Gibco) to prevent glial proliferation, 0.2 mM glutamine, 100 μg/ml of streptomycin, and 1 unit/ml of penicillin (Sigma). Cell viability is determined by Trypan blue exclusion. Cells are plated at 0.1×10^6 cells/well on 48-well plates previously coated with poly-L-lysine (prepared from a 0.01% solution in 150 mM boric acid, pH 8.4) and incubated at 37° in a humidified atmosphere containing 5% CO_2. Cells are rinsed twice with serum-free medium and then covered with 0.3 ml of medium before lipid–DNA complexes are added. To evaluate the effect of the neuronal stage of differentiation on luciferase expression, cells are maintained in culture for different periods of time (from day 0–day 7).

Evaluation of Transgene Expression

Gene expression is usually evaluated in terms of transfection activity, assessed as the total amount of reporter protein produced by the transfected cells per milligram of cell protein. Among other proteins, luciferase gene expression constitutes a suitable and sensitive assay for this purpose. Lipid–DNA complexes are added gently to cells in a volume of 0.2 ml per well. After an incubation for 4 h (in 5% CO_2 at 37°) the medium is replaced with DME-HG–containing 10% FBS, and the cells are further incubated for 24 or 48 h. The cells are then washed twice with phosphate-buffered saline (PBS), and 100 μl of lysis buffer (1 mM dithiothreitol [DTT]; 1 mM ethylenediamine tetracetic acid [EDTA]; 25 mM TRIS-phosphate [pH 7.8]; 8 mM MgCl$_2$; 15% glycerol; 1% (v/v) Triton X-100) is added to each well.

The level of gene expression in the lysates is evaluated by measuring light production by luciferase in a luminometer and using a standard curve for luciferase activity. Because luminescence decreases rapidly, the lysates should either be analyzed immediately after their preparation or stored at −20°. To prevent misunderstanding of the results from artifacts related to variation in cell density, transfection activity is frequently expressed as nanograms of luciferase per milligram of total cell protein. For that purpose, the protein content of the lysates is measured by the Dc Protein Assay reagent (Bio-Rad, Hercules, CA) using BSA as the standard.

To investigate whether the enhancement of transfection resulting from changes in the experimental conditions is due to an increase in the number of cells transfected or only an increase in the levels of gene expression, transfection efficiency should also be evaluated. The latter is defined as the percentage of treated cells that express the transgene. For this purpose, the plasmids encoding β-galactosidase or GFP are complexed with the cationic liposomes, and cells are transfected as described previously. The results are presented as the percentage of cells scored for the expression of the referred proteins. Cells transfected with 1 μg of pCMV.SPORT-β-gal are washed with PBS, fixed in a solution of 2% formaldehyde and 0.2% glutaraldehyde, and stained with a solution containing 5'-bromo-4-chloro-3-indoyl-b-D-galactopyranoside (X-gal) (Invitrogen Corporation, San Diego, CA). The cells are incubated at 37° for 24 h and then examined under a phase-contrast microscope for the development of blue color. The percentage of cells exhibiting β-gal activity is evaluated by counting 1000 cells in duplicate wells (see insert in Fig. 1). In the GFP assay, cells are washed with PBS and detached from plastic by adding 0.5 ml of cell dissociation buffer (Life Technologies, Inc.). They are then mixed with 0.5 ml of DME-HG medium containing 1% of FBS and 1 μg/ml propidium iodide (Sigma Chemical, St. Louis, MO). Green fluorescence is detected with a single argon laser flow cytometer (e.g., Facscan, Becton Dickinson, San Jose, CA). Green Fluorescence protein and propidium iodide are excited at 488 nm, and respective emissions are collected using a 530-nm band-pass filter (30-nm band width) for GFP and a 630-nm band-pass filter (22-nm band width) for propidium iodide. Five thousand events are recorded for each sample. Forward scatter and propidium iodide fluorescence signals are used to gate the cell subset of interest and to eliminate debris, dead cells, and cell aggregates.

Typical results demonstrating the enhancing effect on transfection of both cultured cell lines (HeLa cells) and primary cell cultures (cortical neurons) observed on association of a protein (transferrin) to the lipid–DNA complexes is shown in Fig. 1. The lipid/DNA charge ratio of the complexes plays a major role in the levels of transfection. For each type of protein, the amount to be associated to the complexes should be optimized to achieve the best transfection conditions. The results obtained from a titration experiment on the amount of HSA added to the lipoplexes is illustrated in Fig. 2. An enhancement of transfection is observed as the amount of albumin associated with (1/1) DOTAP:DOPE–DNA complexes is increased up to 32 μg. Doubling this amount leads to a decrease in the level of luciferase gene expression, indicating that 32 μg of HSA is optimal for transfection. In addition to the enhancing effect on transfection described previously, protein-associated lipoplexes are shown not to be

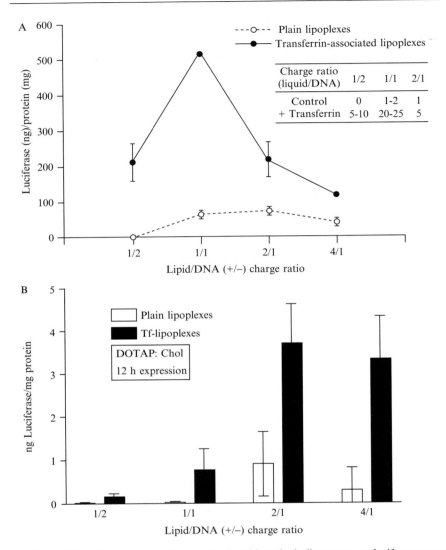

Fig. 1. The effect of associating transferrin with cationic liposomes on luciferase gene expression in (A) HeLa cells. Liposomes composed of DOTAP:DOPE are complexed in the presence or absence of 32 μg of transferrin, with 1 μg of pCMVluc at the indicated theoretical lipid/DNA charge ratios. After incubation for 4 h, the medium is replaced with DME-HG containing 10% FBS, and the cells are further incubated for 48 h. The level of gene expression is evaluated as described in the text. The data are expressed as nanograms of luciferase per milligram of total cell protein (mean ± standard deviation obtained from triplicate wells) and are representative of two independent experiments. Insert, expression of β-gal in HeLa cells

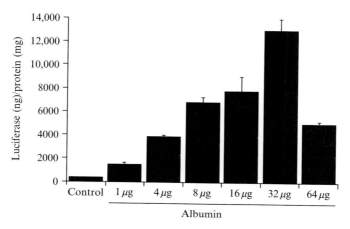

FIG. 2. The effect of the amount of human serum albumin (HSA) complexed with DOTAP:DOPE liposomes on luciferase gene expression in COS-7 cells. The liposomes are complexed in the presence or absence of different amounts of HSA, with 1 μg of pCMVluc to obtain a 1/1 lipid/DNA charge ratio. The protocol for transfection and evaluation of gene expression is described in the text. Reproduced from Simões et al.,[9] with permission.

sensitive to the presence of serum. This not only results in a simplification of the transfection procedure, because the washing steps can be eliminated, but also increases the possibility that these complexes can be used for gene delivery *in vivo*.

Cell Viability Assays

Cationic liposomes can be toxic to cells, depending on the lipid concentration, duration of treatment, cell type, and confluency of the culture.[3,6] To evaluate whether plain or ternary complexes are toxic to cells, cell viability is assessed following transfection by using different assays, either on the basis of cell membrane integrity or cell metabolic activity. The former can be performed by monitoring lactate dehydrogenase release or by scoring viable cells on Trypan blue staining[12] and the latter by measuring cell proliferation. This can be achieved by monitoring the reducing environment of the cell either by quantification of tetrazolium salt reduction

---

(% of blue cells). The liposomes are complexed with 1 μg of pCMV.SPORT-β-gal at the indicated theoretical lipid–DNA charge ratios in the absence (control) or presence of transferrin. The percentage of cells exhibiting β-gal activity is evaluated by counting 1000 cells in duplicate wells (adapted from Simões et al.[6]). (B) Hippocampal neurons are transfected at day zero of differentiation with DOTAP:Chol–DNA complexes with or without transferrin, and luciferase gene expression is evaluated 12 h following transfection.

(MTT-3-(4,5-Dimethyl-2-thiazolyl)-2,5-diphenyl-2H-tetrazolium bromide (Sigma, St. Louis, MO), XTT-2,3-Bis(2-methoxy-4-nitro-5-sulfophenyl)-2H-tetrazolium-5-carboxanilide (Sigma, St. Louis, MO), or MTS-3-(4,5-dimethylthiazol-2-yl)-5-(3-carboxymethoxyphenyl)-2-(4-sulfophenyl)-2 assays)[13] or quantitation of alamar Blue™ reduction.[14] The colorimetric Alamar Blue™ (Trek Diagnostic Systems, Cleveland, OH) assay has the advantage of using a reagent that is soluble, stable in culture medium, and nontoxic, as well as of allowing the determination of viability over the culture period without harming the cells.

In our studies, the Alamar Blue™ assay is applied using the following experimental protocol: 1 ml of 10% (v/v) alamarBlue™ dye in complete DME medium is added to each well 45 h following transfection. After 2.5–4 h of incubation at $37°$, 200 μl of the supernatant is collected from each well and transferred to 96-well plates. The absorbance at 570 nm and 600 nm is measured with a microplate reader (Molecular Devices, Menlo Park, CA). Cell viability (as a percentage of control cells) is calculated according to the formula $(A_{570}-A_{600})$ of treated cells $\times$ 100/ $(A_{570}-A_{600})$ of control cells. It should be noted, nevertheless, that a good correlation can be established between the results obtained using the MTT and the alamarBlue™ assays.

Evaluation of the Mechanisms Underlying the Enhancing Effect of Transfection by Proteins Associated to Lipoplexes

Although based on rational strategies, the actual mechanisms by which proteins promote gene delivery when associated with lipoplexes should be clarified. Different approaches can be used to assess the mode by which protein-associated lipoplexes interact with cells, namely, targeting specificity and competitive inhibition studies and pretreatment of the cells with drugs that interfere with the endocytotic pathway. Protocols used for mechanistic studies involving transferrin-associated lipoplexes are described in the following:

1. To demonstrate whether internalization of transferrin (Tf)-lipoplexes is mediated by specific interaction with Tf receptors, parallel transfection experiments are performed using a protein as similar as possible to Tf but to which the cell surface receptors for Tf do not exhibit affinity. For that purpose, ternary complexes are prepared by associating 32 μg of apotransferrin (iron-depleted transferrin) with the lipid–DNA complexes at different

[13] J. Carmichael, W. G. DeGraff, A. F. Gazdar, J. D. Minna, and J. B. Mitchell, *Cancer Res.* **47,** 936 (1987).
[14] S. A. Ahmed, R. M. J. Gogal, and J. E. Walsh, *J. Immunol. Methods* **170,** 211, (1994).

lipid/DNA $(+/-)$ charge ratios. Complexes are added to cells, and transfection activity is evaluated as described previously.

2. Aiming at blocking Tf receptors, a large excess of free Tf (8 mg/ 0.3 ml of DME-HG medium) is added to cells and incubated for 30 min at $37°$ before the addition of ternary complexes (cationic liposomes, pCMVluc plasmid, Tf), which are incubated with the cells for 1 h always in the presence of the excess of Tf. The medium is then replaced with DME-HG containing 10% FBS, and the cells are further incubated for 48 h before being harvested for luciferase activity measurements. These experimental approaches can be extended to studies involving other proteins, like albumin.

3. To further define the mechanisms involved in the internalization and intracellular fate of the various lipoplexes, before the addition of the lipoplexes, cells are incubated for 30 min at $37°$ in the absence of serum, either with (1) a mixture of antimycin A (1 µg/ml), NaF (10 mM), and NaN3 (0.1%), which, are known to strongly inhibit both receptor-mediated and nonreceptor-mediated endocytosis[15,16]; (2) cytochalasin B (25 µg/ml), a drug that is known to disrupt the microfilament network by inhibiting actin polymerization, thereby blocking phagocytosis and pinocytosis, but not receptor-mediated endocytosis[17–20]; (3) bafilomycin A_1 (125 nM), a specific inhibitor of the vacuolar ATPase proton pump present in the intracellular membrane compartments, thus preventing the acidification of the endocytotic pathway.[21–23] Cells are further incubated for 1 h at $37°$ with the lipoplexes in the presence of the various drugs and then washed once with serum-free medium. The medium is then replaced with DME-HG containing 10% FBS, and the cells are further incubated for 48 h before evaluation of transfection. The viability of the cells transfected in the presence of these agents should be evaluated and compared with that of untreated control cells.

The schematic model presented later illustrates some of the strategies that can be applied to gain insights into the mechanisms by which

[15] K. D. Lee, S. Nir, and D. Paphadjopoulos, *Biochemistry* **32**, 889 (1993).

[16] V. A. Slepushkin, S. Simoes, P. Dazin, M. S. Newman, L. S. Guo, M. C. Pedroso de Lima, and N. Düzgüneş, *J. Biol. Chem.* **272**, 2382 (1997).

[17] J. P. Paccaud, K. Siddle, and J. L. Carpentier, *J. Biol. Chem.* **267**, 13101 (1992).

[18] X. Zhou and L. Huang, *Biochim. Biophys. Acta* **1189**, 195 (1994).

[19] O. Zelphati and F. C. Szoka, Jr., *Pharm. Res.* **13**, 1367 (1996).

[20] H. Matsui, L. G. Johnson, S. H. Randell, and R. C. Boucher, *J. Biol. Chem.* **272**, 1117 (1997).

[21] T. Umata, Y. Moriyama, M. Futai, and E. Mekada, *J. Biol. Chem.* **265**, 21940 (1990).

[22] D. D. Pless and R. B. Wellner, *J. Cell Biochem.* **62**, 27 (1996).

[23] L. Cattani, P. Goldoni, M. C. Pastoris, L. Sinibaldi, and N. Orsi, *Antimicrob. Agents Chemother.* **41**, 212 (1997).

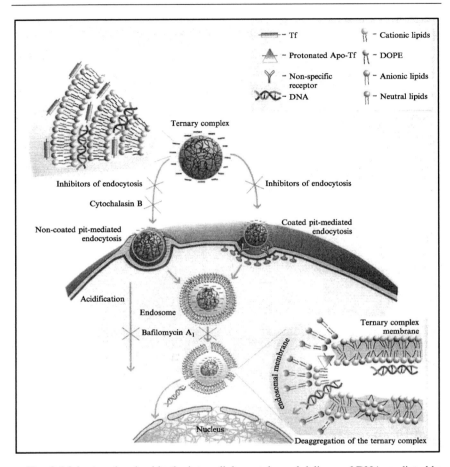

FIG. 3. Main steps involved in the intracellular uptake and delivery of DNA mediated by transferrin (Tf)-lipoplexes. Because of electrostatic interactions, transferrin and DNA associate with cationic liposomes (DOTAP:DOPE), leading to the formation of negatively charged ternary complexes (considering a 1/1 lipid/DNA $[+/-]$ charge ratio) by a still unknown mechanism. Presumably complexes are internalized either by receptor-mediated endocytosis by means of clathrin-coated pits or by phagocytosis by means of uncoated pits, depending on their size. Both of these endocytotic pathways can be inhibited by pretreatment of the cells with inhibitors of endocytosis (mixture of antimycin A, NaF, NaN$_3$), whereas phagocytosis can be selectively inhibited by cytochalasin B. Efficient delivery of DNA into the cytoplasm (from where it can reach the nucleus) seems to be dependent on the acidification of the endosomal lumen, which can be inhibited partially by bafilomycin A$_1$. We speculate that this acidification process triggers a cascade of synergistic effects that would lead to dissociation of DNA from the ternary complexes and to destabilization of the endosomal membrane. Structural changes induced by protonation of apotransferrin would promote dissociation of the complex. It is also possible that, under these conditions, apotransferrin

protein-associated lipoplexes are internalized and mediate intracellular gene delivery.

Albumin has been described as being able to undergo a low pH-induced conformational change, thus acquiring fusogenic properties.[24–26] Therefore, protonation of HSA and its subsequent interaction with the endosome membrane may be involved in the destabilization of the latter, contributing to intracellular delivery of DNA (Fig. 3). This destabilization may then promote the transbilayer movement (flip-flop) of anionic lipids from the cytoplasmic leaflet of the endosome membrane to the luminal leaflet. It is also possible that the conformational change of albumin, involving a reversible expansion of the protein,[25] reinforces the dissociation of the complexes promoted by lipid flip-flop. To investigate this hypothesis, experiments using fluorescent probes, such as 1-aminonaphtalene-8-sulfonic acid (ANS), which allows detection of hydrophobic microenvironments in proteins,[27] can be performed. These studies involve the incubation of 32 μg of albumin in 2 ml of HEPES buffer (adjusted at different pH values) with 23 μM ANS at 37° for 15 min. Conformational changes of the protein, namely, through exposure of hydrophobic domains as a function of the pH, are monitored by the increase in the ANS fluorescence (wavelength ranging from 440–580) and by the blue shift in the emission maximum observed (on excitation at 350 nm). Similar studies can be carried out with other proteins associated with the cationic liposome–DNA complexes. A typical result obtained from experiments using this method is illustrated in Fig. 4. Conformational changes of proteins associated with complexes can also be evaluated by monitoring the tryptophan intrinsic fluorescence.

[24] S. Shenkman, P. S. Araujo, R. Dijkman, F. H. Quina, and H. Chaimovich, *Biochim. Biophys. Acta* **649,** 633 (1981).
[25] L. A. M. Garcia, S. Shenkman, P. S. Araujo, and H. Chaimovich, *Brazil. J. Med. Res.* **16,** 89 (1983).
[26] J. Wilschut and D. Hoekstra, *Chem. Phys. Lipids* **40,** 145 (1986).
[27] J. Ramalho-Santos, R. Negrão and M. C. Pedroso de Lima, *Biosci. Rep.* **14,** 15 (1994).

becomes fusogenic, promoting destabilization of the endosomal membrane, therefore creating favorable conditions for the flip-flop of anionic lipids from the cytoplasmic leaflet of the endosomal membrane. Electrostatic interactions between the anionic lipids present in the inner leaflet of the endosomal membrane and cationic lipids would not only promote deaggregation of the complexes but also facilitate DOPE to undergo a transition from a bilayer to an inverted hexagonal phase, thus acquiring fusogenic properties. All these events would lead to dissociation of DNA from the complexes and to its escape into the cytoplasm (Reproduced from Simões *et al.*,[8] with permission).

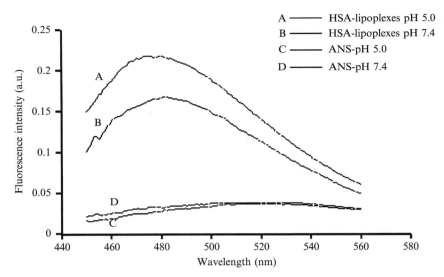

FIG. 4. The effect of pH on the hydrophobicity of HAS-lipoplexes. Emission spectra of ANS in the absence (C and D) and in the presence of HSA-associated lipoplexes (A and B) as a function of pH. As shown, ANS fluorescence is not affected by pH, but an increase in fluorescence intensity is observed on incubation of the probe with albumin-associated complexes. This increase is even more pronounced and accompanied by a blue shift on lowering the pH, indicating that under these conditions protein hydrophobic domains are exposed.

In Vivo Transfection Studies

Because of restrictions in the sample volume that can be injected into mice through the tail vein (approximately 200 μl), complexes should be prepared at considerably higher concentrations than those used for *in vitro* studies. Typically, 100 μg of plasmid DNA pCMVluc (encoding luciferase) is mixed with a solution containing 3.2 mg of protein (either albumin or Tf) and the required amount of cationic liposomes to obtain the desired lipid/DNA (+/−) charge ratio and incubated for 15 min at room temperature. It should be noted that, because of the high concentration of lipoplex components, a large excess of positive charge should be used to obtain small ternary complexes, which is a requirement for intravenous (iv) administration. Alternately, small complexes can be obtained by preparing them in low ionic strength medium (e.g., 5 mol% dextrose). Regarding the liposomal formulation, DOTAP:Chol (1:1) liposomes have been used because of their high stability in biological fluids. Following iv administration into

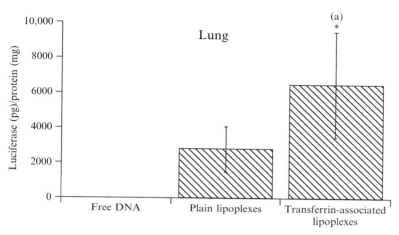

FIG. 5. Levels of luciferase gene expression in the lung followed by iv administration in BALB/c mice of DOTAP:Chol–DNA complexes with or without transferrin.

8-week-old BALB/c mice, animals are killed, and the organs are harvested and homogenized 24 h following injection. Luciferase expression in the supernatant is measured as described previously, and transfection activity is reported as picograms of luciferase per milligram of organ. Similar to what is observed in *in vitro* studies, Tf-associated lipoplexes are able to enhance transfection activity *in vivo* compared with plain lipoplexes. The lung is shown to be the organ in which the highest levels of gene expression are observed (Fig. 5).

Concluding Remarks

The development of new approaches to improve the features of lipoplexes aiming at generating viable alternatives to viral vectors is crucial for gene therapy applications. To achieve such a goal, attempts have been made to confer viral attributes to lipoplexes, namely, through the association of certain proteins or peptides. These strategies are designed taking into consideration the various biological barriers faced by the complexes on their *in vivo* administration, which need to be overcome for gene expression to occur. Whether these improvements result in a system that, although exhibiting satisfactory ability to mediate transfection, would lead to such a complexity that could endanger its versatility and large-scale production or could limit extended/repeated *in vivo* use as a result of immunogenicity, are important questions that remain to be addressed.

[25] Glycosylated Cationic Liposomes for Carbohydrate Receptor–Mediated Gene Transfer

By MAKIYA NISHIKAWA, SHIGERU KAWAKAMI, FUMIYOSHI YAMASHITA, and MITSURU HASHIDA

Introduction

The success of *in vivo* gene therapy relies on the development of a vector that achieves target cell–specific, efficient, and prolonged transgene expression following its *in vivo* application. Nonviral vectors have some advantages in that they are less toxic, less immunogenic, and easier to prepare than viral ones. Following the report of Felgner *et al.*,[1] cationic lipid–containing liposomes have become widely used as reagents to introduce DNA into cells. The addition of cationic lipids to plasmid DNA (pDNA) decreases its negative charge and facilitates its interaction with cell membranes, leading to transgene expression in the cells. In an attempt to increase the efficiency of transgene expression and to reduce cytotoxicity, several kinds of cationic lipids, such as quaternary ammonium detergents, cationic derivatives of cholesterol and diacylglycerol, and alkyl derivative of polyamines, have been developed. Some cationic lipid–DNA complexes have been used in clinical trials for the treatment of cancer and cystic fibrosis.[2]

Although the pDNA–cationic liposome complex (lipoplex) is a useful nonviral vector, it lacks specificity in delivery and transfection. The attachment of a ligand that can be recognized by a specific mechanism would endow a vector with the ability to target a specific population of cells. In the search for macromolecule-based gene delivery systems, several ligands, including asialoglycoproteins, galactose, mannose, transferrin, antibodies, and lung surfactant proteins, have been used to improve the delivery of pDNA to target cells.[3] Therefore, the incorporation of such ligands into cationic liposomes would improve the target-cell specificity of *in vivo* gene transfer by lipoplexes. Here we describe novel glycosylated cationic liposomes, consisting of a glycosylated cationic cholesterol derivative and several helper lipids, which increase the transfection efficiency to cells possessing the corresponding receptors, such as hepatocytes and macrophages.[4]

[1] P. L. Felgner, T. R. Gadek, M. Holm, R. Roman, H. W. Chan, M. Wenz, J. P. Northrop, G. M. Ringold, and M. Danielsen, *Proc. Natl. Acad. Sci. USA* **84,** 7413 (1987).
[2] G. J. Nabel, E. G. Nabel, Z. Y. Yang, B. A. Fox, G. E. Plautz, X. Gao, L. Huang, S. Shu, D. Gordon, and A. E. Chang, *Proc. Natl. Acad. Sci. USA* **90,** 11307 (1993).
[3] M. Nishikawa and L. Huang, *Hum. Gene Ther.* **12,** 861 (2001).

Materials

N-(2-aminoethyl)carbamic acid *tert*-butyl ester, *N*-(4-aminobutyl)-carbamic acid *tert*-butyl ester, *N*-(6-aminohexyl)carbamic acid *tert*-butyl ester, and *N*-[1-(2,3-dioleyloxy)propyl]-n,n,n-trimethylammonium chloride (DOTMA) are purchased from Tokyo Chemical Industry (Tokyo, Japan). Cholesteryl chloroformate is obtained from Sigma Chemicals (St. Louis, MO), dioleoylphosphatidylethanolamine (DOPE) is purchased from Avanti Polar Lipids (Alabaster, AL), and 3β(*N*,*N*,*N*-dimethylaminoethane-carbamoyl)cholesterol (DC-Chol) is synthesized according to the published method.[5]

Synthesis of Glycosylated Cholesterol Derivatives

The following is the protocol for the synthesis of cholesten-5-yloxy-*N*-(4-((1-imino-D-thiogalactosyl-ethyl)amino)butyl)formamide (Gal-C4-Chol) (Fig. 1). Instead of D(+)-galactose, D(+)-mannose and L(−)-fucose are used for the synthesis of cholesten-5-yloxy-*N*-(4-((1-imino-D-thiomanno-syl-ethyl)amino)butyl)formamide (Man-C4-Chol) and cholesten-5-ylox-y-*N*-(4-((1-imino-L-thiofucosyl-ethyl)amino)butyl)formamide (Fuc-C4-Chol), respectively. The length of the spacer (C4) can be controlled by replacement of *N*-(4-aminobutyl)carbamic acid *tert*-butyl ester in step 2 with *N*-(2-aminoethyl)carbamic acid *tert*-butyl ester (C2) or *N*-(6-amino-hexyl)carbamic acid *tert*-butyl ester (C6).

Step 1 Preparation of 2-imino-2-methoxyethyl 1-thioglycopeyranoside

Sugars are converted into their 2-imino-2-methoxyethyl 1-thioglycoside forms by the method of Lee *et al.*[6] To the suspension of D(+)-galactose (7.5 g, 42 mmol) in acetic anhydride is added diluted perchloric acid solution (0.3 ml in 10 ml acetic anhydride) drop by drop while stirring in an ice bath. After the galactose has completely dissolved, the mixture is stirred at room temperature for 3 h to obtain 1,2,3,4,6-penta-*O*-acetyl-galactopyra-noside. Then, the reaction mixture is transferred to ice water, and the 1,2,3,4,6-penta-*O*-acetyl-galactopyranoside is extracted with ethyl acetate. The ethyl acetate phase is washed three times with ice-cold water, dried with magnesium sulfate, and evaporated. The syrup obtained is dissolved in ethyl acetate and applied to a silica-gel column (about 500 ml) and eluted with ethyl acetate. Fractions containing the product are

[4] G. Ashwell and J. Harford, *Ann. Rev. Biochem.* **51**, 531 (1982).
[5] X. Gao and L. Huang, *Biochem. Biophys. Res. Commun.* **179**, 280 (1991).
[6] Y. C. Lee, C. P. Stowell, and M. K. Krantz, *Biochemistry* **15**, 3956 (1976).

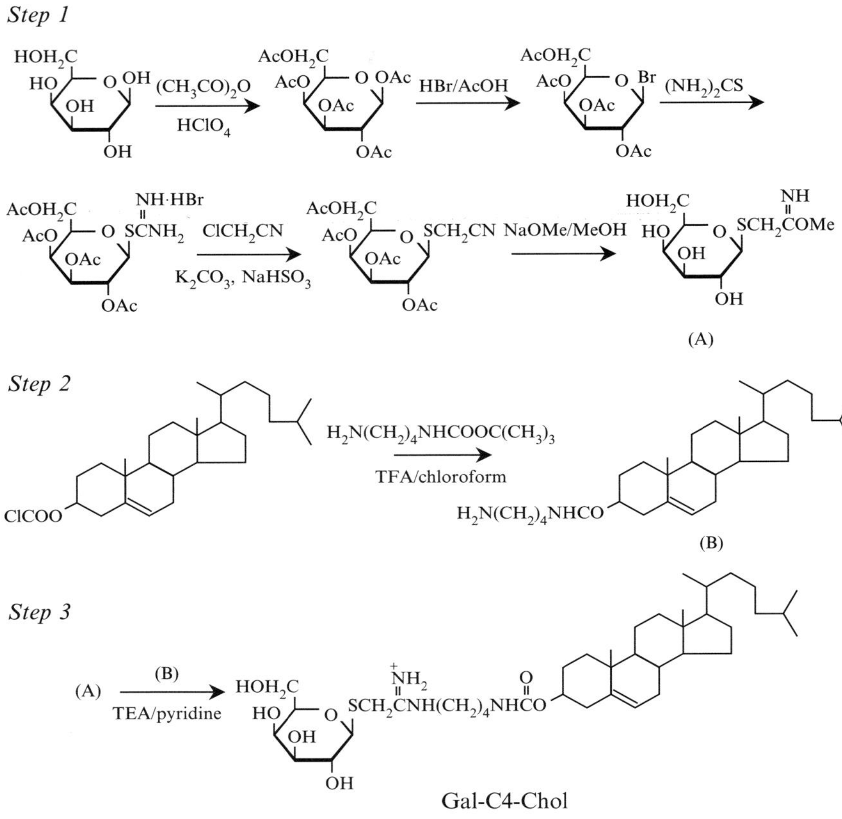

Step 1

(A)

Step 2

(B)

Step 3

Gal-C4-Chol

FIG. 1. Scheme of the synthesis of cholesten-5-yloxy-*N*-(4-((1-imino-D-thiogalactosyl-ethyl)amino)butyl)formamide (Gal-C4-Chol). Abbreviations: Ac, CH₃CO-; Me, CH₃-; TFA, trifluoroacetic acid; TEA, triethylamine.

collected and evaporated to obtain the crystal form of 1,2,3,4,6-penta-*O*-acetyl-galactopyranoside (11 g, 28 mmol). To this compound (6.2 g, 16 mmol) is added 25 ml hydrogen bromide/acetic acid solution (30%) while stirring. After dissolution, the mixture is kept at 4° overnight. Then, the solvent is evaporated to obtain 1-bromo-2,3,4,6-tetra-*O*-acetyl-galacto-pyranoside (4.1 g, 10 mmol). To the syrup of this compound is added thio-urea (0.8 g, 10 mmol) and dry acetone, and the mixture is refluxed for 15 min under nitrogen to obtain 2-*S*-(2,3,4,6-tetra-*O*-acetyl-thiogalactopyr-anosyl)-2-thiopseudourea hydrobromide (4.5 g, 9.3 mmol). This compound and chloroacetonitrile (3.3 g, 44 mmol) are stirred with a 1:1 (v/v) mixture of water and acetone (20 ml) until a nearly clear solution is obtained. To this solution are added potassium carbonate (1.5 g, 11 mmol) and sodium bisulfite (2 g, 20 mmol), and the mixture is stirred for 30 min at room temperature. Then, the mixture is added to an ice-water mixture (80 ml) and stirred for 2 h. The precipitated products are suction-filtered and washed with cold water. The air-dried precipitate is recrystalized from methanol to obtain cyanomethyl 2,3,4,6-tetra-*O*-acetyl-thiogalactopyranoside (CNM-thiogalactopyranoside) (2 g, 5 mmol). Just before conjugation, CNM-thiogalactopyranoside (2 g, 5 mmol) is reacted with 2 ml 0.01 *M* sodium methoxide methanolic solution at room temperature for 24 h. The solvent is then evaporated, and the resultant syrup of 2-imino-2-methoxyethyl 1-thiogalactopyranoside (IME-thiogalactoside) is obtained.

Step 2 Preparation of N-(4-aminobutyl)-(cholesten-5-yloxyl)formamide

Cholesteryl chloroformate (450 mg, 1 mmol) and *N*-(4-aminobutyl)carbamic acid *tert*-butyl ester (1.1 mmol) are reacted in 20 ml chloroform for 24 h at room temperature. A mixed solution of trifluoroacetic acid (2 ml) and chloroform (5 ml) is added dropwise, and the mixture is stirred for 4 h at 4°. After coevaporation of the solvent with toluene, 10 ml hexane is added to the resultant syrup, and the white precipitate formed is collected.

Step 3 Synthesis of Cholesten-5-yloxy-N-(4-((1-imino-D-thiogalactosyl-ethyl)amino)butyl)formamide (Gal-C4-Chol)

IME-thiogalactoside (approximately 2.5 mmol, based on the conversion ratio) is dissolved in 20 ml of pyridine containing 1.1 mmol triethylamine. To this solution is added *N*-(4-aminobutyl)-(cholesten-5-yloxyl) formamide (1 mmol), and the mixture is stirred for 24 h at room temperature. After evaporation, the resultant material is suspended in water, dialyzed against distilled water for 48 h using dialysis tubing (12-kDa cutoff), then lyophilized. The crude product obtained is purified by washing three times with diethylether to obtain Gal-C4-Chol (0.2–0.4 mmol).

Preparation of pDNA–Liposome Complex

Step 1 Preparation of Liposome Formulations

Cationic lipids used are DC-Chol, Gal-C4-Chol, Man-C4-Chol, and DOTMA. The neutral lipids used for the preparation are cholesterol and DOPE. The mixture of lipids is dissolved in chloroform and evaporated to dryness in a round-bottomed flask. Then, the lipid film is vacuum desiccated to remove any residual organic solvent and resuspended in sterile 20 mM 4-(2-hydroxyethyl)-1-piperazineethanesulfonic acid (HEPES) buffer (pH 7.8), phosphate-buffered saline (PBS, pH 7.4), or 5% dextrose. After hydration, the dispersion is sonicated for 5–10 min in a bath sonicator to form liposomes. The liposome formulation is passed through a polycarbonate membrane filter (0.45 μm) or the suspension is extruded through 200-and/or 100-nm pore size polycarbonate membranes using an extruder (Northern Lipids, Vancouver, Canada). Table I summarizes the composition and properties of the liposome formulations used in our previous studies. When hepatocytes are the targets for *in vivo* gene transfer, the size

TABLE I
COMPOSITIONS AND PROPERTIES OF CATIONIC LIPOSOMES

Liposome[a]	Glycosylated (cationic) lipid	Cationic lipid	Neutral lipid	Particle size[b] (nm)	Zeta potential[c] (mV)
DC-Chol/ DOPE (6:4)	—	DC-Chol	DOPE	210.3 ± 21.8 152.1 ± 8.7	39.8 ± 2.9 n.d.
Gal-C4-Chol/ DOPE (6:4)	Gal-C4-Chol	—	DOPE	195.6 ± 15.2 150.1 ± 17.1	34.5 ± 4.6 n.d.
Gal-C4-Chol/ DC-Chol/ DOPE (3:3:4)	Gal-C4-Chol	DC-Chol	DOPE	211.0 ± 13.2 145.1 ± 7.5	37.1 ± 3.5 n.d.
DOTMA/ cholesterol (1:1)	—	DOTMA	Cholesterol	137.7 ± 25.7	n.d.
DOTMA/ Gal-C4-Chol (1:1)	Gal-C4-Chol	DOTMA	—	141.1 ± 5.5	n.d.
DOTMA/cholesterol/ Gal-C4-Chol (1:0.5:0.5)	Gal-C4-Chol	DOTMA	Cholesterol	141.2 ± 6.5	n.d.
DOTMA/DOPE (1:1)	—	DOTMA	DOPE	200.4 ± 8.3	35.6 ± 7.2
Man-C4-Chol/ DOPE (6:4)	Man-C4-Chol	—	DOPE	211.4 ± 30.9	30.5 ± 2.5
Man-C4-Chol/ DC-Chol/ DOPE (3:3:4)	Man-C4-Chol	DC-Chol	DOPE	224.5 ± 24.6	36.8 ± 2.2

[a] The numbers in parentheses represent the weight ratio of each lipid in the formulation.
[b] The particle size of liposomes is measured by dynamic light scattering spectrophotometry (LS-900, Otsuka Electronics, Osaka, Japan).
[c] The zeta potential of liposomes is determined with a laser electrophoresis zeta-potential analyzer (LEZA-500T, Otsuka Electronics, Japan).

of the final lipoplex should be smaller than the fenestrae between the discontinuous endothelial cells of the liver (30–500 nm).

Step 2 Complex Formation with pDNA

To a solution of cationic liposomes is added an equal volume of pDNA solution at room temperature. Then, the complex is agitated rapidly by pipetting it up and down twice, and left to stand for 30 min at room temperature before use. The particle size and zeta potential of the complex are measured using a dynamic light-scattering spectrophotometer (LS-900, Otsuka Electronics, Osaka, Japan) and a laser electrophoresis zeta-potential analyzer (LEZA-500T, Otsuka Electronics), respectively.

In Vitro Transfection to Cells Possessing Asialoglycoprotein Receptors

The aim of targeted gene delivery is to increase the efficiency of transgene expression in target cells, as well as increasing the ratio of the expression in target cells to that in nontarget cells. Therefore, in addition to the transfection efficiency in the target cells, cell type–dependent expression should be investigated to guarantee the usefulness of a targeted nonviral vector.

A human hepatoma cell line, HepG2, is known to possess asialoglycoprotein receptors on its cellular membrane and is sometimes used as a model for hepatocytes. In an attempt to understand the nature and requirements of galactosylated cationic liposome–based gene transfer to hepatocytes *in vivo*, gene transfer to HepG2 cells is investigated.

HepG2 cells are plated on a 6-well cluster dish at a density of 2×10^5 cells/10.5 cm^2 and cultivated in 2 ml Dulbecco's modified Eagle minimum essential medium (DMEM) supplemented with 10% fetal bovine serum (FBS). After 24 h, the cells are incubated with pDNA (0.5 μg/ml) complexed with cationic liposomes in 2 ml OptiMEM I (Gibco BRL, Grand Island, NY) for 6 h. Then, the transfection medium is replaced with the cell culture medium, and the cells are incubated for an additional 42 h. After removal of the culture medium, 200 μl PBS is added, and the cells are scraped off. The cell suspension is then subjected to three cycles of freezing and thawing, and transgene expression in the cells is measured by assaying for luciferase activity in the cell lysate using a kit (Picagene, Toyo Ink, Tokyo, Japan).

Liposome Composition

The transfection efficiency in HepG2 cells is dependent on the type of cationic liposomes used and the mixing ratio of pDNA and liposomes.[7]

Liposome formulations used were DC-Chol/DOPE (6:4 in a molar ratio), Gal-C4-Chol/DOPE (6:4), and Gal-C4-Chol/ DC-Chol/DOPE (3:3:4) liposomes. Among the tested formulations, Gal-C4-Chol/DC-DOPE (3:3:4) liposomes showed the highest transgene expression in HepG2 cells, followed by Gal-C4-Chol/DOPE (6:4) and DC-Chol/DOPE (6:4) liposomes (Fig. 2). This difference might be explained by the amount of $[^{32}P]pDNA$ taken up by the cells with each liposome. When ^{32}P-labeled pDNA ($[^{32}P]pDNA$) in the complexes is traced, radioactivity is significantly ($P <$ 0.01) greater in cells treated with $[^{32}P]pDNA/Gal-C4-Chol/ DC-Chol/$ DOPE (3:3:4) than in cells treated with other complexes. Transgene expression with pDNA–galactosylated liposome complexes is inhibited

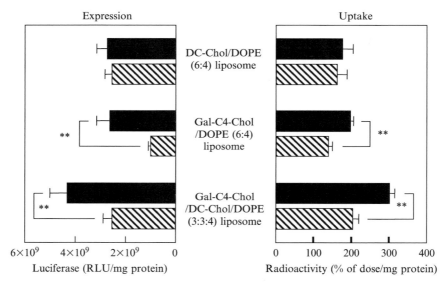

FIG. 2. Transgene expression and uptake of $[^{32}P]pDNA$ by HepG2 cells with various pDNA–cationic liposome complexes. pDNA (1 μg) was mixed with each liposome preparation at various ratios from 1:2.5 to 1:15 (pDNAliposome, wt/wt), and the results with the highest transfection efficiency for each complex are presented (1:5 for DC-Chol/ DOPE [6:4], 1:10 for Gal-C4-Chol/DOPE [6:4], and Gal-C4-Chol/DC-Chol/DOPE [3:3:4]). Cells were transfected with each complex in the absence (closed bars) or presence (hatched bars) of 20 mM galactose. The uptake of the complexes was determined by measuring the radioactivity in the cells 3 h following the application of each $[^{32}P]pDNA$ complex. Each value represents the mean ± S.D. values (N = 3). Statistical analysis was performed by analysis of variance ($^{**}P <$ 0.01).

[7] S. Kawakami, F. Yamashita, M. Nishikawa, Y. Takakura, and M. Hashida, *Biochem. Biophys. Res. Commun.* **252,** 78 (1998).

significantly ($P < 0.01$) by the addition of galactose, but not by mannose, indicating the involvement of asialoglycoprotein receptor–mediated uptake in the expression.

Galactose Density

Studies of the recognition mechanism of naturally occurring sugar chains by asialoglycoprotein receptors using di-, tri-, and tetraantennary oligosaccharides have suggested the importance of a precise geometry of the sugar chains.[8,9] Sugar clusters arranged at the vertices of a triangle with sides of dimensions 15, 22, and 25 Å are preferable for efficient recognition. Therefore, natural or synthetic branched sugar chains with several galact-ose moieties at the nonreducing terminal ends are sometimes used as ligands for targeted delivery to hepatocytes. We have shown previously, however, that the surface density of galactose on galactosylated proteins regulates their *in vivo* recognition by the asialoglycoprotein receptors on hepatocytes.[10] When the calculated average distance of two galactose resi-dues is 30 Å or closer, the uptake rate of galactosylated protein by the liver is high and limited only by the hepatic plasma flow rate.

The density of galactose on the lipoplex can be controlled by changing the mixing ratio of lipids. When the ratio of Gal-C4-Chol/DC-Chol is changed from 2:8 to 8:2, with a constant ratio of the total cholesterol de-rivatives and DOPE at 6:4, transgene expression in HepG2 cells is not affected by the ratio (Fig. 3). Although a precise structure for the pDNA–galactosylated cationic liposome complex has not yet been de-scribed, it can be concluded that the density of Gal-C4-Chol on the surface of the liposomes is sufficient at a molar content of 12% (Gal-C4-Chol/DC-Chol ratio of 2:8).

Effect of Helper Lipids

Endocytosed ligands are usually transported to lysosomes, where they are eventually degraded, and this would limit the efficiency of gene trans-fer.[3] DOPE can destabilize the endocytotic vesicles containing lipoplexes composed of cationic lipid and DOPE, resulting in relatively high transfection efficiency in cultured cells.[11] This has also been shown for targeted transfection with galactosylated cationic liposomes. Figure 4

---

[8] Y. C. Lee, R. R. Townsend, M. R. Hardy, J. Lönngren, J. Arnarp, M. Haraldsson, and H. Lönn, *J. Biol. Chem.* **258**, 199 (1983).

[9] R. R. Townsend, M. R. Hardy, T. C. Wong, and Y. C. Lee, *Biochemistry* **25**, 5716 (1986).

[10] M. Nishikawa, C. Miyazaki, F. Yamashita, Y. Takakura, and M. Hashida, *Am J Physiol.* **268**, G849 (1995).

[11] D. C. Litzinger and L. Huang, *Biochim. Biophys. Acta* **1113**, 201 (1992).

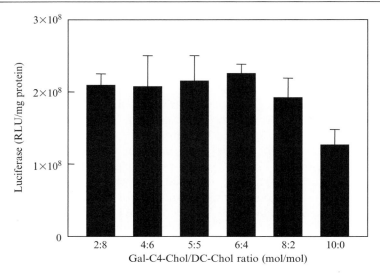

F<small>IG</small>. 3. Transgene expression in HepG2 cells transfected with pDNA/Gal-C4-Chol/DC–Chol/DOPE liposome complexes with different ratios of Gal-C4-Chol and DC-Chol. pDNA (1 μg) was mixed with each liposome preparation at a mixing ratio of 1:5 (wt/wt). Each value represents the mean $\pm$ S. D. values ($N = 3$).

shows the transgene expression in HepG2 cells treated with different pDNA–liposome complexes. When Gal-C4-Chol is added to two cationic liposomal formulations, (i.e., DC-Chol/DOPE (6:4) and DOTMA/cholesterol (1:1)), galactose-specific transgene expression in HepG2 cells is observed.[12] However, the level is about 30-fold greater with Gal-C4-Chol/DC-Chol/DOPE (3:3:4) liposomes than with DOTMA/cholesterol/Gal-C4-Chol (1:0.5:0.5) liposomes. These results suggest that the intracellular sorting or trafficking is also a critical factor for targeted nonviral vectors to achieve efficient transgene expression.

These characteristics of helper lipids on the transgene expression in HepG2 cells are, however, quite different from those observed in whole animals as discussed in the following.

Cell Specificity of Gene Transfer

Although it is not yet known whether transgene expression in nontarget cells leads to serious side effects *in vivo*, such expression would be highly undesirable. In particular, the uptake of pDNA or its complex by immune

---

[12] S. Kawakami, S. Fumoto, M. Nishikawa, F. Yamashita, and M. Hashida, *Pharm. Res.* **17**, 306 (2000).

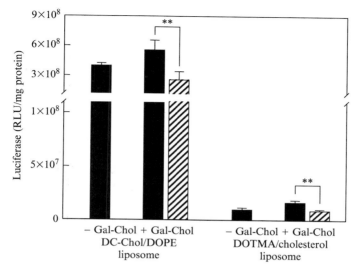

FIG. 4. Transgene expression in HepG2 cells transfected with pDNA–cationic liposome complexes containing DOPE or cholesterol as a helper lipid. pDNA (1 μg) was mixed with each liposome preparation at a charge ratio of 1:2:3. Cells were transfected with each complex in the absence (closed bars) or presence (hatched bars) of 20 mM galactose. Each value represents the mean ± S.D. values ($N = 3$). Statistical analysis was performed by analysis of variance ([**]$P < 0.01$).

cells is a major obstacle to nonviral gene therapy approaches, because the uptake sometimes triggers a severe immune reaction that reduces the level and duration of transgene expression in target cells. High levels of cytokines, such as interferon-γ and tumor necrosis factor-α, are observed after *in vivo* administration of lipoplexes.[13] The immune reaction against pDNA, which is largely due to the unmethylated CpG immunostimulatory motif in pDNA,[14] is amplified by the use of cationic liposomes, probably because of its ability to efficiently deliver pDNA into cells. These cytokines not only cause toxicity to the treated animals but also inhibit transgene expression. Therefore, cell-specific gene transfer is a major goal for constructing glycosylated nonviral vectors.[15]

[13] S. Li, S. P. Wu, M. Whitmore, E. J. Loeffert, L. Wang, S. C. Watkins, B. R. Pitt, and L. Huang, *Am. J. Physiol.* **276,** L796 (1999).
[14] A. M. Krieg, A. K. Yi, S. Matson, T. J. Waldschmidt, G. A. Bishop, R. Teasdale, G. A. Koretzky, and D. M. Klinman, *Nature* **374,** 546 (1995).
[15] M. Nishikawa, M. Yamauchi, K. Morimoto, E. Ishida, Y. Takakura, and M. Hashida, *Gene Ther.* **7,** 548 (2000).

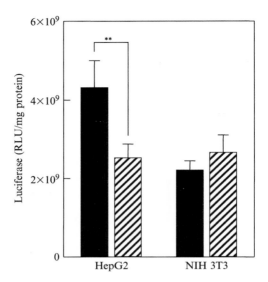

Fig. 5. Transgene expression in HepG2 and NIH 3T3 cells transfected with pDNA/Gal-C4-Chol/DC-Chol/DOPE (3:3:4) liposome complexes. pDNA (1 μg) was mixed with each liposome preparation at a ratio of 1:10 (wt/wt). Cells were transfected with each complex in the absence (closed bars) or presence (hatched bars) of 20 mM galactose. Each value represents the mean ± S.D. values ($N = 3$). Statistical analysis was performed by analysis of variance ($^{**}P < 0.01$).

NIH 3T3 cells are used as control cells lacking asialoglycoprotein receptors. In these cells, DC-Chol/DOPE (6:4) liposomes mediate the highest transgene expression, followed by Gal-C4-Chol/DC-Chol/DOPE (3:3:4) and Gal-C4-Chol/DOPE (6:4) liposomes.[7] Figure 5 shows the transgene expression following transfection with pDNA/Gal-C4-Chol/DC-Chol/ DOPE (3:3:4) liposome complex. Although the addition of 20 mM galactose significantly decreases gene expression in HepG2 cells, it hardly affects expression in NIH 3T3 cells. The ratio of transgene expression in HepG2 cells to that in NIH 3T3 cells is about 2 with a pDNA/Gal-C4-Chol/DC-Chol/DOPE (3:3:4) liposome complex, indicating its selective gene transfer to asialoglycoprotein receptor–positive cells.

Effect of Spacer Length on Gene Transfer

When HepG2 cells are transfected with Gal-C2-Chol/DC-Chol/DOPE (3:3:4), Gal-C4-Chol/DC-Chol/DOPE (3:3:4), or Gal-C6-Chol/DC-Chol/ DOPE (3:3:4) liposomes, transgene expression increases as the spacer length is increased.[7] Because the numbers of galactose units per liposome are almost identical, these results suggest that a longer spacer arm helps the recognition of galactose by the asialoglycoprotein receptors. These results suggest that the galactose moiety is not fully exposed on the surface of the complex when a short spacer arm is used. Although Gal-C6-Chol–containing liposomes show the greatest potential for gene transfer, the yield of Gal-C6-Chol (percent relative to the starting amount of cholesterol) in the synthesis is low compared with Gal-C4-Chol (%). Therefore, it is preferable to use Gal-C4-Chol in the experiments.

In Vivo Transfection by Galactosylated Liposomes

Under *in vivo* conditions, the targeted nonviral vector encounters and interacts with various nontarget cells and serum proteins, as well as the target cells. Such interactions might alter and, in most cases, reduce the targeting potential of the vector. Therefore, delivery systems should be specially suited to *in vivo* gene transfer. For example, nontargeted pDNA–cationic liposome complexes become effective *in vivo* following an increase in the positive charge of the complex compared with the value optimized for *in vitro* transfection.[16]

Effect of Helper Lipids on Biodistribution and Gene Transfer

Figure 6 shows the transgene expression in tissues of mice injected with pDNA–cationic liposome complexes. When administered to mice through a tail vein, pDNA–galactosylated cationic liposome complexes show the highest transgene expression in the lung among the tissues tested, including the liver.[12] These results suggest that the biodistribution of pDNA–galactosylated cationic liposome complexes cannot fully be controlled by its galactose moieties. Changing the administration route could, at least partially, solve this problem. When administered into the portal vein, DOTMA/cholesterol/Gal-C4-Chol (1:0.5:0.5) liposomes show the highest transgene expression in the liver. Although the *in vitro* transfection efficiency to HepG2 cells is higher with DC-Chol/DOPE (6:4) liposomes than with DOTMA/cholesterol/Gal-C4-Chol (1:0.5:0.5) liposomes, this is not the case for *in vivo* transfection following intraportal injection. This could be explained by experimental data showing that some pDNA–cationic liposome complexes interact with erythrocytes and are trapped in the lung capillaries, depending on the composition of the liposomes.[17,18]

[16] Y. P. Yang and L. Huang, *Gene Ther.* **4,** 950 (1997).
[17] F. Sakurai, T. Nishioka, F. Yamashita, Y. Takakura, and M. Hashida, *Eur. J. Pharm. Biopharm.* **52,** 165 (2001).

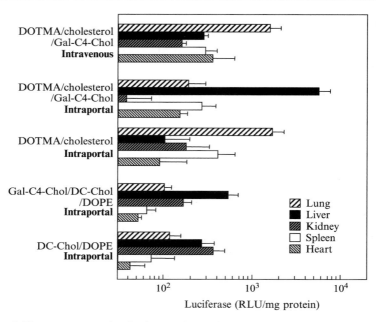

FIG. 6. Transgene expression in tissues of mice injected with pDNA–cationic liposome complexes through a tail or portal vein. pDNA (30 μg) was complexed with various cationic liposomes at a charge ratio of 2.3. At 6 h following injection, tissues were collected and homogenized in a lysis buffer, and the luciferase activity in the supernatant was measured. Each value represents the mean $\pm$ S.D. values ($N = 3$).

Using DOTMA liposomes with different helper lipids, Sakurai et al.[17] reported that transgene expression in the lung was highly dependent on the helper lipid used. DOTMA/cholesterol and plain DOTMA complexes did not induce fusion between erythrocytes, whereas DOTMA/DOPE complexes did.

Compared with DOTMA/cholesterol (1:1) liposomes, which have the highest transgene expression in the lung following intraportal injection, DOTMA/cholesterol/Gal-C4-Chol (1:0.5:0.5) liposomes are taken up extensively and rapidly by the liver. The expression in the liver is fairly specific for hepatocytes (Fig. 7), and it can be inhibited by coadministration of galactosylated bovine serum albumin, a well-known ligand for asialoglycoprotein receptors. These results indicate that pDNA/DOTMA/ cholesterol/

[18] F. Sakurai, T. Nishioka, H. Saito, T. Baba, A. Okuda, O. Matsumoto, T. Taga, F. Yamashita, Y. Takakura, and M. Hashida, *Gene Ther.* **8,** 677 (2001).

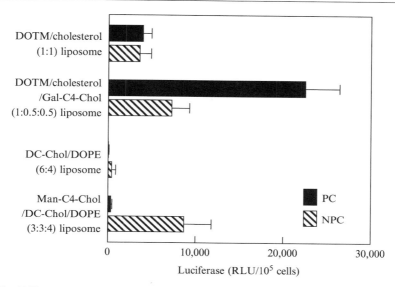

FIG. 7. Transgene expression in liver parenchymal (PC) and nonparenchymal (NPC) cells separated from mice injected with pDNA–cationic liposome complexes. pDNA (50 μg) was complexed either with DOTMA/Chol/Gal-C4-Chol (1:0.5:0.5) liposomes or with Man-C4-Chol/DOPE (6:4) at a charge ratio of 2.3. At 6 h following injection, the liver was perfused with a buffer containing collagenase (type 1A, Sigma), and the liver cells were separated into parenchymal cells (hepatocytes) and nonparenchymal cells by differential centrifugation. Each value represents the mean ± S.D. values ($N = 3$).

Gal-C4-Chol (1:0.5:0.5) liposome complexes can transfect hepatocytes *in vivo* by means of an asialoglycoprotein receptor–mediated mechanism.

Transfection with Mannosylated Liposomes

The activity of mannose receptors has been demonstrated in macrophages, such as Kupffer cells, peritoneal and pulmonary macrophages, and hepatic sinusoidal endothelial cells. Therefore, mannosylated ligands can be used as carriers for these types of cells. Macrophages are primary targets for these vectors because of their importance in various diseases and resistance to gene transfer.

Transfection to Macrophages

Elicited peritoneal macrophages are harvested from mice 4 days after intraperitoneal injection of 1 ml 2.9% thioglycolate medium (Nissui Pharmaceutical, Tokyo, Japan). The washed cells are suspended in RPMI

1640 medium supplemented with 10% heat-inactivated FBS, penicillin G (100 U/ml), and streptomycin (100 μg/ ml). After incubation for 24 h at 37° in 5% CO_2–95% air, nonadherent cells are washed off with culture medium, and the cells are then cultivated for another 48 h.

Although cationic liposomes such as DOTMA/DOPE (1:1) and DC-Chol/DOPE (6:4) can achieve transgene expression in macrophages, the efficiency is very low compared with that in other cells, such as HepG2 cells and NIH 3T3 cells.[19] The addition of Man-C4-Chol to liposome formulations improved transgene expression in macrophages three-fold: Man-C4-Chol/DOPE (6:4) or Man-C4-Chol/DC-Chol/DOPE (3:3:4) liposomes exhibit mannose-sensitive transgene expression. These liposomes are also effective *in vivo* and achieve a relatively high transgene expression in the liver, especially in nonparenchymal cells. Figure 7 shows the transgene expression in liver cells of mice injected with various pDNA–cationic liposome complexes. Parenchymal cells (hepatocytes) and nonparenchymal cells (including Kupffer and endothelial cells) are separated by a collagenase perfusion method. Complexes containing mannose moieties show a significantly higher transgene expression in liver nonparenchymal cells than in parenchymal cells, whereas those with galactose moieties mediate high expression in parenchymal cells.

Tertiary Complex for Improved Gene Transfer

Polyethyleneimine (PEI) is reported to possess a pH-buffering capacity in endosomes and, therefore, destabilizes the vesicles and facilitates the cytoplasmic delivery of endocytosed compounds.[20] In addition, cationic polymers like PEI are known to have a greater ability to condense pDNA when mixed than cationic liposomes.[21] Therefore, in an attempt to increase transgene expression with mannosylated cationic liposomes, a branched PEI with an average molecular weight of about 60,000 is added to the formulation of Man-C4-Chol/DOPE (6:4) liposomes. This results in a two-fold increase in the uptake of [^{32}P]pDNA by macrophages and a sixfold increase in transfection efficiency.[22]

[19] S. Kawakami, A. Sato, M. Nishikawa, F. Yamashita, and M. Hashida, *Gene Ther.* **7,** 292 (2000).
[20] O. Boussif, F. Lezoualch, M. A. Zanta, M. D. Mergny, D. Scherman, B. Demeneix, and J. P. Behr, *Proc. Natl. Acad. Sci. USA* **92,** 7297 (1995).
[21] S. Li, M. A. Rizzo, S. Bhattacharya, and L. Huang, *Gene Ther.* **5,** 930–937 (1998).
[22] A. Sato, S. Kawakami, M. Yamada, F. Yamashita, and M. Hashida, *J. Drug Targeting* **9,** 201 (2001).

Conclusion

Glycosylated cationic liposomes are possible candidates for cell-specific nonviral vectors by means of a receptor-mediated mechanism, although their cell specificity and transfection efficiency need further improvement. The preparation of glycosylated liposomes shown here is much easier than that of asialoglycoproteins as targeting ligands for liposomes and may be less immunogenic. These glycosylated liposomes can be applied not only to gene delivery, but also to targeted delivery of conventional drugs, such as prostaglandin E_1 and probucol.[23–25] In such cases, different criteria for the preparation of glycosylated liposomes are required to obtain efficient drug delivery.

[23] S. Kawakami, C. Munakata, S. Fumoto, F. Yamashita, and M. Hashida, *J. Drug Targeting* **8,** 137 (2000).
[24] Y. Hattori, S. Kawakami, F. Yamashita, and M. Hashida, *J. Controlled Release* **69,** 369 (2000).
[25] S. Kawakami, C. Munakata, S. Fumoto, F. Yamashita, and M. Hashida, *J. Pharm. Sci.* **90,** 105 (2000).

[26] Asialofetuin Liposomes for Receptor-Mediated Gene Transfer into Hepatic Cells

By SALVADOR F. ALIÑO, MARTA BENET, FRANCISCO DASÍ, and JAIME CRESPO

Introduction

Liposomes may be targeted to hepatocytes by covalent coupling of asialofetuin glycoprotein (ASF) onto the liposome surface, using the heterobifunctional reagent *N*-succinimidyl 3-(2-pyridyldithio) propionate (SPDP). Changes in liposome charge can be made by inclusion in the liposome backbone of anionic or cationic lipids (such as phosphatidylserine or DOTAP, respectively), cationic lipopeptides, and cationic polymers. Moreover, cationic polymers are able to condense and compact DNA, which correlates with greater DNA uptake and resistance to DNase digestion. Thus, a number of studies have used cationic polymers, such as polylysine[1,2] or polyethyleneimine (PEI) in the vector composition to improve the efficacy of receptor-mediated gene transfer. Because the nuclear localization signal (NLS) is also a cationic polypeptide, it can interact directly with

[1] S. F. Aliño, M. Bobadilla, J. Crespo, and M. Lejarreta, *Hum. Gene Ther.* **7,** 531 (1996).
[2] J. Crespo, C. Blaya, A. Crespo, and S. F. Aliño, *Biochem. Pharmacol.* **51,** 1309 (1996).

DNA by electrostatic interaction, increasing the nuclear availability of delivered DNA and thereby contributing to increase the DNA expression. With the aim of increasing the transfection efficacy of liver-destined liposomes, we synthesized the dipalmitoylphosphatidyletanolamine-nuclear localization signal lipopeptide (DPPE-NLS), prepared from DPPE covalently coupled to cationic NLS, and subsequently investigated its efficiency in mediating *in vivo* hepatic gene transfer. On the other hand, in this study we use the human α_1-antitrypsin (hAAT) gene for two main reasons: (1) as a general model of liver-destined gene transfer, because hAAT is mainly produced in the liver, secreted into the bloodstream in peak amounts (1.5–3.5 mg/ml) in healthy individuals, and mediates extracellular functions and the fine control of the gene expression of its gene is not critical; (2) as a specific strategy for gene therapy, because hAAT deficiency is a monogenic inherited disease in humans that causes emphysema as result of failure in limiting the enzyme activity of neutrophil elastase, which is able to destroy all the protein connective tissue components of the lung.

Rationale for Asialofetuin-Liposomes

The liver is an attractive organ for gene therapy, because it is the largest gland in the body and the production site for more than 90% of all plasma proteins, including enzymes and most of the factors involved in blood coagulation. It plays a central role in the metabolism of lipids, carbohydrates, and proteins, as well as in the detoxification of exogenous chemicals. The liver is therefore an excellent site for gene transfer in treating a wide variety of diseases that either affect liver function or may be ameliorated by modification of liver metabolism. In addition, the important vascularization and the high rate of plasma protein production (30–60 g/day) indicate that the liver is an ideal organ for rapid delivery of intravenously (iv) injected gene complexes, high level expression of therapeutic genes, and efficient systemic distribution of the resulting therapeutic proteins secreted into the bloodstream.

For DNA delivery, two main classes of systems could be considered: viral and nonviral vectors. Although viral vectors are the most effective procedure for DNA delivery, evidence for clinical effectiveness has been scant in most gene therapy protocols. Besides, viral vectors present additional limitations, such as potential toxicity, restricted targeting of specific cell types, limited DNA-carrying capacity, production and packaging problems, recombination, and high cost.[3,4] On the other hand, although

---

[3] R. G. Crystal. *Science* **270**, 404 (1995).
[4] S. K. Tripathy, H. B. Black, E. Goldwasser, and J. M. Leiden, *Nat. Med.* **2**, 545 (1996).

nonviral vectors presently show low efficiency of DNA delivery into the cells, they offer safety as the main advantage. For this reason, artificial vectors have become desirable to formulate genes as medicines. Nevertheless, the fundamental challenge of improving the efficacy of exogenous DNA transfer into cells remains to be solved. In this respect, it has been described[5] that only a small portion (0.3%) of input DNA is available inside the nucleus after transfection with cationic lipids. These data suggest that the low efficiency of gene transfer mediated by this procedure is due to loss of DNA molecules at each step in the trajectory from point of administration to the nucleus.

A considerable variety of nonviral procedures[6] have been used for enhancing gene delivery to cells, including mechanical (microinjection, pressure, bombardment), electrical (electroporation), and chemical procedures (calcium phosphate, lipids, proteins, dendrimers, polymers). Currently, liposomes are an attractive vehicle for gene transfer and are one of the most commonly used methods in human clinical trials. Liposomal vectors, however, present important limitations, such as lack of targeting, variations arising during preparation, and poor intracellular and nuclear delivery. Furthermore, the structural features of liposome–DNA complexes relevant to gene delivery are poorly understood. Advances have been made with the aim of improving liposome properties for gene transfer, such as liposome targeting and nuclear DNA targeting. Furthermore, our knowledge of lipid–DNA interaction is improving continuously.[7]

The final efficacy of most nonviral vectors involves some of following aspects: (1) DNA complexation and stability, which is related to the physicochemical characteristics of the vector particles (e.g., charge, size, structure, stability); (2) the nature and efficiency of DNA delivery into the cell; and (3) the nuclear availability of DNA. Therefore, to improve the gene transfer efficacy of a new vector, the following major barriers must be overcome: limited stability in biological fluids, low uptake across the plasma membrane, inadequate release of DNA into the cell, and lack of nuclear targeting. Extracellularly, DNA can be protected by liposome encapsulation or complex formation with cationic lipids or polymers.[8–10]

[5] A. R. Holmes, A. F. Dohramn, A. R. Ellison, K. K. Goncz, and D. C. Gruenert. *Pharm. Res.* **16,** 1020 (1999).
[6] D. Luo and M. Saltzman, *Nat. Biotechnol.* **18,** 33 (2000).
[7] M. C. Pedroso de Lima, S. Simões, P. Pires, H. Faneca, and N. Düzgüneş, *Adv. Drug Delivery Rev.* **47,** 277 (2001).
[8] S. F. Aliño, M. Bobadilla, F. J. Unda, M. Garcia-Sanz, E. Hilario, and M. Lejarreta, *J. Microencapsul.* **10,** 163 (1993).
[9] I. Moret, J. E. Peris, V. M. Guillem, M. Benet, F. Revert, F. Dasi, A. Crespo, and S. F. Aliño, *J. Control Release* **76,** 169 (2001).

Pharmacokinetic properties of lipid-based and polymer-based vectors after *in vivo* administration should be studied to select those advantageous for gene transfer,[11–13] mainly when appropriate ligands are covalently coupled on the vector surface to mediate a selective DNA delivery by a receptor-mediated mechanism.[14] Intracellularly, the mechanisms by which DNA avoids endosome degradation,[15–17] as well as DNA protection during the journey from cytosol to the nucleus,[18–19] have been explored actively. Finally, the nuclear envelope is the last barrier for limiting the efficacy of DNA delivery, and in this context, artificial vectors are notably more inefficient than viral vectors for DNA transport to the nucleus. Some synthetic polymers, such as PEI, can promote entry into the nucleus,[20,21] but this property is more limited in other nonviral vectors, including cationic lipids.[22] This limitation could be circumvented, because the addition of NLS peptides has been demonstrated to enhance the nuclear localization of several macromolecules,[23–26] NLS peptides also increase the expression efficacy of delivered genes by use of nonviral vectors such as

[10] V. M. Guillem, M. Tormo, F. Revert, I. Benet, J. García-Conde, A. Crespo, and S. F. Aliño *J. Gene Med.* **4**, 170 (2002).

[11] R. I. Mahato, Y. Takakura, and M. Yashida, *Crit. Rev. Ther. Drug Carrier Syst.* **14**, 133 (1977).

[12] C. W. Pouton, P. Lucas, B. J. Thomas, A. N. Uduehi, D. A. Milroy, and S. H. Moss, *J. Control. Release* **53**, 289 (1998).

[13] S. F. Aliño, J. Crespo, G. Tarrason, C. Blaya, J. Adan, E. Escrig, M. Benet, A. Crespo, J. E. Peris, and J. Piulats, *Xenobiotica* **29**, 1283 (1999).

[14] F. Dasi, M. Benet, J. Crespo, A. Crespo, and S. F. Aliño, *J. Mol. Med.* **79**, 205 (2001).

[15] A. El Ouahabi, M. Thiry, V. Pector, R. Fuks, J. M. Ruysschaert, and M. Vandenbranden, *FEBS Lett.* **414**, 187 (1997).

[16] O. Boussif, F. Lezoualc'h, M. A. Zanta, M. D. Mergny, D. Scherman, B. Demeneix, and J. P. Behr, *Proc. Natl. Acad. Sci. USA* **92**, 7297 (1995).

[17] J. G. Duguid, C. Li, M. Shi, M. J. Logan, H. Alila, A. Rolland, E. Tomlinson, J. T. Sparrow, and L. C. Smith, *Biophys. J.* **74**, 2802 (1998).

[18] R. Lee, *Crit. Rev. Ther. Drug Carrier Syst.* **14**, 173 (1997).

[19] G. F. Ross, M. D. Bruno, M. Uyeda, K. Suzuki, K. Nagao, J. A. Whitsett, and T. R. Korfhagen, *Gene Ther.* **5**, 1244 (1998).

[20] H. Pollard, J. S. Remy, G. Loussouarn, S. Demolombe, J. P. Behr, and D. Escande, *J. Biol. Chem.* **27**, 7507 (1998).

[21] W. T. Godbey, K. K. Wu, and A. G. Mikos, *Proc. Natl. Acad. Sci. USA* **96**, 5177 (1999).

[22] S. F. Aliño, E. Escrig, F. Revert, V. M. Guillem, and A. Crespo, *Biochem. Pharmacol.* **60**, 1845 (2000).

[23] N. Michaud and D. S. Goldfarb, *Exp. Cell Res.* **208**, 128 (1993).

[24] U. Stochaj, M. A. Bossie, K. Van-zee, A. M. Whalen, and P. A. Silver, *J. Cell Sci.* **104**, 89 (1993).

[25] N. L. Allbritton, E. Oancea, M. A. Kuhn, and T. Meyer, *Proc. Natl. Acad. Sci. USA* **91**, 12458 (1994).

[26] X. Leng, V. G. Wilson, and X. L. Xiao, *J. Gen. Virol.* **75**, 2463 (1994).

liposome–DNA complex[27] and polylysine-mediated transfection.[28] In this way, we consider that the NLS peptide-bound liposome–DNA complex could be targeted to the nucleus in a way similar to other macromolecules, thereby contributing to increasing the efficiency of DNA delivery.[14,29,30]

For strategies of liver-targeted gene therapy, we must consider that the continuous endothelium is the major anatomical barrier for liposome access to most organs. Still, the liver sinusoid endothelium offers a unique advantage for gene delivery, because it contains pores with a mean diameter of 100 nm, which would allow small vectors to leave the circulation and reach the hepatocytes. Accordingly, small liposomes are able to cross the natural barrier of the sinusoidal endothelium to deliver the entrapped material to hepatocytes,[31] including recombinant genes.[1,32] On the other hand, we should take into account that hepatocyte possess highly specific receptors that recognize galactose-terminal glycoproteins or asialoglycoproteins[33] (ASGP), facilitating the removal of desialylated glycoproteins from the circulation. By taking advantage of the exclusive localization of the ASGP receptor on hepatocytes, the gene complex could be targeted specifically to the liver.[34,35] In this sense, the ASGP receptor is a good candidate target for liver-targeted gene therapy, considering that a high number of ASGP receptors (100,000–500,000) are present per cell,[36] the receptors are rapidly and constitutively recycled, and a relatively constant number of receptors are available on the cell surface at all times.

Synthesis of Modified Molecules and Covalent Coupling for Targeting

Targeted liposomes permitting internalization by a receptor-mediated mechanism can be prepared easily by use of the bifunctional reagent SPDP (Fig. 1). Covalent bonding on the surface of the liposome of adequate ligands and/or monoclonal antibodies should allow the identification of cell surface receptors with a capacity for internalization. By use of SPDP, it is

[27] A. I. Aronsohn and J. A. Hughes, *J. Drug Targeting* **5,** 163 (1997).
[28] C. K. Chan and D. A. Jans, *Hum. Gene Ther.* **10,** 1695 (1999).
[29] L. J. Branden, A. J. Mohamed, and C. I. Smith, *Nat. Biotechnol.* **17,** 784 (1999).
[30] A. Subramanian, P. Ranganathan, and S. L. Diamond, *Nat. Biotechnol.* **17,** 873 (1999).
[31] G. L. Scherphof, H. H. Spanjer, J. T. Derksen, F. Kuipers, R. J. Vonk, T. Daemen, and F. H. Roerdink, *Biochem. Soc. Trans.* **15,** 345 (1987).
[32] S. F. Aliño, M. Bobadilla, M. García-Sanz, M. Lejarreta, F. Unda, and E. Hilario, *Biochem. Biophys. Res. Commun.* **192,** 174 (1993).
[33] G. Ashwell and A. G. Morell, *Adv. Enzymol.* **44,** 99 (1974).
[34] G. Y. Wu and C. H. Wu, *J. Biol. Chem.* **262,** 4429 (1987).
[35] G. Y. Wu, J. M. Wilson, F. Shalaby, M. Grossman, D. A. Shafritz, and C. H. Wu, *J. Biol. Chem.* **266,** 14338 (1991).
[36] M. Spiess, *Biochemistry* **29,** 10009 (1990).

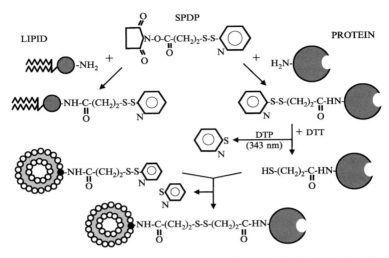

FIG. 1. Synthesis of modified molecules and covalent coupling for targeting. Liposomes containing DPPE or DOPE and proteins are modified with linker molecules (SPDP). Once modified, the reactive groups on the protein are activated by adding DTT, which breaks the disulfide group, releasing DTP. Because DTP shows a maximum absorbance peak at 343 nm, the reaction is monitored by measuring the optical density increments at 343 nm produced by DTP groups released before and after the addition of DTT. In a second step, the modified liposome is mixed with the activated protein resulting in a spontaneous covalent disulfhide linkage between both molecules. The reaction involves the release of DTP, which allows one to evaluate the efficacy by spectrophotometry at 343 nm.

possible to perform a spontaneous covalent coupling reaction between lipids and proteins previously modified in agreement with the general strategy described in Fig. 1. The lipid generally used for modification is phosphatidylethanolamine (usually, DPPE or DOPE). The amino group of the lipid is modified with SPDP (forming PDP-PE) and then is included at 2 mol% in the lipid mixture for liposome preparation, enabling the coupling of molecules containing SH groups. Proteins may also be modified with the same bifunctional reagent. The reactive groups on the protein are activated by rupturing the disulfide group in the presence of dithiothreitol (DTT). The released 2-thiopyridine group (DTP) can be separated by dialysis or exclusion chromatography. The efficacy of the procedure can be determined by spectrophotometry, measuring the absorption at 343 nm before and after DTT and evaluating, by means of the coefficient of extinction, the molar concentration of the released groups and thus the number of modified groups in the protein. Finally, the liposome is mixed with the activated protein, resulting in spontaneous covalent disulfide bonding among the modified PE groups and the protein SH groups. Because the reaction bonding generates the release of DTP, quantification of the latter allows the evaluation of the efficacy of the reaction. The procedures used for the synthesis and/or modification of the lipid and the protein are described in the following.

Synthesis of Pyridyldithiopropionate–Dipalmitoylphosphatidylethanolamine (PDP-DPPE)

The following solutions are prepared: DPPE is dissolved (10 mg/ml) in benzenemethanol (9:1, v/v), SPDP is dissolved (12.5 mg/ml) in methanol, and triethylamine (TEA) is diluted 1:10 in chloroform. The preceding solutions are mixed in a glass tube (DPPE, 2.1 ml; SPDP, 0.9 ml; and TEA, 0.09 ml) and incubated with magnetic stirring for 4 h at room temperature. After the addition of chloroform (3 ml) and 0.2 M TRIS buffer, pH 7.4 (2 ml), the sample is centrifuged at $1000g$ for 10 min. The supernatant is discarded, and the lower phase is washed with 3 ml of distilled, deionized water. After centrifugation at $1500g$ for 30 min, the supernatant is discarded. The remaining sample containing the modified DPPE (DPPE-PDP) is clarified by adding methanol, dried under an N_2 stream, and then lyophilized overnight. The DPPE-PDP is dissolved in benzenemethanol (9:1), and a 50-μl aliquot is dried and resuspended in 950 μl HEPES-buffered saline (HBS: 150 mM NaCl, 20 mM HEPES, pH 7.4). The concentration of modified groups is determined on the basis of the increased ultraviolet absorbance at 343 nm after adding 25 μl 1 M DTT, because of the release of 2-thiopyridine (DTP, extinction coefficient = 8080 M^{-1} cm^{-1}).

Synthesis of Protein-Thiol

Proteins, such as asialofetuin (ASF) or streptavidin (St), are dissolved (1 mg/ml) in HBS, and SPDP (20 mM) is dissolved in ethanol. A 12.5-microliter volume of SPDP solution is mixed with 1 mg of protein and incubated for 30 min at room temperature. The protein-PDP is recovered after desalting the mixture using a PD10 column (Pharmacia Biosystems, Barcelona, Spain) equilibrated in HBS. The ratio of modified amino groups per protein molecule is quantified by spectrophotometry, on the basis of the increased absorbance at 343 nm of the sample caused by DTP release by the addition of 25 μl 1 M DTT. The modified protein-SH (St-SH or ASF-SH) is recovered by desalting on a PD10 column.

Monitoring the Synthesis of Protein–Liposome

The procedure for monitoring the efficiency of protein coupling is based on the capacity to quantify the DTP groups released during synthesis, by means of absorption spectrophotometry at 343 nm, after rupture of the disulfide groups induced by DTT. In this way, we can compare the number of available groups for reaction before and after liposome incubation with the modified protein. This procedure establishes the number of groups that have been consumed after the incubation, indicating indirectly the efficiency of covalent bond formation. The capacity to measure the reactive groups is independent of the amount of sample used for the analysis (Fig. 2). Thus, aliquots of liposome preparation are progressively diluted at a 1:2 ratio with HBS, and absorption (at 343 nm) is then measured before and after the addition of DTT. The results confirm a good linear correlation in all the cases (coefficient > 0.99), indicating that an aliquot of the sample can be sufficient for quantifying the number of reactive groups available on the surface of liposomes with the capacity to form covalent

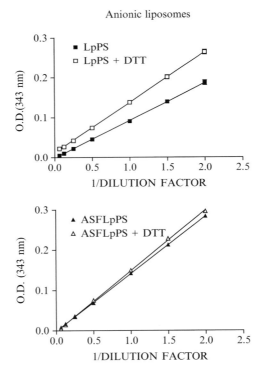

Fig. 2. Evaluation of asialofetuin–liposomes conjugation. Anionic liposomes encapsulating the plasmid pTG7101 are prepared from a lipid mixture containing 2% DPPE-PDP. Samples of liposomes are incubated in the absence (Lp) or presence of SPDP-modified asialofetuin (ASFLp). Samples of liposomes are diluted in HEPES buffer, and the optical density (OD) at 343 nm is measured before and after the addition of DTT (25 mM final concentration).

links with the modified protein. Likewise, DTT does not increase the absorbance at 343 nm when liposomes are incubated previously with modified asialofetuin (asialofetuin-SH), suggesting that available groups on the liposomal surface have been used to form the covalent bond. Also in these cases, the information obtained from samples in different degrees of dilution indicates the existence of a very good correlation among the data, suggesting that an aliquot of the sample can be sufficient to indicate the number of groups that have been consumed after the incubation, thus affording an easy procedure to evaluate the efficiency of covalent bonding between the liposome surface and the molecule of interest for targeting.

Liposome Preparation

To achieve functional gene delivery, liposomes must be able to protect DNA from enzymatic degradation in biological fluids during the journey from the point of administration to the nucleus of target cell, as well as to facilitate DNA transport from the cytoplasm to the nucleus. Anionic[1,2,32,37,38] and cationic[2,39] liposomes have been used for gene transfer. The net negative charge of the former complicates nonspecific interaction with cell membrane surfaces, which also possess a net negative charge. Consequently, they may have major opportunities to be taken up as colloidal suspensions by way of the pinocytic pathway of the hepatocyte, but the entrapment of negatively charged DNA into anionic liposomes leads to low encapsulation efficiency. The positive lipid charge of cationic liposomes facilitates both liposome–DNA and liposome–cell surface interactions,[40] leading to a relatively efficient entry to the target cell. However, the iv administration of cationic liposomes produces extensive tissue distribution because of the ionic interaction with the outer surface of all cell plasma membranes.[39] On the other hand, increased gene bioavailability can be achieved using viral fusogenic proteins to deliver DNA into the cytoplasm of the cell by a liposome-mediated fusion mechanism[41] and encapsulating nuclear proteins to facilitate DNA transport to the nucleus.[41,42] This is a very attractive research topic, because it has been

[37] S. F. Aliño, J. Crespo, M. Bobadilla, M. Lejarreta, C. Blaya, and A. Crespo, *Biophys. Res. Commun.* **204,** 1023 (1994).
[38] R. M. Straubinger, *Methods Enzymol.* **221,** 361 (1993).
[39] N. Zhu, D. Liggitt, Y. Liu, and R. Debs, *Science* **261,** 209 (1993).
[40] J. Y. Legendre and F. C. Szoka, Jr., *in* "Liposomes, New Systems and New Trends in Their Applications" (F. Puisieux, P. Couvreur, J. Delattre, and J. P. Devissaugnet, eds.) pp. 669. Editions de Santé, Paris, 1995.
[41] K. Kato, Y. Kaneda, M. Sakurai, M. Nakanishi, and Y. Okada, *J. Biol. Chem.* **266,** 22071 (1991).

estimated that only 0.1% of the plasmid delivered by liposomes actually reaches the nucleus and is expressed. Therefore, the efficiency of liposome-mediated gene transfer could be increased three orders of magnitude if the gene delivery process leads to 100% gene bioavailability.

Backbone Neutral Liposome

Liposome preparation is based on the previously described dehydration–rehydration method,[43] but we have introduced considerable modifications. A lipid mixture (20 μmol total lipid) of phosphatidylcholine cholesterol (PC:CH, 6:4 molar ratio) dissolved in benzenemethanol (9:1, v/v) containing DPPE-PDP (dipalmitoylphosphatidylethanolamine, 0.2 μmol) is dried under an N_2 stream to form a thin film around the walls of a round-bottomed glass tube. Lipids are dispersed mechanically and then by sonication probe (two cycles of 1.5 min) in 1 ml ultrapure water. The samples are centrifuged at 20,000g for 20 min to remove large aggregates and titanium particles released from the probe. The supernatant fraction is mixed with the DNA (50 μg in 10 mM TRIS, 1 mM EDTA, pH 7.5). The mixture is then frozen in liquid nitrogen and lyophilized overnight. The samples are rehydrated using 4 μl of ultrapure water/μmol of total lipid. After a minimum of 30 min of rehydration at room temperature with gentle agitation, the volume is raised to 0.6 ml with HBS and immediately filtered (five cycles) through consecutive polycarbonate membrane filters of 400 and 50 nm pore diameter. The liposome suspension is then incubated 1 h at 37° with 50 μg/ml DNase I (Sigma Química, Madrid, Spain), and the nonentrapped material is separated from liposomes by exclusion chromatography using a Sepharose 4B column and HBS elution buffer. After the column chromatography, small liposomes encapsulating the DNA are recovered within the supernatant fraction after the ultracentrifugation of the liposome preparation at 10^5 g for 45 min. The liposome encapsulation efficiency is evaluated measuring the DNA/phospholipid (μg/μmole) ratio. The phospholipid concentration is determined by the ammonium ferrothiocyanate technique,[44] and the encapsulated DNA is measured by fluorometry, using the PicoGreen dye (Molecular Probes, Ex: 485, Em: 530). Liposomes entrapping DNA (50 μl) or equivalent volumes of HEPES and standard concentrations of free DNA are vortexed for 1 min with 0.4 ml of chloroform. An aliquot of 0.6 ml HEPES buffer is added, vortexed again, and the aqueous phase after centrifugation at 20,000g for 20 min is collected for DNA quantitation.

[42] Y. Kaneda, K. Iwai, and T. Uchida, *Science* **243,** 375 (1989).
[43] C. Kirby and G. Gregoriadis, *Biotechnology* **2,** 979 (1984).
[44] J. C. M. Stewart, *Anal. Biochem.* **104,** 10 (1980).

Liposomes Containing Anionic Phosphatidylserine

Twenty micromoles of a lipid mixture of PCCHPS (5:4:1) dissolved in benzene/methanol (9:1, v/v) containing DPPE-PDP (0.2 μmol) is dried to form a thin film around the walls of a round-bottomed glass tube. After dispersion of the lipid by a probe sonicator in ultrapure water, the samples are centrifuged (20,000g, 20 min) to remove the large aggregates and the titanium particles released from the probe. The supernatant fraction is mixed with pTG7101 plasmid (50 μg) containing the genomic human α_1-AT gene driven by the natural promoter.[32] The mixture is then frozen in liquid nitrogen and lyophilized overnight. Liposomes are prepared as indicated previously.

Liposomes Containing Cationic N-[1-(2,3-dioleoyloxy)propyl]- N,N,
N-trimethylammonium Methylsulfate (DOTAP)

DOTAP suspension is sonicated and centrifuged (20,000g, 5 min) before use. DOTAP liposomes from supernatant are mixed with the pTG7101 plasmid (100 μg DNA/μmol DOTAP) and incubated to form a DOTAP–pTG7101 complex (Dt-pTG), resulting in a final 3.3:1 ratio of positive/negative charges. Then, a lipid mixture of PC/CH/GM1 (4:4:1, molar ratio) containing the DPPE-PDP (2% of total lipid) is dispersed by sonication probe. The supernatant fraction of centrifuged liposomes (20,000g, 20 min) is mixed with Dt-pTG complex (10% of total lipid) to obtain a homogeneous liposome mixture with optimized DNA–cationic lipid interaction. The mixture is then frozen in liquid nitrogen and lyophilized overnight. Liposomes are prepared as indicated previously.

In the different types of liposomes used by us (anionic liposome containing 10% of phosphatidylserine, cationic liposome containing 10% of DOTAP), 2% of modified DPPE is incorporated for the purpose of preparing asialofetuin-targeted liposomes. The efficiency of the synthetic reaction is evaluated measuring the release of DTP groups in the presence of DTT. In each sample, the released groups were measures in several aliquots obtained from 1:2 serial dilutions. As can be observed in Table I, linear regression analysis yields a good correlation coefficient, supporting the reliability of the procedure. The table shows representative data of optical density at 343 nm (in absence or presence of DTT), as well as the equivalences in molar concentration of released DTP groups. In all the cases, the release of DTP after the incubation with modified ASF is reduced 80%–100%, indicating that covalent linkage among the liposome and ASF took place.

TABLE I

LINEAR REGRESSION ANALYSIS OF ASIALOFETUIN-LIPOSOMES CONJUGATION

Experimental group	Released DTP			Slope	Y-intercept	R
	OD	μM	%			
LpPS						
before DTT	0.187			0.0942 ± 0.0006	-0.0020 ± 0.0006	0.999
LpPS						
after DTT	0.264	9.542	100	0.1264 ± 0.0008	0.0113 ± 0.0008	0.999
ASFLpPS						
before DTT	0.283			0.1424 ± 0.0005	-0.0011 ± 0.0005	0.999
ASFLpPS						
after DTT	0.295	1.509	15.8	0.1510 ± 0.0015	-0.0026 ± 0.0015	0.999
LpDOTAP						
before DTT	0.061			0.0310 ± 0.0005	-0.0012 ± 0.0006	0.998
LpDOTAP						
after DTT	0.111	6.291	100	0.0516 ± 0.0001	0.0087 ± 0.0001	1.000
ASFLpDOTAP						
before DTT	0.215			0.1087 ± 0.0003	-0.0016 ± 0.0003	1.000
ASFLpDOTAP						
after DTT	0.217	0.210	3.3	0.1120 ± 0.0003	-0.0064 ± 0.0003	1.000

Optical density at 343 nm is measured from serially diluted liposome samples (ratio 1:2), in the absence and presence of 25 μl of DTT [1 M]. The concentration of 2-thiopyridine (DTP) molecules released is determined by the following formula: (OD in presence of DTT-OD in the absence of DTT)/8080. Parameters of linear regression analysis like slope, y-intercept, and coefficient of correlation (R) are calculated by GraphPad Prism software.

Liposomes Containing Cationic Nuclear Localization Signal Lipopeptide

DPPE-PDP (0.2 μmol/ml) is resuspended in HBS by probe sonication and incubated (3 h at 37°) in the presence of NLS (150 μg) to synthesize the PE-NLS by covalent reaction between the modified PE-PDP and the SH groups of cysteine in the NLS peptide (nuclear localization signal: Cys-Gly-Tyr-Gly-Pro-Lys-Lys-Lys-Arg-Lys-Val-Gly-Gly). The PE-NLS is mixed with 100 μg of plasmid DNA (30 min, room temperature), incubated with DNase I (100 μg/ml) for 30 min at 37°, and the complex is purified by exclusion chromatography with Sepharose 4B. Because the NLS contains five amino acids with positive charge, the PE-NLS/DNA complex yields a final 1.8:1 ratio of positive/negative charges.

Table II shows the characteristics of the covalent bonding between the NLS peptide and different PDP coupled to PE species (DPPE or DOPE). The results of the reaction are summarized in the table, where the absorbance (343 nm) of a sample mixture of NLS plus PE is shown at time zero

TABLE II
CHARACTERISTICS OF NLS PEPTIDE BINDING TO LIPOSOMES CONTAINING
PDP-DPPE OR PDP-DOPE

| | OD (343 nm) | | [DTP] μM | | |
LpNLS	Time 0	Time 180 min	Initial	Final	Groups DTP released (%)
DPPE	0.0477	0.2067	50	20.0	40.0
	0.0713	0.3173	100	30.4	30.4
DOPE	0.0521	0.1821	50	16.0	32.0
	0.0683	0.2687	100	24.8	24.8

The modified lipids DPPE-PDP or DOPE-PDP, 50 or 100 nM in both, are resuspended in HBS by probe sonication and dilution samples are measured by optical density (OD) at 343 nm before (time 0) or after incubating (time 180 min) with nuclear localization sequence (NLS) to synthesize phosphatidylethanolamine-NLS by covalent reaction. The respective concentration of DTP released from the mentioned samples (before and after incubation with NLS) was evaluated by 343 nm absorbance in presence or absence of DTT. Percentage of groups DTP released refers to availability of reactive groups on lipopeptide.

and after 3 h (when the reaction has reached > 90% of maximum value), verifying that the proportion of groups consumed was 65% and 70% for DPPE-PDP and DOPE-PDP, respectively. In this experiment, the amount of NLS was limited to avoid consumption of the totality of the available groups of modified lipid; thus, at the end of the reaction, each of the preparations shows 30%–35% of the modified lipid to remain intact (i.e., without reacting). These groups will then be used for the purpose of establishing covalent linkage with modified asialofetuin to obtain the corresponding asialofetuin-targeted liposome.

Preparation of Targeted Liposomes

The procedure for covalent linkage of a molecule to the surface of a liposome for facilitating internalization by a receptor-mediated mechanism is the same in all cases, although the strategies to achieve targeting may differ (Fig. 3). We should emphasize two main strategies. The first is based on the direct linkage of the molecule to the surface of liposomes. This procedure is simple, but less versatile. Classically, the type of molecules used are ligands for specific receptors (in our case, ASF for the asialoglycoprotein receptor located on the surface of hepatocytes) or monoclonal antibodies against the receptor. The second strategy is more complex but also more versatile. It consists of coupling an intermediate molecule such as the streptavidin to the liposome surface. Then, biotinylated molecules

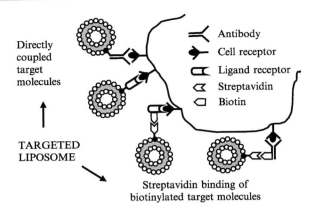

FIG. 3. Strategies for covalent binding of a molecule on the surface of the liposome to mediate internalization by cell receptor. Two main strategies are described: first, direct linkage of the molecule on the liposome surface by an antibody or a ligand receptor recognized by specific cell receptors; the second strategy involves the binding of a bridge molecule such as streptavidin, which is recognized by biotinylated target molecules.

(e.g., biotinylated target element) are easily incorporated by a simple incubation because of the high affinity of biotin for streptavidin. In this sense, it is possible to incorporate biotin-labeled molecules, such as ASF-biotin or antibody-biotin, to direct the liposome against hepatocyte receptors for delivering the entrapped DNA by a receptor-mediated mechanism.

Liposomes Bearing Asialofetuin on Their Surface (ASF–Liposome)

For covalently coupled ASF, liposomes bearing 2% of DPPE-PDP and encapsulating the pTG7101 plasmid are incubated overnight at room temperature in the presence of 100 μg of ASF-SH (82 μl ASF-SH + 20.5 μl 5 M NaCl), are then treated with DNase I (50 μg/ml) for 30 min at 37°, and purified by exclusion chromatography using Sepharose 4B. The liposome fraction is centrifuged (100,000g, 45 min), and the supernatants are used for *in vivo* treatment of mice. To determine the concentration of modified groups, a small aliquot of the sample is incubated with 1 M DTT. The increased absorbance at 343 nm (caused by the release of DTP) indicates the ratio of modified ASF groups. In Fig 4 we see the data corresponding to the absorbance at 343 nm obtained in liposome preparations before and after incubation with modified asialofetuin. The results show that liposome samples from both anionic liposomes (containing PS) or cationic liposomes (containing DOTAP) show increments in absorbance in the presence of DTT, whereas no significant increases are observed

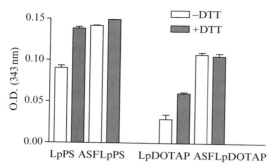

FIG. 4. To determine the concentration of modified groups, a small aliquot of the sample is taken, dried, and incubated with 1 M DTT. The increased absorbance at 343 nm (caused by the release of DTP) indicates the ratio of modified ASF groups.

when liposome samples are incubated previously in the presence of modified asialofetuin. This suggests that all the reactive groups are consumed in the formation of the ASF–liposome covalent link. The efficiency of liposome encapsulation is evaluated as the DNA/phospholipid (microgram/ micromole) ratio in each case. Phospholipid concentrations are determined by the ammonium ferrothiocyanate technique,[44] and the encapsulated DNA is measured by spectrophotometry and fluorimetry (Cytofluor 2350, Millipore, Billerica, MA) using the DNA-binding dye PicoGreen (Molecular Probes, Eugene, DR).

Liposomes Bearing Streptavidin on their Surface (St-Liposome)

For preparation of targeted liposomes, DNA is encapsulated as described in the section on liposome preparation. Liposomes (after DNase I digestion) are mixed with 0.5 mg of modified streptavidin-SH and incubated at room temperature overnight. The protein noncovalently coupled to liposomes, as well as the DTP groups released in the reaction, is separated by exclusion chromatography using Sepharose 4B. The liposomes are centrifuged at 20,000g for 45 min, and the supernatant fraction is recovered and analyzed to determine the encapsulation efficiency. Before treatment, the liposomes are mixed with the appropriate biotinylated molecule such as ligands (e.g., asialofetuin) or monoclonal antibodies to evaluate the ability of streptavidin–liposome to bind biotin-labeled molecules.

In a typical experiment, the appropriate amount of biotinylated antibody is calculated by its ability to block liposome binding to biotin-peroxidase. To study the liposome-binding ability of biotinylated antibodies, 50-μl aliquots of streptavidin-liposomes are incubated with biotin-peroxidase for 30 min at room temperature. Other samples are similarly incubated with

increasing amounts of biotinylated monoclonal antibody (2 μg) before incubation with biotin peroxidase. The mixture is separated by chromatography with Sepharose 4B using a 1.5 × 7-cm column. The elution solvent
is HBS at a flow of 0.3 ml/min, and the fractions are collected every 2 min.
min. The peroxidase activity in every fraction is determined by the addition
of 50 μl/sample of a solution containing o-phenylendiamine (0.4 mg/ml in
0.1 M citrate phosphate buffer, pH 5) plus H_2O_2 (1.5 μl/ml of 30% solution). The reaction is stopped by adding 50 μl/sample of 2 M H_2SO_4, and
the optical density (OD) is measured at 460 nm by spectrophotometry.
The data are expressed as OD percentage per fraction with respect to the
total OD.

Because the bound and unbound biotinylated molecules are separated
by exclusion chromatography (Fig. 5), we expect to find the liposome fraction in the first peak, whereas the small molecules retained in the column
(including the biotinylated peroxidase) emerge as a second peak. In this
sense, the exclusion chromatography of a mixture of streptavidin–liposomes plus biotinylated peroxidase must show in the first peak, the enzymatic activity of biotinylated peroxidase bound to streptavidin–liposomes,
whereas the free fraction of biotinylated peroxidase should appear in the

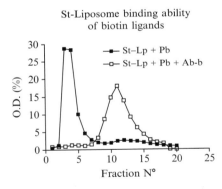

FIG. 5. Ability of streptavidin-liposomes to bind biotinylated molecules. Liposomes from
each fraction obtained after exclusion chromatography are incubated with biotinylated
peroxidase in the absence or presence of biotinylated antibody. The enzymatic peroxidase
activity in the fractions is developed incubating 50-μl aliquots of streptavidin-liposomes
(St-Lp) with biotin-peroxidase (Pb) for 30 min at room temperature or with biotinylated
monoclonal antibody (Ab-b) before incubation with biotin peroxidase. The peroxidase
activity in every fraction is determined by the addition of 50 μl/sample of a solution containing
o-phenylendiamine (0.4 mg/ml in citrate phosphate buffer, pH 5) plus H_2O_2 (1.5 μl/ml of a
30% solution). The reaction is stopped by adding 50 μl/sample of H_2SO_4, and the optical
density (OD) is measured at 460 nm by spectrophotometry. The data are expressed as OD
percentage per fraction with respect to total OD.

second peak. The experimental results agree with the expected data and show that in experiments in which streptavidin–liposomes are mixed with biotinylated peroxidase, 90% of enzymatic activity is found in the first peak. However, when liposomes are incubated previously with an excess of biotinylated antibody, with the purpose of blocking the biotin receptors of streptavidin, 90% of peroxidase activity is found in the second peak. This displacement of enzymatic activity from the first to the second peak indicates that biotinylated molecules bind to streptavidin–liposomes through a specific streptavidin-biotin interaction.

Efficacy of Gene Expression After *In Vivo* Transfection

Gene delivery using nonviral vectors provides a finite period of gene expression and, therefore, a limited synthesis period of the therapeutic product. Although spontaneous integration of DNA into the genome and homologous recombination has been observed *in vitro* with a low frequency, this event has not been observed *in vivo*. Nevertheless, consideration should be given to the possibility that administered DNA can reside transiently in the nucleus of a target cell as a stable extrachromosomal element (episome), where it can be expressed and offer a therapeutic gene product. Obviously, the ultimate efficiency of gene expression is limited by extracellular and intracellular events affecting gene bioavailability, but the efficacy of gene expression is also limited by the intrinsic efficacy of the cassette construct, involving the promoter, enhancer, and stabilizing sequences that greatly influence gene expression and the stability of the gene product. Therefore, one of the major issues in each gene therapy strategy is success in selecting the optimal combination of the gene construct elements. In many cases, such as α_1-antitrypsin deficiency, it is desirable to secure the highest possible level of expression of the transduced gene, mainly by means of one of the cytomegalovirus promoters. This promoter is extremely active in primary hepatocytes,[45] although the activity decreases after a short period following the hepatocellular transplant of transduced cells.[46,47] Therefore, the use of gene constructs under the control of liver-specific promoters,[1,2,32,37,48] which are active in normal

[45] K. P. Ponder, R. P. Dunbar, D. R. Wilson, G. J. Darlington, and S. L. C. Woo, *Hum. Gene Ther.* **2,** 41 (1991).

[46] M. A. Kay, P. Baley, S. Rothenberg, F. Leland, L. Flaming, D. P. Ponder, T. J. Liu, M. Finegold, G. Darlington, W. Pokorny, and S. L. C. Woo, *Proc. Natl. Acad. Sci. USA* **89,** 89 (1992).

[47] R. Sharfmann, J. M. Axelrod, and I. M. Verma, *Proc. Natl. Acad. Sci. USA* **88,** 4626 (1991).

[48] G. Lutfalla, L. Armbrust, S. Dequin, and R. Bertolotti, *Gene* **76,** 27 (1989).

hepatocytes, should be considered for *in vivo* liver-destined gene therapy. In turn, gene persistence within the liver and the duration of gene product expression can be prolonged by performing a partial hepatectomy.[2,37,49] However, it remains to be determined whether the efficacy of nonviral vectors can be increased and whether partial hepatectomy is a necessary condition for long-term expression, because successful gene transfer to the liver with long-term gene expression can be achieved by systemic administration of naked DNA using a hydrodynamic-based procedure.[50] Because it is difficult to apply the partial hepatectomy or hydrodynamic procedures within the clinical context, further work is required to combine the liver-destined nonviral vector approaches with mechanisms for the stable establishment of the therapeutic gene within the liver. We now describe some gene therapy strategies used by us for *in vivo* delivery of the human α_1-AT gene and to evaluate the efficacy of gene expression.

In Vivo *Gene Transfer*

In a typical experiment, C57BL/6 mice ($n = 4$ per group) are cage-housed, and blood samples (200 μl) are taken from the tail vein to pool plasma per group, mixing equal volumes. Two days later, 200 μl of the liposome–DNA complex is injected (0.2-ml volume) into the tail vein with: (1) encapsulated pTG7101 plasmid in anionic or cationic liposomes (100 ng DNA/mouse); (2) encapsulated pTG7101 plasmid in anionic or cationic ASF-liposomes (100 ng DNA/mouse); (3) encapsulated pTG7101 plasmid in ASF targeted (200 ng/mouse); and nontargeted liposomes (200 and 500 ng/mouse) containing the cationic NLS peptide. In all cases, liver regeneration is induced by a partial hepatectomy 1 h after treatment. Murine blood samples (200 μl) are taken at different time points from the tail vein by use of heparinized glass capillaries. After centrifugation, plasma is recovered, and 50 μl/mouse is pooled for each group. The pooled plasma samples are inactivated (55°, 45 min), centrifuged (20,000g, 15 min), and kept at $-20°$ until use for enzyme-linked immuno sorbent assay (ELISA).

Determination of AAT in Mouse Plasma

The assay is performed in 96-well microtitration plates (CoStar, Valencia, Spain), as described previously.[1,37] In a typical ELISA assay, goat anti-hAAT (0.1 μg/well) and goat anti-hAAT peroxidase conjugate (1.5 μg/well) are used as capture and detecting antibodies (Sigma), respectively. Capture antibody (1 μg/ml) diluted in adsorption buffer (0.05 M carbonate, pH 9.6)

[49] C. D. V. Black and G. Gregoriadis, *Biochem. Soc. Trans.* **4**, 253 (1976).
[50] G. Zhang, Y. K. Song, and D. Liu, *Gene Ther.* **7**, 1344 (2000).

is distributed (100 μl/well) in the microplate and incubated at 4° overnight. After three washings (0.05% PBS–Tween-20 (w/v), 200 μl/well), wells are incubated for an additional 2 h at 37° with 1% bovine serum albumin in PBS (200 μl/well) to block nonspecific binding sites. Then, the wells are rinsed five times and 100 μl of hAAT (from 0–30 ng/ml), or the mice plasma samples are added. After incubation (2 h, 37°), wells are rinsed five times and incubated for an equivalent time with goat anti-hAAT peroxidase conjugated (15 μg/ml, 100 μl/well) in PBS-albumin. The enzymatic reaction is induced with o-phenylenediamine (0.4 ng/ml, in citrate-phosphate buffer, pH 5) with 30% H_2O_2 (1.5 μl/ml). The reaction is stopped 2.5 min later by adding 2 M H_2SO_4. The sensitivity of hAAT detection in mouse plasma is 0.2 ng/ml. The results from experiments of *in vivo* hAAT gene transfer, using liposomes and asialofetuin targeted liposomes (Fig. 6), show that asialofetuin–liposomes increase the efficacy of transfection, because a significant increase in the plasma levels of the human protein can be observed. In addition, the increased transfection efficacy mediated by asialofetuin–liposomes is independent of the characteristics of lipid composition of the backbone liposomes (containing anionic PS, cationic DOTAP, or the cationic NLS-DPPE), supporting the idea that liposome targeting with ASF is an effective procedure for increasing the efficiency of liver transfection.

Real-Time Quantitative Polymerase Chain Reaction (qPCR) and Reverse Transcriptase-Polymerase Chain Reaction (qRT-PCR)

The method for PCR and RT-PCR involves isolation of DNA and RNA, respectively, from each liver after transfection with the DNA–liposome complex. The preparation of cDNA is performed using reverse transcriptase and random hexamers. Livers are cut into small pieces and homogenized with an Ultra-Turrax homogenizer (2 × 10 s; 9500 rpm). Subsequently, total RNA and DNA are isolated with a commercial RNA/DNA simultaneous isolation kit (QIAGEN, Valencia, CA) according to the instructions of the manufacturer.

In additional experiments, real time PCR and RT-PCR assays are used for quantification of DNA copies and mRNA transcripts of both the hAAT gene and the mouse α_1-antitrypsin (MUSaat) gene. The real-time system is able to detect PCR products as they accumulate during PCR, thus allowing more accurate and reproducible quantitation of DNA and RNA over a wide (at least five orders) linear range. Reactions are characterized by the point in time during cycling when amplification of a PCR product is first detected rather than the amount of PCR product accumulated after a fixed number of cycles. The higher the starting copy number of the nucleic acid target, the sooner a significant increase in fluorescence is observed. Amplification

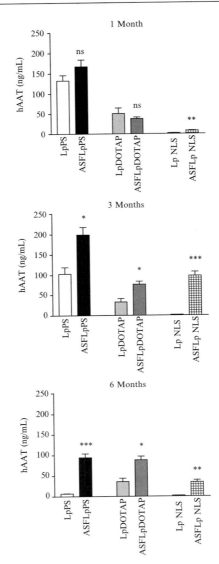

Fig. 6. Effect of vector charge, ASF, and NLS liposomes on hAAT gene transfer. Anionic, cationic, and NLS liposomes were prepared as aforementioned. Mice ($n = 4$ per group) are iv treated (100 ng DNA/mouse) with ASF-targeted and nontargeted liposomes, entrapping the pTG7101 plasmid. Liver regeneration is promoted by partial hepatectomy (PH) 1 h after treatment. Blood samples are collected the first, third, and sixth month after treatment, and aliquots of mouse plasma are pooled and used to measure the plasma levels of hAAT by ELISA testing.

produces increasing amounts of double-stranded DNA, which binds the SYBR Green (Applied Biosystems, Madrid, Spain), resulting in an increase in fluorescence. Amplification plots are produced by plotting the increase in fluorescence versus cycle number.

The method used for obtaining quantitative data of gene amount and relative gene expression is the comparative C_T method as described by the manufacturer (PE-ABI PRISM 7700 Sequence Detection System (Applied Biosystems, Madrid, Spain); User Bulletin #2). The C_T is measured in PCR cycles at which the reporter fluorescent emission increases beyond a threshold level based on the background. Because C_T values are inversely correlated to input target mRNA or DNA levels in the sample, the derived $2^{-\Delta C_T}$ parameter has been proposed as an easy parameter that directly relates the increased values of $2^{-\Delta C_T}$ with increased amounts of PCR product. The precise amount of RNA (based on optical density) and its quality (lack of degradation) is normalized with respect to an endogenous control gene (18S rRNA).

For qPCR, 50 ng of isolated genomic DNA is amplified in a final volume of 25 μl with the SYBR Green Universal PCR Master Mix according to the instructions of the manufacturer (Applera, Foster City, CA). Primers for hAAT and for MUSaat gene quantification are respectively, hAAT-FW: 5′-CATCGTTACAGCCTTTGCAA-3′; hAAT-RV: 5′-AGCCTTCATGGATCTGAGCCT-3′; GeneBank accession No. K02212; MUSaat-FW: 5′-GGCCAGGTTCACAGGATACAAA-3′; MUSaat-RV: 5′-CCTCATTAGCCTCTGGTCTTTTCA-3′ GeneBank accession No. M12585.

For qRT-PCR, isolated total RNA is further purified with the Nucleo-Spin RNA II kit (Macherey-Nagel, Düren, Germany, Cat. No. 740 955 50) according to the instructions of the manufacturer. First-strand cDNA is prepared from 1 μg of RNA with random hexamers using the SYBR Green RT-PCR reagents kit (P/N 4310179; (Applied Biosystems, Madrid, Spain)) according to the manufacturer's instructions. RT-PCR (in a final volume of 50 μl) was performed with one tenth of the resulting first-strand cDNA with primers specific for MUSaat (FW: 5′-CTCTCGGGCATGCTT-GATGT-3′; RV: 5′-AATCCATCAGCAGCACCCAG-3′) and hAAT (hAAT-FW: 5′-GAGGGCCTGAATTTCAACCTC-3′; hAAT-RV: 5′-CCAGGAACTTGGTGATGATAA-3′; GeneBank accession No. X01683 and No. K02212). The hAAT primers were located on both sides of intron 2, allowing differentiation between the amplified product of cDNA and any amplified product derived from plasmidic or genomic DNA.

For both qPCR and qRT-PCR a primer set for 18S rRNA is used as described for the manufacturer (Applera; Pre-Developed Assay Reagents for Gene Expression Quantification; P/N 4310893E). Experiments are

TABLE III

RESULTS OF THE MOLECULAR ANALYSIS AS DETERMINED BY qPCR Y qRT-PCR

	Mouse			Human			
Treatment	RT-PCR $2^{(-\Delta Ct)}$	PCR $2^{(-\Delta Ct)}$	Intrinsic efficacy (mE_i) RT-PCR/ PCR	RT-PCR $2^{(-\Delta Ct)}$	PCR $2^{(-\Delta Ct)}$	Intrinsic efficacy (hE_i) RT-PCR/ PCR	Relative efficacy hEi/mEi
HEPES	0.104	0.012	8.00	—	—	—	—
Naked plasmid	0.088	0.007	12.00	—	—	—	—
NLS(pTG)200	0.112	0.011	10.00	—	—	—	—
NLS(pTG)500	0.116	0.010	11.00	0.04	1.35	0.03	0.003
ASF-NLS(pTG) 200	0.136	0.010	13.00	34.48	12.23	2.82	0.22

Data are obtained as normalized C_T values and presented as $2^{-\Delta Ct}$ values. The intrinsic efficacy of human and mouse genes have been expressed as the RT-PCR/PCR ratios, respectively. In a similar way, the relative efficacy of human gene expression with respect to mouse gene expression is indicated by the ratio between the intrinsic efficacies of human versus mouse genes.

performed in triplicate, and quantitative results are obtained with the ABI-PRISM 7700 Sequence Detection System (Applera).

The results of the molecular analysis are summarized in Table III. Data are obtained as normalized C_T values and presented as $2^{-\Delta C_T}$ values. The intrinsic efficacy of human (hEi) and mouse (mEi) genes has been expressed as the RT-PCR/PCR ratios, respectively. In a similar way, the relative efficacy of human gene expression with respect to mouse gene expression is indicated by the ratio (hEi/mEi) between the intrinsic efficacies of human versus mouse genes. The data confirm the idea that asialofetuin–liposomes offer advantages with respect to efficiency for *in vivo* liver gene transfer. Thus, the hAAT gene expression in transfected mice with untargeted liposomes is confirmed in only one of the four mice (data not shown in the table) and with very low levels. When mice are treated with 2.5-fold higher DNA, the number of animals with hAAT messenger increases, but the amount of gene expression remains very low. However, all animals transfected with asialofetuin–liposomes show hAAT gene expression, and the level of expression is several orders of magnitude higher. These data support the previous observations in the study of plasma levels of hAAT, confirming that the high levels of messenger are related to the high plasma levels of protein in the experimental group of mice transfected with asialofetuin–liposomes. In addition, the expression of the human gene targeted with asialofetuin–liposomes is about 20% of that of the natural mouse α_1-antitrypsin gene.

The amplified qPCR y qRT-PCR products are observed by agarose gel electrophoresis. Fragments are extracted from the gel (QIAGEN), and sequence analysis is performed by direct sequencing with an ABI 377 Genetic Analyzer (Applera). Reaction sequences are done with the same primers used in each specific amplification reaction and the BigDye terminators Kit (Applera). Sequences are analyzed with Factura and SeqEd software (Applera). Sequencing analysis confirms the specificity of the amplified products.

Concluding Remarks

The liver is an excellent organ for gene transfer in treating a wide variety of diseases that affect liver function. In addition, it is an ideal organ for a high amount of expression of therapeutic genes and efficient systemic distribution of the resulting therapeutic proteins secreted into the bloodstream. For strategies of liver-destined gene therapy, we must consider that the liver sinusoid endothelium contains pores with a mean diameter of 100 nm that would allow a small vector to leave the blood circulation and reach the hepatocyte and that the hepatocyte possess highly specific receptors for asialoglycoproteins (ASGP) that are good candidates for specific gene delivery to hepatocytes by a receptor-mediated mechanism.

The preparation of asialofetuin–liposomes targeted to hepatocytes can be made by covalent coupling of asialofetuin glycoprotein (ASF) onto the liposome surface, by use of the heterobifunctional reagent N-succinimidyl 3-(2-pyridildithio) propionate (SPDP). Changes in liposome charge can be made by inclusion on the liposome backbone of anionic or cationic lipids (such as phosphatidylserine and DOTAP, respectively), cationic lipopeptides (such as PE-NLS), and cationic polymers. Because nuclear localization signal (NLS) is also a cationic polypeptide, it can interact directly with DNA by electrostatic interaction, increasing the nuclear availability of delivered DNA and thereby contributing to increase the efficacy or gene transfer.

Acknowledgments

The authors are grateful to Mr. A. Crespo for technical assistance. This work has been partially supported by CICYT projects PB92-0877 and SAF99-0109 and IMTEFA/2000/41.

[27] Fusogenic Polymer–Modified Liposomes for the Delivery of Genes and Charged Fluorophores

By KENJI KONO and TORU TAKAGISHI

Introduction

Fusogenic liposomes have the ability to fuse with biological membranes. These liposomes can introduce their contents into the cytoplasm by fusing with plasma membrane and/or endosomal membrane, and hence these are of great importance as delivery systems of membrane-impermeable molecules with biological activities, such as proteins, genes, and oligonucleotides. Recently, their importance has increased in connection with the demand for nonviral vectors for gene therapy.

Fusogenic liposomes have been prepared by various approaches. One of the most frequently used methods is the use of lipids capable of undergoing a bilayer-to-hexagonal II phase transition such as dioleoylphsophatidylethanolamine (DOPE). pH-Sensitive liposomes that become fusogenic under weakly acidic conditions have been prepared using DOPE in combination with various titratable amphiphiles.[1,1a,2] Also, liposomes with triggerable fusion activity have been prepared by incorporating photopolymerizable lipids[3] or an enzymatically cleavable peptide–lipid conjugate into DOPE membranes.[4]

Another approach to the production of fusogenic liposomes is conjugation of fusogenic molecules to stable liposome membranes. Viral fusion proteins,[5,6] fusion peptides,[7,8] and their analogues[9,10] have been used to give fusogenic activity to stable liposomes.

[1] N. Düzgüneş, R. M. Straubinger, P. A. Baldwin, and D. Papahadjopoulos, *in* "Membrane Fusion" (J. Wilschut and D. Hoekstra, eds.), p. 195. Marcel Dekker, New York, (1991).

[1a] D. C. Litzinger and L. Huang, Biochim. Biophys. Acta **1113,** 201 (1992).

[2] C. J. Chu and F. C. Szoka Jr., *J. Liposome Res.* **4,** 61 (1994).

[3] D. E. Bennet and D. F. O'Brien, *Biochemistry* **34,** 3102 (1995).

[4] C. C. Pak, S. Ali, A. S. Janoff, and P. Meers, *Biochim. Biophys. Acta* **1372,** 13 (1998).

[5] A. Loyter and D. J. Volsky, *in* "Membrane Reconstitution" (G. Poste and G. L. Nicolson, eds.), p. 215, Elsevier, Amsterdam, 1982.

[6] M. Nakanishi, T. Uchida, H. Sugawa, M. Ishiura, and Y. Okada, *Exp. Cell Res.* **159,** 399 (1985).

[7] A. L. Bailey, M. A. Monck, P. R. Cullis, *Biochim. Biophys. Acta* **1324,** 232 (1997).

[8] C. Puyal, L. Maurin, G. Miruel, A. Bienvenüe, and J. Philippot, *Biochim. Biophys. Acta* **1195,** 259 (1994).

[9] K. Vogel, S. Wang, R. J. Lee, J. Chmielewski, and P. S. Low, *J. Am. Chem. Soc.* **118,** 1581 (1996).

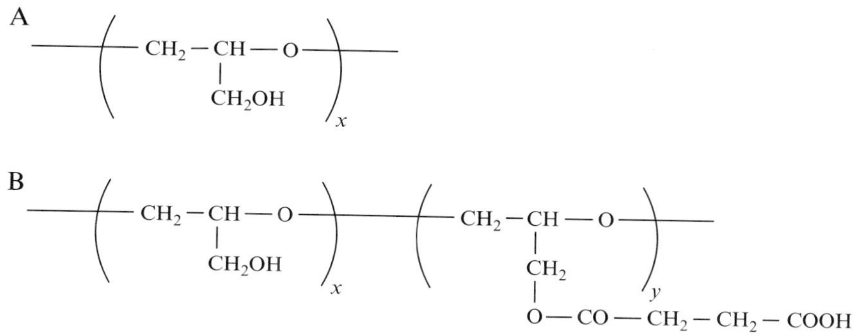

A

$$\left(\!\!-\!\!\begin{array}{c} CH_2-CH-O- \\ | \\ CH_2OH \end{array}\!\!\right)_{\!x}$$

B

$$\left(\!\!-\!\!\begin{array}{c} CH_2-CH-O- \\ | \\ CH_2OH \end{array}\!\!\right)_{\!x}\!\!\left(\!\!-\!\!\begin{array}{c} CH_2-CH-O- \\ | \\ CH_2 \\ | \\ O-CO-CH_2-CH_2-COOH \end{array}\!\!\right)_{\!y}$$

Fig. 1. Structures of poly(glycidol) (A) and SucPG (B).

We have attempted to obtain fusogenic liposomes with synthetic polymers derivatized from poly(glycidol) (Fig. 1). Poly(glycidol) has a main chain structure similar to that of poly(ethylene glycol) (PEG), which is a well-known fusogenic polymer. Each repeating unit of poly(glycidol) possesses a hydroxyl group, to which various functional moieties, such as carboxyl and amino groups, can be connected. Introduction of these functional groups will affect interaction of the polymer main chain with liposome membranes. Thus, we designed succinylated poly(glycidol) (SucPG) (Fig. 1) as a pH-sensitive polymer that exhibits fusogenic activity in a pH-dependent fashion. Succinylated poly(glycidol) hardly enhances fusion of liposomes at neutral pH, but promotes it at acidic pH.[11] In addition, surface modification with this polymer gives stable liposomes with the ability to fuse under weakly acidic conditions.[11] Here, we describe experimental methods for the preparation of SucPG-modified liposomes and their application to cytoplasmic delivery of calcein[12] and plasmid DNA.[13]

Synthesis of Succinylated Poly(Glycidol)

Succinylated poly(glycidol) is synthesized according to the scheme shown in Fig. 2. Poly(epichlorohydrin) is first converted to poly(glycidol), and then poly(glycidol) is succinylated by the reaction with succinic anhydride. The acid–base titration for SucPG shows that the charge density

[10] A. L. Bailey, M. A. Monck, and P. R. Cullis, *Biochim. Biophys. Acta* **1324**, 232 (1997).
[11] K. Kono, K. Zenitani, T. Takagishi, *Biochim. Biophys. Acta* **1193**, 1 (1994).
[12] K. Kono, T. Igawa, and T. Takagishi, *Biochim. Biophys. Acta* **1325**, 143 (1997).
[13] K. Kono, Y. Torikoshi, M. Mitsutomi, T. Itoh, N. Emi, H. Yanagie, and T. Takagishi, *Gene Ther.* **8**, 5 (2001).

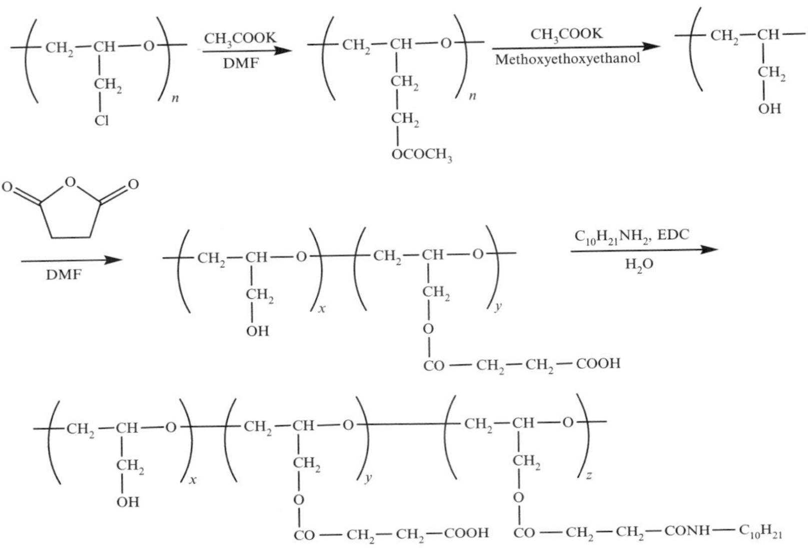

Fig. 2. Synthesis of SucPG coupled to n-decyl groups.

of this polymer changes between pH 3.4 and 9.0, and the apparent pKa is 5.4.

Previous studies have shown that hydrophilic polymers can be fixed on liposome membranes by conjugation of hydrophobic moieties to the polymer.[14–16] Thus, n-decyl groups are incorporated in to the polymers as anchoring moieties for the fixation of the polymer onto liposome membranes.

Synthesis of Poly(glycidol)

Poly(glycidol) is synthesized according to the method of Cohen[17] by use of poly(epichlorohydrin) as the starting material. Although poly(epichlorohydrin) is obtained by polymerization of epichlorohydrin in the presence of a catalyst, such as aluminum alkyls,[17] this polymer is commercially available. Procedures of conversion of poly(epichlorohydrin) to poly(glycidyl acetate) and its conversion to poly(glycidol) are described in detail in

[14] M. Takada, T. Yuzuriha, K. Katayama, K. Iwamoto, and J. Sunamoto, *Biochim. Biophys. Acta* **802**, 237 (1984).
[15] H. Ringsdorf, B. Schlarb, and J. Venzmer, *Angew. Chem. Int. Ed. Engl.* **27**, 113 (1988).
[16] N. Oku, S. Shibamoto, F. Ito, H. Gondo, and M. Nango, *Biochemistry* **26**, 8145 (1987).
[17] H. L. Cohen, *J. Polym. Sci., Polym. Chem. Ed.* **13**, 1993 (1975).

Ref. 17. Poly(epichlorohydrin) (20 g) is dissolved in dimethylformamide (330 ml) under nitrogen, and dry potassium acetate (40 g) is added to the solution and stirred for 4 h at 150°. Then, the solution is added to distilled water to afford a white precipitate of poly(glycidyl acetate). The precipitate is dissolved in acetone, reprecipitated in diethyl ether, and dried under vacuum. The obtained poly(glycidyl acetate) (18 g) and potassium acetate (40 g) are dissolved in methyl carbitol (200 ml) and stirred for 1.5 h at 140° under nitrogen. The solution is dialyzed against distilled water and freeze dried. The obtained rubbery solid is dissolved in methanol, precipitated in acetone, and dried under vacuum.

We have observed that poly(glycidol) obtained by the conversion reactions exhibited a smaller molecular weight than expected from the molecular weight of the poly(epichlorohydrin) before the reaction.[18] Partial cleavage of the polymer chain may have taken place during this reaction.

Synthesis of Succinylated Poly(Glycidol)

Succinylated poly(glycidol) is prepared easily by the reaction of poly (glycidol) with succinic anhydride according to the following procedure. Poly(glycidol) (10.5 g) is dissolved in 120 ml dimethylformamide at 60° with stirring under nitrogen. When the solution is complete, succinic anhydride (28.4 g) is added, and the mixture is stirred for 6 h at 60°. The reaction mixture is cooled, precipitated with diethyl ether, and dried under vacuum overnight. Then, the crude polymer is dissolved in an aqueous NaHCO$_3$ solution, dialyzed against water, and freeze dried. This procedure gives complete succinylation of poly(glycidol). Partial succinylation of the polymer can be achieved by controlling the ratio of succinic anhydride to poly(glycidol) during the reaction.

n-Decylamine is coupled to some carboxyl groups of SucPG by means of the following procedure. SucPG (4.5 g) is dissolved in water, and the pH of the solution is adjusted to 5 by adding 0.1 N HCl. n-Decylamine (0.31 g) and 1-ethyl-3-(3-dimethylaminopropyl) carbodiimide (EDC, 0.38 g) are added to the solution at 4°, and the solution is stirred overnight at 4°, and then at room temperature for 2 days. The reaction mixture is freeze dried, dissolved in methanol, and applied on a Sephadex LH-20 columun, eluting with methanol. Fractions that contain the polymer are collected and evaporated to remove methanol. The polymer is dissolved in an aqueous NaHCO$_3$ solution, washed three times with chloroform, dialyzed against water, and freeze dried.

[18] K. Kono, M. Iwamoto, R. Nishikawa, H. Yanagie, and T. Takagishi, *J. Control. Release* **68,** 225 (2000).

Liposome Preparation

Liposomes modified with hydrophilic polymers can be obtained either by incubating the anchor-bearing polymer with liposomes or by using a mixture of the anchor-bearing polymer and lipids for the preparation of liposomes.[19] The former gives liposomes with the polymer fixed on the outer leaflet of the liposome membrane, whereas the latter gives liposomes with the polymer fixed on both surfaces of the membrane. These preparations produce different results.[20]

Preparation of Succinylated Poly(Glycidol)–Modified Liposomes Encapsulating Calcein

Liposomes with SucPG fixed on both sides of the membrane are prepared by the following procedure. Egg yolk phosphatidylcholine (EYPC) (7 mg) and SucPG (3 mg) are dissolved in methanol (5 ml) in a 20-ml round-bottom flask. Methanol is removed by evaporation of the solution to afford a thin film of a mixture of EYPC and SucPG. The film is further dried under vacuum for more than 3 h. Then, the membrane is dispersed in 2 ml of an aqueous calcein solution (200 mM, pH7.4) and sonicated for 30 min with an ultrasonic disruptor to produce liposomes composed of SucPG and EYPC. The dispersion can be extruded through a polycarbonate membrane with a pore size of 50 or 100 nm to prepare the liposomes. The liposome suspension is applied on a Sephadex G-75 column, eluting with phosphate-buffered saline (PBS) to remove free calcein. Calcein-loaded unmodified EYPC liposomes are prepared by the same procedure, except that a dry EYPC film is used instead of the dry film of a mixture of EYPC and SucPG.

The SucPG-modified liposomes are stable at neutral pH but are destabilized under acidic conditions. As a result, their contents are released, and fusion takes place.[11] Because carboxylate groups of SucPG become protonated under acidic conditions, the polymer interacts with the liposome membrane probably through hydrophobic interactions and hydrogen bond formation, resulting in the destabilization of the liposomes.

Cytoplasmic Delivery of Calcein Mediated by Succinylated Poly(Glycidol)–Modified Liposomes

In our laboratory CV1 cells (an African green monkey kidney cell line) are used, because their interaction with liposomes has been well characterized.[21–23] The experimental procedure is as follows. CV1 cells

[19] K. Kono, *Adv. Drug Delivery Rev.* **53**, 307 (2001).
[20] H. Hayashi, K. Kono, and T. Takagishi, *Bioconj. Chem.* **10**, 412 (1999).

are grown on precleaned coverslips for 24 h in Dulbecco's modified Eagle medium (DMEM) supplemented with 10% fetal bovine serum (FBS). The cells are washed three times with PBS containing 0.36 mM CaCl$_2$ and 0.42 mM MgCl$_2$ (PBS-CM) and incubated with the calcein-loaded liposomes suspended in PBS-CM for 3 h at 37° (concentration of EYPC, 1.5 mM). After the incubation, the cells are washed with PBS-CM and viewed under an epifluorescence microscope with an excitation filter set that produces excitation in the range of 470–500 nm and allows observation of fluorescence emission in the range of 515–540 nm, with a long wave pass dichroic mirror and barrier filter. Fig. 3 (upper and middle) shows typical examples of fluorescence micrographs of the cells treated with SucPG-modified and SucPG-unmodified EYPC liposomes encapsulating calcein. The cells treated with the unmodified EYPC liposomes display weak and vesicular fluorescence of calcein. In contrast, the cells incubated with the SucPG exhibit strong and diffuse fluorescence of calcein, indicating that calcein molecules are transferred into the cytoplasm.

Fusion of Succinylated Poly(Glycidol)–Modified Liposomes with CV1 Cells

Resonance energy transfer between N-(7-nitrobenz-2-oxa-1,3-diazol-4-yl)phosphatidylethanolamine (NBD-PE) and N-(lissamine rhodamine B sulfonyl)phosphatidylethanolamine (Rh-PE) is used frequently for the detection of the liposome–liposome fusion and the liposome–cell fusion.[24a] This fusion assay is based on the fact that the resonance energy transfer efficiency from NBD-PE to Rh-PE varies depending on their concentration in the membrane. When liposomes containing NBD-PE and Rh-PE are mixed with fluorescent lipid-free liposomes and their fusion takes place, energy transfer efficiency between these fluorescent lipids decreases. Although an increase in fluorescence intensity of NBD-PE on liposome fusion is used to detect occurrence of fusion,[24,24a] it is also useful for the detection of fusion to monitor the ratio of the NBD-PE fluorescence intensity to the Rh-PE fluorescence intensity.[25,26] We used this fusion assay for the detection of fusion between liposomes and the endosomal and/or

[21] R. T. Fraley, S. Subramanni, P. Berg, and D. Papahadjopoulos, *J. Biol. Chem.* **255**, 10431 (1980).

[22] R. M. Straubinger, K. Hong, D. S. Friend, and D. Papahadjopoulos, *Cell* **32**, 1069 (1983).

[23] K.-D. Lee, K. Hong, and D. Papahadjopoulos, *Biochim. Biophys. Acta* **1103**, 185 (1992).

[24] D. K. Struck, D. Hoekstra, and R. E. Pagano, *Biochemistry* **20**, 4093 (1981).

[24a] D. Hoekstra, and N. Düzgünes, *Methods Enzymol.* **220**, 15 (1993).

[25] K. Kono, S. Kimura, and Y. Imanishi, *Biochemistry* **29**, 3631 (1990).

[26] K. Kono, H. Nishii, and T. Takagishi, *Biochim. Biophys. Acta*, **1164**, 81 (1993).

lysosomal membranes.[12] Occurrence of fusion can be recognized directly from the change of the color of fluorescence observed in the cell. The experimental procedure is outlined in the following.

The liposomes containing NBD-PE and Rh-PE are prepared by means of the preceding method using a mixture of EYPC, NBD-PE, and Rh-PE (94:5:1, mol/ mol/mol) as the membrane lipid. CV1 cells are grown on coverslips for 24 h in DMEM supplemented with 10% FBS. The cells are washed three times with PBS-CM and incubated with the labeled liposomes suspended in PBS-CM for 3 h at 37° (concentration of EYPC, 1.5 mM). After the incubation, the cells are washed with PBS-CM and viewed with an epifluorescence microscope with an excitation filter set that produces excitation in the range of 470–500 nm and allows observation of fluorescence emission above 515 nm. Typical examples of fluorescence micrographs of CV1 cells treated with SucPG-modified and SucPG-unmodified EYPC liposomes are shown in Fig. 3 (lower). The cells treated with the unmodified EYPC liposomes display vesicular, red fluorescence, namely fluorescence of Rh-PE, indicating that most liposomes are still intact in endosomes or lysosomes. However, vesicular, yellow fluorescence, namely fluorescence of NBD-PE, is seen in the cells treated with SucPG-modified liposomes, indicating the occurrence of fusion between the liposomes and endosomal and/or lysosomal membranes. In addition, the percentage of the liposomes fused in the cells can be estimated by comparing fluorescence spectra of the treated cells and the liposomes.[12]

Preparation of Gene Delivery Systems Using Succinylated Poly(Glycidol)–Modified Liposomes

Fusogenic liposomes are expected to deliver DNA into cells efficiently by fusing with the plasma membrane and/or endosomal membrane. One of the major problems for their use as gene delivery systems is the difficulty of efficient encapsulation of genes by the liposomes. To solve this problem, complexation of pH-sensitive liposomes with cationic polymer–plasmid DNA complex has been examined.[27]

On the other hand, it is well known that cationic liposomes associate with DNA efficiently by means of electrostatic interactions and form complexes, which are termed lipoplexes. Although the lipoplexes per se are used frequently for gene delivery, their transfection ability is not very high. It is considered that a large fraction of lipoplexes taken up by cells are

[27] R. J. Lee and L. Huang, *J. Biol. Chem.* **271**, 8481 (1996).

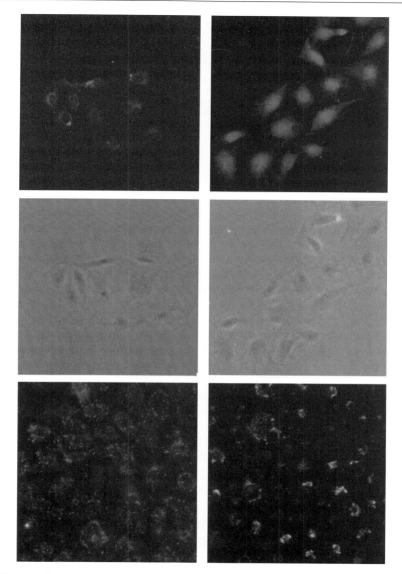

Fɪɢ. 3. Micrographs of CV1 cells treated with plain EYPC liposomes (left) and SucPG-modified EYPC liposomes (right). (Upper and middle) Fluorescence and phase-contrast micrographs of CV1 cells treated with the calcein-loaded liposomes. (Lower) Fluorescence micrographs of CV1 cells treated with the liposomes containing NBD-PE and Rh-PE. (See color insert.)

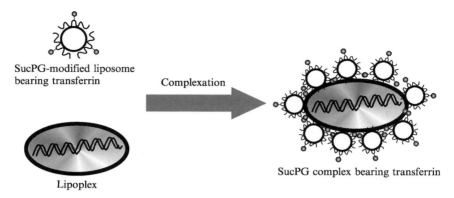

Fig. 4. Preparation of SucPG complexes bearing transferrin.

trapped in endosomes and eventually degraded in lysosomes.[28,29] If some mechanism that promotes fusion of lipoplexes with endosomes is conferred to lipoplexes, an improvement of their transfection activity is expected. Thus, we designed complexes of SucPG-modified liposomes and lipoplexes (which are termed SucPG complexes here).[13] Because SucPG-modified liposomes have a negatively charged surface and lipoplexes can be prepared to have a positive charge, they readily form a complex through electrostatic interactions (Fig. 4). SucPG complexes with a negatively charged surface exhibit very low transfection activity against HeLa and K562 cells, but conjugation of transferrin to SucPG complexes increases their transfection activity significantly, indicating that these are taken up by the cell through the specific interaction between transferrin and its receptor on the surface of the target cells. Although a variety of cationic lipids can be used for the preparation of transferrin-conjugated SucPG complexes, an example, in which 3β-(N-(N',N'-dimethylaminoethane)carbamoyl) cholesterol (DC-chol) is used as a cationic lipid, is described here.

Preparation of Succinylated Poly(Glycidol) Complexes

Succinylated poly(glycidol)–modified liposomes are prepared by suspending a mixture of SucPG (3 mg) and egg yolk phosphatidylcholine (7 mg) in PBS (2 ml) according to the preceding procedure. The liposome

[28] A. Singhal and L. Huang, *in* "Gene Therapeutics" (J. A. Wolf, ed), p. 118, Birkhäuser, Boston, 1993.

[29] J. Zabner, A. J. Fasbender, T. Moninger, K. A. Poellinger, and M. J. Welsh, *J. Biol. Chem.* **270**, 18997 (1995).

suspension is extruded through a polycarbonate membrane with a pore size of 50 nm. 1-Ethyl-3-(3-dimethylaminopropyl)carbodiimide (0.7 mg) is added to the liposome suspension at pH 6.0 and stirred for 2 h at $0°$. Then, transferrin (3 mg) is added to the liposome suspension and kept at $4°$ overnight. The liposome suspension is adjusted to pH 7.4 by adding 1 N NaOH, and 0.5 M ferric citrate (5 μl) is added to the suspension. The transferrin-conjugated SucPG-modified liposomes are applied to a Sepharose 4B column eluting with PBS at $4°$ to remove free transferrin and other impurities. Lipoplexes of DC-chol–DOPE liposomes and plasmid DNA are prepared according to the method of Gao and Huang.[30] A thin film of a mixture of DC-chol (0.34 mg) and DOPE (0.6 mg) is prepared by evaporation of a solution of these lipids in chloroform and subsequent drying under vacuum for more than 3 h. The membrane is dispersed in PBS (5 ml) and sonicated for 2 min using a bath-type sonicator. Plasmid DNA (2 μg) is dissolved in 20 mM TRIS-HCl (100 μl), added to the DC-chol–DOPE liposome suspension (100 μl), and incubated for 10 min in an ice bath to generate a lipoplex. A given volume (0–60 μl) of a suspension of SucPG-modified liposomes (0.3 mM) is added to the lipoplex suspension, and the mixed suspension is allowed to stand for 10 min in an ice bath.

Transfection of HeLa Cells Mediated by Succinylated Poly(Glycidol) Complexes

Transfection of HeLa cells is carried out according to the following procedure. The cells are seeded in 1 ml of DMEM supplemented with 10% FBS in 24-well culture plates at 5×10^4 cells/well the day before transfection. The cells are washed with PBS and then covered with DMEM supplemented with 10% FBS (1 ml). The SucPG complex containing plasmid DNA (0.5–2 μg) is added gently to the cells and incubated for 4 h at $37°$. Then, the cells are rinsed with PBS, covered with DMEM containing 10% FBS, and incubated at $37°$ for 12–40 h before the assay for gene expression.

Figure 5 shows luciferase activity of HeLa cells treated with transferrin-attached or transferrin-free SucPG complexes with varying succinylated unit/DC-chol ratios. Zeta potential measurements show that the SucPG complex with the succinylated unit/DC-chol ratio of 0.5 is electrically neutral. Therefore, both of the SucPG complexes with or without transferrin exhibit higher transfection activity than lipoplexes as long as these SucPG complexes are positively charged. However, when the complexes are

---

[30] X. Gao and L. Huang, *Biochem. Biophys. Res. Commun.* **179,** 280 (1991).

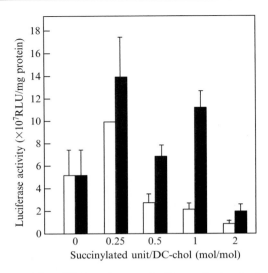

FIG. 5. Luciferase activity of HeLa cells treated with transferrin-attached (solid bars) and transferrin-free (open bars) SucPG complexes. The cells were treated with SucPG complexes containing 2 μg plasmid DNA in serum-free medium.

negatively charged, only transferrin-conjugated complex shows high activity. The SucPG complex with the succinylated unit/DC-chol ratio of 2 shows a low transfection activity, probably because the transferrin-attached SucPG liposomes, which do not form the SucPG complex, coexist in the sample, and these liposomes inhibit the binding of the ligand-bearing SucPG complex to the receptor of the cell.

Although SucPG complexes prepared using DC-chol are described here, other cationic lipids can be used as the component of SucPG complexes. For example, SucPG complexes prepared using dimethyldioctadecylammonium bromide exhibit a high transfection activity. In contrast to lipoplexes, which interact with cells through nonspecific charge interaction, the ligand-bearing SucPG complexes with negatively charged surfaces interact with cells by means of the specific interaction with its receptor and deliver genes into cells efficiently. In addition, SucPG complexes exhibit lower cytotoxicity and higher serum resistance than the parent lipoplexes. These advantages imply that SucPG complexes and their relevant systems may be useful as nonviral vectors for gene therapy.

[28] Liposome-Mediated Gene Delivery: Dependence on Lipid Structure, Glycolipid-Mediated Targeting, and Immunological Properties

By Renat Zhdanov, Elena Bogdanenko, Alexey Moskovtsev, Olga Podobed, and Nejat Düzgüneş

Introduction

The targeted delivery of therapeutic genes to specific tissues and organs remains one of the most challenging problems of gene and gene-cellular therapy.[1–8] According to Inder Verma, "there are three major problems in gene therapy: delivery, delivery, and again delivery (of genes)." The vehicles most commonly used for gene delivery to cells, both viral[9–11] and non-viral,[12–15] are far from being perfect, especially with respect to targeted delivery.[16] Improved delivery systems must exploit elements of

[1] P. L. Felgner and G. M. Ringold, *Nature* **337,** 387 (1989).

[2] D. Lasic, "Liposomes in Gene Delivery", CRC Press, Boca Raton, 1997.

[3] R. I. Zhdanov, A. N. Somov, and S. N. Viryasov, *in* "Development of Recombinant Immunomodulating Preparations for Gene and Immunotherapy of Cancer", p. 37, Scientific reviews, Obolensk, 1996 (in Russian).

[4] M. Strauss and J. A. Barrenger, eds., "Concepts in Gene Therapy." De Gruyter, Berlin, 1997.

[5] K. Xantopoulos, ed., "Gene Therapy." NATO ASI Series, Series H Cell Biology, vol. 105, Springer-Verlag, Berlin, Heidelberg, 1998.

[6] A. V. Kabanov P. L. Felgner, and L. W. Seymour, eds., "Self-Assembling Complexes for Gene Delivery." John Wiley & Sons, London, 1999.

[7] R. I. Zhdanov and A. I. Archakov (eds), *Voprosy Med. Khim (Problems Med. Chem)* **46,** 1–150 (2000) (in Russian).

[8] R. I. Zhdanov and E. S. Severin, eds., *Voprosy Biol. Med. Farm. Khimii (Problems Biol. Med. Pharm. Chem.)* **4,** 1999 (in Russian).

[9] C. P. Hodgson, ed., "Retro-Vectors for Human Gene Therapy," Springer, New York, Berlin, 1996.

[10] M. G. Kaplit and A. D. Loewy, eds., "Viral Vectors. Gene Therapy and Neuroscience Applications." Academic Press, San Diego, 1995.

[11] K. I. Berns and C. Giraud, eds., "Adeno-Asociated Virus (AAV) Vectors in Gene Therapy." Springer, Berlin, Heidelberg, 1996.

[12] R. I. Zhdanov, N. G. Kutsenko, and V. I. Fedchenko, *Voprosy Med. Khimii* **43,** 3 (1997) (in Russian).

[13] O. V. Podobed and R. I. Zhdanov, *Voprosy Biol. Med. Farm. Chem.* **4,** 7 (1999) (in Russian).

[14] M. C. Pedroso de Lima, S. Simões, P. Pires, H. Faneca, and N. Düzgüneş, *Adv. Drug Deliv. Rev.* **47,** 294 (2001).

[15] R. I. Zhdanov, O. V. Podobed, and V. V. Vlassov, *Bioelectrochem.* **58,** 53 (2002).

both viral and non-viral systems and should include at least the following: (a) the address group—most likely an oligosaccharide having an affinity to cell surface lectins; (b) a polycationic core (possibly a synthetic polymer) for DNA condensation and preferably coated with poly(ethyleneglycol) for prolonged residence time in the bloodstream; and (c) minor protein/peptide components, to facilitate nuclear localization and targeting for enhancing the system's efficiency.[17–20] At the same time non-viral methods based on physical techniques, such as electroporation,[21,22] are growing in importance, even in clinical practice.

The purpose of this chapter is to outline the methodological aspects of gene delivery by a novel set of lipidic transfection reagents including cationic lipids,[23–29] glycolipids,[24,30,31] pH-sensitive and neutral lipids[32,33]

[16] W. F. Anderson, *Nature. Therapeutic horizons, (Suppl.)* **392,** 25 (1998).

[17] R. I. Zhdanov, *Voprosy Biol. Med. Farm. Chem.* **4,** 3 (1999) (in Russian).

[18] R. I. Zhdanov, N. V. Semenova, and A. I. Archakov, *Voprosy Med. Khimii,* **46,** 223 (2000) (in Russian).

[19] R. I. Zhdanov and G. G. Krivtsov, *Voprosy Med. Khimii* **46,** 246 (2000) (in Russian).

[20] N. Düzgüneş, C. Tros de Ilarduya, S. Simões, R. Zhdanov, K. Konopka, and M. C. Pedroso de Lima, *Curr. Med. Chem.* **10,** 1213 (2003).

[21] R. I. Zhdanov, R. S. Khusainova, G. R. Ivanitskyi, and A. S. Borisenko, *Voprosy Biol. Med. Farm. Chem.* **1,** 10 (2000) (in Russian).

[22] L. Heller, M. J. Jaroszeski, D. Coppola, C. Pottinger, R. Gilbert, and R. Heller, *Gene Ther.* **7,** 826 (2000).

[23] R. I. Zhdanov, N. G. Kutsenko, O. V. Podobed, O. A. Buneeva, T. A. Tsvetkova, D. N. Konevets, and V. V. Vlassov, *Doklady Akad. Nauk* **361,** 695 (1998).

[24] R. I. Zhdanov, O. V. Podobed, I. O. Konstantinov, A. I. Petrov, D. A. Gaichevsky, Yu, V. Sviridov, D. N. Konevets, V. V. Vlassov, A. Haberland, M. Bottger, and H. Haller, *J. Gene Med.* **1** (Suppl.), 88 (1998).

[25] I. D. Konstabtinova, I. P. Ushakova, G. A. Serebrennikova, *Rus. J. Bioorgan Chem.* **19,** 845 (1993).

[26] M. A. Maslov, E. V. Sycheva, N. G. Morozova, and G. A. Serebrennikova, *Rus. Chem. Bull.* **48,** 545 (1999).

[27] R. I. Zhdanov, O. V. Podobed, N. G. Kutsenko, O. A. Buneeva, T. A. Tsvetkova, G. A. Serebrennikova, I. D. Konstantinova, and M. A. Maslov, *Doklady Akad. Nauk* **362,** 557 (1998).

[28] A. A. Moskovtsev, A. Elkady, E. Yu. Filinova, N. K. Vlasenkova, A. A. Kaloshin, A. A. Sokolovskaya, R. I. Zhdanov, A. A. Churilov, M. A. Skripnikova, A. Larin, Yu. L. Sebyakin, and D. Yu. Blokhin, *Bioterapevticheskii Zhurnal* **3,** (2003) (in press, in Russian).

[29] A. Elkady, Y. Sebyakin, M. Gallyamov, A. Moskovtsev, G. Bischoff, R. Zhdanov, and A. Khokhlov, *in* "Micro- and Nanostructures of Biological Systems" (G. Bischoff and H.-J. Hein, eds), p. 53. Shaker Verlag, Aachen, 2003.

[30] R. P. Evstigneeva, Y. L. Sebyakin, E. V. Kalinina, and E. V. Kazakova, *Doklady Akad. Nauk* **325,** 959 (1992).

[31] R. P. Evstigneeva, E. V. Kazakova, and Y. L. Sebyakin, *Doklady Akad. Nauk* **323,** 495 (1992).

and to describe a new method for the estimation of the influence of these compounds on the complement system.[34] Both *in vitro* (cell culture)[23,27,28,35] and *in vivo*[36–39] gene delivery methods are described. The lipids include the pH-sensitive/neutral structures I and II;[32,33,40] the cationic lipids III, IV, V, and VI;[27] the oligocationic molecules VII, VIII, and IX;[23] the dicationic XII;[28,29] the glycocationic molecule XI;[41] and the glycolipid X[42] (Fig. 1).

Cationic Lipids and Glycolipids

We have employed several novel cationic, dicationic, and glycocationic lipids and glycolipids, as well as pH-sensitive lipids. Figure 1 shows the chemical structures of these new groups of compounds.

1. pH-sensitive hydrophobic compounds—glycyrrhizin, GL (I) and α-tocopherol ester of succinic acid, TSA (II) (both commercially available from Sigma, St. Louis, MO).

2. New cationic lipids with different distances between the polar and hydrophobic moieties—*rac*-[(2,3-dioctadecyloxypropyl)(2-hydroxyethyl)]-dimethylammonium chloride, CLIP1 (III), *rac*-[2-(2,3-dihexadecyloxypro-

[32] R. I. Zhdanov, O. A. Buneeva, O. V. Podobed, N. G. Kutsenko, T. A. Tsvetkova, and T. P. Lavrenova, *Voprosy Med. Khimii* **43**, 212 (1997) (in Russian).

[33] R. I. Zhdanov, O. S. Dyabina, A. A. Moskovtsev, R. S. Khusainova, and G. R. Ivanitskiy, *Rus. J. Cytology* **42**, 280 (2000).

[34] L. V. Kozlov, R. I. Zhdanov, V. A. Guzova, O. V. Podobed, Yu. V. Sviridov, E. V. Bogdanenko, V. I. Schvets, O. S. Dyabina, and A. S. Ermakov, *Cytobios* **106 S1,** 67 (2001).

[35] R. I. Zhdanov, O. V. Podobed, A. S. Borisenko, G. G. Krivtsov, Y. V. Sviridov, A. A. Moskovtsev, V. V. Vlasov, and N. Düzgüneş, *Mol. Therapy* **1** (**5**, part 2), S223 (2000) (abstract).

[36] D. Kovalenko, R. Shafei, A. Borisenko, I. Zelenina, M. Semenova, O. Samuilova, and R. Zhdanov, *Rus. J. of Genetics* **32**, 1299 (1996).

[37] R. I. Zhdanov, A. A. Moskovtsev, A. S. Ermakov, O. V. Podobed, E. V. Bogdanenko, D. D. Bronova, Yu. V. Sviridov, D. N. Konevets, and V. V. Vlasov, *Mol. Therapy* **1**, S341 (2000) (abstract).

[38] E. V. Bogdanenko, Yu. V. Sviridov, A. A. Moskovtsev, and R. I. Zhdanov, *Voprosy Med. Khimii* **46**, 229 (2000) (in Russian).

[39] Y. V. Sviridov, R. I. Zhdanov, O. V. Podobed, T. A. Tsvetkova, I. O. Konstantinov, and E. V. Bogdanenko, *Cytobios* **106 S1**, 7 (2001).

[40] R. I. Zhdanov, F. M. Venanzi, A. Amici., C. Petrelli, P. Moretti, F. Petrelli, O. V. Podobed, T. P. Lavrenova, T. A. Tsvetkova, Y. V. Sviridov, I. O. Konstantinov, and O. V. Bohdanenko, *J. Gene Med.* **1**, (Suppl.), 89 (1998) (abstract).

[41] R. I. Zhdanov, N. G. Morozova, Z. Alshoaibi, M. A. Maslov, G. A. Serebrennikova, O. V. Podobed, A. Haberland, and M. Boettger, *ChemBioChem.* (submitted).

[42] R. I. Zhdanov, Yu. L. Sebyakin, E. V. Bogdanenko, Yu. V. Sviridov, and R. P. Evstigneeva, *ChemBioChem.* (submitted).

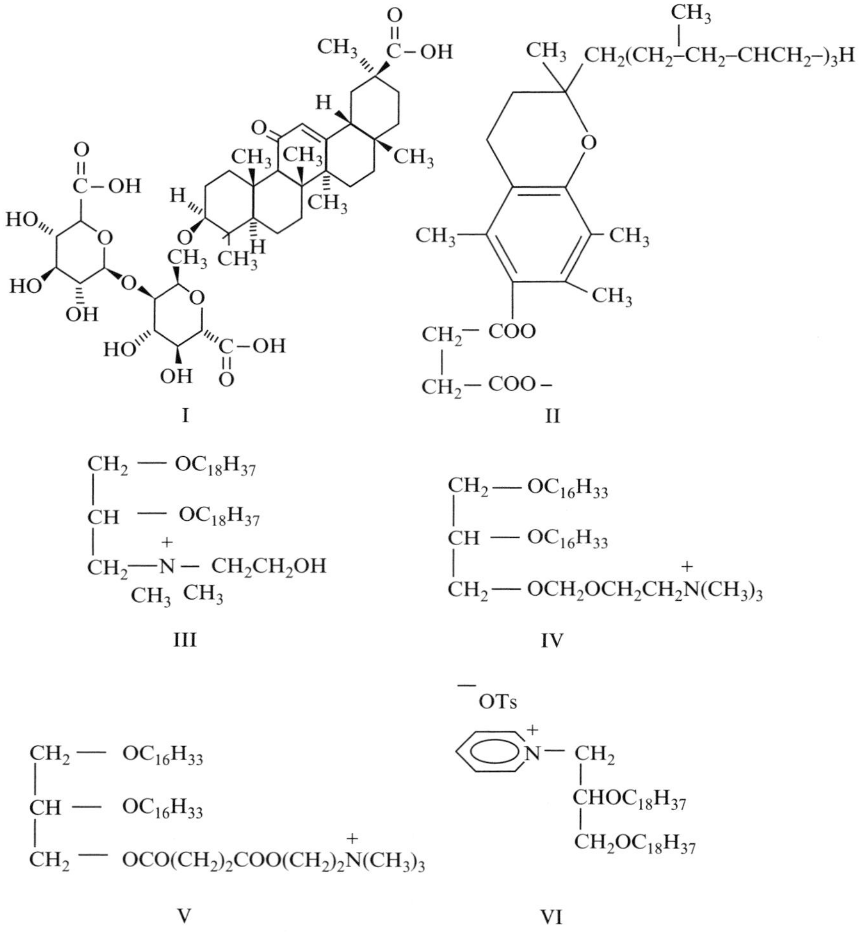

Fig. 1. (*continued*)

pyloxymethyloxy)ethyl] trimethylammonium bromide, CLIP6 (IV), *rac*-[2((2,3-dihexadecyloxypropyl-oxysuccinyloxy) trimethylammonium chloride, CLIP9 (V), and *rac*-N-[2,3-di(octadecyloxy)propyl]pyridinium p-toluenesulfonate, CLIP (VI). Glycerolipids (CLIPs) are synthesized by Prof. G. A. Serebrennikova's group (Moscow)[25–27] from the corresponding dialkylglycerides by treatment with alkyl/arylsulfochloride, followed by substitution with the corresponding amino compound.

3. New hydrophobic oligocations: di-[2-N-(3-aminopropyl)-amino-ethyl]-[2-N-(3-cholesteroyloxycarbonylaminopropyl)-aminoethyl]-

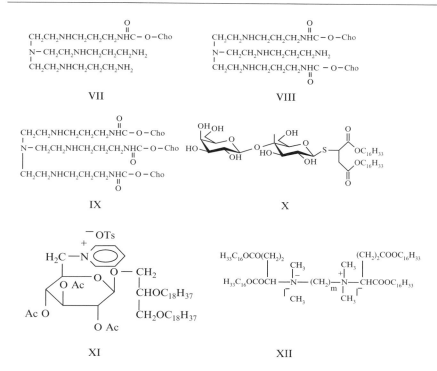

FIG. 1. Chemical structures of new cationic lipids.

amine, MONOCHOLENIM (VII), di-[2-N-(3-cholesteroyloxycarbo-nylaminopropyl)-aminoethyl]-[2-N-(3-aminopropyl)-aminoethyl]amine, DICHOLENIM (VIII), and tri-[2-N-(3-cholestroyloxycarbonylamino-propyl)-aminoethyl]-amine, TRICHOLENIM (IX). The CHOLENIM compounds are synthesized by Prof. V. V. Vlassov and co-workers (Novosibirsk), using acylation of tris[2-N-(3-aminopropyl)-aminoethyl]-amine with, respectively, one (VII), two (VIII), or three (IX) equivalents of cholesterolchloroformate (the latter amine is prepared by addition of tris(2-aminoethyl)amine to acrylonitrile followed by hydrogenation of the product at Renei nickel).[23]

4. New glycolipids: lactosylated lipid—dihexadecyl-2-β-D-thiolactosyl-succinate, LacS (X) synthesized by Dr. Y. L. Sebyakin (Moscow) by addition of thio-lactose to dipalmitoyl ester of succinic ester, followed by treatment of the product with hydrazine hydrate;[30,31,42] and rac-1,2-dioctadecyl-3-O-(2,3,4-tri-O-acetyl-6-deoxy-6-pyridinium-β-D-glucopyra-nosyl)-glycerol p-toluene-sulfonate, GLYCOCLIP (XI), synthesized by

the glycosylation of *rac*-1,2-dioctadecylglycerol[25] with 6-O-(4-toluenesul-fonyl)-2,3,4-tri-O-acetyl-α-D-glucopyranosyl bromide (in the presence of Hg(CN)$_2$ and HgBr$_2$) by Prof. G. A. Serebrennikova and colleagues.[41]

5. Novel dicationic lipids based on glutamic acid derivatives—(N'-dimethyl-N''-dimethyl)-bis-(dihexadecylglutamate)butan or hexane (DEGA, XII, m = 4 or 7).[28,29] These dicationic lipids represent amphiphilic dimeric derivatives of L-glutamic acid, containing as mono-mers two diester and two quaternary ammonium moieties, conjugated to each other (4 to 7 methylene moieties) with an aliphatic spacer, synthesized by Dr. Y. L. Sebyakin.

Egg yolk phosphatidylcholine (PC), cholesterol (both from Fluka, Buchs, Switzerland), cardiolipin (CL) (Khar'kov, Ukraine),[43] and DOPE (Avanti Polar Lipids, Alabaster, AL)[44–46] are used as helper lipids to form liposomes in this work and are reagent grade. Lipids are stored at −80°. DOTAP is obtained from Avanti Polar Lipids. Lipofectin (Invitrogen, USA) and Dosper (Boehringer Mannheim, Mannheim, Germany) are stable suspensions at 1 mg/ml and are used as control transfection agents.

Liposome Preparation and Characterization

Liposomes are prepared by a modification of the reverse-phase evapor-ation procedure (*method A*),[47,48] hydration of a dried lipid thin film followed by sonication (*method B*), or injection of solvent solution of lipids to aqueous solution (*method C*). Glycyrrhizin (I) is used as the ammonium salt and is dissolved in the mixture water/ethanol/ether (1:1:1), but PC and

---

[43] It should be noted that cardiolipin preparations purchased from the Khar'kov's plant for production of phospholipids are quite stable under appropriate conditions of storage and can show one TLC spot even after 20 years (Y. Loesch, Martin -Luther University, Halle, Germany, personal communication).

[44] The oleic acid derivative DOPE may not be an optimal helper lipid for lipofection of cells. Results of computer simulation of DNA-fatty acids complexes suggest that linoleic acid seems to bind to DNA with higher energy gain and form more stable complexes. Therefore, dilinoleylphosphatidylethanolamine appears to be a more effective helper lipid compared to DOPE: V. A. Struchkov, E. P. Dyachkov, N. B. Strazhevskaya, R. I. Zhdanov, and P. N. Dyachkov, *Doklady Biochem. Biophys.* **381,** 554 (2001).

[45] R. I. Zhdanov, E. P. Dyahkov, V. A. Struchkov, N. B. Strazhevskaya, and P. N. Dyachkov, *Rus. Chem. Bull.* **52,** (2003) (in press).

[46] R. I. Zhdanov, E. P. Dyahkov, V. A. Struchkov, N. B. Strazhevskaya, and P. N. Dyachkov, *Doklady Biochem. Biophys.* **400,** (2003) (in press).

[47] F. Szoka, Jr. and D. Papahadjopoulos, *Proc. Natl. Acad. Sci. USA* **75,** 4194 (1978).

[48] N. Düzgüneş, J. Wilschut, K. Hong, R. Fraley, C. Perry, D. S. Friend, T. L. James, and D. Papahadjopoulos, *Biochim. Biophys. Acta* **732,** 289 (1983).

compound II (TS) are well dissolved in chloroform. To prepare liposomes according to *method A*, a diethyl ether or methylene chloride solution containing a mixture of lipids (cationic lipids and helper lipids) is added by drops (at an average rate of about 1 ml/min) to a sterilized saline or dextrose solution (prepared with milliQ water, Millipore, Bedford, MA) in a round-bottom flask at 50–60° while shaking. The mixture is subjected to controlled vacuum in a rotary evaporater, resulting in "boiling." In the case of CHOLENIM compounds, glycolipids X and XI in CH_2Cl_2, this procedure should be repeated until a homogeneous, opalescent solution without phase separation is obtained.

In *method B* the lipid solution in organic solvent (chloroform, ethanol) is evaporated under reduced pressure (at 30–50°) until a dry lipid film is formed. The lipid film is dried *in vacuo* for 2–3 h to remove solvent traces. The lipid film is hydrated in sterilized milliQ water, saline, or a 5% aqueous dextrose solution[64] to give a final concentration of up to 20 mM total lipid. The hydrated film is rotated in a water bath at a temperature above the lipid phase transition temperature, usually at 50° for 30–40 min.[48] After overnight storage in the cold, the suspension is sonicated at low frequency in a bath sonicator (Avanti) for 30 min. In certain cases the liposome mixture is subjected to 4–5 freeze-thawing cycles. To achieve a uniform size distribution, the liposome suspension is then subjected to extrusion through polycarbonate filters with 50 nm pore size (Avestin, Ottawa, Canada).

Liposomes containing the dicationic lipid DEGA are prepared according to *method B*, whereas liposomes containing the oligocationic lipid CHOLENIM are prepared according to *method C*. Compounds VII, VIII, and IX are first mixed with glycerol/water/ethanol (34:40:26), and the mixture is sonicated at 0° for 10 min.[23] Liposomes prepared from mixtures of cationic lipids with helper lipids are usually more stable and effective in transfection, which is the case with liposomes containing the lipids III-VI and the dicationic DEGA.[28]

Quality control. Nitrogen (or argon) is passed through sterilized liposomal preparations for 10 min. Liposomes are stored under these gases at 4° and are used within 1–2 weeks of preparation. Phospholipids are checked by thin layer chromatography for the appearance of *lyso*-forms. Liposomal phospholipids for transfection should be characterized by an oxidative index value (OD_{233}/OD_{215}) no higher than 0.1 (after dissolving the liposomes in ethanol).

Extrusion. In some cases, liposomal preparations are extruded through 100 or 200 nm polycarbonate filters using the Liposofast (Avestin, Canada) extruder, which leads to no loss of lipids. For liposomes extruded through a 100 nm filter, the average size estimated by negative-stain electron microscopy is 85 ± 15 nm. The preparation of dicationic (DEGA) liposomes via extrusion is problematic, leading to loss of lipid material. This is most likely due to the aggregation of liposomes composed of DEGA-7, 11, and 21.

Size. Size estimation is usually carried out by dynamic laser light scattering, using a Nicomp 380 Submicron Particle Sizer (Santa Barbara, CA) and a program for polydispersed systems. Lipofectin preparations (Invitrogen) are characterized by this method as particles with a size distribution around 250 nm, and supercoiled plasmid DNA preparations (5–7 kb) are characterized by a size of 8–10 nm. Micelles of CHOLENIM preparations (VII-IX) are in the range of 100–200 nm in size (assessed by negative stain electron microscopy). The use of atomic force microscopy (Device Nano-Scope IIIa, Digital Instruments, Santa Barbara, CA) for size determination may not be adequate, because this technique usually gives smaller values (due to the drying of liposomes). A number of good quality papers are devoted to the study of the structure of DNA/liposomes complexes, including cationic[49–56] and amphiphilic[57–61] ones. The results of these studies are important for the creation of new liposomal preparations and corresponding lipoplexes, and for a study of their physicochemical properties.

[49] H. Gershon, R. Ghirlando, S. B. Guttman, and A. Minsky, *Biochemistry* **32,** 7143 (1993).

[50] B. Sternberg, F. L. Sorgi, and L. Huang, *FEBS Lett.* **356,** 361 (1994).

[51] J. Gustafsson, G. Arvidson, G. Karlsson, and M. Almgren, *Biochim. Biophys. Acta* **1235,** 305 (1995).

[52] Y. Tarakhovsky, R. Khusainova, A. Gorelov, K. Dawson, and G. Ivanitsky, *FEBS Lett.* **390,** 133 (1996).

[53] J. O. Radler, I. Koltover, T. Salditt, and C. R. Safinya, *Science* **275,** 810 (1997).

[54] S. J. Eastman, C. Siegel, J. Tousignant, A. E. Smith, S. H. Cheng, and R. K. Scheule, *Biochim. Biophys. Acta* **1325,** 41 (1997).

[55] V. Oberle, U. Bakowsky, I. S. Zuhorn, and D. Hoekstra, *Biophys. J.* **79,** 1447 (2000).

[56] A. E. Regelin, S. Fankhaenel, L. Gurtesch, C. Prinz, G. von Kiedrowski, and U. Massing, *Biochim. Biophys. Acta* **1464,** 151 (2000).

[57] R. I. Zhdanov and V. V. Kuvichkin, *in* "New Developments in Lipid-Protein Interactions and Receptor Function" (J. A. Gustafsson and K. W. A. Wirtz, eds.), NATO ASI Series Vol. 246, p. 249. Plenum, New York, 1993.

[58] R. I. Zhdanov, L. A. Volkova, V. V. Kuvichkin, and A. I. Petrov, *Appl. Magn. Reson.* **7,** 115 (1994).

[59] R. I. Zhdanov, L. A. Volkova, L. G. Artemova, and V. V. Rodin, *Appl. Magn. Reson.* **7,** 146 (1994).

[60] V. V. Kuvichkin, S. M. Kuznetsova, V. I. Emel'yanenko, A. I. Petrov, and R. I. Zhdanov, *Biophysics (Moscow)* **44,** 630 (1999).

[61] F. M. Venanci, R. Zhdanov, C. Petrelli, P. Moretti, A. Amici, and F. Petrelli, *in* "Liposomes, Nineties and Beyond. Book of Abstracts" G. Gregoriadis and A. Florence (eds), Abstracts 122 and 123. London, 1993.

Reporter Genes and DNA

Plasmid DNA (pCMV-SPORT-β-gal (7.2 kb) (GIBCO-BRL, Grand Island, NY), pEGFP-N1 (4.7 kb) (Clontech, Palo Alto, CA), and pCSEAP and pQE-LacZ (a gift from Dr. A. Borisenko, Moscow) contain 90% of supercoiled form (based on agarose gel electrophoresis), and the D_{260}/D_{280} ratio is 2.5. pCMV-Luc encodes the American firefly luciferase gene under the control of the cytomegalovirus promoter (a gift of Dr. M. Boettger, MDC, Berlin). The plasmids are grown in preparative scale in transformed *E. coli* culture and isolated, purified, and analyzed according to common procedures.[62,63] Ultracentrifugation in a CsCl gradient is used at the final stages of purification.

[^{14}C]-tymidine labeled DNA ([^{14}C]-DNA) (7,000–10,000 cpm/μg) is isolated from rat Gasser ganglion neuroma cells (RGGN-1), which are grown in a medium containing [^{14}C]-tymidine (56 mCi/mmol) and used at a concentration of 5 μCi/ml, using the conventional phenolic method.[62,63] [^{14}C]-DNA is fragmented for 15 min using impulse sonication (with 30 sec rest after every minute of sonication at $0°$), resulting in 4–6 kb [^{14}C]-DNA fragments (determined by electrophoresis).

Lipoplex Preparation

To form lipid-DNA complexes suitable for gene delivery, a solution of plasmid or marker[^{14}C]-DNA is added to one of the liposome suspensions (I-XII) at various ratios (usually at DNA/liposome ratio 1:5, w/w). The mixture is shaken carefully and incubated for 30 min at room temperature. For preparing pH-sensitive or neutral liposomes a 2 M solution of magnesium or calcium chloride is added to the incubation mixture up to 15–50 mM.[21,32,33] To transfect cell cultures the necessary amount of medium with serum (or without) is added to the mixture, and 9% sodium chloride aqueous solution is added until the 0.9% concentration required for administration to animals is obtained.

Lipoplex preparations based on lipofectin and other common cationic lipids usually have a diameter of approximately 400 nm.[64] In general, but not without exceptions, the larger the lipoplex size (up to 400 nm) the

[62] T. Maniatis, E. F. Fritsch, J. Sambrook, "Molecular Cloning" Vol. 1, Cold Spring Harbor Laboratory Press, Cold Spring Harbor, New York, 1997.

[63] F. M. Ausubel, R. Brent, R. E. Kingston, D. D. Moore, I. G. Seidman, J. A. Smith, and K. Struhl, *in* "Current Protocols in Molecular Biology", Suppl. 17, Unit 9.4.1.13. Wiley, New York, 1991.

[64] N. S. Templeton, D. D. Lasic, P. M. Frederik, H. H. Strey, D. D. Roberts, and G. N. Pavlakis, *Nat. Biotechnol.* **15,** 647 (1997).

higher the efficacy of *in vitro* transfection, and the smaller the lipoplexes (up to 50 nm) the higher the efficacy of *in vivo* transfection.[64]

CHOLENIM (VII to IX)-based lipoplexes are prepared by adding a CHOLENIM solution in a glycerol/water/ethanol (34:40:26) mixture (micellar preparation) to a plasmid DNA solution [(DNA/lipid ratio, 1:3, 1:1 or 3:1) (30 min incubation) (*method C*)], followed by extrusion of the lipoplex preparation (only for *in vivo* experiments) through 400 nm, 200 nm, and 100 nm filters. [64][14C]-DNA–based lipoplexes are administered by a single intraperitoneal injection (either in *vena porta* or in renal artery) of 20–50 μg [14C]-thymidine DNA in 200 μl saline.

While preparing DEGA liposome-based lipoplexes, it was found that the incubation time used for mixing liposomes and plasmid DNA could be an important parameter with respect to the efficiency of transfection. An optimal incubation time for many cases is up to 2 h. In addition to the size, the shape of lipoplex particles is important for efficient transfection (Fig. 2). DEGA-4 lipoplexes have a toroid shape, as visualized by atomic force microscopy, while DEGA-7 lipoplexes appear as rods (Fig. 2). Calculations indicate that the micellar phase is predominant in the former case and the hexagonal phase in the latter. The shape of lipoplexes appears to be important in lipofection (*vide infra*).

The required lipoplexes are prepared from liposomes I-XII and reporter plasmids. Prior to employing the lipoplexes for transfection, it is necessary to study the influence of the corresponding liposomes on cell line growth and on DNA synthesis in these cells within 24 or 48 h using conventional techniques. If the liposomes at the concentrations used only slightly influence (if at all) the growth of particular cell lines, they are then used in transfection experiments. We have used liposomes formed from lipids III

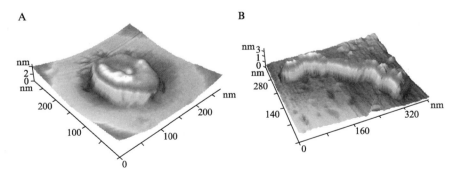

Fig. 2. Atomic force microscopy-images of lipoplexes composed of (A) DEGA-3 liposomes and pEGFP-N1, and (B) DEGA-7 liposomes and pCMV-SPORT-β-gal. The lipid/DNA ratio was 7.5:1 (w/w). (See color insert.)

to IX for transfection of HeLa and RGGN-1 cells by the alkaline phosphatase gene (pCSEAP). The RGGN-1 and PC-12 cells have been transfected
by the β-galactosidase gene (plasmids pQE-LacZ and pCMV-SPORT-β-
gal), using liposomes formed from agents VII–IX. Lipofection of CHO
cells has been achieved using liposomes based on the glycocationic lipid
XI, in various combinations with helper lipids and the CMV-Luc plasmid.
The GFP gene (pEGFP-N1 plasmid) has been used for testing gene transfer efficiency against a number of cultured cells (293, MCF-7) with the aid
of liposomes formed fom dicationic DEGA lipids. The protocols for these
experiment are given later.

Cell Culture

Cell lines are cultured in medium supplemented with 10% fetal bovine
serum and 50 μg/ml of gentamicin, in a CO_2-incubator (Flow, UK) under
5% of carbon dioxide at 37°. For HeLa, L929, and 293 (human embryonic
kidney) cells, DMEM is used as the medium, and for SKOV-3, MCF-7
(human mammary gland carcinoma), RGGN-1 (rat Gasserian ganglion
neuroma[65]), PC-12 (rat adrenal pheochromocytoma) RPMI-1640 is
employed.[66]

Cytotoxicity of Lipids and Liposomes: Influence on Cell Growth

Potential cytotoxicity of liposomes is determined by two common
techniques: standard MTT test using PC-12 or MCF-7 cell lines[67] after 48 h
incubation in the presence of liposome preparations (*method D*), or by
testing the influence of liposomes on cell growth 24 or 48 h after their addition to the culture medium, followed by counting living cells (*method E*).[23,27]

Method D

Method D is based on the conversion of 3-[(4,5-dimethylthiazol-2-yl)-2,
5,-diphenyltetrazolium bromide] (MTT) (Sigma, St. Louis, MO) to a dark
blue dye, formasan, in living cells as a result of the action of mitochondrial
dehydrogenases on the reagent. MTT is dissolved in phosphate-buffered
saline (PBS) or other medium without indicator at a concentration of
5 mg/ml, sterilized by filtration (through a 0.22 μm filter) and stored at
4°. Phenol red-free 199 medium (Sigma) is used to carry out the reaction.

[65] A. P. Avtsin, L. I. Kondakova, A. S. Khalanskii, G. P. Polyakova, and N. V. Chudinovskaya, *Cytology (U.S.S.R.)* **31,** 97 (1989).

[66] R. L. P. Adams. "Cell Culture for Biochemists." Elsevier/North-Holland, Amsterdam, 1980.

[67] H. Tada, O. Shito, K. Kuroshima, M. Koyama, and K. Tsukamoto, *J. Immunol. Methods* **13,** 157 (1986).

The culture medium is removed from 96-well plates, and cell monolayers are washed with Hank's balanced salt solution (HBSS), 100 μl of 199 medium and 50 μl of 0.5% MTT solution are added and the mixture is incubated for 3 h at 37°. Formasan crystals formed are then dissolved for 15–30 min by adding 150 μl of 25% aqueous SDS solution to every well. The mixture is pipetted a few times to obtain a homogeneous mixture, and the optical density at 540 nm is measured.

Method E

Method E is based on the binding of crystal violet dye to total cellular proteins. Dye binding is measured spectrophotometrically after incubation and removal of non-bound crystal violet by washing. Cells in 24-well plates (U-bottomed) (2×10^5 cells/well) are stained with 0.2% crystal violet in 2% aqueous ethanol. Cells are then washed with water, and the dye is eluted with 10% acetic acid. The amount of cells is calculated from the optical density at 595 nm,[68] using a calibration curve based on the number of cells counted by microscopy. *Method D* is found to be more precise when making large numbers of measurements.

The Effect of Lipids and Liposomes on DNA Synthesis

The effect of liposomal preparations on DNA synthesis is determined by the incorporation of an isotope label (^{14}C-thymidine) into growing cells (RGGN-1) 24, 48, or 72 h after addition of liposomes to the culture medium.[23,27,66] Two μCi of ^{14}C-thymidine (56 mCi/mmol) («IZOTOP», Russia) are added to the cells in a 24-well plate. After 4 h, cells are washed with cold Hank's medium and then fixed with ethanol/acetic acid mixture (9:1) overnight to remove the unbound isotope label. Then cells are subjected to lysis overnight with 0.3 M KOH. Following neutralization, Bray's solution is added at a 1:5 volume ratio, the radioactivity is counted using a Rackbeta scintillation counter, and results are expressed in cpm per 10^5 cells (values could be also expressed as dpm).

Interaction of Liposomes or Lipoplexes with Cells

The study of affinity of liposomes/lipoplexes to cells and cell membrane and their accumulation by cells is carried out with fluorescence microscopy using the lipophilic fluorescent dye, coumarin, according to the following procedure. Cells are removed from the cultural flask, seeded on coverslips in 6-well plate with an initial density 2.10^5 cells per well, and then incubated for 24 h in common growth medium (37°, 5% CO_2). Cells on coverslips are

[68] A. E. Medvedev, B. B. Fuchs, and A. L. Rachmilevich, *Biomed. Sci.* **1**, 261 (1991).

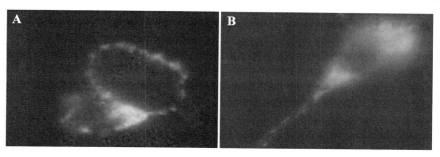

FIG. 3. Interaction of coumarin-labeled DEGA-3 liposomes with MCF-7 cells. The incubation times were (A) 5 min (fluorescence at the cell perimeter) and (B) 15 min (fluorescence at the cell perimeter, with accumulation in cytoplasmic compartments) (1250×). (See color insert.)

washed twice with cultural medium (without serum), and transferred (under non-sterile conditions) to microscope slides with a drop of growth/ culture medium containing fluorescent-labeled liposomes or lipoplexes without serum on the surface. Images are recorded by photography or time-lapse video microscopy using a Zeiss Axioskop 20 equipped with an FITC filter (Fig. 3).

Lipofection Procedures

Gene transfer into cultured eukaryotic cells is carried out using DNA-liposome complexes according to a standard protocol.[69] Cells are seeded in the corresponding growth medium (DMEM, RPMI 1640, OPTIMEM) containing 10% fetal bovine serum at 10^4, 5×10^4, and 10^5 cells per well in 96-, 48-, and 24-well plates, respectively.[69]

The culture medium is removed after 24 h, the cell monolayer is washed with medium (without serum), and lipoplexes suspended in the same medium are added. Lipofection is usually carried out in medium (without serum) to avoid complexation between lipoplexes and plasma proteins and thus decrease transfection efficacy, although the presence of serum does not influence lipofection in some cases. Cells are incubated for 4–5 h at 37° and 5% CO_2, and an equal volume of growth medium containing 20% of fetal bovine serum is added, followed by incubation for 3 days. In the case of suspension cultures (Jurkat), cells are plated at twice the initial density of adherent cells (e.g., 2×10^4 cells per well of a 96-well

[69] O. V. Abakumova, O. V. Podobed, T. A. Tsvetkova, I. V. Yakusheva, T. A. Moskvitina, L. I. Kondakova, D. C. Navasardyantz, and A. E. Medvedev, *J. Neural. Transmit. (Suppl.)* **52**, 87 (1998).

plate). The subsequent stages of work do not differ from the experiments with adherent cell cultures.

Expression of Alkaline Phosphatase in Cells

Lipoplexes [e.g., pCSEAP (2 μg) mixed with PC (65 mol. %)/CLIP1, CLIP6, or CLIP9 (10 mol.%)/lipid III or IV (25 mol. %) (20 μg)] are added to each well (2 × 10^5 RGGN-1 cells). Cells are incubated for 5 h (37°, 5% CO_2), then an equal volume of DMEM medium with 20% of serum is added, and cells are incubated for 3 days.[23,27]

Alkaline phosphatase expression is determined colorimetrically using 4-nitrophenylphosphate-bis-(2-amino, 2-ethyl-1,3–propanediol) (Serva, Heidelberg, Germany) as a substrate. The activity of the enzyme is expressed in relative units, 1 RU being defined as an increase of 0.1 OD units at 405 nm. The activity is normalized to the number of cells in each well and determined after staining with crystal violet.

Expression of β-Galactosidase in Cells Determined with the X-Gal Reagent

The expression of the β-galactosidase gene is measured by the production of a blue colored product following incubation with X-gal reagent, 5-bromo, and 4-chloroindolyl-β-D-galactopyranoside.[70,71] The medium is removed from wells, the cell monolayers are washed with Hank's medium and fixed for 4 min at 4° with 0.25% glutaraldehyde and 2% formaldehyde in PBS buffer (0.14 M NaCl, 0.02 M potassium-phosphate buffer, pH 7.4). The monolayers are washed again with Hank's medium, the developing solution is added, and the cells are incubated for 12–36 h at 37°. The developing solution is prepared immediately before use in PBS buffer and contains 5 mM potassium ferricyanide, 5 mM potassium ferrocyanide, 2 mM magnesium chloride, and 0.2% X-gal reagent (the reagent is first dissolved in a small volume of dimethylformamide). After removing the colored solution the cells are washed with PBS and the percentage of blue-coloured cells is assessed.

Expression of β-Galactosidase in Cells Determined with the N-Gal Reagent

Chlorophenol red-β-D-galactopyranoside (N-gal) can be used as an alternative for substrate for β-galactosidase.[70] Cells are seeded at a density of 5 × 10^4 per well in 96-well plates. Following transfection and an appropriate

[70] J. H. Felgner, R. Kumar, C. N. Sridhar, C. J. Wheeler, Y. J. Tasi, R. Border, P. Ramsey, M. Martin, and P. Felgner, *J. Biol. Chem.* **269**, 2550 (1994).
[71] D. J. Weiss, D. Liggitt, and J. G. Clark, *Hum. Gene Ther.* **8**, 1545 (1997).

incubation period, the medium is removed without disrupting the cell monolayer, and lysis solution (0.1% Triton X-100, 0.25 M Tris-HCl, pH 8.0) is added to every well. Cells are then frozen at $-70°$ and thawed at room temperature for 10 min. The reaction is carried out in phosphate buffer (pH 8.0) containing N-gal reagent (1 mg/ml), 1 mM magnesium sulfate, 10 mM potassium chloride, 50 mM mercaptoethanol, and 0.5% bovine serum albumin at $37°$ until color is developed (15 min to 24 h). The enzyme in samples is quantified from a calibration curve generated from a standard enzyme sample (Sigma, cat. no. 9031-11-2), subtracting the amount of enzyme in control samples without the N-gal reagent.

Expression of Luciferase in Cells[41,72–74]

The plasmid pCMV-Luc and H1 histone are mixed in buffer at a protein to DNA weight ratio of 1:20, in the presence of CaCl$_2$ (final concentration 2 mM) at room temperature. The protein is added gradually to the DNA under gentle agitation in order to avoid precipitation. After mixing, the samples are shaken for at least 15 min. Higher ratios do not result in higher transfection efficiency. The final complex volume is 100 μl. About 2×10^5 CHO cells per well in 24-well plates (Falcon) are grown in RPMI 1640 medium with L-glutamine (Serva) supplemented with 10% FBS. The culture medium is removed before transfection and the cells are washed twice with RPMI medium. The protein-DNA complexes are first added to 0.9 ml medium containing 10% FBS and 8 mM CaCl$_2$. This transfection mixture is then added to the cells. Some experiments are performed in DMEM medium (Invitrogen/Gibco, Carlsbad, CA) because of its higher inherent Ca^{2+} concentration compared with RPMI (1.79 and 0.42 mM, respectively). After a 4 h incubation under 4.5% CO$_2$ at $37°$, the transfection mixture is removed and replaced by 1 ml RPMI 1640 containing 10% FBS. After incubation of about 30 h the cells are harvested and the luciferase activity is measured by the Promega assay system using a luminometer (Berthold Luma). Transfection with Lipofectin is performed according to the manufacturers recommendations. Two μg DNA and 3 μl Lipofectin are both diluted separately in 100 μl Optimem medium. After gently mixing both solutions at room temperature and incubation for 15 min, the mixture is diluted with 0.8 ml Optimem medium and is added to the washed cells.

[72] S. V. Zaitsev, A. Haberland, A. Otto, V. I. Vorobyev, H. Haller, and M. Bottger, *Gene Ther.* **4**, 586 (1997).

[73] M. Bottger, S. V. Zaitsev, A. Otto, A. Haberland, and V. I. Vorobyev, *Biochim. Biophys. Acta* **1395**, 78 (1998).

[74] A. Haberland, T. Knaus, S. V. Zaitsev, R. Stahn, A. Mistry, C. Coutelle, H. Haller, M. Bottger, *Biochim. Biophys. Acta* **1445**, 21 (1999).

Expression of Green Fluorescent Protein in Cells

To test the lipofection efficiency, light, and fluorescent microscopy of transfected cells is carried out by using a Zeiss AXIOSKOP-20 fluorescent microscope (Jena, Germany), equipped with an image analysis system and FITC/TRITC filters for double fluorescent labeling. Images are processed by a numerical system based on a Sony color-integrating 3CCD videocamera and a workstation for non-linear videomontage (P4, 1.5 GHz, FRGR Matrox Meteor, Dorval, Quebec, Canada). KS 100 (Zeiss, Jena, Germany), Adobe Photoshop, and Ulead Media Studio programs are used for this purpose.

Transfection efficiency is determined by using fluorescence microscopy (in our case, for the plasmid pEGFP-N1) by counting the green fluorescent cells 24–48 h after lipofection. Transfection efficiency is counted as the percentage of cells that express fluorescent protein (Fig. 4). Plasmids with genes expressing red and blue fluorescent protein are also available.

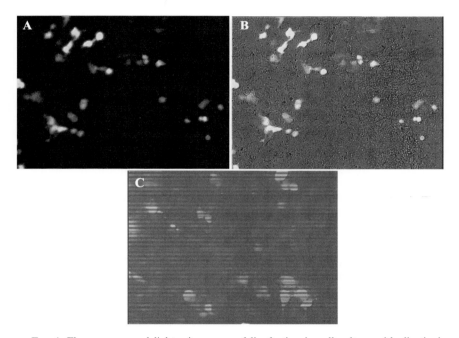

FIG. 4. Fluorescence and light microscopy of lipofection in cell culture with dicationic DEGA-3 and DEGA-7 lipoplexes. (A) 293 cells transfected with DEGA-3/pEGFP-N1 lipoplexes (fluorescence microscopy, 300×); (B) 293 cells transfected with DEGA-3/pCMV-SPORT-β-gal lipoplexes (light microscopy, 300×); (C) 293 cells transfected with Lipofectin[R]/pEGFP-N1 lipoplexes (fluorescence microscopy, 300×). (See color insert.)

Transfection Efficiency Determined by Flow Cytometry

Flow cytometric analysis is carried out by using a FACScan or FACS-Calibur (Becton Dickinson, Franklin Lakes, NJ) instrument, equipped with an air-cooled argon laser (wavelength: 488 nm) and the CELL Quest program.[75] MCF-7 or SKOV-3 cells transfected with a plasmid expressing GFP are removed from their substrate 24 h after lipofection, using Trypsin-EDTA or non-enzymatic cell dissociation solution (Sigma), and are washed twice with PBS buffer. Transfected human macrophages are washed and detached from plastic by adding cell dissociation buffer (GIBCO/Invitrogen, Carlsbad, CA). Then they are mixed with DME-HG medium containing 1% FBS and 1 μg/ml propidium iodide (Sigma) to assess cell viability. The GFP fluorescence is detected using a 530 nm band-pass filter (30 nm band width) and propidium iodide is monitored using a 630 nm band-pass filter (22 nm band width).

Cells remain viable during the measurement. Forward laser light scattering is useful in estimating the size and shape of cells, while side scattering at 90° to the laser beam indicates structural peculiarities of various cell populations. The corresponding discrimination window (gate) is set in the cytogram in which the main cell pool is selected to avoid small (debris) and big (aggregates) particles, based on the forward and side scatter data. The fluorescence intensity is given by the X axis of the histogram, while the number of events (number of cells) is given by the Y axis. From 5 to 10 thousand events are accumulated for every sample, depending on the aims of the study. The wavelength for green fluorescence (FITC) is set in the FL1 channel (515–545 nm).

Measurements are carried out with a low flow speed. The percentage of transfected cells is counted based on the ratio of the squares of two peaks in the FL1 channel, which reflects: (1) the presence of transfected cells in the population that have synthesized GFP and (2) non-transfected cells with autofluorescence only. For example, the efficiency of gene transfer in MCF-7 cells, using lipoplexes based on DEGA (XII, m = 3) liposomes was 19 ± 2% and gene transfer efficiency for lipofectin was 27 ± 2%.

Dependence of Lipofection Efficiency on Lipid Structure (Summary and General Comments)

We have described the protocols of lipofection of cultured cells and various methods for testing the efficiency of gene transfer, using lipoplexes based on lipids I–IX, XI, and XII. Each of the reporter genes we have

[75] S. Simões, V. Slepushkin, E. Pretzer, P. Dazin, R. Gaspar, M. C. Pedroso de Lima, and N. Düzgüneş, *J. Leukoc. Biol.* **65,** 270 (1999).

described has particular advantages. The plasmid encoding firefly luciferase appears to be the most convenient reporter for the majority of experiments to evaluate the efficiency of transfection. The use of plasmids pCSEAP and pCMV-SPORT-β-gal, which encode alkaline phosphatase and β-galactosidase, respectively, is less efficient because the cells contain such endogenous enzymes. Bacterial alkaline phosphatase and the enzyme from eukaryotes differ in their pH optimum. In the case of pCMV-SPORT-β-gal there is a noticeable background of colored indigogenic reaction. Plasmids containing the gene for fluorescent proteins (green or red), (in our case, pEGFP-N1), are convenient from the viewpoint of subsequent testing of expression: the availability of a fluorescent microscope allows one to rapidly test the results of transfection. The possibility of using flow cytometry to detect GFP expression is another major advantage.

The effect of differences in lipid structure on the efficacy of gene transfer into eukaryotic cells *in vitro* and *in vivo* have been studied.[70,76–78] The results of transfection of a variety of cell lines using lipoplexes containing pCSEAP, pCMV-SPORT-β-gal, pFGFP-N1 (or pCMV-Luc), and liposomes composed of novel cationic lipids (III–VI and VII–IX), glycolipids (XI), and helper lipids (PC, DOPE, DOTMA), are presented in Table I.

The efficiency of pH-sensitive liposomes formed of lipids I or II and PC as a helper lipid in delivering the Lac Z plasmid to RGGN-1 cells is not high and comparable with that of Ca-phosphate transfection. We should note that RGGN-1 cells are quite difficult to transfect with any of the non-viral systems used. A series of liposomes formed of PC and cationic lipids III–V, which represent a molecular ruler with varying distances between the hydrophobic and polar moieties, were used to transfect HeLa cells with pCSEAP, and RGGN-1 cells with pQE-LacZ. The efficiency of these vectors varied for different cell lines. Liposomes formed of CLIP IV containing a spacer of 6 chemical bonds, were the most effective against HeLa cells. The efficiency of these systems was not very high because PC was used as a helper lipid in the study.

A new group of hydrophobic oligocations (CHOLENIMs)[23] exhibit higher transfection efficiency. DICHOLENIM VIII appears to be relatively more effective in gene transfer to RGGN-1 cells, compared to CHOLENIMs VII and IX. The ability of lipoplexes prepared from MONO-, DI-, and TRICHOLENIM oligocations (VII–IX) to deliver a functional (reporter) gene to PC-12 cells was studied in detail using

[76] P. L. Katsel and R. J. Greenstein, *Biotechnol. Ann. Rev.* **5,** 197 (2000).

[77] N. Düzgüneş, S. Simões, P. Pires, and M. C. Pedroso de Lima, *in* "Polymeric Biomaterials" (S. Dumitriu, ed.), p. 943, Marcel Dekker, New York, 2002.

[78] Y. Liu, L. C. Mounkes, H. D. Liggitt, C. S. Brown, I. Solodin, T. D. Heath, and R. J. Debs, *Nat. Biotechnol.* **15,** 167 (1997).

pCMV-SPORT-β-gal. Table I presents the results of the study of dependence of the gene transfer efficiency in PC-12 cells on the DNA:oligocation ratio. For the MONOCHOLENIM-containing lipoplexes the highest efficiency of expression (187 pg/10^5 cells) is attained at a DNA/MONOCHOLENIM ratio of 1:0.5. With DICHOLENIM the highest expression (113 pg/10^5 cells) is also exhibited at the same ratio. The ratio 1:1.5 produces the lowest level of expression (14 pg/10^5). The efficiency of transfection depends on the transfected cell type. The results of transfection of the same plasmid against RGGN-1 cells by the use of CHOLENIM-containing lipoplexes differ from those obtained for PC-12 cells. The highest efficiency of reporter gene expression in these cells is substantially lower than with PC-12. Thus, with the DNA/MONOCHOLENIM lipoplexes at the ratios of 2:1 and 1:1, the gene expression levels are 32 and 35 pg/10^5 cells, respectively.

Liposomes composed of a cationic glycolipid containing glucose and helper lipid are not cytotoxic and have little influence on DNA synthesis in rat glial cells. Transfer of pCMV-Luc with cationic glycoliposomal formulations (GLYCOCLIP/DOPE) into CHO cells is much more effective than with the corresponding cationic liposomes V and is higher in combination with cholesteroyl oligocation (see Table I). Transfection of CHO cells with MONOCHOLENIM VII/DOPE/glycocationic lipid XI at the DNA/cationic lipid ratio of 1:1.5 mediates relatively high gene expression compared to lipoplexes containing cationic lipid VI or compound VII. The levels are also comparable to that obtained with Lipofectin and Dosper. The increased proportion of DOPE (with the same DNA/cationic lipid ratio of 1:1.5) had no significant influence on transfection efficiency. When compared with the results with MONOCHOLENIM alone, there was a decrease in gene expression.

The efficiency of gene transfer to 293 and MCF-7 cells with a series of liposomes formed from the new dicationic DEGA lipids [with different distances between 2 cationic moieties (n = 3 or 4)] is relatively high and comparable with that of Lipofectin. It is interesting to note that DEGA-7 (n = 7), containing 7 methylene groups between two cationic moieties, is not active in gene transfer, which could be related to the size and shape of the corresponding lipoplex (see Fig. 2). The transfection efficiency of lactosylated lipid X in PC-12 cells is comparable to that of CHOLENIM. The lactose moiety of lipid X may facilitate targeted gene delivery.

In Vivo Targeted Gene/DNA Delivery

Liposomes formed of pH-sensitive (I, II) cationic (VII–IX), and glyco (X) lipids are used for delivery in vivo (in mice) of β-galactosidase, GFP

TABLE I

TRANSFECTION ACTIVITY OF LIPOPLEXES BASED ON NOVEL CATIONIC AND GLYCOLIPIDS

Lipid/liposomal preparation	Charge of the lipid	Method of preparation	Plasmid	Lipid/DNA ratio, w/w	Cell line	Efficiency of lipofection
I	−3	A	pQE-Lac Z	5	RGGN-1	98%
II	−1	A	pQE-Lac Z	5	RGGN-1	50%
Control		Ca-phosphate	pQE-Lac Z	5	RRGN-1	100%
Control		Ca-phosphate	pCSEAP	5	HeLa	100%
III§	+1	B	pCSEAP	5	HeLa	50%
IV§	+1	B	pCSEAP	5	HeLa	96%
V§	+1	B	pCSEAP	5	HeLa	65%
VII	+2	C	pCMV-SPORT-β-gal (N-gal)	3; 1; 0.33	PC-12	105 ng/10^6 cells 187 36
VII	+2	C	pCSEAP	1.5	RGGN-1	0.4 RLU
VIII	+1	C	pCMV-SPORT-β-gal (N-gal)	3; 1; 0.33	PC-12	100 113 14
VIII	+1	C	pCSEAP	1.5	RGGN-1	1 RLU
IX	0	C	pCMV-SPORT-β-gal (N-gal)	3; 1; 0.33	PC-12	56.6 31.3 75
IX	0	C	pCSEAP	1.5	RGGN-1	0.45 RLU
VII/XI	+1.5	A	pCMV-Luc	5	CHO	1.10^8 RLU
VI#	+1	A	pCMV-Luc	5	CHO	3.10^3
XI#	+1	A	pCMV-Luc	5	CHO	5.10^6
Control	+1	Lipofectin	pCMV-Luc	5	CHO	2.10^8
Control		Dosper	pCMV-Luc	5	CHO	2.10^8
XIIa¶	+2	B	pEGFP-N1	2.5	293, MCF-7	19%
XIIb¶	+2	B	pEGFP-N1	2.5	293, MCF-7	0
Control	+1	Lipofectin	pEGFP-N1	5	293, MCF-7	25%
X	0	A	pCMV-SPORT-β-gal (N-gal)	5	PC-12	92 ng/10^6 cells

§ Efficiency compared to Ca-phosphate transfection (100%) as a control (Relative Units./10^6 cells).

In relative luminescence units.

¶ Number of transgene-expressing (fluorescent) cells, compared to Lipofectin efficiency using either direct counting of transfected cells or a flow cytometer.

reporter genes, and [^{14}C]-thymidine DNA. Most of the papers devoted to *in vivo* gene delivery report studies of gene expression following intravenous administration via the tail vein in animals. This method has its own peculiarities and has not been attempted in human experimentation.[79] Lipoplex administration into the portal vein or the renal artery are of interest as alternative methods and will be described in this section. Intraperitoneal injection of liposomes is not discussed in detail in this chapter because it is carried out according to commonly used protocols.[80]

Animal Experiments

Experiments to study the tropism of ^{14}C-DNA/polycationic complexes and gene delivery *in vivo* are carried out with male ICR mice (inbred stock) or male NMRI mice (outbred stock), which are obtained from the "Svetlye Gory" nursery (Russian Academy of Medical Sciences). Mice that are 10–12 weeks old are maintained in plastic cages, each containing 5 animals under conventional conditions. The animals are given free access to food such as full-ration granulated feed. The average body weight of one animal is 30–40 g. Mice are euthanized (or killed) by cervical dislocation according to the instructions of the American Physiological Society, 1995. Lipoplexes in saline are injected into mice intraperitoneally or through the liver portal vein.

Avertin Preparation for Anesthesia of Mice

Safe and effective narcosis for general anesthesia is achieved by using the following two components: the solid 2,2,2-tribromoethanol (Avertin, Aldrich, 25 μg) is dissolved in the second component tert-amyl alcohol (155 μl). Before injection, 100 μl of this solution is diluted into 5 ml isotonic saline kept in a boiling water bath ("Mice *in vivo* Protocol," University of Colorado Medical Center).

Tissue Distribution after Administration of [^{14}C] DNA

The tissue distribution of [^{14}C]-DNA in male NMRI mice organs is studied after a single intraperitoneal, intravenous, or intraarterial injection of 20–50 μg [^{14}C]-thymidine labeled DNA in 100–200 μl of saline. The mice are euthanized 24 h after injection, and the inner organs are removed, homogenized, and undergo alkaline lysis in 0.6 N KOH at 37° for 24 h with

[79] G. Cichon and M. Strauss, *in* "Concepts in Gene Therapy" (M. Strauss and J. A. Barrenger, eds), p. 268, De Gruyter, Berlin, 1997.
[80] M. Monk, ed. "Mammalian Development: A Practical Approach," IRL Press, Oxford, 1997.

shaking. The lysate is neutralized with 0.6 N $HClO_4$. The radioactivity of the acid precipitated fraction is detected by liquid scintillation using Bray toluene scintillation liquid (scintillation solution—10^6) by means of a Rack-beta scintillation counter. Detection of DNA content in the organs of animals is carried out according to Schmidt and Thanhauser's general technique modified by Fleck and Munro.[81] The results are presented as cpm/mg DNA and also as a fraction of the radioactive label in the given organ relative to the total radioactivity recorded in all organs studied, normalized to the mass of the organ: (cpm in organ/total cpM recorded)/organ mass. Most data are given as the percentage of the total $[^{14}C]$-thymidine labeled DNA detected in the organs observed (% equivalent dose/g). In all cases the results are the means of 3 or 4 series of independent experiments, together with the standard deviations.

Administration of Lipoplexes via the Portal Vein[79]

For injections of lipoplexes formed of either lipids VII–X or $[^{14}C]$-DNA into the liver portal vein Avertin is injected intraperitoneally or subcutaneously at a dosage noted by the manufacturer. The standard dosage used in our laboratory is 20 μl of the solution described per g of body mass. This method has proven to be the most convenient, particularly since it does not cause serious complications even if accidentally overdosed. After the mouse falls asleep, it is placed on its back on the operating table and its legs are constricted with Scotch tape. The fur is removed along the white midline of the belly using pincers and the bare skin is sterilized with 70% aqueous ethanol. The small transversal fold of the skin in the middle of abdominal white line of the belly is held with the pincers, and an incision is made that does not damage the muscles of the abdominal wall. The incision is then enlarged upward (to the swordlike appendix of the thoracic bone and downwards) to the additional sex gland (of a male) or bladder (of a female). The same cut is made along the mid-line of the muscular wall. Then the mouse is covered with gauze over the fur and over the operating field on the surface of the table to the right and left of the body. The cut on the gauze must correspond to the cut on the abdominal wall. The intestines are removed carefully and placed on the saline-soaked operational field by using pincers, the branches of which are wrapped with gauze to prevent damage to the intestines. The intestines are positioned in such a way that the portal vein area appears outside. Then the intestines are placed on the right-hand side and are turned from left to right because the portal vein is positioned within/inside the mesentery between the loops of the

[81] A. Fleck and N. H. Munro, *Biochim. Biophys. Acta* **55,** 571 (1962).

intestines. The portal vein is the main vessel leading to the liver. To prevent the intestines from "running away" back to the abdominal cavity, they are slightly fixed with the pincers in the turned position.

The lipoplexes in saline are taken up into a glass, sterilized capillary (flame-stretched glass with a sharp tip of 50–100 μm in diameter). The portal vein is punctured with this capillary, and the lipoplexes (formed on the basis of lipids VII–X) are administered into it by use of a mouth pipette. Then the capillary is pulled out carefully, the puncture wound is plugged with cotton, and the intestines are placed back into the abdominal cavity. If the bleeding is considerable, it should be stopped with 1–2 drops of noradrenaline solution for injections. The intestines are wetted with physiological saline if the administration of lipoplexes is delayed. If the mouse wakes up during the operation, it is possible to administer injectional anesthetic drops into its abdominal cavity or to allow the mouse to inhale ether using cotton balls. The latter is preferable because in the first method narcosis could be easily overdosed. Care should be taken not to expose the eyes of the animal to ether. After administration of lipoplexes, the muscles of the abdominal wall are sewed up with an atraumatic needle and silk thread, and the skin is fixed with metal sutures to avoid gnawing by the animal. The inner suture is powdered with a wide spectrum antibiotic (e.g., penicillin or ampicillin) and the frequency of the sutures is increased towards the abdomen to avoid the slipping of the intestines between them. The mouse is transferred to an empty cage with clean litter and food. The mouse should be kept warm before it wakes up completely, otherwise it may be lost because of hypothermia.

Administration of Lipoplexes to the Renal Artery

For administration of the lipoplexes into the renal artery,[82,83] the rules for dissection of animals, putting in stitches, and antiseptics, remain unchanged, and the excision of skin and abdominal wall is carried out similarly. For right-handed individuals, working on the left renal artery is more convenient. Prior to lipoplex administration, the intestines are shifted to the left, and the left kidney area is liberated in the vicinity of the peritoneal artery. The artery is ligated because its puncture without ligation causes heavy bleeding. The renal artery is punctured using the stretched, sharpened, sterile capillary containing the lipoplexes, and the latter are infused under pressure. Lipoplexes are administered between the ligature and the kidney, then the capillary is removed carefully and the ligature is

[82] N. Tomita, J. Higaki, R. Morishita, K. Kato, H. Mikami, Y. Kaneda, and T. Ogihara, *Biochem. Biophys. Res. Commun.* **186,** 129 (1992).
[83] J. Wagner, H. Madry, and R. Reszka, *Nephrol. Dial. Transpl.* **10,** 1801 (1995).

taken out. When the renal circulation is restored, the intestines are placed back into the peritoneal cavity. The remaining procedures are essentially the same as those described for the portal vein. After the operations the mice are kept separately for 24 h in a warm environment with the same nutrition regimen. Intraarterial administration of lipoplexes to the kidney is a much more complicated procedure, and the animals bear it with greater difficulty, than intravenous injection.

Histological Sectioning and Testing for Exogenous β-Galactosidase Activity with X-Gal Reagent

Organs of mice transfected with liposomes (formed of lipids I, II, VII–X) and control organs are frozen at $-80°$ immediately after they are removed from mice. Sections (25 μm) are prepared from frozen tissue pieces using a cryostatted microtome and the 25 μm section is placed on a slide. Two hundred μl of PBS (pH 7.5), containing X-gal reagent (6 mg/ml), magnesium sulfate (1 mM), 4 mM $K_4[Fe(CN)_6]$ and 4 mM $K_3[Fe(CN)_6]$, are placed on every slide. X-gal (6 mg) reagent is previously dissolved in a minimal amount of dimethylsulfoxide. The slides are placed into a moisture chamber, which is then placed in an incubator at $37°$. The time of color development is kept constant (30–50 min). Then these slides are fixed with 2.5% glutaraldehyde for 2 h at $4°$. The slides are stained again with hematoxylin for visualization of cellular structures (Karachhi's modification)[84] and dehydrated by washing in 70%, 96%, and 100% aqueous ethanol, ethanol/o-xylene (1:1), and o-xylene. A drop of Canadian balsam is added, and the sample is covered with a coverslip (Fig. 5).

Testing for β-Galactosidase Activity with N-Gal Reagent in Tissues

To measure exogenous β-galactosidase activity *in vivo*, organs frozen at $-80°$ are placed in a homogenizer with PBS (pH 8.0) containing 1 mM magnesium sulfate and 10 mM potassium chloride. The samples are homogenized in a melting ice bath. The N-gal reagent (at a final concentration 1 mg/ml) and mercaptoethanol (150 mM) are added to the homogenized samples collected in Eppendorf tubes. The mixture is vortexed and placed in a $37°$ water–bath for 30 min. The mixture is then centrifuged, and the supernatant is collected to tubes and kept in the cold. Spectrophotometric 3 ml quarz cuvettes are filled with PBS (pH 8.0) and 200 μl of the reaction mixture is added to the cuvettes. The optical density at 580 nm is measured using a standard spectrophotometer. As controls, samples with N-gal are

---

[84] Yu. S. Chentsov, "Practicum on General Cytology", Moscow State University Press, Moscow, 1986 (in Russian).

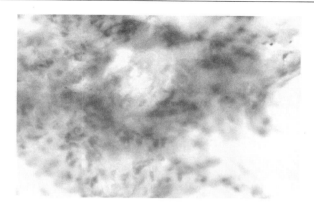

FIG. 5. *In vivo* transfection by DICHOLENIM/p-CMV-SPORT-β-gal lipoplexes follow-ing portal vein injection in mice: Expression of the *Lac Z* gene in the lung (light microscopy, 400×). (See color insert.)

placed in an ice bath. Enzyme activity is expressed in international units and is normalized to 1 *g* of tissue.

Dependence of the Biodistribution of Lipoplexes on Lipid Structure

Gene expression detected by X-gal[85–89] in the lung is shown in Fig. 5. The tissue distribution of [^{14}C]-DNA/GL or TSA lipoplexes is evaluated 24 h after administration of a small dose.[39] This is the time required to reach the stationary phase, when the expression of the *Lac Z* transgene probably reaches its optimum.[88] [^{14}C]-DNA, when given alone, is not detected in any organ 1–1.5 h after injection. In the case of [^{14}C]-DNA-PC/GL lipoplexes, the maximal [^{14}C]-DNA level is observed in the kidneys and liver at 2 h and for the spleen and heart at 4 h post-injection. The maximal content of the label for both types of lipoplexes (PC/GL and PC/TSA) is observed in the intestines (50% dose equiv/g), and in the spleen (30% dose equiv/g). The lowest [^{14}C]-DNA content is found in the heart and lungs. The most noticeable result is the change of [^{14}C]-DNA content in

[85] Y. Liu, L. C. Mounkes, H. D. Liggitt, C. S. Brown, I. Solodin, T. D. Heath, and R. J. Debs, *Nat. Biotechnol.* **15,** 167 (1997).

[86] V. Budker, V. Gurevich, J. Hagstrom, F. Bortsov, and J. A. Wolff, *Nature Biotechnol.* **14,** 760 (1996).

[87] J. Zabner, A. J. Fasbender, T. Moninger, K. A. Poellinger, and M. Welsh, *J. Biol. Chem.* **270,** 18997 (1995).

[88] S. Osaka, H. Tsui, and H. Kiwada, *Biol. Pharm. Bull.* **17,** 940 (1994).

[89] S. Ishida, Y. Sakiya, and Z. Taira, *Biol. Pharm. Bull.* **17,** 960 (1994).

the liver and kidneys depending on the lipid composition of the liposomes used (namely 4% and 10% in the liver and kidneys, respectively) for the PC/GL lipoplexes and 15% and 6% in the case of the PC/TSA-lipoplexes. Thus, the [^{14}C]-DNA content in the liver is noticeably higher for PC/TSA lipoplexes than for PC/GL lipoplexes. The lower accumulation in the liver with PC/GL lipoplexes is surprising, since glycyrrhizin is known to facilitate liposome targeting to liver hepatocytes.[89] Calcium ions (< 50 mM) are added into the system to form PC/GL (or PC/TSA) lipoplexes to enhance DNA condensation, based on previous studies with non-cationic, amphiphilic liposomes.[21,32,33,36]

The biodistribution of MONO- and DICHOLENIM-based lipoplexes was studied using ICR (or NMRI) mice. There are no differences in the tissue distribution of [^{14}C]-DNA upon intraperitoneal administartion of lipoplexes based on all three CHOLENIM oligocations; most of the label is distributed between the spleen and intestines.

With MONOCHOLENIM-based lipoplexes, most of the [^{14}C]-DNA is found in the spleen and heart (25% and 20%, respectively), 24 h after infusion into the portal vein. Smaller percentages are found in the intestines, lungs, and plasma (10%, 10%, and 5%, respectively). For [^{14}C]-DNA/DICHOLENIM lipoplexes (1:1), the percentages of [^{14}C]-DNA in the liver, spleen, heart, and lungs are essentially the same (20% in each). Smaller amounts of [^{14}C]-DNA are localized in the kidneys and intestines (5% in each). The effect of the DNA/oligocation ratio on the distribution of labeled DNA in mice was also investigated using DICHOLENIM lipoplexes. With [^{14}C]-DNA/DICHOLENIM (2:1) lipoplexes the uptake of [^{14}C]-DNA was 10% higher in the liver and 5% higher in the spleen and heart, while it was reduced by more than half in the lungs (down to 15%). Thus, with the 1:1 DNA/oligocation ratio in the complex, the maximal inclusion of the label is observed in the lungs and liver, and with the 2:1 ratio in the liver and heart.

Upon administration of DNA/DICHOLENIM lipoplexes to the kidney artery, the localization of the label in the operated kidney was increased no more than two-fold, compared with the control (intact) kidney. The low level of label in the kidney is probably explained by the fact that the capillary network of the kidney is not extensive, blood flow is very fast, and blood easily enters into the venous system. The distribution of the [^{14}C]-DNA complex in other organs on renal artery injection was analogous to that observed on intravenous injection. The maximal label localization was in the lungs with various intravenous administration techniques.

In histochemical preparations prepared upon *in vivo* gene delivery of the pCMV-SPORT-β-gal-PC/DICHOLENIM lipoplexes, β-galactosidase expression is mostly observed in endothelial cells of lung vessels and in the cells most proximal to these vessels (see Fig. 5). In cases where the liposomes based on PC and lactosylated diglyceride (Lac S (X)) are used, the expression of the reporter gene is mostly registered in liver cells, vascular endothelial cells, and the immediate vicinity of the latter.

Immunological Properties of Liposomes: Influence on the Complement System

While carrying out transfection *in vivo*, it is essential that liposomes and their complexes with DNA (lipoplexes) not cause any undesirable immunological reactions in the organism. There are several procedures for testing for the activation of the complement system.[90,91] Here, we present a procedure for assessing the influence of lipoplexes on the complement system specially developed for this purpose.[34] As a typical non-viral system, we have chosen liposomes formed of the oligocationic lipid DI-CHOLENIM, VIII, and also two helper lipids—PC and cardiolipin (CAL) The choice of cardiolipin (along with PC) is determined by the special significance of cardiolipin, not only as an important plasma lipid but also as one of the key lipids of chromatin,[92–95] with a significant role in the functioning of the genome and in effective transcription.

A simple test-system is first developed for detecting the ability of effectors (lipoplexes) to activate the complement system in an antibody-independent manner, to serve as acceptors of nascent C4b and to inhibit formation of the key enzyme of complement, C3-convertase. The effect on the complement system of plasmid DNA (pCMV-SPORT-β-gal) and negatively charged CAL or neutral phosphatidylcholine PC liposomes and their lipoplexes is studied. It is revealed that PC vesicles do not affect the complement system, while CAL vesicles manifest low activation. The influence of plasmid DNA (and its lipoplex based on PC liposomes) on the complement system is very low. PC/pDNA lipoplexes (143 μg/ml) acts on the complement system similar to 5.36 μg/ml heat-aggregated IgG (agg) (the level of non-pathological ruptures), whereas CAL/pDNA lipoplexes

[90] J. Marjan, Z. Xie, and D. V. Devine, *Biochim. Biophys. Acta* **1192**, 35 (1994).

[91] C. Planck, K. Mechtler, F. C. Szoka Jr, and E. Wagner, *Hum. Gene Ther.* **7**, 1437 (1996).

[92] R. I. Zhdanov, V. A. Struchkov, O. S. Dyabina, and N. B. Strazhevskaya, *Cytobios* **106**, 55 (2000).

[93] V. A. Struchkov, N. B. Strazhevskaya, and R. I. Zhdanov, *Bioelectrochem.* **56**, 195 (2002).

[94] V. A. Struchkov, N. B. Strazhevskaya, and R. I. Zhdanov, *Bioelectrochem. (A special issue devoted to DNA-bound membranes and lipids, R. I. Zhdanov and T. Hianik, eds.)* **58**, 23 (2002).

[95] R. I. Zhdanov, N. B. Strazhevskaya, A. R. Jdanov, and G. Bischoff, *J. Biomol. Struct. Dyn.* **20**, 232 (2002).

(143 μg/ml) act at a level similar to 10.7 μg/ml IgG (agg). Thus, weak activation of the complement system with CAL lipoplexes (and with an even weaker activation with PC lipoplexes) confirms the usefulness of neutral and negatively charged lipoplexes in gene therapy protocols. The technique can also be used for testing the influence of injectable gene therapy vectors on the complement system.

Reagents

Sheep red blood cells (E) are used between 10–30 days after blood collection and guinea pig serum is used as the complement. Reagent R4 (complement without C4 activity) is prepared by treating the complement with hydrazine hydrate. Sensitized erythrocytes (EA) [1.5×10^8 cells/ml] and veronal-buffered saline (pH 7.2) containing optimal concentrations of Ca^{2+} and Mg^{2+} ions (VBS^{2+}) are prepared by standard techniques.[96,97]

For preparation of human IgG (agg), gamma-globulin (10 mg/ml in VBS^{2+}) is heated to 63° for 30 min, centrifuged for 15 min at 3000g (after cooling), and subjected to gel chromatography on Sephadex G-200. The first fraction (high molecular mass) is then collected. The protein is concentrated (e.g., by ultrafiltration) and dialyzed against VBS^{2+} buffer. The concentration of immunoglobulin is determined in terms of optical density at 280 nm.

PC (or CAL/DICHOLENIM) liposomes are prepared from the lipid mixture using the reverse-phase evaporation technique.[47,48,98] Liposomal preparations are extruded through a 200 nm polycarbonate filter using a syringe extruder (Liposofast, Avestin, Ottawa, Canada). The lipoplexes are prepared by mixing liposomal preparations with plasmid DNA followed by a 30 min incubation.

Testing Anti-Complement Activity

VBS^{2+} buffer (25 μl) is placed in every well of a 96-well plate (U-bottomed). Immediately, 25 μl of guinea pig complement are added to all wells of column 1. The complement is diluted serially in two-fold steps from left to right along each row (25 μl are transferred from column

---

[96] L. V. Kozlov, Yu. I. Krylova, V. P. Chikh, and N. N. Molchanova, *Rus. J. Bioorg. Chem.* **8,** 652 (1982).

[97] R. A. Harrison and P. J. Lachmann, *in* "Handbook of Experimental Immunology" 4th Ed., Vol. 1, Immunochemistry, D. M. Weir and L. A. Herzenberg (eds). Blackwell Science Publications, Boston, 1986.

[98] N. Düzgüneş, *Methods Enzymol.* **367,** 23 (2003).

TABLE II
ANTI-COMPLEMENT ACTIVITY OF HEAT-AGGREGATED IgG

Row number	IgG (agg.) (mg/ml)	Number of wells with 50% lysis[*]	A[§]
1	0 (Control of complement)	6	
2	0 (Control of system)	10	0
3	1.2	0	10
4	0.6	3	7
5	0.3	5	5
6	0.15	7	3
7	0.075	8	2
8	0.0375	9	1

[*] The data are based on at least 10 experiments.
[§] Anti-complement activity.

1 to column 2 and mixed using a pipette column 2 to column 3, and so on). After that IgG (agg) solution (5 μl) is added to each well in rows 3 to 8 (row 3 receives 5 μl of the 1.2 mg/ml solution; row 4 is diluted twice, and so on up to the eighth row, to which the IgG (agg) solution (5 μl) is added at a concentration of 0.0375 mg/ml). The R4 reagent (5 μl) is added to each well of the plate from rows 2 to 8. The first 2 rows of the plate are controls. After incubation at 37° for 1 h, EA (20 μl) are added to each well and the plate is incubated again at 37° for 1 h with shaking. Unlysed erythrocytes are pelleted by centrifugation at 1,000g for 5 min. Then the well numbers, in which nearly half of the lysis occurs, are determined visually for each row. Anti-complement activity (A) is determined as the difference between the number of wells with half the lysis in the control series (the second row) and the number of wells in the series with known IgG (agg) concentrations.[99,100] The results obtained are listed in Table II. The anti-complement activity of the lipoplex preparation is carried out in the same way. Thus, the preparation sample is added to the solution instead of IgG (agg). The anti-complement activity of all preparations is expressed in terms of the IgG (agg) concentration with the same level of activity (Table III).

[99] L. V. Kozlov and T. V. Lebedeva, *Rus. J. Bioorg. Chem.* **24,** 350 (1998).
[100] L. V. Kozlov, V. M. Lakhtin, T. G. Skorokhodova, T. N. Batalova, B. B. Shojbonov, V. L. D'yakov, V. A. Guzova, and N. S. Matveevskaya, *Rus. J. Bioorg. Chem.* **26,** 817 (2000).

TABLE III

ANTI-COMPLEMENT ACTIVITY OF LIPOSOMES AND LIPOPLEXES

Number of row	Preparation[*]	Number of well with 50% lysis	A[§]	Activity in terms of IgG (agg)[*] (mg/ml)
1	0 (Control of complement)	6	—	—
2	0 (Control of system)	10	0	0
3	CAL/VIII vesicles (1 mg.ml)	8	2	0.075
4	PC/VIII vesicles (1 mg/ml)	10	0	0
5	Plasmid pCMV-SPORT-β-gal (1 mg/ml)	9	1	0.0375
6	3 vol of number 3 + 1 vol of plasmid	8	2	0.075
7	3 vol of number 4 + 1 vol of plasmid	9	1	0.0375

[*] The initial concentration of liposomal preparations (effectors) is given in all cases. The actual concentration of effectors in experimental media is 7-fold lower. Thus, it is necessary to compare 2 actual concentrations expressed as effector (μg), which is equivalent to heat-aggregated IgG. The initial 0.075 mg/ml concentration corresponds to 10.7 μg/ml of actual concentration in the wells and 0.0375 mg/ml corresponds to 5.36 μg/ml. The data are based on at least 10 experiments.

[§] Anti-complement activity.

To test the influence of liposomes (or any other effectors) on the complement system these compounds are incubated with guinea pig complement R4 reagent (with C4 activity depleted). After incubation, the residual functional activity of the C4 component is determined by measuring hemolysis of sheep erythrocytes sensitized with rabbit antibodies added to the system during the second incubation. The reduction of C4 functional activity compared with the control (where the effector under study is absent) may have been due to: (1) activation of the classical pathway of complement by the effector and the resulting consumption of the C4 component during the first incubation and (2) inhibition of binding of the nascent C4b component to erythrocytes (e.g., formation of C3 convertase) by the effector during the second incubation.

In the first case the effector is bound to the C1 complement component, activating complement in an antibody-independent manner. In the second case the effector acts as a nucleophilic agent reacting with the thioester bond exposed on nascent C4b. This process inhibits the binding of the C4b component to erythrocytes, the formation of C3-convertase, and the final lysis of erythrocytes. In addition, there is a third possibility for action on the complement system, namely inhibition of formation of C5 convertase as a result of inhibition of nascent C3b. It was shown by Kozlov et al.[98,99] that the specificity of binding nascent C3b and C4b components is almost the same, and thus the effectors binding to the C3b component should also bind the C4b component.

Human IgG (agg) is chosen as the standard for evaluation of anti-complement activity. Its ability to bind the C1 component and activate the classical process is well known. Moreover, immunoglobulin IgG may bind the nascent C4b and C3b components.[98,99] Experimental data on the anti-complement activity of IgG (agg) are shown in Table II. Column 2 gives the initial concentrations of IgG solutions used for each experimental row. For each row of the double solutions of the C4 component (its source being guinea pig serum) the number of cells showing 50% lysis of the erythrocytes served as the measure of residual functional activity of the C4 component (column 3). Anti-complement activity (column 4) is shown as the difference between the number of wells with 50% lysis for the control experiment where the effector is absent, and the experimental one. The data in Table II show that the anti-complement activity correlates well with the concentration of IgG (agg), testifying to the suitability of the method developed.

Table III shows values of the anti-complement activity of PC or CAL liposomes, plasmid DNA, and lipoplexes prepared from them. Column 4 indicates anti-complement activity expressed in terms of the concentration of IgG (agg) with equivalent activity. CAL-liposomes (1 mg/ml) possess anti-complement activity corresponding to the activity of 0.075 mg/ml (final concentration: 10.7 μg/ml in the well; see Table III) of aggregated IgG (row 3). PC liposomes do not have complement inhibitory activity up to 1 mg/ml. These results are in accordance with data on activation of the classical pathway by negatively charged liposomes containing CAL and lack of activation by neutral (PC and DPPC) liposomes. Plasmid DNA, pCMV-SPORT-Lac Z, displays weak complement inhibitory activity. Activity of the plasmid DNA (1 mg/ml) is equivalent to the activity demonstrated by 0.0375 mg/ml (5.36 μg/ml in the well) of IgG (agg).

Lipoplexes containing CAL liposomes and pDNA (1 mg/ml) have anti-complement activity corresponding to 0.075 mg/ml (10.7 μg/ml in the well) of IgG (agg) (row number 6). The anti-complement activity of lipoplexes formed of PC liposomes and pDNA is lower and equivalent to 0.0375 mg/ml (5.36 μg/ml) of IgG (row number 7). It should be noted that the initial concentration of liposomal preparations, which is equivalent to the action of 0.0375 mg/ml of IgG (actually 7-fold lower in experimental media), corresponds to the concentration of IgG (agg) or other immunogenic complexes capable of activating complement in circulation without causing pathological lysis. The normal value of such immune complexes in plasma of healthy donors (n = 11) is 6.3 ± 1.5 μg/ml. The anti-complement activities of plasmid DNA and PC-lipoplexes (5.36 μg/ml) are lower. Therefore, PC-lipoplexes should not cause any pathological disturbances. Our results show that in the case of the CAL-lipoplexes, the cardiolipin component provides the main anti-complement activity, while in the case of PC-lipoplexes it is plasmid DNA that contributes to this activity.

The study of the action of phospholipid/CHOLENIM liposomes and lipoplexes on the complement system demonstrates that lipoplexes based on neutral (PC) liposomes show very low inhibition of the complement system, equivalent to 5.36 μg/ml of IgG (agg), and lipoplexes based on negatively charged (CAL) liposomes which shows low activation, equivalent to 10.7 μg/ml of IgG (agg). This result supports the view that lipoplexes formed of neutral liposomes/micelles should be used in the development of injectable non-viral gene delivery systems.

Concluding Remarks

We have described, in detail, various methods used in gene transfer *in vitro* and *in vivo*, and in evaluating the effect of lipoplexes on the complement system. We have also presented a series of novel lipids with a range of transfection activities. The choice of lipids and DNA/lipid ratios will depend on the actual application of the gene transfer technology to a particular therapy. Lipoplexes that mediate high levels of transgene activity in one tissue may not be desirable for gene delivery in another tissue. It should also be considered that achieving high transfection activity may be associated with high toxicity (or untoward effects) on the complement system. Thus, the choice of a non-viral vector for gene delivery *in vivo* will depend on weighing its advantages and limitations in a particular application.

Acknowledgments

The authors wish to honor the memory of Yury Vladimirovich Sviridov, MSc, who passed away July 12, 2001; he contributed to the animal experiments in our laboratory. The authors are greatful to Prof. Dr. Alexander Archakov (Institute of Biomedical Chemistry, Moscow) for his helpful discussions and for leading the Russian gene transfer and therapy program between 1996 and 2000. We wish to thank Dr. M. Boettger and Dr. A. Haberland for help in experiments on GLYCOCLIP lipoplex transfer against CHO cells, Dr. L. I. Kondakova and Dr. O. Yu. Abakumova for the gift of RGGN-1 cell, and Mrs. T. Tsvetkova for assistance in cell culture. The authors are indebted to Prof. Dr. V. V. Vlassov (Novosibirsk Institute of Bioorganic Chemistry) for the synthesis of CHOLENIM preparations, Prof. Dr. A. R. Khokhlov (M. V. Lomonosov Moscow State University) and Dr. D. Yu. Blokhin

(N. N. Blokhin National Oncological Center, Moscow) for their support, help, and discussions; to Prof. Dr. G. A. Serebrennikova and Dr. Yu. L. Sebyakin (MITHT) for the gift of cationic and glycolipids; to Dr. Ashraf Elkady (M. V. Lomonosov Moscow State University) for AFM images of DEGA liposomes and lipoplexes; and to Anastasia Shmyrina (Institute of Biomedical Chemistry, Moscow) for her help with manuscript processing. The work from our laboratories was supported by the following grants: Grant No. 08.01.01.01 of "National Priorities in Medicine and Healthcare" program of the Russian Ministry of Science and Industry (1996–2001), Russian Foundation for Basic Research No 98-04-49042, research and equipment grants from the A. von Humboldt Foundation (Bonn, Germany), and a grant from the Branch of Chemistry and Material Sciences of the Russian Academy of Sciences, 2002–2005.

[29]   Phage DNA Transfer into Liposomes

By OLIVIER LAMBERT and JEAN-LOUIS RIGAUD

Introduction

The transfer of the phage genome into bacteria during phage infection is a very efficient process. Most of our knowledge comes from the studies of the *Escnerchia coli*–tailed phages T4, T5, T7, and lambda (for reviews see Refs. 1–3). Infection of the host cell by different kinds of bacteriophages mostly follows a similar scheme and occurs in three steps: (1) first, the phage binds reversibly to the cell surface, (2) then, it recognizes and binds irreversibly to an outer membrane receptor, (3) third, following binding to its receptor, the phage releases its DNA, which crosses the bacterial envelope linearly, base pair after base pair.[4] Some phages require an energized membrane and/or additional components for DNA release,[5] whereas for others the interaction with their cognate receptors is sufficient to trigger DNA ejection.[6,7] In *E. coli*, the mere interaction of phage T5 with its outer membrane receptor, the ferrichrome transporter FhuA, is sufficient to trigger the release of the DNA (121,000 bp, 40 μm in length) from the phage

[1] B. Dreiseikelmann, *Microbiol. Rev.* **58**, 293 (1994).

[2] L. Letellier and P. Boulanger, *Biochimie* **71**, 167 (1989).

[3] L. Letellier, L. Plancon, M. Bonhivers, and P. Boulanger, *Res. Microbiol.* **150**, 499 (1999).

[4] E. Goldberg, L. Grinius, and L. Letellier, *in* "Molecular Biology of Bacteriophage T4." (J. D. Karam, J. W. Drake, and K. N. Kreuzer, eds.), p. 347. ASM Press, Washington, DC, (1994).

[5] R. Hancock and V. Braun, *J. Bacteriol.* **125**, 409 (1976).

[6] P. Boulanger, M. le Maire, M. Bonhivers, S. Dubois, M. Desmadril, and L. Letellier *Biochemistry* **35**, 14216 (1996).

[7] D. J. McCorquodale and H. R. Warner, *in* "The Viruses" (H. Fraenkel-Conrat and R. R. Wagner eds.), p. 232. Plenum Press, New York, 1975.

capsid into the surrounding medium in a few seconds. No extra component is needed to induce the DNA release. Therefore, reconstituting the membrane receptor FhuA into liposomes should allow the *in vitro* analysis of the mechanism of phage infection (i.e., phage binding) and the subsequent transport of the phage genome into lipid vesicles.

In this chapter, we describe a recent *in vitro* strategy developed in our laboratory using FhuA-reconstituted liposomes as a simple system for analyzing phage DNA delivery into liposomes (Fig. 1). We will present different methods to analyze DNA transfer with a special emphasis on cryotransmission electron microscopy, which has been demonstrated as a powerful tool. Finally, it has to be stressed that these receptor-containing proteoliposomes also represent an attractive model system to study the mechanism of DNA condensation and, importantly, may play a role as alternative gene vectors.

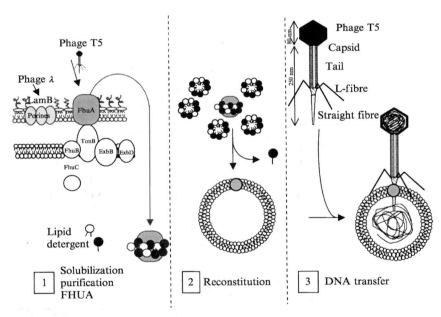

Fig. 1. A strategy to study the mechanism of phage infection. (1) Solubilization and purification of FhuA, the receptor for T5 phage in the outer membrane of *E. coli*. (2) Reconstitution of FhuA into proteoliposomes. To the detergent solubilized protein, an excess of lipids and detergent is added, leading to mixed lipid-protein-detergent micellar solutions. Detergent removal from these micellar solutions leads to the production of well-sealed FhuA-containing proteoliposomes. (3) T5 phages are mixed with FhuA-containing proteoliposomes. Once T5 phage binds to reconstituted FhuA, it transfers its genome into the proteoliposome.

Preparation of Proteoliposomes Containing FhuA, the T5 Phage Receptor

Purification of FhuA

The protein of the *E. coli* outer membrane targeted by phage T5 is FhuA, a multifunctional membrane protein. It is the receptor for the ferric siderophore ferrichrome-iron, for the peptide antibiotics albomycin and microcin 25, for the toxin colicin M, as well as for bacteriophages T5, T1, UC-1, and ϕ 80 (for a recent review, see Ref. 8). X-ray crystallography has revealed that FhuA is a hollow 22-stranded antiparallel β-barrel that encloses a channel with an inner diameter of approximately 2.5 nm.[9,10] Three large surface-exposed hydrophilic loops constrict the entrance of the channel, with L4, the farthest protruding loop, crucial for the binding of T5 phages.[11,12] A globular *N*-terminal hydrophilic domain made up of approximately 160 amino acids, known as the plug or cork, obstructs the channel on the periplasmic face.

Hexahistidine-tagged FhuA protein is overproduced and purified from a porin-deficient *E. coli* strain HO830FhuA.[12] Cells are grown to stationary phase, harvested by centrifugation, and subjected to enzymatic lysis.[13] The outer membrane is solubilized in the presence of 0.1% *N*-*N*-dimethyldodecylamine *N*-oxide (LDAO) and the hexahistidine-tagged FhuA protein purified at room temperature using a Ni^{2+}-NTA Superflow (Qiagen, Valencia, CA) column. Pooled fractions are dialyzed against a buffer containing 20 mM TRIS-HCl (pH 8), 150 mM NaCl, 0.05% LDAO, 0.02% NaN_3, and finally concentrated to 2 mg/ml using Amicon (Millipore, USA) YM-10 ultrafiltration membranes.

Reconstitution of FhuA into Proteoliposomes

Lipids of highest purity are used to produce liposomes by the reverse-phase evaporation method.[14] Typical preparations contain 16 mg/ml of a phospholipid mixture made of egg phosphatidylcholine and egg phosphatidic acid (9:1 M/M). After complete evaporation of the organic solvent, the

[8] G. Moeck and J. Coulton, *Mol. Micobiol.* **28,** 675 (1998).

[9] A. D. Ferguson, E. Hofmann, J. W. Coulton, K. Diederich, and W. Welte, *Science* **282,** 2215 (1998).

[10] K. P. Locher, B. Rees, R. Koebnick, A. Mitschler, L. Moulinier, J. P. Rosenbusch and D. Moras, *Cell* **95,** 771 (1998).

[11] R. Koebnik and V. Braun, *J. Bacteriol.* **175,** 826 (1993).

[12] G. S. Moeck, P. Tawa, H. Xiang, A. A. Ismail, J. L. Turnbull, and J. W. Coulton, *Mol. Microbiol.* **22,** 459 (1996).

[13] K. Hantke, *Mol. Gen. Genet.* **182,** 288 (1981).

[14] J. L. Rigaud and D. Levy, *Methods Enzymol.* **372,** 65 (2003).

liposomes are sequentially extruded through 0.4- and 0.2-μm polycarbonate membranes before use. For proteoliposome reconstitution, 25 μl of these preformed liposomes are diluted to a concentration of 4 mg lipid/ml in a buffer containing 20 mM TRIS HCl, pH 8, 150 mM NaCl, and solubilized in the presence of 30 mM β-octyl glucoside. After complete solubilization, 5 μl of the purified, detergent-solubilized FhuA is added to give a lipid/protein ratio of 40 (w/w) in a total volume of 100 μl. Following 1-h incubation for micellar equilibration, the detergent is removed in approximately 4 h by three successive additions of 5 mg of Biobeads (Bio-Rad, USA) at room temperature.[15]

For experiments dealing with the study of DNA condensation, the proteoliposomes are prepared as described previously, except that liposomes are made of only egg phosphatidylcholine to avoid any interaction between tetravalent spermine-4HCl and negatively charged phospholipids.

Preparation of T5 Phages

T5 phage is composed of a capsid and a flexible noncontractile tail made of four substructures: a long proteinaceous helical structure, a conical part containing several copies of the protein pb5 involved in the specific binding to FhuA, a thin central straight fiber made of several copies of the protein pb2 involved in phage adsorption and DNA transfer through the membrane, and three L-shaped tail fibers (Fig. 1).

T5stamN5 (T5) phages are produced, in *E. coli* Fsuβ^+, a permissive host. Phage stocks are prepared and purified by polyethylene glycol precipitation followed by a CsCl step gradient.[16] They are resuspended in 10 mM TRIS-HCl, pH 7.2, 150 mM NaCl, at a final titer of 1×10^{13} phages/ml.

To analyze the binding of the phages to their receptor, T5 phages (10^{11} particles) are added to 30 μl of FhuA-containing proteoliposomes, such that the ratio of the phage particles to FhuA molecule is approximatively 1:100. In all experiments, DNAse (1 μg) and MgSO$_4$ (5 mM) are added to digest the DNA occasionally released from broken capsids in the external medium.

Methods to Analyze DNA Injection into Liposomes

Several approaches can be used to analyze the release of DNA following binding of the phages to their receptors incorporated in the membrane of a proteoliposome. These include radioactive labeling, spectroscopic methods, and electron microscopy analysis.

[15] J. L. Rigaud, D. Levy, G. Mosser G, and O. Lambert, *Eur. Biophys. J.* **27,** 305 (1998).
[16] M. Bonhivers, A. Ghazi, P. Boulanger, and L. Letellier, *EMBO J.* **15,** 1850 (1996).

Radiolabeling Methods

T5 phages containing radioactively labeled DNA can be prepared by adding (2-^{14}C) uridine at the time of infection of *E. coli* cells. These ^{14}C-labeled DNA T5 phages are mixed for 1 h at 37° with FhuA-containing proteoliposomes in a buffer containing DNAse and Mg^{2+}. After centrifugation of the liposome-phage complexes in an airfuge, no radioactivity can be detected in the supernatant. When the pelleted liposomes are resuspended and the proteoliposomes disrupted by addition of solubilizing detergent concentrations, 20%–30% of the total DNA radioactivity is recovered in the supernatant after DNAse incubation, which is interpreted as an incomplete but effective DNA transfer inside the liposomes[17] (Fig. 2).

Radiolabeling methods have been used in the early phage–receptor studies. Such experiments are difficult to handle and, in addition, require additional centrifugation steps and further solubilization of the pelleted liposomes to determine the amount of DNA that has been entrapped into the vesicles. Furthermore, the information that can be retrieved from such experiments remains essentially qualitative.

Spectroscopic Methods

Release of the DNA into FhuA-containing proteoliposomes can be followed from the increase of the fluorescence of YO-PRO 1, a dye that intercalates between the free DNA base pairs (Fig. 2). To this end, proteoliposomes are prepared in the presence of 4 μM YO-PRO 1 and resuspended in a cuvette also containing 4 μM of the fluorescent probe and the desired amount of phage T5. On proteoliposome addition, the fluorescence measured at 509 nm (λ_{ex}: 491 nm) increases slowly after a lag of 2–3 min and reaches a steady state in about 15 min. Addition of DNAse and Mg^{2+} decreases the fluorescence, immediately and partially indicating that part of the DNA has been ejected in the external medium. Total solubilization of the proteoliposomes leads to a new decrease in fluorescence that corresponds to DNA entrapped inside the liposomes before solubilization. Thus, such experiments indicate the transport of part of the DNA into the vesicles and are consistent with entrapment of 10%–20% of the DNA inside the vesicles. The simplest explanation is that DNA has been transferred only to the largest liposomes that represent a small proportion of the preparation.[18]

A similar approach using another intercalator, ethidium bromide, has been applied for measuring the kinetics of DNA injection from

[17] F. Tosi, B. Labedan, and J. Legault-Démare, *J. Virol.* **50**, 213 (1984).
[18] L. Plançon, M. Chami, and L. Letellier, *J. Biol. Chem.* **272**, 1686 (1997).

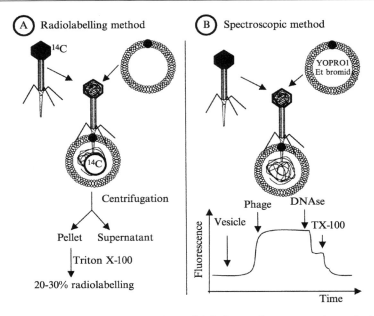

FIG. 2. Analysis of DNA transfer by radiolabeling and spectroscopic methods. (A) Radiolabeling method: T5 phages whose DNA has been ^{14}C-labeled are added to FhuA-containing proteoliposomes. After DNA transfer, the phage–proteoliposomes complexes are isolated by centrifugation. Then, the transferred ^{14}C labeled DNA is released from the proteoliposomes by detergent solubilization. (B) Spectroscopic method: T5 phages are mixed with proteoliposomes containing an entrapped fluorescent dye (YOPRO1 or ethidium bromide). The kinetics of DNA transfer into the proteoliposome, following phage addition, are followed continuously by changes in fluorescence intensity. Once a steady-state has been reached, the addition of external DNAse allows one to determine the amount of DNA ejected in the external medium. Total solubilization of the proteoliposomes by Triton X-100 allows one to determine the amount of DNA that has been transferred into liposomes.

bacteriophage λ into LamB-containing liposomes. LamB-proteoliposomes are prepared in a medium supplemented with 1 mM ethidium bromide, and nonencapsulated solute is removed on a Sephadex G-25 spin column. These proteoliposomes are resuspended in a cuvette, and the changes in fluorescence, induced by addition of bacteriophage λ, are continuously monitored at 598 nm (λ_{ex}: 295 nm). As the DNA injected by the bacteriophages interacts with the entrapped fluorophore, the fluorescence is shown to increase very rapidly reaching a steady-state after 1 min. The rates and the extents of DNA injection into liposomes are shown to be related to the bacteriophage and receptor concentrations.[19] Interestingly, the level of this

[19] S. L. Novick and J. D. Baldeschwieler, *Biochemistry* **27,** 7919 (1988).

fluorescence steady-state corresponds to injection by approximately 20% of the phages present in the sample.

Electron Microscopy Analysis of DNA Release

Radioactive methods provide a quantitative estimate of the efficiency of phage DNA transfer into liposomes. Fluorescence spectroscopy combines the advantage of fast data acquisition and high sensitivity and is well suited for kinetic studies. However, it does not give information about the efficiency of phage binding, the proportion of phage that has injected their genome into liposomes, the complete or partial DNA release from the capsid, the distribution of injected DNA among the liposome population, and the state of DNA compaction inside the liposome. Only morphological analysis by electron microscopy of the phage–proteoliposome complexes can answer these important questions.

Negative Staining

Negative staining consists of the deposition of the proteoliposome–phage complex suspension on a carbon-coated grid and staining with 1% uranyl acetate (Fig. 3). Although this technique consumes a small amount of material, is rapid, and is easy to perform, it is limits in reliability it can provide because of artefacts of dehydration and staining. Indeed, after adsorption of the suspension to a grid and negative staining, liposomes appear collapsed, distorted, and often rearranged (Fig. 4A,B). In addition, because liposomes are not permeable to the stain, it is not possible to visualize clearly a delineated lipid bilayer and to directly visualize DNA that may be present within the liposomes after ejection from the phage capsid. Thus, negative staining can only be used for controling the overall aspect of the preparation.

Nevertheless negative staining can be used to visualize bacteriophages and their interaction with the membrane of proteoliposomes containing their receptor. Also important is that full capsids of bacteriophages that contain DNA can be distinguished readily from empty capsids. Indeed, filled capsids have a characteristic icosahedral shape, whereas empty heads appear broken and retain a lower amount of stain. Thus, it is possible to visualize the content of the capsid heads and determine the percentage of the heads that are full, empty, or partially empty. For example, it has been shown, in the case of λ phage, that 75% of phages bound to their receptor appear to have ejected all or part of their DNA at 37°, whereas at 4° most of the phages are still bound to the liposome, with their heads full of DNA.

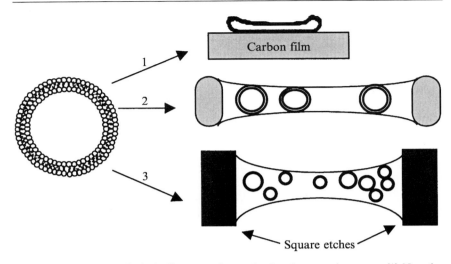

FIG. 3. Different methods for liposome observation by electron microscopy. (1) Negative staining: A suspension of liposomes is deposited on a continous carbon film and stained with 1% uranyl acetate. Dehydration and collapse on the grids have dramatic effects on the morphology of the liposome. (2) Cryo-TEM on perforated carbon support: A suspension of liposomes is deposited on a perforated carbon support. After blotting the excess solution, the thin layer of liposome suspension formed within the hole is rapidly frozen in liquid ethane. The frozen–hydrated liposomes embedded in amorphous ice exhibit a spherical shape. (3) Cryo-TEM using the bare grid method: In the absence of a carbon support, the layer of liposome suspension fills the squares of the grid after blotting the excess solution.

Freeze-Fracture

Freeze-fracture is considered a powerful method to study the morphology of a liposome preparation.[20] A drop of the liposome suspension is frozen rapidly into a cryogen (liquid ethane for example), and the sample is fractured *in vacuo* at $-110°$. Then the surface of the sample is shadowed unidirectionally with platinum/carbon at a $45°$ tilt angle and carbon-coated. The replica is observed at room temperature in the microscope. Because the liposomes are fractured usually in the middle of the lipid bilayer, the information that can be retrieved from this technique is related to the lipid layer itself (i.e., to the lamellarity and the homogeneity in size distribution). Importantly, compared with other electron microscopy techniques, freeze-fracture allows visualization of the incorporation of transmembrane proteins that appear as small bumps on the fracture faces. Because the fracture is located between the two leaflets, however, molecules encapsulated into

[20] T. Gulik-Krzywicki, *Biol. Cell.* **80,** 161 (1994).

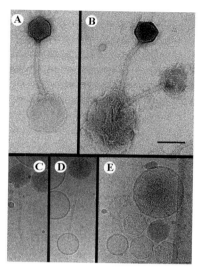

FIG. 4. Phage DNA transfer into FhuA-containing proteopliposome. (A and B) Negative staining electron micrographs of T5 phages bound to a proteoliposome before (A) and after DNA transfer (B). The capsid filled with DNA keeps its icosahedral shape, whereas the empty one appears broken. (C–E): Cryo-TEM electron micrographs of T5 phage (C), T5 phage bound to a proteoliposome before (D), and after DNA transfer (E). Scale bar: 100 nm.

the liposomes or interacting with the outer leaflet of the liposomes are not visible or poorly visible, respectively.

The freeze-fracture method has been applied to analyze FhuA-containing proteoliposomes.[18] It allows a clear description of the morphology of the reconstituted vesicles (size distribution, unilamellarity). In addition, the efficiency of protein reconstitution is demonstrated by the presence of transmembrane particles randomly distributed on the fracture surfaces. Unfortunately, no information concerning the presence of DNA inside the proteoliposome after T5 binding to their receptor can be retrieved by this method.

Cryo-Transmission Electron Microscopy

Cryo-transmission electron microscopy (cryo-TEM), as described in detail by Dubochet *et al.*,[21] is a technique well suited for the observation of biological specimens in a state close to their native state. Dehydration and deformation of biological specimens that represent the main limitation of other conventional electron microscopy methods have been overcome

[21] J. Dubochet, M. Adrian, J. Chang, J. C. Homo, J. Lepault, A. W. Mc Dowall, and P. Schulz, *Q. Rev. Biophys.* **21,** 129 (1988).

by embedding the specimen in amorphous ice and by maintaining the specimen *in vacuo* well below $-135°$ during the observation (Fig. 3). Thus, this technique avoids the artefacts of staining and drying procedures and permits observation of relatively undistorted samples. Over the past 20 years, this technique has provided important information over a large variety of structural biology fields. One of the fields that has greatly benefited from cryo-TEM is the "liposome" field. Cryo-TEM has been applied widely to the studies of the morphology of vesicular and liposomal dispersions, phospholipid phase transitions, spherical micelles, and liposomes at different solubilization steps with various detergents, or during the reconstitution processes by detergent removal.[14,22-24]

The preparation and observation of frozen-hydrated biological particles involve the following operations: (1) forming a thin layer of the suspension; (2) cooling it into the vitreous state; (3) transferring it into the microscope without rewarming below the devitrification temperature ($T_d = -135°$); and (4) observing it below T_d using a low dose of electrons to preserve the structure of the specimen.

For preparation of the unstained, frozen, hydrated liposome specimen, a 5-μl droplet of the liposome suspension (at 0.5–5 mg lipid/ml) is deposited on an electron microscopy grid. After removing the excess solution with a filter paper, the thin layer of water is plunged rapidly into liquid ethane at $-178°$ using a CPC station (Leica, Rueil-Malmaison, France). The grid, kept under liquid nitrogen, is then transferred into a cryoholder (Gatan, Pleasanton, CA and München, Germany) maintained at $-170°$, which, in turn, is introduced into the EM for observation at $-170°$. To limit beam damage, the specimens are observed with a limited amount of electrons using the low-dose technique. Images are recorded at a magnification of 45,000, with a defocus of 1 μm and 1 second exposure time, which corresponds to a dose of less than 10 $e^-/\text{Å}^2$.

For cryo-TEM analysis, different carbon-coated support grids can be used: (1) A continuous supporting carbon film rendered hydrophilic by glow discharge in air. Use of these grids may cause some flattening artefacts because of adsorption of the material onto the supporting film. (2) The holey carbon film (commercially available or prepared by different methods) is the support used widely for liposome vitrification. This method consists of mounting the specimen across the holes of a hydrophilic carbon film (size of the holes 1–5 μm). As with other grids, a drop of solution is deposited on the grid and the specimen frozen immediately after removing most of the liquid with blotting paper. The stabilizing effect of

[22] P. K. Vinson, Y. Talmon, and A. Walter, *Biophys. J.* **56,** 669 (1989).
[23] D. P. Siegel, W. J. Green, and Y. Talmon, *Biophys. J.* **66,** 402 (1994).
[24] O. Lambert, D. Levy, J. L. Ranck, G. Leblanc, and J. L. Rigaud, *Biophys. J.* **74,** 918 (1998).

the perforated support ensures that most of the holes become filled with an adequate layer of suspension, avoiding deformation of the liposomes now present in the thin layers of water filling the holes of the perforated carbon film (Fig. 3). The original random distribution of particles is generally well preserved when the unsupported vitrified layer of suspension is much thicker than the particles. When the thickness of the layer is close to the dimension of the particles, the interactions with the surface become noticeable. (3) An alternative way of preparation is referred as the bare grid method.[25] In this method, a drop of solution is deposited on a clean, uncoated, 200–600 mesh, copper specimen supporting grid. After removal of the excess liquid, the grid is frozen quickly. On most grids, few squares are filled with a film of adequate thickness (less than 3000 Å), allowing electron microscopy observations.

In principle, a wide range of buffer compositions can be used to prepare frozen-hydrated specimens. However, high salt concentrations, greater than 200 mM, can lead to a grainy appearance. High concentrations of sugar or glycerol do not disturb the preparation but lead to a drastic decrease of the image contrast, rendering the observation difficult.

Cryo-TEM Analysis of Phage DNA Transfer into FhuA-Containing Proteoliposomes

Observation of Proteoliposomes

The images of ice-embedded biological objects exhibit a low contrast and correspond to a projection of these objects. FhuA-containing liposomes appear as round shells delineated clearly by a 4–5 nm lipid bilayer (Fig. 4). Most of them have a diameter ranging from 70–150 nm.

Two kinds of artefacts may arise from incorrect specimen preparation: (1) Vesicle invagination is thought to occur on evaporation during specimen preparation. Dubochet et al.[21] reported that cryo-TEM specimens of lipid vesicles in 100 mM NaCl prepared with intentional evaporation show invaginated structures, whereas specimens prepared without evaporation do not show the osmotic effect. If such invaginations are systematically observed during specimen preparation, a controlled environment of the chamber of vitrification (temperature and humidity) can be used to limit the evaporation of the water from the droplet placed on the grid before freezing. (2) A second type of distortion may arise from the mechanical stress caused by the fluid flows on the grid when the sample is blotted to form the very thin liquid film. The shear force of this flow squeezes vesicles against the carbon support, where they align with each other.

[25] P. M. Frederik, M. C. Stuart, P. H. Bomans, and W. M. Busing, *J. Microsc.* **153,** 81 (1989).

Cryo-TEM provides information on the size, the homogeneity, the morphology, and the lamellarity of the proteoliposomes. However, membrane proteins incorporated in the membrane are hardly visible, except for proteins with a large extramembrane part. Compared with freeze-fracture, cryo-TEM requires a 10-fold lower concentration of liposome. In practice, blotting can be achieved easily by touching the droplet with a filter paper while removing the liquid excess from the opposite side of the droplet; this facilitates the retention of a larger amount of material present in the holes.

Observation of T5 Phages

Phage morphology can be visualized clearly by cryo-TEM. The icosahedral capsids of isolated T5 bacteriophages (80 nm in diameter) are filled with a dark gray striated material similar to that found in the capsids of phages T4 and T7 and corresponding to densely packed DNA. The long flexible noncontractile tail and its four substructures are also visible clearly: a long proteinaceous helical structure (190 nm in length, 8 nm in diameter), a conical part containing the protein pb5, a thin central straight fiber (50 nm in length, 2 nm in diameter), and three L-shaped tail fibers (3 nm in diameter) (Fig. 4C).

Observation of Proteoliposome–Phage Complexes and DNA Transfer

In the presence of FhuA-containing proteoliposomes, phages bind to the proteoliposomes by their distal tail. The straight fiber is clearly visible at the surface of the vesicle, and, in some pictures, part of it has apparently passed through the bilayer (Fig 4D). Control experiments show that T5 phages do not bind to pure liposomes, demonstrating the specificity of phage–FhuA interaction.

Some capsids of phages attached to the proteliposomes are full of DNA (Fig. 4D), but many other capsids are only partially filled with DNA or totally empty, as shown by their lower electron densities (Fig. 4E). Phages with partially or completely empty capsids are always associated with proteoliposomes that contain DNA (dark gray material) originating from the capsid. Interestingly, control experiments demonstrate that phages Φ 80 can bind to FhuA-containing proteoliposomes but are unable to eject their DNA. This is consistent with the fact that DNA ejection from Φ 80 requires, in addition to its receptor FhuA, other cell envelope components that are not present in the reconstituted system.

Noteworthy, the genome can be entrapped in proteoliposomes of different sizes and even in proteoliposomes as small as the phage capsids.[25] The amount of entrapped DNA seems increased with the size of the proteoliposomes and with the number of phages bound to their receptors. More than one complete genome is transferred into vesicles of 150 nm in diameter (Fig. 4E), and several genomes are observed in vesicles of 250-nm diameter.[26] DNA, whose concentration inside the vesicles reaches values as high as 130 mg/ml, always occupies the entire internal volume of the proteoliposomes. Although densely packed, it never appears condensed into peculiar structures. In addition, although the DNA occupies all space available within the vesicle, it does not disturb the morphology of the vesicle, indicating that DNA does not interact with the lipids. Furthermore, the proteoliposomes remain impermeant to large molecules, because the entrapped DNA is protected from digestion of externally added DNAse.

Condensation of DNA into Proteoliposome

To increase the amount of DNA that can be transferred from bacteriophages into proteoliposomes, condensing agents can be encapsulated inside the proteoliposomes to induce a higher packing of DNA strands. We have focused on polycations, and more specifically on polyamines that modify the electrostatic interactions between DNA segments, either through neutralization of their charges and/or through mediation of attractive forces.[27,28] DNA condensation has been observed previously by a variety of techniques that detect changes in polymer size or chirality. This includes total intensity and dynamic laser light scattering, sedimentation, viscometry, linear optical dichroism, and circular dichroism.[29,30] Also, various electron microscopy approaches have been used to get information about the size and morphology of the condensates formed. Striking toroidal condensates have been characterized by electron microscopy[31], following addition of polyamines to very dilute aqueous solutions of DNA.

The process of spermine-induced DNA condensation, following its release from T5 phages bound to isolated FhuA molecules in a detergent-containing solution, can be analyzed by cryo-TEM. DNA condensation depends on the concentration of spermine present in solution. In the presence of 1–4 mM spermine, DNA strands released from the phage capsid appear as long disentangled filaments, which are progressively condensed into very long bundles up to 100 μm in length, longer than the 40-μm T5 phage genome (Figs. 5A,B). In the presence of 5 mM spermine, DNA

[26] O. Lambert, L. Plançon, L. Letellier, and J. L. Rigaud, *Mol. Microbiol.* **30,** 761 (1998).

[27] V. A. Bloomfield, *Biopolymers* **31,** 1471 (1991).

[28] G. S. Manning, *Q. Rev. Biophys.* **11,** 179 (1978).

[29] V. A. Bloomfield, *Biopolymers* **54,** 168 (2000).

[30] V. A. Bloomfield, *Curr. Opin. Struct. Biol.* **6,** 334 (1996).

[31] D. K. Chattoraj, L. C. Gosule, and J. A. Schellman, *J. Mol. Biol.* **121,** 327 (1978).

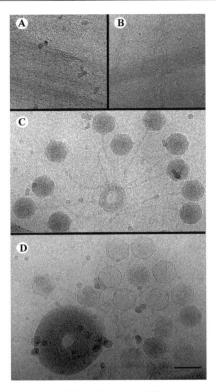

FIG. 5. Phage DNA transfer in bulk solution at different spermine concentrations. T5 phages in the presence of detergent-solubilized FhuA molecules release their genome in bulk solution. In the absence of spermine, released DNA strands appear as long disentangled filaments. In the presence of 3 mM spermine (A), released DNA strands appear roughly aligned or condensed into bundles. In the presence of 4 mM spermine (B), the bundles formed by DNA strands look more compact, extending over several tens of microns in length. In the presence of 5 mM spermine (C and D), the released DNA strands are condensed into typical toroidal structures. At the early stages of DNA release (C), small toroids start to be formed from DNA strands of several phages, and after 5 min (D), huge toroidal structures are obtained. Scale bar: 100 mm.

strands appear organized into typical toroidal structures. A kinetic analysis indicates that, after 1–2 min incubation following addition of T5 phages to solubilized FhuA, small annular structures (70–100 nm outer diameter and 30–40 nm inner diameter) can be formed (Fig. 5C). After 5 minutes incubation, these annular structures are transformed into very large toroidal structures, rather homogenous in size with an outer diameter of 300 nm (Fig. 5D). The formation of such large toroids occurs very rapidly and does not require full DNA release from the phages.

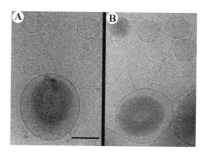

FIG. 6. Phage DNA transfer into proteoliposomes containing encapsulated spermine. Cryo-TEM electron micrographs of T5 phage bound to proteoliposomes prepared in the presence of 50 mM spermine. As opposed to what is observed in Fig. 4 in the absence of spermine, DNA strands transferred into spermine-containing proteoliposomes appear condensed into a single toroidal structure. Scale bar: 100 nm.

It has been possible to produce such condensates in FhuA-proteolipo-somes[32] and to analyze their formation by cryo-TEM. To this end, proteo-liposomes are reconstituted in the presence of different amounts of spermine. Importantly, proteoliposomes are reconstituted using only neutral phosphatidylcholine to avoid any interaction between tetravalent spermine and phospholipids. In the presence of 50 mM spermine-contain-ing liposomes, DNA strands transferred from the phage capsids appear condensed into characteristic toroidal structures, which occupy a volume smaller than the internal volume of the proteoliposome (Fig. 6). Increasing the number of DNA strands transferred into proteoliposomes does not change the number of toroids in the liposomes but increases significantly their size. For example, a toroid composed of two DNA strands ejected from two phages (Fig. 6A) and three different phages (Fig. 6B) has an outer diameter of 150 and 160 nm, respectively (Figs. 6A, 6B), whereas a toroid composed of six DNA strands has a diameter of about 210 nm.[32] The size increases up to 230 nm when about 10 phages have transferred their DNA into one liposome.[32]

In conclusion, it can be stressed that a single toroid can be formed within each liposome, even when several phages have transferred their DNA. Overall, the striking feature of our experiments is that, because of the progressive release of DNA from the phage capsid, the mechanism of toroid formation is fundamentally different from that in the classical stud-ies in which highly dilute, naked DNA is condensed by direct addition of polycations. As a consequence, our method leads to toroids of arbitrary

[32] O. Lambert, W. Gelbart, L. Letellier, and J. L. Rigaud, *Proc. Natl. Acad. Sci. USA* **97**, 7248, (2000).

size; when the linear DNA is released, base pair after base pair, from the capisd, it enters the spermine-containing proteoliposome and progressively wraps into a single toroid that grows in size until the entire DNA strand has been transferred. A single toroid can collect DNA strands that have been delivered either independently or in a synchronized manner.

A Cryo-Electron Tomography Study of Phage DNA Transfer into FhuA-Containing Proteoliposomes

For a better understanding of the mechanism of DNA transfer into liposome, the information should be retrieved not only from convent-ional two-dimensional (2D) electron micrographs but also from three-dimensional (3D) images of the phage–proteoliposome complexes. With recent improvements in automated data collection equipment and image analysis software, the determination of 3D images of radiation-sensitive complexes embedded in vitreous ice have been possible using electron tomography.[33,34] The principle of electron tomography consists of the determination of the 3D structure of a single object from a series of 2D images of this object tilted at various angles. The data collection involves tilting of the specimen in the electron microscope, typically more than ±70 degrees, with 2 degree tilt increments.

Several 3D reconstructions of phage–proteoliposome complexes have been calculated and allowed us to analyze at a molecular level the inter-action of T5 phages bound to their receptors (Fig. 7). The resolution of the 3D reconstructions demonstrates unambiguously that, after binding to its receptor, the straight fiber of the phage crosses the lipid bilayer, undergoes a major conformational change, allowing the transfer of its double-stranded DNA into the proteoliposome. Taking into account the crystal structure of the receptor FhuA, we are able to conclude that FhuA is only used as a docking site for the phage and that the straight fiber of the phage tail acts like an "injection needle" creating a passageway at the periphery of FhuA, through which the DNA crosses the membrane.

Conclusion

Besides the important consequences of our study on the molecular mechanisms of bacterial phage infection, the experimental approach using reconstituted proteoliposomes has important applications in many bio-logical contexts. It represents an attractive model system to study the

[33] W. Baumeister, R. Grimm, and J. Walz, *J. Trends Cell Biol.* **9,** 81 (1999).
[34] J. Böhm, O. Lambert, A. S. Frangalis, L. Letellier, W. Baumeister, and J. L. Rigaud, *Cur. Biol.* **11,** 1168 (2001).

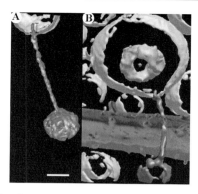

Fig. 7. Three-dimensional reconstructions of phage DNA transfer into FhuA-containing proteoliposome. Isosurface representations of the 3D reconstruction of the phage bound to the proteoliposome before (A) and after DNA transfer (B). (A) T5 phage (icosahedral capsid and long tail) bound to a proteoliposome (upper left of the panel). The top of the vesicle has been cropped to reveal the tip of the phage visible inside the vesicle. (B) The proteoliposome contains DNA condensed into a toroidal structure as a result of the presence of spermine. The phage attached to the vesicle is visible in the lower part of the image. The tail of the phage bends over a bar in the holy film, and its tip crosses the membrane of the proteoliposome. Scale bar: 50 nm.

mechanisms of DNA condensation, and our studies could be extended to DNA condensation by other polyamines or globular, cationic proteins, such as histones, to mimic the cellular environment in pressure-controlled conditions.

From another point of view, the approach can be important for gene therapy applications. These DNA-containing proteoliposomes are conceptually very different from the DNA–cationic lipid complexes classically used for gene delivery. The DNA is truly entrapped and condensed within unilamellar liposomes. Interaction between DNA and lipids is unlikely, because the liposomes retain their initial morphology. The structure of DNA-containing liposomes is stable. DNA-containing proteoliposomes produced in this way could serve as alternative vehicles for the transfer of foreign genomes into eukaryotic cells because: (1) as opposed to cationic liposomes, disruptive electrostatic interactions between lipids and DNA, and between liposomes and cell membranes, are unlikely in our system; (2) the entrapped DNA is of far larger size than the plasmids commonly used; (3) DNA can be strongly condensed as demonstrated in this study and can be thereby better protected from degradation; (4) the lipid bilayer comprising the "membrane" of the encapsulated DNA can be functionalized easily to optimize targeting. The lipid composition of the vesicles can be altered to contain different ratios of neutral to charged lipids, glycosylated lipids, or even cholesterol. Furthermore, other proteins can be co-reconstituted with the phage receptor and components or entrapped within the liposomes to favor the gene delivery process or to protect the DNA from degradation. Also, the technique might be improved by using phage lambda, which is used extensively as a cloning vector and in recombinant technology and whose receptor, the LamB protein, has been reconstituted functionally into liposomes.

Acknowledgments

We acknowledge J. Böhm and W. Baumeister for their fruitful collaboration on electron tomography. We have benefited through this work with the extensive collaboration of L. Letellier and her collaborators. We also thank W. Gelbart and D. Levy for stimulating discussions.

[30] The Hemagglutinating Virus of Japan–Liposome Method for Gene Delivery

By YASUFUMI KANEDA, SEIJI YAMAMOTO, and KAZUYA HIRAOKA

Introduction

Toward the success of human gene therapy, numerous viral and nonviral (synthetic) methods of gene transfer have been developed,[1,2] but each has its own limitations as well as advantages. Therefore, to develop *in vivo* gene transfer vectors with high efficiency and low toxicity, several groups have attempted to overcome the limitations of one vector by combining them with the strengths of another.

Our basic concept is the construction of novel, hybrid-type liposomes with functional molecules inserted into them.[3,4] On the basis of this concept, DNA-loaded liposomes were fused with ultraviolet (UV) inactivated hemagglutinating virus of Japan; Sendai virus (HVJ) to form HVJ-liposomes (approximately 400–500 nm in diameter). Those viral liposomes bind to cell surface sialic acid receptors and fuse with cell membrane to directly introduce DNA into the cytoplasm without degradation. The HVJ–liposomes can encapsulate DNA smaller than 100 kb. RNA, oligodeoxynucleotides (ODN), proteins, and drugs can also be enclosed and

[1] R. C. Mulligan, *Science* **260**, 926 (1993).
[2] F. D. Ledley, *Hum. Gene Ther.* **6**, 1129 (1995).
[3] Y. Kaneda, *Biogenic Amines* **14**, 553 (1998).
[4] Y. Kaneda, Y. Saeki, and R. Morishita, *Mol. Med. Today* **5**, 298, (1999).

delivered to cells. Recently, we explained by use of fluorescence resonance energy transfer (FRET) that the degradation of ODN in the process of delivery to the cytoplasm is inhibited by HVJ-liposomes but not with simple cationic liposomes.[5]

HVJ-liposomes are useful for *in vivo* gene transfer.[6] When HVJ–liposomes containing the LacZ gene were injected directly into one rat liver lobe, approximately 70% of cells expressed LacZ gene activity, and no pathological hepatic changes were observed.[7] One advantage of HVJ–liposomes is allowance for repeated injections. Gene transfer to rat liver cells was not inhibited by repeated injections. After repeated injections, anti-HVJ antibody generated in the rat was not sufficient to neutralize HVJ–liposomes. Cytotoxic T cells recognizing HVJ were not detected in rats transfected repeatedly with HVJ-liposomes.[7] The safety of HVJ–liposomes has been tested and evaluated in monkeys.[8]

To increase the efficiency of gene delivery by HVJ–liposomes, we investigated the lipid components of liposomes[9] and concluded that the most efficient gene expression occurred with liposomes consisting of a phosphatidylcholine, phosphatidylethanolamine, sphingomyelin, phospatidylserine, and cholesterol at the molar ratio of 13.3:13.3:13.3:10 and 50, respectively. The lipid components of the liposomes are very similar to the HIV envelope and mimic the red blood cell membrane.[10] We called the liposomes HVJ–AVE liposomes (i.e., HVJ-artificial viral envelope liposomes). With HVJ–AVE liposomes, gene expression in heart, liver, and muscle was 5–10 times greater than that observed with various nonviral gene transfer methods such as conventional HVJ–liposomes, cationic-lipid–mediated lipofection and naked DNA injection.[4]

Another improvement was construction of cationic-type HVJ–liposomes using cationic lipids. Of the cationic lipids, positively charged DC-cholesterol (DC)[11] has been the most efficient for gene transfer. For luciferase expression, HVJ–cationic DC liposomes were 100 times more

[5] N. Nakamura, D. A. Hart, C. B. Frank, L. L. Marchuk, N. G. Shrive, N. Ota, K. Taira, H. Yoshikawa, and Y. Kaneda, *J. Biochem.* **129**, 755 (2001).
[6] V. J. Dzau, M. Mann, R. Morishita, and Y. Kaneda, *Proc. Natl. Acad. Sci. USA* **93**, 11421 (1996).
[7] T. Hirano, J. Fujimoto, T. Ueki, H. Yamamoto, T. Iwasaki, R. Morishita, Y. Sawa, Y. Kaneda, H. Takahashi, and E. Okamoto, *Gene Ther.* **5**, 459 (1998).
[8] N. Tsuboniwa, R. Morishita, T. Hirano, J. Fujimoto, S. Furukawa, M. Kikumori, A. Okuyama, and Y. Kaneda, *Hum. Gene Ther.* **12**, 469 (2001).
[9] Y. Saeki, N. Matsumoto, Y. Nakano, M. Mori, K. Awai, and Y. Kaneda, *Hum. Gene Ther.* **8**, 1965 (1997).
[10] R. Chander and H. Schreier, *Life Science* **50**, 481 (1992).
[11] K. Goyal and L. Huang, *J. Liposome Res.* **5**, 49 (1995).

efficient than conventional HVJ–anionic liposomes.[9] However, HVJ–cationic liposomes were not appropriate for gene transfer to liver, kidney, heart, and muscle, but they were much more effective for gene transfer to tumor masses or disseminated cancers[12,13] in animal models compared with anionic-type HVJ–AVE liposomes. Therefore, HVJ-anionic and cationic liposomes can complement each other, and each liposome should be used for proper targeting.

Materials and Instrumentation

Chromatographically, pure bovine brain phosphatidylserine-sodium salt (PS) (No. 83032L) is obtained from Avanti Polar Lipids Inc. (Alabaster, AL), and other lipids such as dioleoyl-L-alpha-phosphatidy-lethanolamine (DOPE) (P 5078), sphingomyelin (Sph) (S 0756), egg yolk phosphatidylcholine (PC) (P 2772), DC-Cholestrol (DC-chol) (C 2832), and cholesterol (Chol) (C 8667) are from Sigma (St. Louis, MO). All the lipids are stored at $-20°$. Polypeptone (Pancreatic Digest of Casein) (No. 394-00115) is obtained from Wako (Osaka, Japan). EDTA-3Na (ED3SS), sucrose (S 7903), trizma base (T 1503), chloroform (C 2432), NaCl (S 3171), KCl (P 9333), and dimethylsulfoxide (DMSO) (D 8779) are from Sigma.

The procedures require 50-ml conical tubes (Becton-Dickinson, Lincoln Park, NJ), 35-ml centrifuge tubes (Beckman Instruments, Tokyo, Japan), ultracentrifuge tubes (Hitachi, Tokyo, Japan) and cellulose acetate membrane filters (0.45 μm, No. 190-2545, and 0.20 μm, No. 190-2520 (Nalgene Co., Rochester, NY).

For preparing lipid mixtures, glass tubes 24 mm in diameter and 12 cm long were custom-made (Fujiston 24/40, Iwaki Glass Co. Ltd., Tokyo, Japan), but similar sterilized tubes resistant to chloroform are available. The fresh glass tubes are immersed in saturated KOH-ethanol solution for 24 h, rinsed with distilled water, and heated at $180°$ for 2 h before use. A rotary evaporator (Type SR-650, Tokyo Rikakikai Inc., Tokyo, Japan), vacuum pump with a pressure gauge (Type Asp-13, Iwaki Glass Co. Ltd., Tokyo, Japan), water bath (Thermominder Jr 80, TAITEC, Saitama, Japan), vortex mixer (Scientific Industries, Bohemia, NY), and water bath shaker (Thermominder, TAITEC, Saitama, Japan) are used, but similar instruments are also available from other manufacturers for this purpose. For preparation of plasmid DNA, an endotoxin-free column (Qiagen Inc., Germany) is recommended.

---

[12] T. Otomo, S. Yamamoto, R. Morishita, and Y. Kaneda, *J. Gene Med.* **3,** 345 (2001).
[13] T. Miyata, S. Yamamoto, K. Sakamoto, R. Morishita, and Y. Kaneda, *Cancer Gene Ther.* **8,** 852 (2001).

For purifying HVJ, a low-speed centrifuge (05PR-22, Hitachi, Tokyo, Japan), a centrifuge with JA-20 rotor (J2-HS, Beckman Instruments), and a photometer (Spectrophotometer DU-68, Beckman Instruments) are needed. A UV cross-linker (Spectrolinker XL-1000, Spectronics Co.) is used for inactivation of HVJ. For purifying HVJ–liposomes, an ultracentrifuge with RPS-40T rotor (55P-72, Hitachi, Tokyo, Japan) is used.

Procedures for Preparation of Hemagglutinating Virus of Japan–Liposomes

Preparation of Hemagglutinating Virus of Japan

Preparation of Hemagglutinating Virus of Japan in Eggs

REAGENTS.

1. Polypeptone solution (1% polypeptone, 0.2% NaCl, pH 7.2): To make 500 ml, 5 g of polypeptone and 1 g of NaCl are solubilized in distilled water, the pH is adjusted to 7.2 by adding aliquots of 1 M NaOH, and the total volume is brought to 500 ml with distilled water. The solution is sterilized by autoclaving and stored at 4°.

2. BSS (137 mM NaCl, 5.4 mM KCl, 10 mM TRIS-HCl pH7.6): Eight grams of NaCl, 0.4 g of KCl, and 1.21 g of Trizma base are dissolved in distilled water, the pH is adjusted to 7.6 with aliquots of 1 M HCl and the total volume is brought to 1 l with distilled water. The solution is sterilized by autoclaving and store at 4°.

3. Seed of HVJ: Aliquots (100 μl) of the best seed of HVJ (Z strain) in 10% DMSO are stored at −80°.

METHODS.

1. The seed is thawed rapidly and diluted 1000 times with poly peptone solution. The diluted seed should be kept at 4° before proceeding to the next step.

2. Embryonated eggs are observed under illumination in a dark room, and an injection point is marked at about 0.5 mm above the chorioallantoic membrane. The eggs are disinfected with tincture of iodine and punctured at the point marked.

3. The diluted seed (0.1 ml) is injected into each egg using a 1-ml disposable syringe with a 26-gauge needle. The needle should be inserted vertically so as to stab the chorioallantoic membrane.

4. After inoculation of the seed, the hole punctured on the egg is covered with melted paraffin. Then the eggs are incubated for 4 days at 36.5° in sufficient moisture.

5. The eggs are chilled at 4° for more than 6 h before harvesting the virus.

6. The egg shell is removed partially, and the chorioallantoic fluid is aspirated with a 10-ml syringe (18-gauge needle) and placed in an autoclaved bottle. The fluid should be kept at 4° to avoid freezing. The virus is stable in the fluid at least for 3 months. Steps 2, 3, and 6 can be carried out at room temperature.

Purification of Hemagglutinating Virus of Japan from Chorioallantoic Fluid

REAGENTS.

1. BSS is prepared as described previously.

METHODS.

1. Two hundred milliliters of the chorioallantoic fluid is transferred into four 50-ml disposable conical tubes and subjected to centrifugation at 3000 rpm (1000g) for 10 min at 4° in a low-speed centrifuge.

2. Then, the supernatant is aliquoted into six tubes (Beckman JS-20) and centrifuged at 15,000 rpm (27,000g) for 30 min at 4°.

3. About 5 ml of BSS is added to the pellet in one of the tubes, and the materials are kept at 4° overnight.

4. The pellets are gently suspended, collected in two tubes, and centrifuged as described in step 2. The resultant pellet in each tube is kept at 4° in 5 ml of BSS for more than 8 h.

5. The pellets are suspended gently and centrifuged at 3000 rpm in a low-speed centrifuge.

6. The supernatant is removed to an aseptic tube and stored at 4°.

7. The virus titer is determined by measuring the absorbance at 540 nm of the 10 times–diluted supernatant using a photometer. An optical density at 540 nm corresponds to 15,000 hemagglutinating units (HAU), which is well correlated with fusion activity. The supernatant as prepared previously usually shows 20,000–30,000 HAU/ml. A virus solution prepared aseptically maintains fusion activity for 3 weeks.

Preparation of the Lipid Mixture

METHODS

1. Dry reagents of DOPE (12.2 mg), Sph (11.5 mg), and Chol (23.8 mg) are dissolved in 3870 μl of chloroform. The PC choloform solution (130 μl) is added to the 3870 μl lipid solution. This 4000 μl of lipid solution is called a basal mixture for liposomes. The master mixture is

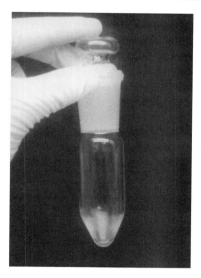

FIG. 1. After evaporation, a thin lipid film is formed inside a glass tube.

ready to prepare anionic or cationic liposomes (described later) or can be stored at $-20°$ after infusing nitrogen gas over the solution.

2. For preparation of an anionic or cationic lipid mixture, 10 mg of PS on 6 mg DC–Chol is added to the basal mixture, respectively.

3. The lipid solution of 0.5 ml is aliquoted into eight glass tubes; each tube contains 10 mg of lipids. The tubes are kept on ice or $-20°$ under nitrogen gas before evaporation. The lipid solution should be evaporated as soon as possible.

4. The tube is attached to a rotary evaporator. The tube should be immersed in a $40°$ water bath at the tip.

5. The organic solvent is evaporated in the rotary evaporator under vacuum. The lipids are dried up for about 10 min. (Fig. 1).

Preparation of Hemagglutinating Virus of Japan–Liposomes

The procedure for the preparation of HVJ–liposomes is illustrated in Fig. 2.

Preparation of Hemagglutinating Virus of Japan–Liposomes Containing DNA

REAGENTS.

1. BSS, prepared as described previously.

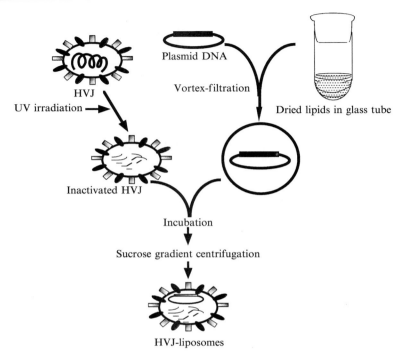

Fig. 2. Preparation of HVJ–liposomes and fusion-mediated gene transfer. DNA-loaded liposomes are fused with UV-inactivated HVJ to form HVJ–liposomes. HVJ–liposomes bind cell surface sialic acid receptors, degrade the receptor, and associate with the lipids in the cell membrane to induce membrane fusion. By fusion of the envelope of HVJ–liposomes with the cell membrane, DNA in the HVJ–liposomes can be introduced directly into the cytoplasm.

2. Plasmid DNA: Plasmid DNAs are purified by a column procedure. The preparations are dissolved in TE solution, BSS, or water. The final concentration of DNA should be more than 1 mg/ml, and stored at $-20°$.

3. Sucrose solutions: To prepare 30% and 60% (w/v) sucrose solutions, 150 g of sucrose is solubilized in BSS and the solution bright up to a total volume to 500 ml and 250 ml, respectively. The solution is sterilized by autoclaving and stored at $4°$.

METHODS.

1. Plasmid DNA (200 μg) in 200 μl is added to a lipid mixture in the glass tube prepared as described previously, and agitated intensely by vortexing for 30 s. followed by incubation at $37°$ for 30 s. This cycle is repeated eight times. By this method, plasmid DNA up to 20 kb is

encapsulated at an efficiency of 10–30% in anionic liposomes or 50–60% in cationic liposomes.

2. For preparing sized unilamellar liposomes, the liposome suspension is filtered through a 0.45-μm pore size cellulose acetate filter and then through a 0.2-μm filter. Sizing by an extruder with polycarobanate filters is better for preparing sized liposomes.

3. In the meantime, HVJ is inactivated and keep on ice. HVJ (15,000 HAU) is added to the liposome suspension and the tube is incubated on ice for 5 to 10 min. Then, the sample is incubated at 37° for 1 h, with shaking (120/min), in a water bath.

4. One milliliter of 60% sucrose and 7 ml of 30% sucrose are added to a centrifuge tube, and the HVJ–liposome mixture is overlaid on top (Fig. 3a). The HVJ–liposome complexes are separated from the free HVJ by sucrose density gradient centrifugation at 62,000g for 90 min at 4°.

5. The conjugated liposomes just above the 30% layer are collected gently (Fig. 3b). Free HVJ is concentrated at the layer between 30% and 60% (Fig. 3b). The final volume of the HVJ–liposome suspension should be approximately 1 ml.

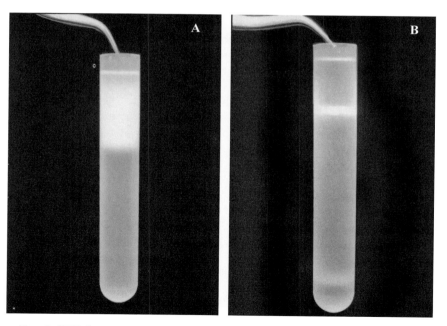

FIG. 3. HVJ–liposomes and free HVJ can be separated by sucrose-density gradient centrifugation. HVJ–liposome mixture (white layer) was added to a sucrose layer consisting of 1 ml of 60% and 7 ml of 30% sucrose solution before centrifugation (A). The upper white band is the conjugate of HVJ and liposomes, and the lower is free HVJ (B).

Gene Transfer by Hemagglutinating Virus of Japan–Liposomes

Transfer of DNA into Cultured Cells

METHODS.

1. HVJ–cationic liposomes should be used for *in vitro* gene transfer, because HVJ–cationic liposomes are approximately 100 times more efficient in gene transfer to cultured cells than HVJ–anionic liposomes.[8]

2. One hundred microliters of the HVJ–cationic liposome suspension is added to 2×10^6 cells (6-well plate) in a serum-containing culture medium.[14]

3. The cells are incubated with the liposomes at 37° for 2 h. Then the medium is replaced with fresh culture medium, and the culture continued.

Gene Transfer In Vivo *by Hemagglutinating Virus of Japan HVJ–Liposomes*

METHODS.

1. For gene transfer to tissues, HVJ–anionic liposomes are recommended. The liposomes are useful for gene transfer to liver, skeletal muscle, heart, lung, artery, brain, spleen, eye, and joint space of rodents, rabbits, dogs, sheep, and monkeys. For example, to introduce DNA into rat liver, 2–3 ml of HVJ–anionic liposomes is injected into the portal vein with a 5-ml syringe with a butterfly-shaped needle[15,16] or directly into the liver under the perisplanchnic membrane using a 5-ml syringe with a 27-gauge needle.[17,18] For gene transfer into rat kidney, 1 ml of anionic HVJ–liposome solution is injected into the renal artery.[19,20] For gene transfer into the rat carotid artery, a lumen of a segment of the artery is filled with 0.5 ml anionic HVJ–liposome complex for 20 min at room temperature using a cannula.[21]

[14] T. Nishikawa, D. Edelstein, X. L. Du, S. Yamagishi, T. Matsumura, Y. Kaneda, M. A. Yorek, D. Beebe, P. J. Oates, H-P. Hammes, I. Giardino, and M. Brownlee, *Nature* **404,** 787 (2000).

[15] Y. Kaneda, K. Iwai, and T. Uchida, *Science* **243,** 375 (1989).

[16] Y. Kaneda, K. Iwai, and T. Uchida, *J. Biol. Chem.* **264,** 12126 (1989).

[17] K. Kato, M. Nakanishi, Y. Kaneda, T. Uchida, and Y. Okada, *J. Biol. Chem.* **266,** 3361 (1991).

[18] N. Tomita, R. Morishita, J. Higaki, S. Tomita, M. Aoki, T. Ogihara, and Y. Kaneda, *Gene Ther.* **3,** 477 (1996).

[19] N. Tomita, J. Higaki, R. Morishita, K. Kato, Y. Kaneda, and T. Ogihara, *Biochem. Biophys. Res. Comm.* **186,** 129 (1992).

[20] Y. Isaka, Y. Fujiwara, N. Ueda, Y. Kaneda, T. Kamada, and E. Imai, *J. Clin. Invest.* **92,** 2597 (1993).

[21] R. Morishita, G. Gibbons, Y. Kaneda, T. Ogihara, and V. J. Dzau, *J. Clin. Invest.* **91,** 2580 (1993).

2. For gene transfer to tumor masses or disseminated tumors, direct injection of cationic HVJ–liposomes (0.1 ml–0.5 ml) is recommended.

3. Repetitive transfection *in vivo* is successful in cancer masses (Yamamoto and Kaneda: manuscript in preparation), liver,[7] and skeletal muscle (Fig. 4).

The Storage of Hemagglutinating Virus of Japan–Liposomes

Reagents

1. DMSO

Methods

1. DMSO is added to HVJ–liposome suspension at a final concentration of 10%.

2. The mixture is stored immediately in a freezer (below $-20°$).

3. Before use, frozen HVJ–liposomes should be thawed rapidly in a water bath at $37°$.

4. Once thawed, all the samples should be used up for gene transfer.

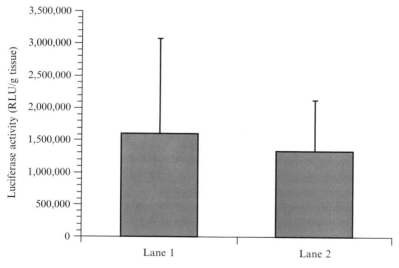

Fig. 4. Repetitive transfection of HVJ–AVE liposomes to mouse skeletal muscle. One hundred microliters of saline (lane 1) or empty HVJ–AVE liposomes (lane 2) was injected into the left and right deltoid muscles of C57BL/6 mouse on days 0 and 7, respectively, and HVJ–AVE liposomes containing the luciferase gene were injected into the left pretibial muscle of the same mouse on day 21. Twenty-four hours after the third injection, luciferase activity in the muscle was measured. The mean and SD was shown. There was no significant difference between lane 1 and lane 2.

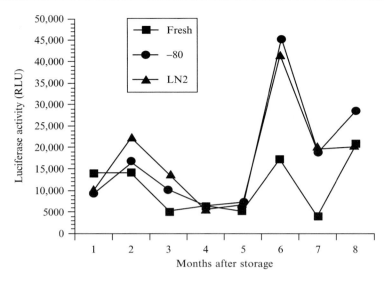

FIG. 5. Stability of HVJ–AVE liposomes. After preparation of HVJ–AVE liposomes containing the luciferase gene, aliquots were stored at $-80°$ or $-170°$ in liquid nitrogen after adding DMSO at a final concentration of 10%. At each time point, the aliquots were thawed and transferred to cultured BHK-21 cells (2×10^6 cells). Twenty-four hours after transfection, luciferase activity was measured. As a control, freshly prepared HVJ–AVE liposomes containing luciferase gene were transferred to BHK-21 cells, and luciferase activity was measured similarly.

Notes

1. UV-inactivated HVJ can be stored for more than 6 months in 10% DMSO at $-80°$ or $-170°$ (in liquid nitrogen).

2. The lipids can be stored, after evaporation, at $-20°$ under nitrogen gas for 1 month.

3. HVJ–liposomes stored as described previously can maintain gene transfer activity for more than 8 months in 10% DMSO at $-80°$ or $-170°$ (in liquid nitrogen) (Fig. 5).

4. The most important component of HVJ–liposomes is the fusion activity of the HVJ envelope. Hemagglutination ability should be checked frequently by hemagglutination of chick red blood cells.[22]

5. Reconstituted fusogenic liposomes can be prepared using isolated fusion proteins derived from HVJ instead of inactivated whole viral particles.[23]

[22] Y. Okada and J. Tadokoro, *Exp. Cell Res.* **26,** 98 (1962).
[23] K. Suzuki, H. Nakashima, Y. Sawa, R. Morishita, H. Matsuda, and Y. Kaneda, *Gene Ther. Reg.* **1,** 65 (2000).

Concluding Remarks

HVJ–liposomes were constructed by combining DNA-loaded liposomes with a fusogenic envelope derived from HVJ. The advantage of fusion-mediated delivery is the protection of molecules in endosomes and lysosomes from degradation. HVJ–liposomes have not been shown to induce significant cell damage *in vivo*. Repetitive transfection is successful *in vivo* because of the low immunogenicity of HVJ–liposomes. Numerous gene therapy strategies using the HVJ–liposome system have been successful in animals. However, one limitation of this system is that the fusion activity of HVJ–liposomes is approximately 2% of that of HVJ itself because of the reduced density of fusion proteins on HVJ–liposomes. Another limitation is that vector production in large scale is difficult because of the complicated vector system using two distinct vesicles, HVJ particles, and liposomes. To overcome these disadvantages, we have recently developed a simpler and more effective vehicle, HVJ envelope vector, by direct incorporation of DNA into an inactivated HVJ particle without liposomes. This vector has also been useful for drug delivery both *in vitro* and *in vivo*.

[31] Enhancement of Retroviral Transduction by Cationic Liposomes

By Krystyna Konopka

Introduction

Background

Successful gene therapy requires (1) the safe and efficient introduction of genetic material into the target cell and (2) the expression of the transgene at therapeutic levels.[1] In the case of retroviral vectors, the viral particles encapsulate a modified genome carrying a therapeutic gene in place of viral genes. Transduction is defined as the abortive (nonreplicative) infection that introduces functional genetic information expressed from the recombinant vector into the target cell.

Recombinant retroviruses are the most studied and widely used vectors for preclinical and clinical gene therapy trials.[2–4] The advantages of retroviral gene transfer are a relatively high efficiency of expression and stable

---

[1] V. F. I. Van Tendeloo, C. Van Broeckhoven, and Z. N. Berneman, *Leukemia* **15,** 523 (2001).

chromosomal integration. The disadvantages include the inability to infect nondividing cells,[5] transcriptional repression,[6] and a potential of insertional mutagenesis.[7] Because only a fraction of cells pass through mitosis at any given time, this severely reduces the application of retroviral vectors in gene therapy to selected targets *ex vivo* such as lymphocytes and hematopoietic progenitor cells. The size of the gene insert is limited, because retroviruses are capable of accommodating approximately 8–10 kb of genetic material.[8]

Further limitations include the slow binding of the virus to specific receptors, a rapid decay of the virus, low concentrations of viral stocks, and the presence of inhibitors (proteoglycans and soluble virus envelope proteins) in viral stocks. Retroviral transduction is usually performed in the presence of a polycation, polybrene (hexadimethrine bromide), to reduce electrostatic interference between the virus and the target cell.

Commericially available cationic liposome preparations are used for the delivery of DNA and RNA into mammalian cells, but expression is episomal and transient. In most cases, cationic liposomes are used for transfection with plasmid vectors.[9,10] We have reported previously that the infectivity of human immunodeficiency virus type 1 (HIV-1) is enhanced *in vitro* by liposomes composed of the positively charged synthetic lipid *N*-2,3-(dioleyloxy) propyl]-*N,N,N*-trimethylammonium chloride (DOTMA).[11–13] The enhancing effect of DOTMA liposomes on HIV-1 infectivity is CD4-dependent; DOTMA liposomes do not mediate HIV-1 infection and replication in CD4⁻ cell lines.[12] A significant enhancement of transduction efficiency has been observed using retrovirus particles complexed with cationic liposomes.[14] In this chapter, we outline the procedures of retroviral transduction and its enhancement by cationic liposomes.

[2] M. A. Kay, J. C. Glorioso, and L. Naldini, *Nat. Med.* **7**, 33 (2001).

[3] D. Miller, *Curr. Top. Microbiol. Immunol.* **158**, 1 (1992).

[4] W. F. Anderson, *Nature (London)* **392**, 25 (1998).

[5] D. G. Miller, M. A. Adam, and A. D. Miller, *Mol. Cell Biol.* **10**, 4239 (1990).

[6] P. M. Chalita and D. B. Kohn, *Proc. Natl. Acad. Sci. USA* **91**, 2567 (1994).

[7] K. T. Smith, A. J. Shepherd, J. E. Boyd, and G. M. Lees, *Gene Ther.* **3**, 190 (1996).

[8] M. Verma and R. K. Naviaux, *Curr. Opin. Genet. Dev.* **1**, 54 (1991).

[9] P. L. Felgner, T. R. Gadek, M. Holm, R. Roman, H. W. Chan, M. Wenz, J. P. Northrop, G. M. Ringold, and M. Danielsen, *Proc. Natl. Acad. Sci. USA* **84**, 7413 (1987).

[10] R. W. Malone, P. L. Felgner, and I. M. Verma, *Proc. Natl. Acad. Sci. USA* **86**, 6077 (1989).

[11] K. Konopka, B. R. Davis, C. E. Larsen, D. R. Alford, R. J. Debs, and N. Düzgüneş, *J. Gen. Virol.* **71**, 2899 (1990).

[12] K. Konopka, L. Stamatatos, C. E. Larsen, B. R. Davis, and N. Düzgüneş, *J. Gen. Virol.* **72**, 2685 (1991).

[13] K. Konopka, E. Pretzer, B. Plowman, and N. Düzgüneş, *Virology* **193**, 877 (1993).

[14] C. P. Hodgson and F. Solaiman, *Nat. Biotechnol.* **14**, 339 (1996).

Retroviral Vectors

Retroviral vectors are packaged into infectious particles after introduction into a packaging cell line containing the necessary structural genes (*gag, pol,* and *env,* coding respectively for structural proteins, reverse transcriptase/integrase, and envelope proteins) but lacking the packaging signal (ψ, psi) required for encapsidation of vector RNA into virions. The recombinant vector containing the therapeutic gene is largely deleted of all viral genes but maintains the packaging signal and the long terminal repeats (LTR) required for viral integration and transcription. This split construct design ensures that the producer cell line generates replication-incompetent retroviral particles that are capable of only one round of infection.[2] The risk of replication-competent retrovirus (RCR) has been further decreased by placing the *gag-pol* and *env* genes on two plasmids that can be introduced separately into the packaging cell line.[15] The problem of the LTR overlap between the vector components has been solved by using heterologous promoters and polyadenylation signals in the retroviral constructs.[16]

Most retroviral vectors are derived from Moloney murine leukemia virus (MoMuLV). The viral envelope glycoproteins determine the host range of retroviral particles through their interactions with receptors on target cells.[17] The ecotropic virus (MoMuLV-E) displays the ecotropic envelope protein and binds to a cationic amino acid transporter (ATRC1) expressed only in rodent cells.[18] MoMuLV vectors have been pseudotyped with the amphotropic envelope protein, allowing transduction of a wide variety of species, including human cells. The receptor for the amphotropic vector is Pit-2 (previously designated as Ram-1), a sodium-dependent phosphate transporter.[17] MoMuLV-based vectors have been also pseudotyped with the gibbon ape leukemia virus (GALV) envelope protein, which uses as a receptor the phosphate transporter Pit-1. Moreover, MoMuLV-based vectors have been pseudotyped with the envelope G protein of the vesicular stomatitis virus (VSV-G) that interacts with ubiquitous membrane receptors such as phosphatidylserine. Most human cells, including bone marrow cells and lymphocytes, express receptors for amphotropic retroviruses. The vast distribution of these receptors seems to increase the vector host range from zebrafish to humans.[19]

[15] D. Markowitz, S. Goff, and A. Bank, *J. Virol.* **62,** 1120 (1988).
[16] H. M. Temin, *Hum. Gene Ther.* **1,** 111 (1990).
[17] D. Miller, D. G. Miller, J. V. Garcia, and C. M. Lynch, *Meth. Enzymol.* **217,** 581 (1993).
[18] H. Wang, M. P. Kavanaugh, R. A. North, and D. Kabat, *Nature (London)* **352,** 729 (1993).
[19] T. Friedmann and J. K. Yee, *Nat. Med.* **2,** 276 (1995).

Retroviral Transduction

Vectors derived from retroviruses have been used widely for gene transfer, although parameters that control the kinetics of infection are still understood poorly. Infection is initiated by binding of the virus to the specific cell surface receptor. The interactions between viral envelope and the receptor trigger envelope-mediated fusion, followed by release of the viral RNA genome and viral proteins into the cytoplasm.[20] The viral RNA is reverse transcribed into double-stranded DNA, which is transported to the nucleus and integrated randomly in the host genome. Once integrated, the viral DNA may be transcribed, leading to the synthesis of therapeutic proteins that are normally not expressed by the target cell. Although integration does not guarantee stable expression of the transduced gene, it is an effective way to maintain the genetic information in self-renewing tissue and in the clonal outgrowth of a stem cell.

Limitations in Retrovirus–Cell Interactions

Low transduction efficiency in human cells and rapid virus inactivation by human serum significantly limit the use of retroviral vectors.[21] The low efficiency is due to (1) the slow rate of virus binding to target cells, (2) the rapid decay of the virus, and (3) the low concentration of the virus in retroviral stocks. Retroviruses lose spontaneously their ability to transfer genes into cells with a half-life of 5.5–7.5 h.[22,23] Retrovirus binding is diffusion-limited. Because of their large size (100 nm), retroviruses diffuse only $\sim$300 μm in one half-life, and, as a result, most viruses lose their infectivity before they reach the surface of target cells. Chuck et al.[24] have shown that in a static transduction system, most viral particles are degraded before they interact with cells. Wang et al.[25] observed the depletion of Mo-MuLV-E virus from the target cell medium. The multiplicity of infection (MOI) is dependent on the virus concentration but is independent of the number of target cells. The authors concluded that binding of the virus is the limiting step in the transduction process. Transduction with an amphotropic MoMuLV vector is also dependent on the virus concentration rather than on the virion-cell multiplicity.[26] The Brownian motions of retroviral

[20] D. Miller, *Proc. Natl. Acad. Sci. USA* **93,** 11407 (1996).

[21] D. Jolly, *Cancer Gene Ther.* **1,** 51 (1994).

[22] S. T. Andreadis and B. O. Palsson, *J. Virol.* **71,** 7541 (1997).

[23] J. M. Le Doux, H. E. Davis, J. R. Morgan, and M. L. Yarmush, *Biotechnol. Bioeng.* **63,** 654 (1999).

[24] A. S. Chuck, M. F. Clarke, and B. O. Palsson, *Hum. Gene Ther.* **7,** 1527 (1996).

[25] H. R. Wang, R. Paul, R. E. Burgeson, D. R. Keene, and D. Kabat, *J. Virol.* **65,** 6468 (1991).

[26] J. R. Morgan, J. M. Le Doux, R. G. Snow, R. G. Tompkins, and M. L. Yarmush, *J. Virol.* **69,** 6994 (1995).

particles in the medium impose a significant limitation for infection of murine NIH 3T3 fibroblasts.[27] The association rate constant is lower than the calculated limitation imposed by viral diffusion, suggesting that binding rather than encounter is rate-limiting.[28]

Several modifications that may improve the interaction between viral particles and target cells are used in transduction protocols. These include the use of polycations, such as polybrene or protamine sulfate, cationic liposomes, polyamines, recombinant fibronectin fragment (CH296), calcium precipitation, centrifugation, and the transwell flow-through system.[14,29–40]

Protocols for Transduction with Retrovirus–Liposome Complexes

Materials

Retroviral vectors
Cells of interest with appropriate cell culture equipment and media
Cationic liposomes
β-gal Staining Kit (Invitrogen Corporation, San Diego, CA)
Toxicity assays

Retroviral Vectors

Retroviral vectors are produced using packaging cell lines, such as $\psi 2$, PA317, Bing, PG13, and FLYRD18. Amphotropic PA317 cells are used

[27] A. S. Chuck and B. O. Palsson, *Hum. Gene Ther.* **7,** 743 (1996).

[28] H. Yu, N. Soong, and W. F. Anderson, *J. Virol.* **69,** 6557 (1995).

[29] K. Toyoshima and P. K. Vogt, *Virology* **38,** 414 (1969).

[30] Y. Kaneko and A. Tsukamoto, *Cancer Lett.* **105,** 39 (1996).

[31] Y. Kaneko and A. Tsukamoto, *Cancer Lett.* **107,** 211 (1996).

[32] W. P. Swaney, F. L. Sorgi, A. B. Bahnson, and J. A. Barranger, *Gene Ther.* **4,** 1379 (1997).

[33] J. Dybing, C. M. Lynch, P. Hara, L. Jurus, H.-P. Kiem, and P. Anklesaria, *Hum. Gene Ther.* **8,** 1685 (1997).

[34] C. D. Porter, K. V. Lukacs, G. Box, Y. Takeuchi, and M. K. L. Collins, *J. Virol.* **72,** 4832 (1998).

[35] M. Themis, S. J. Forbes, L. Chan, R. G. Cooper, C. J. Etheridge, A. D. Miller, H. J. F. Hodgson, and C. Coutelle, *Gene Ther.* **5,** 1180 (1998).

[36] A. G. Wu, X. Liu, A. Mazumder, J. A. Bellanti, and K. R. Meehan, *Hum. Gene Ther.* **10,** 977 (1999).

[37] H. Liu, Y. Hung, S. D. Wissink, and C. M. Verfaillie, *Leukemia* **14,** 307 (2000).

[38] J. J. Song, J.-H. Kim, H. Lee, E. Kim, J. Kim, Y. S. Park, J. Ahn, N.-C. Yoo, J. K. Roh, and B. S. Kim, *Oncol. Reports* **7,** 119 (2000).

[39] H. Lee, J. J. Song, E. Kim, C.-O. Yun, J. Choi, B. Lee, J. Kim, J. W. Chang, and J.-H. Kim. *Gene Ther.* **8,** 268 (2001).

[40] C. D. Porter, *J. Gene Med.* **4,** 622 (2002).

frequently to make retroviral vectors for gene therapy clinical trials. The amphotropic viral envelope glycoprotein, expressed by PA317 cells, permits vectors to infect cells of many species, including humans. The BAG virus is a low-to-medium-titer retroviral vector that is used often for testing gene delivery systems. The producer cells PA317/BAG produce the BAG vector containing an *Escherichia coli lacZ* gene (encoding β-galactosidase), driven by the MoMuLV LTR, and a bacterial neomycin phosphotransferase gene (*neo*) driven by an internal simian virus 40 (SV40) early gene promoter. The expression of the BAG vector can be detected either by β-galactosidase assay, using 5-bromo-4-chloro-3-indolyl-β-D-galacopyranoside (X-gal) as a substrate or by G418 (neomycin analogue) selection. The producer cells are grown to near confluence in RPMI 1640 or in Dulbecco's modified Eagle medium (DMEM) containing 10% fetal bovine serum (FBS) (Gibco-BRL Life Technologies, Inc., Gaithersburg, MD). The viral supernatant is harvested by centrifugation (3000g for 5 min) or by filtration through 0.45- or 0.22-μm pore-size filters (Millipore Corporation, Bedford, MA). The titer of viral supernatants is determined usually on murine NIH 3T3 fibroblasts (CRL 1658; ATCC, Manassas, VA), using serial dilutions of the virus in the absence or presence of polybrene.[14,34–36]

Cells

Cells of choice are cultured according to standard protocols. MoMuLV-based vectors pseudotyped with the amphotropic envelope proteins allow transduction of a wide variety of human cells, such as bone marrow CD34$^+$ cells,[33,37] T lymphocytes,[36] and a number of established cell lines (Table I).

Cationic Liposomes

Cationic liposomes provide a simple means of introducing DNA and other polynucleotides into eukaryotic cells in a process generally referred to as "lipofection." The positive charge of cationic lipids facilitates their association with negatively charged nucleic acids and cell membrane components. Gene delivery by means of cationic liposomes has been used extensively both *in vitro* and *in vivo* and constitutes a viable alternative to viral vectors.[9,10,41–45] A direct gene transfer protocol using a liposome–DNA complex has been approved for injection into solid tumors in

[41] S. Simões, P. Pires, N. Düzgüneş, and M. Pedroso de Lima, *Curr. Opin. Mol. Ther.* **1**, 147, (1999).

[42] J. Zabner, A. J. Fasbender, T. Moninger, K. A. Poellinger, and M. J. Welsh, *J. Biol. Chem.* **270**, 18997 (1995).

TABLE I
CELLS USED FOR RETROVIRAL TRANSDUCTION IN THE PRESENCE OF CATIONIC LIPOSOMES

Name	Cell line species	Type	References
PLC/PRF/5	Human	Hepatoma	31
SK-Hep1	Human	Hepatoma	38,39
Hep3B	Human	Hepatoma	38,39
HepG2	Human	Hepatoma	35
YCC-2, SNU-1	Human	Gastric cancer	38
YCC-1	Human	Gastric cancer	39
U-87MG, U251-N	Human	Glioma	38,39
U343, C6	Human	Brain cancer	39
MCF-7	Human	Breast cancer	38,39
HeLa	Human	Cervical cancer	38,39
HCT-116	Human	Colon cancer	38
H-460	Human	Lung cancer	38,39
HT1080	Human	Fibrosarcoma	14
TE671	Human	Rhabdomyosarcoma	34,40
H-Meso-1	Human	Mesothelioma	34
K-562	Human	Leukemia	37
TF-1[a]	Human	Leukemia	32
KMT2[b]	Human	Hematopoietic	33
NIH 3T3	Mouse	Fibroblast	14,34,35,40

[a] Human myeloerythroid $CD34^+$ leukemia cell line dependent on the presence of granulocyte-macrophage colony-stimulating factor or interleukin-3.
[b] $CD4^+$ cell line spontaneously immortalized from cord blood.

patients.[44] The efficiency of lipofection is strongly influenced by the liposome formulation, the lipid/DNA charge ratio, and target cells.[41,43,46]

The cationic liposome formulations that have been used to enhance retroviral transduction are summarized in Table II. The commercially available lipid formulations are obtained from the following sources: Lipofectin and LipofectAMINE from Gibco-BRL, DOTAP from Sigma (St. Louis, MO) or Boehringer Mannheim Corporation (Indianapolis, IN), PerFect Lipid Kit from Invitrogen, Tfx-50 from Promega Inc. (Madison, WI), Lipid 67 from Genzyme (Framingham, MA), Cytofectin from Glen Research (Sterling, VA), T-MAG and DC-9-12 from Sogo Pharmaceutical

[43] N. Düzgüneş, C. Tros de Ilarduya, S. Simões, R. I. Zhdanov, K. Konopka, and M. C. Pedroso de Lima, *Curr. Med. Chem.* **10**, 1213 (2003).
[44] G. J. Nabel, A. Chang, A. E. G. Nabel, G. Plautz, B. A. Fox, L. Huang, and S. Shu, *Hum. Gene Ther.* **3**, 399 (1992).
[45] H. Gershon, R. Ghirlando, S. B. Guttman, and A. Minsky, *Biochemistry* **32**, 7143 (1993).
[46] P. Hawley-Nelson, V. Ciccarone, G. Gebeyehu, and P. L. Felgner, *Focus* **15**, 73 (1993).

TABLE II
CATIONIC LIPIDS USED TO ENHANCE RETROVIRAL TRANSDUCTION

Lipids	Molar ratio	Commercial name	References
DOTMA[a]/DOPE[b9]	1:1	Lipofectin	14,31,32,35
DOSPA[c]/DOPE[44]	3:1	Lipofectamine	32–40
DOSPA/DOPE	1:1	—	14
DC-Chol[d]/DOPE[47]	3:2	—	32,34,38–40
DOTAP[e]			14,31,34,38–40
DMRIE[f]/DOPE	1:1		35,37
Pfx1 to Pfx8		PerFect Lipid Kit	37
Tfx[g]/DOPE		Tfx-50	34
Lipid 67/DOPE			34
CDAN[h](B198)/DOPE	3:2		34
ACHx[i](CJE52)/DOPE	3:2		34
ACO[j](B130)/DOPE	3:2		34
CTAP[k](B232)/DOPE	3:2		34
Cytofectin			37
T-MAG[l]/DLPC[m]/DOPE	1:2:2		31
T-MAG/DSPC[n]/DOPE	1:2:2		31
DC-912L[o]/DSPC/DOPE	1:2:2		31
T-MAG/DDBA[p]/DOPE	1:2:2		31
DC-912L/DDBA/DOPE	1:2:2		31
T-MAG/DMPC[r]/DOPE	1:2:2		31
T-MAG/DPPC[s]/DOPE	1:2:2		31
DOTMA[t]/PE[u]	9:1		13

[a] DOTMA: N-[1-(2,3 dioleyloxy)-propyl]-N,N,N-trimethylammonium chloride.

[b] DOPE, dioleoylphosphatidylethanolamine.

[c] DOSPA, 2,3-dioleyloxy-N[2(sperminecarboxamido)ethyl]-N,N-dimethyl-1-propanaminium trifluoroacetate.

[d] DC-Chol, $3\beta[N$-$(N',N'$-dimethylaminoetane) = carbamoyl]cholesterol.

[e] DOTAP, N-[1-(2,3-dioleyloxy)propy]N,N,N-trimethylammonium methylsulfate.

[f] DMRIE, 1,2-dimyristyloxypropyl-3-dimethyl-hydroxyethyl ammonium bromide.

[g] Tfx, of N,N,N',N'-tetramethyl-N,N'-bis(2-hydroxyethyl)-2,3-di(oleoyloxy)-1,4-butanediammonium iodide.

[h] CDAN, N^1-cholesteryloxycarbonyl-3,7-diazanonane-1,9-diamine.

[i] ACHx, 3-aza-N^1-cholesteryl-oxycarbonylhexane-1,6-diamine.

[j] ACO, 4-aza-N^1-cholesteryl-oxycarbonyloctane-1,8.

[k] CTAP, $N^1$5-cholesteryloxycarbonyl-3,7,12-triazapentadecane-1,15-diamine.

[l] T-MAG, N-(α-trimethylammonium acetyl)-didodecyl-O-glutamate chloride.

[m] DLPC, dilauroyl L-α-phosphatidycholine.

[n] DSPC, distearoyl-L-α-phosphatidycholine.

[o] DC-912L, $2C_{12}$-L-Glu-ph-C_4N^+.

[p] DDBA, dimethyldioctadecylammoniumbromide.

[r] DMPC, dimyristoyl-phosphatidycholine.

[s] DPPC, dipalmitoyl-phosphatidycholine.

[t] DOTMA, of N-[2,3-(dioleyloxy) propyl]-N,N,N-trimethylammonium chloride.

[u] PE, phosphatidyethanolamine.

[47] X. Gao and L. Huang, *Biochem. Biophys. Res. Commun.* **179,** 280 (1991).

Co. (Tokyo, Japan). For the studies reported from our laboratory, DOT-MA was provided by N. Dyson (formerly Syntex Research, Palo Alto, CA). DOTMA–DOPE liposomes are available currently as Lipofectin (Gibco-BRL). If not mentioned otherwise, lipids are obtained from Avanti Polar Lipids (Alabaster, AL).

Basic Protocol for Transduction with Virus–Liposome Complexes

Adherent target cells are plated into 12- or 24-well plates at 5×10^4 or 2.5×10^4 cells/well, respectively, and incubated overnight at 37°. The liposome preparations are diluted in either OptiMEM or serum-free DMEM (Irvine Scientific, Santa Clara, CA). Liposome–virus complexes (virosomes) are prepared by mixing viral supernatant (serum containing medium) with an appropriate concentration of the liposome reagent at a ratio of 1:1 (v/v). The virus–liposome mixtures are incubated in sterile polystyrene tubes for 30–45 min at room temperature before adding to cells. Medium is aspirated from the target cells and replaced with the virosome complex in a volume of 0.1–0.5 ml per well. Cells incubated with virus plus medium alone serve as controls. The cells are incubated with virosomes for 4 h at 37° and then refed with fresh medium.[38] Alternately, the cells are incubated with virosomes in serum-free DMEM or serum containing DMEM for 24 h, and the medium is changed to DMEM/10% FBS for an additional 24 h before X-gal staining and G418 drug selection (400–600 μg/ml) (Gibco-BRL).[14] Colonies are counted after 10–14 days of G418 selection.

Alternative Protocols for Transduction with
  Virus–Liposome Complexes

*Transduction of Mouse and Human T Lymphocytes with
  PA317/BAG Supernatant*

T lymphocytes from human marrow or spleens of mice are isolated using R&D columns (R&D Systems, Minneapolis, MN). The prestimulated lymphocytes are suspended with the virus–lipid mixture in 25-cm² flasks at 10^6 cells/ml, supplemented with interleukin(IL)-2 and incubated at 37° for 12 h. Six hours after transduction, 10 ml of supernatant is replaced with a similar supernatant supplemented with IL-2. Twelve hours after transduction, the cells are washed twice with phosphate-buffered saline (PBS) and cultured in complete medium. Seventy-two hours after transduction, the cells are stained for β-galactosidase activity by use of X-gal staining.[36]

Transduction of K562 Leukemic Cells and Primary Bone Marrow
Cells with the Amphotropic Retroviral Vector G1Na, Packaged in
PA317-Producing Cells

K562 cells (CCL 243; ATCC) are maintained in DMEM/20% FBS. Transduction is performed with serum-containing viral supernatant, with or without cationic lipid (10 μl/1 ml viral supernatant), recombinant fibronectin fragment (CH296), and the transwell flow-through system alone, or in combination. The transduced cells are harvested 18 h later and replated in Iscove's modified DMEM medium (IMDM) containing methylcellulose (Fisher, Chicago, IL) at a final concentration of 1.12% and 30% FBS with or without G418 (1 mg/ml) for 12 days. Transduction efficiency is determined by the number of colonies in G418-containing media divided by the number in the absence of G418.[37]

One thousand to 10,000 normal CD34$^+$ or chronic myelogenous leukemia (CML) CD34$^-$ cells are prestimulated for 48 h with low-phosphate serum-free medium supplemented with 20 ng/ml Flt-L (Immunex, Seattle, WA), 20 ng/ml stem cell factor (Amgen, Thousand Oaks, CA), and 2 ng/ml IL-3 (R&D Systems). On days 3 or 4, cells are transduced with low-phosphate serum-free viral (PA317/G1N) supernatant in the absence or in the presence of some cytokines and cationic lipids (10 μl/1 ml viral supernatant). Transduced cells are cultured for 2 weeks in the clonogenic methylcellulose medium containing 30% FBS, erythropoietin, (3 UI/ml; Amgen), and the supernatant of the bladder carcinoma cell line 5637 (7.5%), with or without G418 (1 mg/ml). Transduction efficiency is determined as for K562 cells.[37]

Transduction of Brain Tumor Cells with VSV-G Pseudotyped Retrovirus
293T/G/GP/LacZ

The brain tumor cells, U251-N, U87-MG, U343, or C6, are plated at 2.5×10^4 per well in 48-well plates. For liposome treatment, equal volumes of the virus supernatant, 293T/G/GP/LacZ, and the liposome solution at a designated concentration are mixed; 200 μl of the mixture is added to the cells and incubated for 4 h at 37°, then the cells are refed with fresh media. Two days after infection, the transduction efficiency is determined by X-gal staining.[39]

Although these protocols have been published, to obtain optimal enhancement of transduction it is advisable to optimize the cationic lipid/virus ratio for the particular cell line being used.

Monitoring Retroviral Transduction

The BAG vector is detected either by β-galactosidase assay or by identification of G418-resistant colonies containing integrated virus. Of these two methods, β-gal staining is more useful, because it provides rapid visual determination of the relative number of transduced cells (transduction efficiency), whereas the *neo* gene is more often used to evaluate the number of integrated proviruses.[14]

The β-gal staining is usually performed within 48–72 h after infection. The cells are washed with PBS, fixed in the fixative solution containing 2% formaldehyde and 0.2% glutaraldehyde, and stained with an X-gal solution according to the manufacturer's protocol. The cells are examined under a phase-contrast microscope for the development of blue color. Blue stain begins to appear within 2 h. Complete color development occurs after overnight incubation at 37°. Transduction efficiency, designated also as a transduction unit, is defined as the average number of blue (β-gal producing) cells per $100\times$ field multiplied by magnification factors, plate size, and dilution of the infectious stock.[38,39]

Human Immunodeficiency Virus 1 Infection in the Presence of DOTMA Liposomes

Monocytic THP-1 cells[48,49] (TIB-202; ATCC) are maintained at 37° in medium consisting of RPMI 1640 (Irvine Scientific) supplemented with 10% FBS (Sigma), penicillin (50 units/ml), streptomycin (50 μg/ml), and L-glutamine (2 mM). Cells are split 1:5, and fresh medium is added every 3 to 4 days. THP-1 cells (0.1 ml of 0.9×10^6/ml) are incubated with HIV-1$_{IIIB}$, at 10 ng p24/10^6 cells, in the presence of DOTMA–PE liposomes (2.5 or 20 μM) for 2 h at 37° and washed three times to remove unbound virus. DOTMA–PE (transphosphatidylated from egg phosphatidylcholine) liposomes are prepared by reverse-phase evaporation followed by extrusion through polycarbonate membranes of 100-nm pore diameter.[50] Cells are then resuspended in RPMI/10% FBS at 2.25×10^5 cells/ml and cultured in 12-well culture plates (4 ml/well). Half of the culture medium is replaced with fresh medium every 3 or 4 days.[13] The production of p24 antigen in cell-free supernatants is measured by enzyme-linked immunosorbent assay (ELISA).[11] p24 ELISA plates are

[48] S. Tsuchiya, M. Yamabe, Y. Yamaguchi, Y. Kobayashi, and T. Konno, *Int. J. Cancer* **26,** 171 (1980).
[49] S. Tsuchiya, Y. Kobayashi, Y. Goto, H. Okumura, S. Nakae, T. Konno, and K. Tada, *Cancer Res.* **42,** 1530 (1982).
[50] N. Düzgüneş, *Meth. Enzymol.* **367,** 23 (2003).

obtained from the AIDS Vaccine Program (NCI-Frederick Cancer Research and Development Center, Frederick, MD).

Because DOTMA is not readily available, commercially available cationic liposome preparations, such as Lipofectin Reagent, LipofectAMI-NE^{TM} Reagent (Gibco-BRL) or Escort (DOTAP/DOPE) (Sigma) may be used.

Cell Viability Assay

It is important to realize that cationic lipids may have toxic effects on cells. For example, 8 of 11 lipids tested with K562 cells cause significant cell death.[37] The number of viable cells used for experiments is usually determined by Trypan blue exclusion. Cell viability following treatment with cationic liposomes and virus may be quantified by a modified Alamar Blue assay.[51,52]

One milliliter of 10% (v/v) Alamar Blue dye (Trek Diagnostic Systems, Inc., Cleveland, OH) in the appropriate medium is added to each well. After incubation for 2–4 h at 37°, 200 μl of the supernatant is collected from each well and transferred to 96-well plates. The absorbance at 570 nm and 600 nm is measured with a microplate reader. Cell viability (as a percentage of mock-treated control cells) is calculated according to the formula $(A_{570} - A_{600})$ of test cells $\times 100/(A_{570} - A_{600})$ of control cells. After removal of the Alamar Blue/medium mixture, fresh growth medium is added, and cells are returned to the incubator. Thus, the Alamar Blue assay allows determination of viability over the culture period without the detachment of adherent cells. A good correlation is observed between the Alamar Blue assay and Trypan blue staining.[52]

Concluding Remarks

Enhancement of Retroviral Transduction by Cationic Liposomes

Complexation of retroviral vectors with cationic liposomes results in an increased transduction efficiency. The level of enhancement is strongly dependent on target cells, retroviral vectors, and liposomal formulations. Some cationic liposomes are able to enhance only one virus pseudotype and not another. Generally, a higher cationic liposome concentration is needed to obtain maximum titer for amphotropic than for ecotropic viruses.[35]

[51] R. D. Fields and M. V. Lancaster, *Am. Biotechnol. Lab.* **11,** 48 (1993).
[52] K. Konopka, E. Pretzer, P. L. Felgner, and N. Düzgüneş, *Biochim. Biophys. Acta* **1312,** 186 (1996).

Among the many types of liposomes used, lipofectamine seems to be the best,[14,32,33,35,37,38] although the level of enhancement may differ significantly. For example, in HT1980 cells transduced with the PA317/BAG vector, the effectiveness at optimal concentrations, is lipofectamine > lipofectin > DOTAP, resulting in 60-fold, 37-fold, and 5-fold increases in transduction efficiency, respectively.[14] In TF-1 cells transduced with the amphotropic MFG-lacZ retroviral vector, the effectiveness at optimal concentrations, is lipofectamine > DC-Chol–DOPE > lipofectin, resulting in 11.9-fold, 6.2-fold and 1.2-fold increases, respectively.[32] The effect of lipofectamine might be linked to the number of positive charges per molecule (3^+/mole), allowing it to neutralize the negative charges present on the cell membrane and viral envelope more efficiently compared with DOTAP and DC-Chol–DOPE (1^+/mole).[38] Lipofectamine promotes significant transduction with the PA317/BAG vector in human and murine T lymphocytes prestimulated with IL-2. More than 70% of lipofectamine-treated human T lymphocytes turn blue after X-gal staining compared with 10% in the polybrene-treated samples. Control samples are negative after X-gal staining.[35] Lipofectamine enhances also transduction of K562 cells and primary $CD34^+$ cells with the amphotropic retroviral vector by 3-fold and 10-fold, respectively.[37] Complexation of GALV pseudotyped retroviral vector with lipofectamine increases transduction efficiency of $CD34^+$ cells without a requirement for extended *in vitro* culture, or cocultivation with producer cell lines.[33] A combination of cationic lipids with the recombinant fibronectin fragment, CH296,[37] or with centrifugation[32] further improves transduction efficiency.

Receptor Dependence of Transduction Enhancement by Cationic Liposomes

It has been proposed that enhancement of retroviral transduction by cationic liposomes depends on the specific envelope–receptor interaction.[12,14,34,38] The VSV-G–pseudotyped retrovirus interacts with ubiquitous membrane receptors such as phosphatidylserine, instead of a specific cell surface protein.[19] This virus transduces efficiently a variety of cancer cells, especially brain tumor cells both *in vivo* and *in vitro*.[39] It is interesting to note that the transduction efficiency does not increase significantly when the 293T/G/GP/LacZ (VSV-G envelope protein) retrovirus is complexed with lipofectamine, DOTAP, or DC-Chol–DOPE.

In our study, the enhancing effect of DOTMA liposomes on HIV-1 infection was CD4-dependent.[12] DOTMA liposomes do not mediate HIV-1 infection and replication in the $CD4^-$ human Burkitt's lymphoma cell line, Raji, and in K562 leukemic cells, although they enhance significantly binding of HIV to CD4$^-$ KG1 acute myelogenous leukemia cells and Raji cells (K. Konopka, unpublished results, 1990). These results suggest that cationic liposomes modulate the charge associated with the viral envelope and the cell membrane, facilitating interaction between the envelope glycoprotein and its specific receptor, rather than by providing an alternative pathway of the viral entry.

Recently, Porter[40] has examined the effect of lipofectamine, DOTAP and DC-Chol–DOPE on the envelope/receptor-independent transduction in TE671 cells. The binding of retrovirus to target cells was determined by Western blot analysis for cell-associated capsid. Cationic liposomes enhance transduction efficiency proportional to the enhancement of binding of the virus to target cells. The increase in virus binding for the infectious enveloped viruses is similar to that observed for the noninfectious viruses lacking specific envelope proteins (designated "nonenveloped viruses"). The liposome-mediated transduction by enveloped viruses of cells, in which the receptor is blocked, or cells lacking the specific receptor, is equivalent to that shown by the nonenveloped viruses. The post-entry events of nuclear entry and integration for enveloped and nonenveloped viruses are indistinguishable. It is likely that cationic liposomes affect the initial steps of virus binding and/or membrane fusion followed by release of the capsid into the cytoplasm. Thus, the entry of nonenveloped particles complexed with liposomes may be dependent on the intrinsic fusogenicity of liposomes. In contrast, polybrene enhances binding of viruses lacking envelope proteins but does not enhance transduction, indicating that binding is not sufficient for transduction.

In conclusion, although several questions concerning the mechanism of enhancement of retroviral transduction by cationic liposomes remain to be answered, these liposomes are useful tools for increasing retrovirus-mediated gene transfer.

[32] Gene Targeting *In Vivo* with Pegylated Immunoliposomes

By WILLIAM M. PARDRIDGE

Introduction

Targeting therapeutic genes to tissues in cell culture or *in vivo* has been performed in the past with one of three different technologies: (1) viral vectors, (2) cationic liposomes, and (3) plasmid DNA complexed to conjugates of a polycationic protein and a receptor ligand (asialoglycoproteins, transferrin, or folic acid). Viral vectors, such as adenovirus or herpes simplex virus, generate inflammatory responses as a result of the preexisting immunity of virtually all humans to either virus.[1,2] The single injection of either virus into animal or human brain results in a local, inflammatory reaction leading to demyelination.[3-12] The principal nonviral form of gene delivery uses complexes of cationic liposomes and DNA. However, the *in vivo* application of cationic liposomes is limited by the aggregation of

---

[1] U. Herrlinger, C. M. Kramm, K. S. Aboody-Guterman, J. S. Silver, K. Ikeda, K. M. Johnston, P. A. Pechan, R. F. Barth, D. Finkelstein, E. A. Chiocca, D. N. Louis, and X. O. Breakefield, *Gene Ther.* **5,** 809 (1998).

[2] Y. Stallwood, K. D. Fisher, P. H. Gallimore, and V. Mautner, *Gene Ther.* **7,** 637 (2000).

[3] M. S. Lawrence, H. G. Foellmer, J. D. Elsworth, J. H. Kim, C. Leranth, D. A. Kozlowski, A. L. M. Bothwell, B. L. Davidson, M. C. Bohn, and D. E. Redmond, Jr., *Gene Ther.* **6,** 1368 (1999).

[4] A. P. Byrnes, J. E. Rusby, M. J. A. Wood, and H. M. Charlton, *Neuroscience* **66,** 1015 (1995).

[5] M. M. McMenamin, A. P. Byrnes, H. M. Charlton, R. S. Coffin, D. S. Latchman, and M. J. A. Wood, *Neuroscience* **83,** 1225 (1998).

[6] C. M. Kramm, N. G. Rainov, M. Sena-Esteves, M. Chase, P. A. Pechan, E. A. Chiocca, and X. O. Breakefield, *Hum. Gene Ther.* **7,** 291 (1996).

[7] J. G. Smith, S. E. Raper, E. B. Wheeldon, D. Hackney, K. Judy, J. M. Wilson, and S. L. Eck, *Hum. Gene Ther.* **8,** 943 (1997).

[8] R. A. Dewey, G. Morrissey, C. M. Cowsill, D. Stone, F. Bolognani, N. J. Dodd, T. D. Southgate, D. Klatzmann, H. Lassmann, and M. G. Castro, *Nat. Med.* **5,** 1256 (1999).

[9] M. J. Driesse, A. J. P. E. Vincent, P. A. E. Sillevis Smitt, J. M. Kros, P. M. Hoogerbrugge, C. J. J. Avezaat, D. Valerio, and A. Bout, *Gene Ther.* **5,** 1122 (1998).

[10] M. J. Driesse, M. C. Esandi, J. M. Kros, C. J. J. Avezaat, Ch. J. Vecht, C. Zurcher, I. van der Velde, D. Valerio, A. Bout, and P. A. E. Sillevis Smitt, *Gene Ther.* **7,** 1401 (2000).

[11] M. J. A. Wood, H. M. Charlton, K. J. Wood, K. Kajiwara, and A. P. Byrnes, *Trends Neurosci.* **19,** 497 (1996).

[12] K. Kajiwara, A. P. Byrnes, Y. Ohmoto, H. M. Charlton, M. J. A. Wood, and K. J. Wood, *J. Neuroimmunology* **103,** 8 (2000).

cationic liposome–DNA complexes in physiological saline. The cationic lipid–DNA formulations do not aggregate in water. However, once the salt content is raised to the physiological level, the complexes become electrically neutral and aggregate into multimicron structures.[13–16] Therefore, when cationic liposome–DNA complexes are injected intravenously, large aggregates form immediately and are deposited in the first vascular bed, the pulmonary circulation. Gene expression in the lung is several log orders of magnitude greater than gene expression in peripheral tissues such as the liver or spleen, and there is no gene expression in brain following the intravenous administration of cationic liposome–DNA complexes.[17–19] Aggregation also occurs with complexes of polycations and DNA; in this approach, the polycationic protein substitutes for the cationic lipid. However, aggregation in saline is observed, and there is preferential expression of the exogenous gene in the pulmonary circulation.[20] The lung is targeted by gene delivery systems composed of polycationic lipids or proteins, because the pulmonary circulation is the very first vascular bed immediately distal to an intravenous injection, and aggregates are trapped in the lung microcirculation. The instability of the gene formulation in serum has also been observed for conjugates of a receptor ligand and plasmid DNA complexed to a polycationic protein such as polylysine, and this instability is not altered by pegylation of the polylysine.[21] Presumably, the electrostatic interactions that bind the anionic DNA to the cationic protein are disrupted by serum proteins *in vivo*. Each of the traditional approaches for gene targeting *in vivo* have distinct advantages. However, each approach also has significant disadvantages that prevent the widespread application of the gene targeting technology *in vivo*.

An alternative approach to targeting therapeutic genes to tissues *in vivo* is the use of pegylated immunoliposomes (PIL), which are depicted in Fig. 1A. The PIL formulation is similar to a viral vector in that the DNA

[13] H. Matsui, L. G. Johnson, S. H. Randell, and R. C. Boucher, *J. Biol. Chem.* **272,** 1117 (1997).
[14] R. I. Mahato, A. Rolland, and E. Tomlinson, *Pharm. Res.* **14,** 853 (1997).
[15] C. Plank, M. X. Tang, A. R. Wolfe, and F. C. Szoka, Jr., *Hum. Gene Ther.* **10,** 319 (1999).
[16] T. Niidome, N. Ohmori, A. Ichinose, A. Wada, H. Mihara, T. Hirayama, and H. Aoyagi, *J. Biol. Chem.* **272,** 15307 (1997).
[17] L. G. Barron, L. S. Uyechi, and F. C. Szoka, Jr., *Gene Ther.* **6,** 1179 (1999).
[18] K. Hong, W. Zheng, A. Baker, and D. Papahadjopoulos, *FEBS Lett.* **400,** 233 (1997).
[19] G. Osaka, K. Carey, A. Cuthbertson, P. Godowski, T. Patapoff, A. Ryan, T. Gadek, and J. Mordenti, *J. Pharm. Sci.* **85,** 612 (1996).
[20] S. M. Zou, P. Erbacher, J. S. Remy, and J. P. Behr, *J. Gene Med.* **2,** 128 (2000).
[21] D. Y. Kwoh, C. C. Coffin, C. P. Lollo, J. Jovenal, M. G. Banaszczyk, P. Mullen, A. Phillips, A. Amini, J. Fabrycki, R. M. Bartholomew, S. W. Brostoff, and D. J. Carlo, *Biochim. Biophys. Acta.* **1444,** 171 (1999).

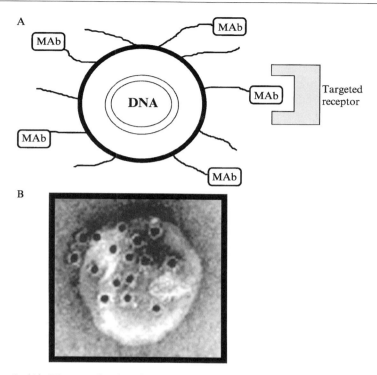

FIG. 1. (A) Diagram showing the supercoiled circular double-stranded plasmid DNA encapsulated in the interior of a pegylated immunoliposome (PIL). There are approximately 3000 strands of polyethylene glycol of 2000-Dalton molecular weight, designated PEG^{2000}, attached to the liposome surface, and 1%–2% of the PEG strands are conjugated with a targeting monoclonal antibody (MAb). For gene targeting to the brain, a MAb directed against the transferrin receptor (TfR) has been used in rats[22,23] and mice,[24] and an antibody directed against the human insulin receptor (HIR) has been used for drug targeting to human cells.[37] The targeted receptor could be the TfR, the HIR, or some other receptor. The targeting MAb must be an endocytosing antibody that enters the cell on binding to an exofacial epitope on the receptor. The MAb binding site on the receptor should be spatially removed from the binding site of the endogenous ligand, so as to not interfere with endogenous transport. (B) Transmission electron microscopy of a PIL. The mouse MAb tethered to the tips of the PEG^{2000} are bound by a conjugate of 10 nm gold and an anti-mouse secondary MAb. The gold particles are bound to the MAb conjugated to the tip of the PEG strand, which projects from the surface of the liposome. The position of the gold particles illustrates the relationship of the PEG-extended MAb and the liposome. Original magnification = 245,000.

is housed in the interior of a nanocontainer, and the PIL formulation is similar to the receptor-mediated approach in that the gene is targeted *in vivo* to a specific receptor. In the PIL formulation, a nonviral double-stranded supercoiled plasmid DNA is packaged in the *interior* of a neutral or slightly anionic liposome.[22–24] Packaging the DNA in the interior of the liposome renders the DNA resistant to the ubiquitous endonucleases that are present *in vivo* and that rapidly degrade exposed DNA.[25] The liposome is prepared from a mixture of neutral lipid, such as 1-palmitoyl-2-oleoyl-*sn*-glycerol-3-phosphocholine (POPC), cationic lipid, such as didodecyldimethylammonium bromide (DDAB), and anionic lipid, such as distearoylphosphatidylethanolamine (DSPE) conjugated to polyethylene glycol (PEG). A small amount of cationic lipid (2–3% of the total lipid) is included in the lipid formulation to stabilize the anionic DNA. The net charge of the liposome is anionic, because there is an excess of DSPE-PEG relative to DDAB (see later).

The encapsulation of DNA in the interior of a conventional liposome would not yield significant gene expression *in vivo*, because this structure would be immediately coated with serum proteins following intravenous injection. The protein-coated liposome would be rapidly removed from blood by cells lining the reticuloendothelial system (RES).[26] The blood transit time can be prolonged, and uptake by the RES can be minimized by pegylation of the surface of the liposome, wherein 2000–3000 strands of PEG of varying molecular weight (2000–5000 Da) are conjugated to the liposome lipids.[27] The 2000-D PEG is typically used and is designated PEG^{2000}, and the strands of PEG projecting from the surface of the liposome are depicted in Fig. 1A. However, DNA encapsulated in the interior of pegylated liposomes would not be specifically targeted to tissues. Organ targeting is accomplished by tethering a receptor-specific ligand to the tips of 1–2% of the PEG strands.[28] The targeting ligand might be an endogenous peptide, modified protein, or a peptidomimetic monoclonal antibody (MAb). The conjugation of the targeting MAb to the tips of the liposome-anchored PEG strands is demonstrated in Fig. 1B. The PIL was

[22] N. Shi and W. M. Pardridge, *Proc. Natl. Acad. Sci. USA* **97,** 7567 (2000).

[23] N. Shi, R. J. Boado, and W. M. Pardridge, *Pharm. Res.* **18,** 1091 (2001).

[24] N. Shi, Y. Zhang, C. Zhu, R. J. Boado, C. Zhu, and W. M. Pardridge, *Proc. Natl. Acad. Sci. USA* **98,** 12754 (2001).

[25] K. W. C. Mok, A. M. I. Lam, and P. R. Cullis, *Biochim. Biophys. Acta.* **1419,** 137 (1999).

[26] D. V. Devine and J. M. J. Marjan, *Crit. Rev. Ther. Drug Carrier Sys.* **14,** 105 (1997).

[27] D. Papahadjopoulos, T. M. Allen, A. Gabizon, E. Mayhew, K. Matthay, S. K. Huang, K. D. Lee, M. C. Woodle, D. D. Lasic, and C. Redemann, *Proc. Natl. Acad. Sci. USA* **88,** 11460 (1991).

[28] J. Huwyler, D. Wu, and W. M. Pardridge, *Proc. Natl. Acad. Sci. USA* **93,** 14164 (1996).

visualized with electron microscopy after binding to the PIL-extended MAb a secondary anti-mouse IgG that was conjugated with 10-nm gold particles.

The MAb binds an exofacial epitope on the targeted receptor (Fig. 1A), and this triggers receptor-mediated endocytosis of the PIL.[29] The targeting of genes to the intracellular compartment of cells following intravenous injection is a "two-barrier" gene-targeting problem. The PIL must traverse both the capillary endothelial membrane (first barrier) and then cross the plasma membrane of the tissue cell (second barrier). For gene targeting to the brain, the targeting ligand binds to a receptor (e.g., the transferrin receptor [TfR] or the insulin receptor) that is present on both the first (capillary) and second (tissue cell) barriers.[29,30] Binding to the capillary endothelial receptor triggers receptor-mediated transcytosis of the PIL across the microvascular endothelial barrier. Binding of the targeting ligand to the receptor on brain cell membranes then triggers receptor-mediated endocytosis of the PIL into the target brain cell subsequent to transport across the microvascular endothelium.

Gene targeting to tissues such as liver or spleen, which have highly porous sinusoidal capillary beds, is a one-barrier gene-targeting problem, because the PILs freely cross the sinusoidal barrier.[23] The limiting barrier is endocytosis across the plasma membrane of the parenchymal cell in liver or spleen. However, in tissues such as heart or brain, which have continuous endothelial barriers, the exodus of the PIL from the capillary compartment to the organ interstitial space is minimal in the absence of targeting of the PIL across the endothelial barrier. The endothelial barrier is tightest in the brain, and the brain capillary wall constitutes the blood–brain barrier (BBB). Gene targeting in the brain is accomplished by using a MAb to the TfR, which is expressed on both the BBB and the brain cell membrane (BCM).[22] The anti-TfR MAb is an endocytosing antibody and binds an exofacial epitope on the TfR, and this binding triggers receptor-mediated transcytosis across the endothelial barrier and receptor-mediated endocytosis into brain cells.[31] In contrast to the brain, there is minimal TfR on the capillary endothelium in heart or kidney. Consequently, the PIL does not escape the capillary compartment in the heart or kidney, and there is no gene expression in these organs when the PIL is targeted to the TfR.[22–24]

[29] J. Huwyler, J. Yang, and W. M. Pardridge, *J. Pharmacol. Exp. Ther.* **282,** 1541 (1997).

[30] J. Huwyler and W. M. Pardridge, *J. Neurochem.* **70** 883 (1998).

[31] W. M. Pardridge, "Brain Drug Targeting; The Future of Brain Drug Development." Cambridge University Press, Cambridge, 2001.

TABLE I
SPECIES-SPECIFIC PEPTIDOMIMETIC MONOCLONAL ANTIBODIES FOR
GENE TARGETING TO THE BRAIN

Species to be targeted	Targeting ligand	Reference
Mouse	8D3 rat MAb to mouse transferrin receptor	33
Rat	OX26 murine MAb to rat transferrin receptor	32
Rhesus monkey	83–14 murine MAb to human insulin receptor	34
Human	Genetically engineered chimeric HIRMAb	35

MAb, monoclonal antibody; HIR, human insulin receptor; HIRMAb, MAb against HIR.

Peptidomimetic MAb targeting ligands tend to be species-specific, and a panel of targeting MAbs has been developed for gene targeting to the brain in different species (Table I). For gene targeting to rat brain, the OX26 murine MAb to the rat TfR is used.[32] However, the OX26 MAb is not active in mice.[33] Gene targeting to mice, including transgenic mice, is accomplished with the rat 8D3 MAb to the mouse TfR.[33] Gene targeting to the primate brain can be achieved with the murine 83–14 MAb to the human insulin receptor (HIR).[34] The HIRMAb cross-reacts with the insulin receptor at both the human BBB and the BBB of Old-World primates, such as Rhesus monkeys. The HIRMAb does not react with the insulin receptor of New-World primates, such as squirrel monkeys, because of the reduced genetic similarity between humans and New-World primates.[34] The murine HIRMAb cannot be used in humans because of immunological reactions in humans to proteins of mouse origin. However, a genetically engineered chimeric HIRMAb has been produced, and this chimeric HIRMAb has an affinity for the HIR that is identical to that of the original murine MAb.[35] The chimeric HIRMAb is avidly transported across the primate BBB *in vivo*, with 2% of the injected dose (ID) being delivered to the primate brain *in vivo* following a single intravenous injection. The chimeric HIRMAb could be used to target therapeutic genes to the brain of humans.

One goal of gene therapy is the widespread expression of the exogenous gene in the targeted organ following noninvasive administration. Because gene targeting technology was not developed, therapeutic genes were

[32] W. M. Pardridge, J. L. Buciak, and P. M. Friden, *J. Pharmacol. Exp. Ther.* **259,** 66 (1991).
[33] H. J. Lee, B. Engelhardt, J. Lesley, U. Bickel, and W. M. Pardridge, *J. Pharmacol. Exp. Ther.* **292,** 1048 (2000).
[34] W. M. Pardridge, Y. S. Kang, J. L. Buciak, and J. Yang, *Pharm. Res.* **12,** 807 (1995).
[35] M. J. Coloma, H. J. Lee, A. Kurihara, E. M. Landaw, R. J. Boado, S. L. Morrison, and W. M. Pardridge, *Pharm. Res.* **17,** 266 (2000).

delivered to the brain with neurosurgical approaches such as craniotomy. In addition to being highly invasive, craniotomy is not a useful approach to gene delivery to the brain, because the effective treatment volume following direct injection into the brain is only the volume at the tip of the injection needle or <1 mm^3. Instead, what is desired is the widespread expression of a therapeutic gene throughout the brain or any organ, and this can only be accomplished by targeting the therapeutic gene through the organ capillary bed. In the human brain, there are 400 miles of capillaries, and the surface area of the brain endothelial barrier is about 20 m^2. Therefore, if an exogenous gene is targeted through the capillary wall, there is immediate distribution of the gene throughout the entire organ volume.

Another goal of gene targeting is tissue-specific expression of the exogenous gene following intravenous administration. One might anticipate that the only way the expression of an exogenous gene might be restricted to a particular organ is to physically inject the gene into the organ. However, tissue-specific gene expression can be achieved with noninvasive intravenous routes of administration through the combined use of (1) gene-targeting technology and (2) tissue-specific gene promoters.[24] Many expression plasmids are driven by the SV40 promoter, which is widely expressed in most tissues. The intravenous injection of a β-galactosidase expression plasmid that is driven by the SV40 promoter, and packaged in the interior of OX26-PIL, results in the expression of the exogenous gene in multiple TfR-rich organs, including brain, liver, and spleen.[22–24] There is no measurable gene expression in TfR-poor organs, such as the kidney or heart.[22–24] The SV40 promoter was then replaced with a brain-specific promoter, taken from the 5'-flanking sequence (FS) of the human glial fibrillary acidic protein (GFAP) gene. The β-galactosidase gene, under the influence of the GFAP promoter, was packaged in the interior of 8D3-PIL and injected intravenously in mice.[24] Under these conditions, there was expression of the exogenous gene only in the brain, with no detectable gene expression in peripheral tissues including the liver or spleen. A gene can be delivered to peripheral tissues rich in the targeted receptor, but there will not be significant expression if the gene is driven by a tissue-specific promoter that is not transcriptionally active in a given tissue. This promoter will be activated only by specific trans-acting factors, which are expressed in a tissue-specific pattern. Therefore, the limiting factor in achieving tissue-specific gene expression *in vivo* with the PIL gene–targeting technology is the choice of the promoter-driving gene expression and the tissue specificity of that promoter.

Before using the gene targeting technology described in the following, the investigator needs to obtain access to the following items that are not readily commercially available. First, if a peptidomimetic MAb is used as the targeting ligand (Table I), 10 to 100-mg quantities of the MAb may be required. This necessitates that the hybridoma secreting the MAb be available and that milligram quantities of the MAb be produced by either propagating liters of hybridoma-conditioned media or by the ascites method. Second, the targeted gene may be a commercially available reporter gene such as luciferase or β-galactosidase. However, if the targeted gene is a therapeutic gene, appropriate expression plasmids must be designed. Important elements in the design of the expression plasmid are the tissue specificity of the 5'-promoter or 3'-enhancer elements and the activity of elements that promote persistence of the gene by means of a extrachromosomal replication.[31] Third, a bifunctional PEG derivative is required (Fig. 2A), and these often must be obtained by custom synthesis.[22,28]

Methods

Maxiprep of Plasmid DNA

A typical PIL formulation starts with encapsulation of 150 μg of plasmid DNA, and this amount of plasmid DNA can be routinely isolated with the QIAfilter Plasmid Maxi kit from Qiagen, Inc. (Valencia, CA). The *Escherichia coli*, or other suitable host, that has been transformed with the plasmid DNA is removed from the freezer, and an aliquot is used to inoculate 150 ml of LB medium. The liquid culture is incubated at 37° with vigorous shaking (200 rpm) for 20–24 hours using a rotary water-bath shaker. The bacterial pellet is harvested in 250-ml plastic bottles at 6000 g for 15 min at 4° and is resuspended in 10 ml of ice-cold resuspension buffer P1 (0.05 M TRIS, pH 8.0, 10 mM EDTA). The P1 buffer contains 100 μg/ml of RNase A. Ten milliliters of lysis buffer P2 (0.2 M NaOH, 1% SDS) is added, and the suspension is mixed gently but thoroughly and incubated at room temperature for 5 min. Vortexing should be avoided, because this could shear the DNA. Ten milliliters of cold neutralization buffer P3 (3 M potassium acetate, pH 5.5) is added to the lysate, and the lysate is poured into the barrel of the QIAfilter cartridge and incubated at room temperature for 10 min. This 10-min incubation at room temperature is necessary for optimal performance of the QIAfilter maxi cartridge. The cartridge should not be agitated at this time. Ten milliliters of equilibration buffer QBT (0.75 M NaCl, 50 mM MOPS, pH 7.0, 15% isopropanol, 0.15% Triton X-100) is added and allowed to elute from the column by gravity. This is filtered until all lysate is passed from the QIAfilter cartridge, one should not apply force. Approximately 25 ml of lysate is

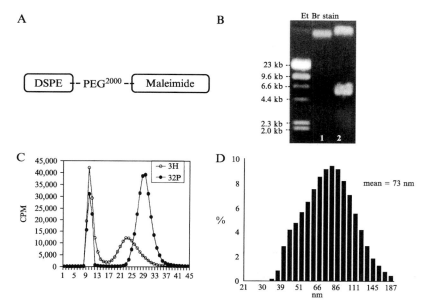

FIG. 2. (A) The structure of the bifunctional PEG^{2000} that contains a distearoylphosphatidylethanolamine (DSPE) moiety at one end, for incorporation into a liposome surface, and a maleimide moiety at the opposite tip of the PEG strand, to enable conjugation to a thiolated monoclonal antibody. (B) DNA encapsulated within pegylated liposomes before (lane 2) and after (lane 1) nuclease treatment is resolved with 0.8% agarose gel electrophoresis followed by ethidium bromide (Et Br) staining. DNA molecular weight size standards are shown on the left side. Before nuclease digestion, approximately 50% of the DNA associated with the pegylated liposome is bound to the exterior of the liposome (lane 2), and this is quantitatively removed by the nuclease treatment (lane 1).[22] (C) The conjugation of the MAb to the pegylated liposome carrying the encapsulated plasmid DNA following nuclease digestion is demonstrated by Sepharose CL-4B gel filtration chromatography. A trace amount of the encapsulated plasmid DNA was radiolabeled with ^{32}P, and a trace amount of the targeting MAb was radiolabeled with ^{3}H. This study shows comigration in the first peak off the column of the MAb, which is conjugated to the PEG^{2000} strand, and the plasmid DNA, which is encapsulated in the interior of the liposome. The unconjugated MAb and the exteriorized, digested DNA elute from the column at later elution volumes.[37] (D) The mean diameter of pegylated immunoliposomes with encapsulated plasmid DNA is 73 nm as determined by quasielastic light scattering.[22]

generally recovered after filtration. The Qiagen-tip is washed with wash buffer QC (1 M NaCl, 50 mM MOPS, pH 7.0, 15% isopropanol) and the DNA is with eluted 15 ml of elution buffer QF (1.25 M NaCl, 50 mM MOPS, pH 8.5, 15% isopropanol), the eluate is collected in 50-ml Nalgene Oak Ridge centrifuge tubes (polycarbonate centrifuge tubes should not be used). The DNA is precipitated by adding 10.5 ml (0.7 volumes) of room temperature 2-propanol to the eluted DNA. This is centrifuged at 15,000 g for 30 min at $4°$ and the supernatant is carefully decanted. All solutions should be kept at room temperature to minimize salt precipitation, although centrifugation is carried out at $4°$ to prevent overheating of the sample. The DNA pellet is washed with 5 ml of room temperature 70% ethanol and centrifuged at 15,000g for 10 min at $4°$. The supernatant is carefully decanted the pellet is air-dried for 5–10 min, and the DNA is dissolved in a suitable volume of TE buffer (0.01 M TRIS, pH = 8.0, 0.1 mM ethylenediamine tetraacetic acid [EDTA]). The concentration of the DNA is determined by measurement of the A_{260}. An A_{260} of 1.0 is equivalent to a DNA concentration of 50 μg/ml. Overdrying the pellet will make the DNA difficult to redissolve. DNA dissolves best under alkaline conditions and is not easily dissolved in acidic buffers because of protonation of the phosphate groups. A typical Maxiprep procedure should generate 750–1250 μg of DNA, which is enough for several PIL preparations.

Labeling of Plasmid DNA by Nick Translation

A small aliquot of the plasmid DNA should be ^{32}P-labeled by nick translation, so that the encapsulation of the DNA in the interior of the PIL can be confirmed and quantitated. The Nick Translation kit is obtained from Life Technologies, Inc. (Gaithersburg, MD) and contains the DNA polymerase I–DNaseI mixture, deoxynucleotide triphosphates (dNTP) minus deoxycytidine triphosphate (dCTP), and stop buffer. Sephadex-G25 (fine) spin columns are obtained from Roche (Indianapolis, IN). To a 1.5-ml microfuge tube, 5 μl of the dNTP mixture, 1 μl of plasmid DNA (1–2 μg), 4 μl (40 μCi) ^{32}P-dCTP, and 35 μl of distilled water are added. Following the manufacturer's instructions, the DNA polymerase I/DNase I is added, the mixture is incubated for 60 min at $15°$, and the reaction is stopped with addition of EDTA stop buffer. The unincorporated nucleotides are removed with a G25 spin column, and the percent of radioactivity that is precipitated by trichloroacetic acid (TCA) should be measured. An aliquot of radioactivity is removed, 2.5% bovine serum albumin (BSA) is added as a carrier, and the DNA is precipitated by the addition of 10% cold TCA. More than 95% of the radioactivity should be precipitated by TCA. If this is not observed, there may have been problems with the gel filtration spin column and removal of unreacted radiolabeled dCTP. There can be degradation of the plasmid DNA during this procedure, and the TCA precipitability will not detect partial degradation. Plasmid DNA degradation is best assessed with agarose gel electrophoresis (see later) and film autoradiography.

Liposome Formation and DNA Encapsulation

POPC (MW = 760 D) and DDAB (MW = 631 D) are obtained from Avanti Polar Lipids (Alabaster, AL). The DSPE-PEG^{2000} (MW = 2748 D) is obtained from Shearwater Polymers as a catalog item. A bifunctional PEG^{2000} derivative with a DSPE moiety at one terminus and a maleimide (MAL) moiety at the other end is custom synthesized by Shearwater Polymers (MW = 2955 D). If stored properly, the DSPE-PEG^{2000}–MAL reagent is stable for at least a year. The LipoFast™-Basic extruder and the 400-, 200-, 100-, and 50-nm pore-sized polycarbonate membranes are obtained from Avestin (Ottawa, Canada). A total of 20 μmol of lipids is typically used for a PIL preparation involving the encapsulation of 150–250 μg of plasmid DNA. To a test tube containing 1 ml of chloroform, 97 μl of 146 mg/ml of POPC in chloroform (18.6 μmol), 60 μl of 10 mM DDAB (0.6 μmol, 3% final concentration), 100 μl of 16.4 mg/ml DSPE-PEG^{2000} (0.6 μmol, 3% final concentration), and 120 μl of 5 mg/ml DSPE-PEG^{2000}–maleimide (0.2 μmol, 1% final concentration) are added. Although DSPE is an electrically neutral lipid, DSPE-PEG or DSPE-PEG-maleimide are anionic lipids. The total concentration of anionic lipids in the formulation is 4%, and the total concentration of cationic lipids is 3%, so the net charge of the liposome is anionic. Lipids are stored under a nitrogen atmosphere. Following mixing of the lipids, the chloroform is evaporated under a stream of nitrogen while it is vortexed to produce a thin-layer lipid film. This should be performed continuously without interruption, and the lipid film should dry well and sit at room temperature for at least 30 min before going to the next step. Then, 0.2 ml of 0.05 M TRIS, pH = 7.4 is added to produce a total lipid concentration of approximately 100 mM. This solution is vortexed three to four times and sonicated for 2 min with a bath sonicator; it is vortexed again for approximately 1 min. Then, 150 μg of plasmid DNA and 2 μCi of ^{32}P-DNA in a total volume of approximately 300 μl is added, and the total ^{32}P radioactivity is determined. This CPM is designated "A" and is used for subsequent calculations (see later). An ethanol–dry ice bath is prepared, and the lipids undergo a freeze–thaw cycle for a total of 10 times with placement in the ice bath for 5 min followed by thawing at 40° for 1.5–2 min. Before the freeze–thaw cycles, if air bubbles are visible in the lipid solution, the solution should be allowed to sit, or an additional two to three freeze–thaw cycles should be performed, so as to eliminate the air bubbles, because these inhibit DNA encapsulation. At this point, 0.05 M HEPES, pH = 7.0, is added to a final volume of 500 μl (40 mM lipid), and the solution is mixed well by tapping. Two stacked 400-nm polycarbonate filters are inserted into a washed extruder and the liposome and DNA mixture is forced through the extruder; the extrusion is repeated seven times. Then, the mixture is forced five times through the extruder containing two stacked 200-nm polycarbonate filters. Finally, the lipids should be forced three to four times through two stacked polycarbonate filters with a pore size of 100 nm. Depending on the results of the liposome sizing experiments, it may be preferable to extrude the solution through an additional two stacked polycarbonate filters with a pore size of 50 nm, so that the mean diameter of the PIL is less than 100 nm. There is a tradeoff, because smaller size liposomes are presumably transcytosed easier but encapsulate DNA less efficiently. A compromise is achieved by preparing liposomes with diameters in the range of 75–100 nm. At this point, approximately half of the DNA has been interiorized in the liposome, and half remains absorbed to the surface of the liposome. The exterior bound DNA can interfere with conjugation of the thiolated targeting ligand to the maleimide moiety on the DSPE-PEG^{2000}–MAL. Therefore, the exteriorized DNA is removed by nuclease digestion.[36]

Nuclease Digestion

Pancreatic endonuclease I is obtained from Sigma (St. Louis, MO), and exonuclease III is obtained from GIBCO-BRL. The extruded pegylated liposome–DNA mixture is added to a 1.5-ml microfuge tube followed by addition of 3 μl (6 units) of DNAse I, 0.5 μL (33 units) of exonuclease III, and 5 μl of 500 mM MgCl$_2$ to yield a final MgCl$_2$ concentration of 5 mM. The solution is incubated at 37° for 60 min, and the nuclease reaction is stopped by the addition of 20 μl of 0.5 M EDTA, pH = 8.0 (to yield a final EDTA concentration of 20 mM). The ^{32}P should be counted (designated "B"). The ratio of the ["B"/"A"] radioactivity allows for calculation of the loss of DNA (and lipid) because of the dead volume of the extruder. The targeting ligand or MAb should be thiolated in parallel with the nuclease digestion step, and the PIL should be formed by overnight conjugation by mixing the nuclease-treated pegylated liposomes and the thiolated MAb. The efficacy of the nuclease treatment can be assessed with 0.8% agarose gel electrophoresis and ethidium bromide staining (Fig. 2B).

Monoclonal Antibody Thiolation

To a glass tube, 3.0 mg of MAb (300 μl of a 10 mg/ml solution or 20 nmol of MAb) is added followed by addition of 2 μCi of either ^{3}H-labeled MAb or ^{125}I-labeled MAb (10–20 μl of a stock solution of 100–200 μCi/ml) (The MAb can be labeled with ^{125}I and chloramine T or

[36] P. A. Monnard, T. Oberholzer, and P. Luisi, *Biochim. Biophys. Acta.* **1329,** 39 (1997).

labeled with ^{3}H-N-succinimidyl propionate). An appropriate volume of 0.15 M sodium borate, pH = 8.5, 0.1 mM EDTA, is added to yield a final sodium borate concentration of approximately 0.05 M. Then, 12.2 μl of 9.1 mg/ml of fresh Traut's reagent (2-iminothiolane) is added. The solution is incubated at room temperature for 60 min, and the ^{3}H or ^{125}I radioactivity is counted (designated "X"). The unreacted Traut's reagent is removed from the thiolated MAb using a Centriprep-30 concentration filter (Amicon, Millipore, Billerica, MA). The mixture is added to the concentrator, and the volume is brought to 20 ml with 0.05 M HEPES, pH = 7.4, and 0.1 mM EDTA (HE buffer), and the volume is reduced to 2 ml by centrifugation at 2700 rpm for approximately 60 min at 4°. This cycle is repeated one more time with a final reduction of the volume to approximately 1 ml. An aliquot should be removed for counting radioactivity to determine whether there is significant loss of the MAb during the Centriprep-30 buffer exchange. The loss of MAb at this step does not alter calculations of MAb conjugation, because the latter is based on the specific activity of the MAb, which is the μCi of radiolabeled MAb per milligram of unlabeled MAb added at the beginning of the MAb thiolation. The thiolated MAb is transferred to a small bottle that can be capped with a rubber stopper, and the liposome–DNA mixture is added to the thiolated MAb and mixed gently. The bottle is capped with an air-tight stopper, and the air phase is flushed with nitrogen and rocked slowly overnight at room temperature.

Pegylated Immunoliposome Purification with Gel Filtration Chromatography

The next day the conjugated PIL is applied to a 1.5 × 20-cm column of Sephacryl CL-4B and eluted with 0.05 M HEPES, pH = 7.0, and 1-ml fractions are collected with a flow rate of 1 ml/min controlled by a chromatography pump. For all tubes off the column, take 100 μl and count in 10 ml of Ultima-gold (Packard) for either ^{3}H/^{32}P or ^{125}I/^{32}P in a liquid scintillation counting (LSC) spectrometer. The DNA is labeled with ^{32}P, and the MAb is labeled with either ^{125}I or ^{3}H. For ^{3}H/^{32}P counting, count ^{3}H in a window of 0–16 keV and count ^{32}P in a window of 16–1700 keV; there is no ^{3}H spillover in the ^{32}P channel, and there is about 2% spillover of ^{32}P into the ^{3}H channel. This spillover correction should be performed when needed. For simultaneous counting of ^{125}I/^{32}P, count ^{125}I in a window of 0–100 keV and count ^{32}P in a window of 100–1700 keV; there is no ^{125}I spillover in the ^{32}P channel, and there is about a 26% spillover of ^{32}P into the ^{125}I channel, and this spillover correction should be performed. For either the radiolabeled DNA or the radiolabeled MAb, there will be two peaks off the CL-4B column. The first peak is the PIL elution, and there should be an exact comigration of the radiolabeled DNA encapsulated in the interior of the PIL and the radiolabeled MAb conjugated to the tips of the PEG strands (Fig. 2C). At a later elution time, the unconjugated radiolabeled MAb should be detected followed by degraded DNA and free nucleotides generated in the nuclease treatment step (Fig. 2C). The DNA and MAb radioactivity in the first (PIL) peak should be determined, and the DNA counts are designated "C" and the MAb radioactivity is designated "Y." It is also useful to check the TCA precipitability of the PIL, and this radioactivity should be more than 95% precipitable by TCA. The diameter of the PIL can be determined with quasielastic light-scattering methods[27] and should be ≤100 nm (Fig. 2D).

Calculations

The two principal parameters that are used to assess the quality of the PIL formulation are (1) the amount of DNA encapsulated in the interior of the PIL and (2) the number of MAb molecules conjugated to the PIL. A typical DNA encapsulation is 20%, and 35–65 MAb molecules are usually conjugated per PIL. For calculation of the number of MAb molecules conjugated per PIL, it is assumed that there are 100,000 lipid molecules in a 100-nm diameter liposome.[28] Given this parameter, it can be calculated that there are 1.2×10^{14} liposome particles in 20 μmol of lipid. The mole of conjugated MAb is determined from the total CPM in the first peak off the CL-4B column (Y) divided by the MAb-specific activity (CPM/mol). The MAb-specific activity is the total CPM in fraction X divided by 20 nmol of initial MAb. The number of MAb molecules conjugated in peak "Y" is divided by the number of liposomes (1.2×10^{14}) and corrected for the loss of lipid in the dead space of the extruder (Z), where

$$Z = 1 - (B/A)$$

The loss of lipid with the extruder step is assumed to be equal to the loss of DNA. The percent of DNA encapsulated in the liposomes is determined from ["C"/"A"] × 100, and the fractional encapsulation is multiplied by the total micrograms of initial DNA (e.g., 150 μg) to determine the total micrograms of encapsulated DNA. If the number of MAb molecules conjugated per liposome is <10–15, and/or the percent of DNA encapsulation is <10%, the synthesis of the PIL should probably be repeated.

Sterilization of Pegylated Immunoliposome for Tissue Culture Experiments

If the PIL is used for tissue culture, the TRIS and HEPES buffers should be sterilized by extrusion through a 0.22-μm Millipore (Billerica, MA) filter. All the plastic and glass tubes and pipet tips should be autoclaved before the experiment. Once the PIL is ready for addition to cultured cells, the PIL is sterilized by extrusion through a 0.22-μm Millipore filter unit (SLGV R25 LS, Millipore Corp.). The recovery of either the DNA or the MAb is >95% after filtration sterilization.[37] If this sterilized PIL is reapplied to the CL-4B column, all of the DNA and MAb elute in the first PIL peak, indicating the Millipore filtration does not disrupt the PIL.[37]

Organ Luciferase Measurements

At various days after a single intravenous injection of the PIL carrying a luciferase expression plasmid, the animal is killed, and the brain or other organs are removed. The organs (about 0.5 g) are homogenized in 2 ml of lysis buffer (0.1 M potassium phosphate, pH = 7.8, 1% Triton X-100, 1 mM dithiothreitol, and 2 mM EDTA) using a Polytron homogenizer. The homogenate is centrifuged at 14,000 rpm for 10 min at 4°, and the supernatant is subsequently frozen for later use for measurements of tissue luciferase activity with a luminometer (Biolumat LB 9507, Berthold, Bundoora, Victoria, Australia); 100 μl of reconstituted luciferase substrate (Promega, Madison, WI) is added to 20 μl of tissue extract. The peak light emission is measured for 10 s at 20° and recorded as relative light units (RLU). Recombinant luciferase (Promega) is assayed in parallel to establish a standard curve, and the standard curve is used to convert the RLU into picograms of luciferase activity. The protein content in the tissue extract is measured with the bicinchoninic acid (BCA) protein assay reagent (Pierce Chemical Co.), and the final organ luciferase activity is expressed as picograms of luciferase per milligram of tissue protein. For studies in mice, the pGL3-control luciferase expression plasmid driven by the SV40 promoter is obtained from Promega and encapsulated in the interior of 8D3-PIL. Adult male BALB/c mice (25–30 g) are injected intravenously with the 8D3-PIL at a dose of 5 μg plasmid DNA per mouse. Organ luciferase is measured at 48 and 72 h after a single intravenous injection (Table II). These data show expression of the luciferase gene in TfR-rich organs of the mouse, such as brain, liver, spleen, and lung, with background levels of gene expression in TfR-poor organs, such as heart or kidney. The luciferase levels at 72 h in the TfR-rich organs range from 50–75% of the luciferase level at 48 h after intravenous injection (Table II). At 72 h after intravenous administration, the level of luciferase expression in brain and spleen is comparable and is about one-third the level in liver (Table II).

The luciferase expression levels in Table II demonstrate the marked differences in organ specificity of gene expression using the PIL technology,

[37] Y. Zhang, H. J. Lee, R. J. Boado, and W. M. Pardridge, *J. Gene Med.*, **4**, 183 (2002).

TABLE II
ORGAN LUCIFERASE ACTIVITY IN THE MOUSE AFTER A SINGLE
INTRAVENOUS INJECTION OF pGL3-CONTROL PLASMID DNA
ENCAPSULATED IN 8D3 PEGYLATED IMMUNOLIPOSOMES[24]

	Picograms luciferase/mg protein	
Organ	48 h	72 h
Brain	0.76 ± 0.09	0.50 ± 0.10
Heart	0.015 ± 0.001	0.018 ± 0.004
Liver	3.1 ± 0.5	1.4 ± 0.1
Spleen	1.1 ± 0.1	0.52 ± 0.11
Lung	0.74 ± 0.06	0.34 ± 0.09
Kidney	0.0092 ± 0.0009	0.0082 ± 0.0003

Data are mean $\pm$ SE ($n = 5$ mice per time point). Mice were
administered a single intravenous injection of 5 μg/mouse of pGL3
plasmid DNA encapsulated in 8D3-PIL.
PIL, pegylated immunoliposome.

compared with either viral vectors or cationic liposomes. With the PIL
gene targeting approach, the gene expression in the lung is comparable to
brain and is less than gene expression in spleen or liver. The opposite is
observed with cationic liposome–DNA formulations, wherein >99% of
gene expression is observed in the lung,[17,18] <1% in liver or spleen, and
0% in brain.[19] Gene expression in the lung is principally in the pulmonary
endothelial compartment[38] as a result of precipitation of the aggregated
cationic liposome–DNA mixture in the lung circulation. The intravenous
injection of adenoviral vectors in the mouse results in >90% of the exogenous gene expression in the liver because of rapid entrapment of adenovirus
by liver cells.[39] In contrast, with the PIL drug-targeting technology, an
exogenous gene is expressed in organs in a pattern that is predicted
from the specificity of the targeting MAb and the tissue distribution of
the targeted receptor. Luciferase expression in organs such as liver or
spleen reflects the fact that these are TfR-rich organs. Expression in liver
or spleen is not due to nonspecific uptake of the PIL by the RES cells in these
organs. For example, when a nonspecific rat IgG is used in lieu of the rat 8D3
MAb and the exogenous gene is encapsulated in rat IgG-PIL and injected
into mice, there is no gene expression in liver, spleen, or brain.[24] Similarly,

[38] H. E. J. Hofland, D. Nagy, J. J. Liu, K. Spratt, Y. L. Lee, O. Danos, and S. M. Sullivan,
Pharm. Res. **14,** 742 (1997).
[39] K. R. Zinn, J. T. Douglas, C. A. Smyth, H. G. Liu, Q. Wu, V. N. Krasnykh, J. D. Mountz,
D. T. Curiel, and J. M. Mountz, *Gene Ther.* **5,** 798 (1998).

when the OX26 MAb is replaced by a mouse IgG_{2a} isotype control antibody and IgG_{2a}-PIL is prepared, there is no expression of the exogenous gene in liver or spleen in the rat (Fig. 3). These results indicate the PIL is targeted to tissues on the basis of the receptor specificity of the targeting ligand, with minimal nonspecific uptake of the PIL by the RES.[23]

The organ luciferase activity in the mouse following the administration of 8D3-PIL is obtained with the intravenous injection of only 5 μg DNA/ 25 g mouse, which is 200 μg/kg (Table II). A similar level of luciferase activity is observed in rat tissues, with the pGL3-control plasmid packaged in OX26-PIL and administered at an intravenous dose of 5 μg/250 g rat, which is a dose of only 20 μg/kg. Because of the inefficiency of cationic liposomes *in vivo*, it is necessary to inject large amounts of DNA, ranging from 30–100 μg per mouse[17,18] or doses up to 5000 μg/kg. The administration of such large amounts of DNA can cause inflammation even with non-viral gene delivery systems, owing to the inflammatory response to large doses of exogenous DNA.[40] The PIL gene-targeting approach allows for tissue expression of exogenous genes following the administration of relatively low doses of systemic DNA ranging from 20–200 μg/kg. This dosage may be reduced further with advances in plasmid DNA formulation that enable persistence of gene expression.[31] The inclusion in the plasmid of the Epstein Barr nuclear antigen (EBNA)-1 gene element allows for persistence of extrachromosomal replication in human, but not rodent, cells.[37] The administration of luciferase expression plasmids carrying the EBNA-1 gene to human cells with the HIRMAb-PIL gene targeting system yields cell luciferase levels that are >100 pg/mg protein.[37] The level of luciferase gene expression in cultured cells with the HIRMAb-PIL gene-targeting system is comparable to that obtained with lipofectamine.[37]

β-Galactosidase Histochemistry

An alternative reporter gene other than luciferase is *Escherichia coli β-*galactosidase, and expression plasmids encoding for *E. coli β*-galactosidase are readily available commercially. The advantage of using *β*-galactosidase as a reporter gene is that histochemistry can be used to identify the cellular location of gene expression (Fig. 3). At various times after a single intravenous injection of a *β*-galactosidase gene encapsulated in the interior of the PIL, organs are removed and rapidly frozen in powdered dry ice and dipped in Tissue-tek OCT embedding medium. Then, 18-μm frozen sections are prepared on a cryostat and stored at $-70°$. The sections are

[40] J. Norman, W. Denham, D. Denham, J. Yang, G. Carter, A. Abouhamze, C. L. Tannahill, S. L. D. MacKay, and L. L. Moldawer, *Gene Ther.* **7**, 1425 (2000).

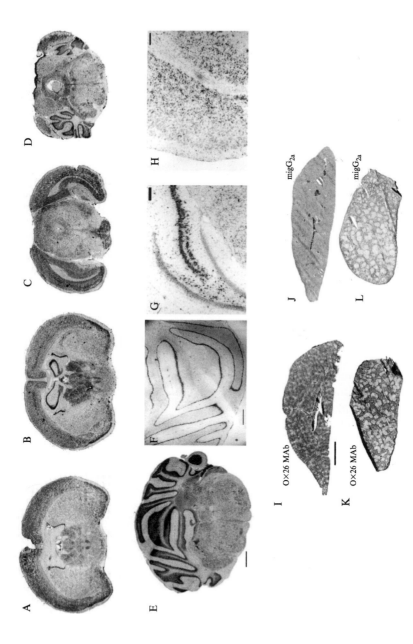

FIG. 3. β-Galactosidase histochemistry of mouse (A–H) and rat (I–L) tissues 48 h after a single intravenous injection of a β-galactosidase expression plasmid encapsulated in the interior of pegylated immunliposomes (PIL) conjugated with either the 8D3 (A–H) or OX26 (I–L) MAb.[23,24] (A–D) Coronal sections through the mouse brain at the level of the septostriatum, the rostral diencephalon, the rostral

fixed at room temperature in 0.5% glutaraldehyde/0.1 M phosphate buffer (pH = 7.0) for 5 min. The β-galactosidase histochemistry is performed with 5-bromo-4-chloro-3-indoyl-β-D-galactoside (X-gal). A kit for X-GAL histochemistry is available from Invitrogen. The slides are developed with the X-gal chromagen overnight at 37° in a humidified chamber. Before coverslipping, the air-dried slides are scanned with a 1200 dpi UMAX flatbed scanner with transluminator and cropped in Adobe Photoshop 5.5 with a G4 Power Macintosh. Scanned images are shown in Fig. 3 (A–E, I–L). After scanning, the slides are coverslipped for light microscopy (F–H of Fig. 3).

The β-galactosidase histochemistry assay should be performed with care, because it is possible to have both false-negative and false-positive results. False-negative results are obtained when there is no blue histochemical product following the overnight X-gal incubation. It is essential to use only polypropylene tubes and pipet tips with the X-gal histochemical assay. The X-gal will precipitate on polystyrene surfaces and yield false-negative results. If care is not taken to maintain the pH at 7.0, false-positive results can be obtained if the overnight incubation is performed under conditions of acidic pH. There is β-galactosidase–like enzyme activity in the mammalian lysosome that is activated at pH 4–5.[41,42] Lysosomal activity will give a false histochemical product under acidic pH conditions. False-positive results can be eliminated by maintaining the pH at 7.0 and can be tracked by performing parallel histochemistry on tissues obtained from control or uninjected mice. The endogenous β-galactosidase–like activity in rat kidney is very high and is readily detectable even at pH 7.0. Therefore, control mouse kidney should be examined in all β-galactosidase assays to confirm that there are no false-negatives results in the histochemical assay. If there is no substantial β-galactosidase histochemical activity in both the cortex and medulla of control rat or mouse kidney, the assay was

---

[41] D. J. Weiss, D. Liggitt, and J. G. Clark, *Histochem. J.* **31**, 231 (1999).
[42] J. Sanchez-Ramos, S. Song, M. Dailey, F. Cardozo-Pelaez, C. Hazzi, T. Stedeford, A. Willing, T. B. Freeman, S. Saporta, T. Zigova, P. R. Sanberg, and E. Y. Snyder, *Cell Transplant.* **9**, 657 (2000).

---

mesencephalon, and the caudal mesencephalon, respectively. Expression of the exogenous gene throughout the brain is shown. (D) A counterstained scanned image of the mouse cerebellum. Light microscopy of selected regions of the mouse brain are shown for the cerebellum (F), the hippocampus (G), and the temporoparietal cortex (H). The β-galactosidase gene is expressed widely throughout the rat liver (I) and spleen (K) following the administration of OX26 PIL. However, if the OX26 MAb is replaced on the pegylated liposome by a mouse IgG$_{2a}$ isotype control, there is no β-galactosidase gene expression in liver (J) or spleen (L). The magnification bars are 600 μm, 250 μm, 70 μm, and 70 μm in E–H, respectively. The magnification bar in I is 2.1 mm. The only section that was counterstained is E. (See color insert.)

not performed properly. A greater histochemical reaction product is obtained with the use of 18-μm sections than with 5-μm sections, presumably because there is a > threefold more β-galactosidase gene product in the thicker sections. β-Galactosidase histochemistry has been performed in rats to examine the persistence of the β-Galactosidase transgene expression following a single intravenous injection of the PIL formulation. The level of β-galactosidase gene expression as determined by either histochemistry or by Southern blotting decreases approximately 50% in the rat 6 days following a single intravenous injection of a β-galactosidase plasmid driven by the SV40 promoter encapsulated in OX26-PIL.[23]

Southern Blotting

Expression of the plasmid DNA in tissues can also be confirmed by Southern blotting.[23] Genomic DNA is isolated from tissues using the Genomic Isolation kit from Qiagen, and a typical yield is approximately 1.3 μg of DNA per mg of tissue. The absorbance at 260–280 nm should be determined for quality control of the isolate. Then, 10-μg DNA aliquots are digested with 15 units of *EcoRI* for 1 h at 37° to remove the insert from the plasmid. The sample is resolved with agarose gel electrophoresis followed by blotting to a GeneScreen-plus membrane, which is then hybridized with ^{32}P-radiolabeled plasmid. Following washing, the membrane is applied to the Kodak X-Omat Blue X-ray film for 3 h at room temperature. During the agarose gel, the migration of xylene cyanol (XC) and bromophenol blue (BPB) tracking dyes are observed. The XC and BPB dyes migrate near the 4.4 and 0.6 kb DNA sizing standards, respectively.

Confocal Microscopy

The intracellular fate of the plasmid DNA following uptake of the PIL can be followed with confocal microscopy following the initial conjugation of fluorescein or an alternative fluorochrome to the plasmid DNA. For the production of fluorescein-conjugated DNA, 15 μg of plasmid DNA is labeled with fluorescein-12-dUTP using the Nick translation kit (Roche) and purified by lithium chloride–ethanol precipitation per the manufacturer's instructions. Then, 15 μg of fluorescein-labeled plasmid DNA is incorporated in a 10-μmol lipid preparation of PIL. Gene targeting to human cells is achieved with the HIRMAb (Table I). The fluoro-DNA packaged in the HIRMAb-PIL can be targeted to U87 human glioma cells in tissue culture.[37] The exogenous gene that is conjugated with the fluorescein is an expression plasmid producing antisense RNA directed against the human epidermal growth factor (EGF) receptor mRNA. The HIRMAb-PIL carrying the fluoresceinated DNA is added to U87 human glioma cells in tissue culture following sterilization of the PIL preparation as described previously. The cells are plated on coverslips at the bottom of wells of a cluster dish, and the glioma cells are incubated for either 3 or 24 h with a total of 1 μg/dish of the fluoresceinated plasmid DNA encapsulated in the interior of HIRMAb-PIL. The incubations are performed in MEM with 10% FBS. The presence of serum does not interfere with cell uptake of the PIL, and this contrasts with the cationic lipid–DNA formulation. The activity of cationic lipids such as lipofectamine is blocked by serum, owing to the binding of the cationic lipid by serum proteins. At the end of the incubation, the media is removed by aspiration, the cells are washed with cold buffer, and fixed with 1 ml/well of 10% formalin in PBS at 4° for 20 mins. The formalin is aspirated, the cells are washed with PBS, and the coverslips containing the adhered cells are removed from the cluster dish and mounted to glass slides. Confocal microscopy is performed with the LSM 5 PASCAL microscope (Zeiss, Jena, Germany) with an argon laser. The intracellular delivery of plasmid DNA encapsulated in the PIL is demonstrated by confocal microscopy (Fig. 4). After 3 h of incubation, the fluorescein-conjugated DNA, delivered to the cell with the HIRMAb-PIL, is distributed throughout the cytoplasm and is also observed within intranuclear vesicles (Fig. 4A). After 24 h of incubation, the fluorescein-conjugated DNA is primarily sequestered within the nucleus (Fig. 4B). These confocal microscopy studies confirm that the plasmid DNA is delivered to the nucleus with the PIL gene targeting system.

A   B

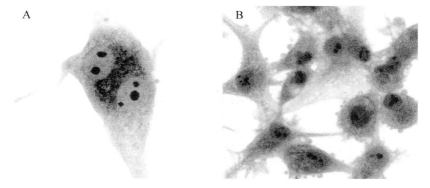

Fig. 4. Confocal microscopy of human U87 glioma cells following a 3-h (A) or 24-h (B) incubation of fluorescein-conjugated plasmid DNA (fluoro-DNA) encapsulated in HIRMAb-pegylated immunoliposomes.[37] The gray-scale confocal image was inverted in Photoshop, so that the localization of the fluoro-DNA is shown by the black areas. There is primarily cytoplasmic accumulation of the fluoro-DNA at 3 h, whereas the fluoro-DNA is largely confined in the nuclear compartment at 24 h. Fluoro-DNA entrapped within intranuclear vesicles is visible at both 3 and 24 h.

Summary

The PIL gene-targeting technology is based on an advanced molecular reformulation of the therapeutic gene (Fig. 1A). This gene-targeting technology is derived from the merger of multiple and disparate disciplines, including liposome technology, pegylation technology, monoclonal antibody technology, and molecular biology. The PIL gene-targeting technology enables the widespread expression of an exogenous gene throughout the target organ (Fig. 3) following a noninvasive, intravenous injection of a nonviral formulation. The targeting specificity of the PIL is strictly a function of the specificity of the targeting ligand or MAb conjugated to the PIL (Fig. 1A, Table I). The specificity of the tissue expression of the exogenous gene is derived from the combined influences of the specificity of the targeting ligand and the tissue specificity of the promoter that is placed at the 5'-end of the therapeutic gene. With the combined use of gene-targeting technology and tissue-specific gene promoters, it is possible to have tissue-specific gene expression widely throughout the target organ following an intravenous injection of the therapeutic gene.[24] The PIL gene targeting technology has thus far been used only for transient or extrachromasomal gene expression *in vivo* using plasmid vectors that do not integrate into the host genome. Conversely, certain viral vectors such as retroviruses or adeno-associated virus stably integrate into the host genome. Nevertheless, it is possible to trigger stable integration of an exogenous gene into genomic host DNA with nonviral expression vectors by incorporating into the plasmid certain transposons. These are 1.6–1.7 kb terminal inverted repeats, and these transposons enable genomic integration of the exogenous gene without the use of viruses.[43] The incorporation of transposon elements into expression plasmids may allow for stable integration of the exogenous gene in the host genome following the intravenous administration of a nonviral plasmid DNA that is encapsulated in a pegylated immunoliposome.

Acknowledgments

The author is indebted for many valuable discussions to Drs. Ruben Boado, Yun Zhang, Ningya Shi, Hwa Jeong Lee, and Frederic Calon. This work was supported by a grant from the UC Davis–MIND Institute and by a grant from the U. S. Department of Defense-Neurotoxin Program.

[43] Z. Izsvak, Z. Ivics, and R. H. Plasterk, *J. Mol. Biol.* **302,** 93 (2000).

[33] Liposome-Mediated Cytokine Gene Delivery to Human Tumor Xenografts

By NEJAT K. EĞILMEZ and RICHARD B. BANKERT

Introduction

Vaccination of cancer patients with cytokine gene–modified tumor cells has been shown to promote the development of antitumor immune responses.[1] In these studies gene modification is commonly achieved by introduction of cytokine genes into patient tumor cells in culture, which are then administered back to the same patient as a vaccine. *Ex vivo* gene transfer requires the establishment of autologous cell lines from patient tumors, which is inefficient, labor intensive, and expensive. An alternative approach involves the direct introduction of cytokine genes to established tumors *in vivo*. *In vivo* gene modification can be achieved either by viral or nonviral methods such as lipofection.[1,2] Liposome-mediated gene delivery has been evaluated extensively because of the simplicity of the technology and has been shown to induce successful antitumor immune responses despite low *in vivo* transfection efficacy.[3,4] More recently, modified liposome–DNA complexes that target membrane receptors have been used to obtain higher transfection efficiencies *in vivo*.[5,6]

Most preclinical studies involving liposome-mediated cytokine gene transfer have been performed in murine tumor models.[7] Although these studies have provided a wealth of data, their relevance to human tumors has not been fully established. The human tumor/SCID mouse xenograft model provides a system in which gene transfer protocols can be evaluated in human tumors *in vivo* before clinical use.[8] The ability to establish and grow different human tumors in SCID mice provides the opportunity to evaluate gene-transfer protocols in a variety of human tumors. This is important, because transfection efficiency of a given formulation is dependent

---

[1] G. Dranoff, *J. Clinical Oncol.* **16,** 2548 (1998).

[2] J. A. Roth and R. J. Cristiano, *J. Nat. Cancer Inst.* **89,** 21 (1997).

[3] M. Nishikawa and L. Huang, *Human Gene Ther.* **12,** 861 (2001).

[4] G. H. Yoo, M-C Hung, G. Lopez-Berestein, S. LaFollette, J. F. Ensley, M. Carey, E. Batson, T. C. Reynolds, and J. L. Murray, *Clin. Can. Res.* **7,** 1237 (2001).

[5] L. Xu, K. F. Pirollo, W-H. Tang, A. Rait, and E. H. Chang, *Hum. Gene Ther.* **10,** 2941 (1999).

[6] H. Hasegawa, M. Shimada, Y. Yonemitsu, T. Utsunomiya, T. Gion, Y. Kaneda, and K. Sugimachi, *Cancer Gene Ther.* **8,** 252 (2001).

[7] G. Parmiani, M. Rodolfo, and C. Melani, *Hum. Gene Ther.* **11,** 1269 (2000).

[8] R. B. Bankert, N. K. Egilmez, and S. D. Hess, *Trends Immunol.* **22,** 386 (2001).

on tumor type. Moreover, this model represents a convenient assay for correlating cytokine gene-transfer efficacy with *in vivo* antitumor activity for selected human cytokines. For example, human interleukin-2 (IL-2) is cross-reactive with murine cells and induces the antitumor activity of SCID mouse natural killer cells. Thus, the human tumor xenograft/SCID mouse model has been used to optimize *in vivo* gene-transfer strategies involving human IL-2 expression plasmids.[9,10] A further improvement of the model involves co-engraftment of human tumors and human PBL into SCID mice, which allows the evaluation of species-specific cytokines such as IL-12.[11]

Required Materials

Animals

Male or female CB17 scid/scid mice, 6- to 8-weeks-old, are obtained from Taconic Labs (Germantown, NY).

Reagents

Plasmid DNA and liposomes: Many different mammalian expression plasmids have been used for cytokine gene transfer.[1,2] Plasmid is stored at 4° in TE buffer (10 mM TRIS HCl, 1 mM ethylenediamine tetraacetic acid [EDTA], pH 8). Numerous commercial or noncommercial liposome formulations have been described and are readily available.[3]

Culture medium: Dulbecco's modified Eagle medium (DMEM) + F12 nutrient mixture + 10% fetal bovine serum + 100 units/ml penicillin + 100 μg/ml streptomycin (Invitrogen-Life Technologies, Rockville, MD).

Phosphate-buffered saline: 137 mM NaCl, 2.7 mM KCl, 4.3 mM Na$_2$HPO$_4$:7H$_2$O, 1.4 mM KH$_2$PO$_4$, pH 7.3).

Tumor disaggregation cocktail; five milligrams Collagenase A (Boehringer-Mannheim, Germany) + 0.2 mg DNAse I (Sigma Chemical Co., St. Louis, MO) dissolved in 10 ml of culture medium, filter sterilized.

Ficoll-Paque: Research grade (Pharmacia Biotech, Uppsala, Sweden).

[9] N. K. Egilmez, R. Cuenca, S. J. Yokota, F. Sorgi, and R. B. Bankert, *Gene Ther.* 3, 607 (1996).
[10] P. R. Clark, A. T. Stopeck, S. E. Parker, and E. M. Hersh, *Cell. Immunol.* 204, 96 (2000).
[11] S. D. Hess, N. K. Egilmez, J. Shiroko, and R. B. Bankert, *Cancer Gene Ther.* 8, 371 (2001).

Trypan blue: 0.1% trypan blue in PBS.

Fixative: PBS containing 2% formaldehyde and 0.05% glutaraldehyde (can be stored up to 1 week at 4°).

Stain solution: 2 mM MgCl$_2$, 5 mM potassium ferricyanide, 5 mM potassium ferrocyanide in PBS, stored at 4° in the dark for up to 1 week.

Substrate: 20 mg/ml X-gal (5-bromo-4-chloro-3-indolyl-β-D-galacto-pyranoside, Sigma) dissolved in dimethyl formamide, stored at −20° in a brown bottle.

Substrate/stain solution: 1 ml of substrate is added to 20 ml of stain solution before use.

Human tumor cell lines: We use primary human lung carcinoma cell lines established from patient samples. Numerous human tumor cell lines are available from the American Type Culture Collection (ATCC, Manassas, VA).

Supplies

Engineer's calipers, 5-ml sterile round-bottom culture tubes, micro pipettors and tips, 1-ml disposable pipettes, small surgical forceps and scissors, scalpel, 30-mm Petri dishes, 15-ml conical-bottom Falcon tubes, Pasteur pipettes, hemacytometer.

Equipment

Laminar flow hood, compound microscope, bench-top centrifuge, rotating mixer, 37°/5% CO$_2$ tissue culture incubator, electric shaver.

Procedures

1. *Induction of human tumor xenografts in SCID mice.* All procedures are performed in a laminar flow hood to maintain sterility and to prevent infection of SCID mice. Mice are injected subcutaneously with 5–10 × 10^5 human tumor cells in 0.1 ml of DMEM. The skin in the ventral caudal midline area is shaved with an electric shaver, scrubbed with 70% alcohol, painted with pevidone-iodine (Betadine), and the cells are injected with a 28.5-gauge needle attached to a 0.5-ml insulin syringe. Mice are monitored, and tumors are measured twice a week with engineer's calipers until the tumor diameter reaches 4–5 mm.

2. *Preparation of liposome–DNA complexes.* It is important to optimize the liposome to DNA ratio before therapy studies, because the optimal ratio is different for each liposome formulation and for each tumor type.[12,13] Moreover, optimization of lipid/DNA ratios should be performed

in vivo, because *in vitro* and *in vivo* transfection optima differ.[12,13] We routinely test lipid (nanomoles) to DNA (μg) ratios in the range of 10:1 to 0.5:1, in which the amount of DNA is kept constant and the lipid quantity is altered. In a standard optimization assay, 10 μg of plasmid DNA is complexed with different amounts of liposomes and injected into 4- to 5-mm tumors in a final volume of 50 μl per tumor. The preparation of the liposome–DNA complexes for a single group of 10 mice in such an experiment is described as follows:

a. Plasmid DNA is prepared at a stock concentration of 1–2 mg/ml in DMEM. A volume of stock DNA equivalent to 120 μg (enough for 12 mice) is removed and diluted to 0.3 ml in DMEM in a sterile 5-ml round-bottom polystyrene culture tube (for example for a 1 mg/ml stock, dilute 0.12 ml of stock into 0.38 ml of DMEM). As many DNA tubes as the number of different ratios to be tested, must be prepared.
b. Appropriate amounts of the selected liposome formulation are distributed to 5-ml tubes, and the final volume is brought up to 0.3 ml with DMEM in each tube. For example, using a 2 mM stock of Lipofectamine (Invitrogen-Life Technologies, Rockville, MD), for lipid/DNA ratios of 10:1, 5:1, 2:1, and 0.5:1, 60, 30, 12, and 3 μl of stock liposome are used, respectively.
c. The DNA (0.3 ml/tube), is added dropwise, with a 1-ml sterile disposable pipette into the liposome solution while gently mixing by hand. Once all the samples are mixed, injections are started. The injections should be performed within 30 min of complex formation.

3. *In vivo gene transfer.* The injections are performed with a 28.5-gauge needle attached to a 0.5-ml syringe. Tumors are scrubbed with 70% ethanol, and 50 μl of lipid–DNA complex is injected directly into the tumor. For maximum dispersal of the complexes within the tumor, the needle is inserted into the center of the tumor nodule and about one third of the solution is injected. The needle is then pulled back slightly without exiting the tumor, reinserted to one side, and one third of the solution is injected. The remaining solution is then injected to the opposite side of the tumor nodule.

[12] N. K. Egilmez, Y. Iwanuma, and R. B. Bankert, *Biochem. Biophys. Res. Commun.* **221,** 169 (1996).
[13] P. R. Clark, A. T. Stopeck, M. Ferrari, S. E. Parker, and E. M. Hersh, *Cancer Gene Ther.* **7,** 853 (2000).

4. *Analysis of transfection efficiency.* Transfection efficiency can be determined by evaluating the transcript levels of a given marker expressed by the plasmid, using reverse transcriptase-polymerase chain reaction (RT-PCR) analysis of total tumor RNA or bioactivity assays of tumor lysates if the marker is an enzyme.[14,15] If the reagents are available, staining of tissue sections for marker activity (either enzymatic or antibody staining) can also be performed.[9,14] Alternately, transfection efficiency can be quantified by staining single-cell suspensions prepared from enzymatically disaggregated tumors for the selected marker, which is more accurate than other approaches for determining efficacy on a per cell basis.[9,12] We have used β-galactosidase as a marker on our plasmids to determine the percent transfection efficacy in single-cell suspensions. The protocol is described as follows:

    a. Preparation of single cell suspensions. Mice are killed 24 h after lipofection, and the tumors are removed surgically. All mouse tissue is carefully dissected away, and the tumor nodule is weighed. The tumor is minced into small (1–2 mm) pieces in a 30-mm Petri dish with a sterile scalpel in 1 ml of tumor disaggregation cocktail. Tumor pieces are then transferred to a 15-ml conical-bottom Falcon tube with a sterile Pasteur pipette, and 0.02 ml of disaggregation cocktail is added per milligram of tumor. The tube is then sealed and incubated at 37° for 2–3 h with continuous mixing. Approximately 70%–90% of the tumor will be digested into single cells after a 2- to 3-h incubation. Remaining large pieces of tumor are allowed to settle, and 2.5-ml aliquots of the supernatant are collected and layered over 5-ml Ficoll cushions in 15-ml Falcon tubes. A small aliquot is saved for determining yield and viability. The tubes are centrifuged at 1000g for 15 min at room temperature with the brake off, and the interface containing the viable cells is collected (2–3 ml) and brought up to 15 ml with PBS in a fresh 15-ml tube. The cells are centrifuged at 500g for 15 min and are washed with 15 ml PBS once more. The cells are then suspended in 5 ml of PBS, and the number of viable cells is

[14] G. N. Hortobagyi, N. T. Ueno, W. Xia, S. Zhang, J. K. Wolf, J. B. Putnam, P. L. Weiden, J. S. Willey, M. Carey, D. L. Branham, J. Y. Payne, S. D. Tucker, C. Bartholomeusz, R. G. Kilbourn, R. L. De Jager, N. Sneige, R. L. Katz, P. Anklesaria, N. K. Ibrahim, J. L. Murray, R. L. Theriault, V. Valero, D. M. Gershenson, M. W. Bevers, L. Huang, G. Lopez-Berestein, and M. C. Hung, *J. Clin. Oncol.* **19,** 3422 (2001).
[15] V. Braiden, A. Ohtsuru, Y. Kawashita, F. Miki, T. Sawada, M. Ito, Y. Cao, Y. Kaneda, T. Koji, and S. Yamashita, *Hum. Gene Ther.* **11,** 2453 (2000).

determined by Trypan blue exclusion (an aliquot of the cells is mixed 1:1 with Trypan blue, and the cells are counted using a hemacytometer).

b. Staining of cells for β-galactosidase activity. Cells are suspended at a concentration of 1×10^7 cells/ml in 2 ml of fixative. After a 5-min incubation at room temperature, they are brought up to 15 ml in PBS and are centrifuged at 500 g for 10 min. The cells are then washed with PBS once more, resuspended in 2 ml of substrate/stain solution, and then incubated at 37° for 3 h.

c. Transfection efficiency. Ten microliters of the cell suspension is loaded into the counting chamber of the hemacytometer, and the cell concentration is determined. The entire chamber is then scanned under the 10× objective of the microscope for blue-staining cells. Percent positive cells is determined by the following formula:

$$\frac{\text{Number of blue-staining cells}}{\text{Number of viable cells in 10 } \mu\text{l of sample}} \times 100 = \text{transfection}$$

A Sample Experiment

The results from an experiment, in which the efficacy of different liposome formulations for *in vivo* gene transfer was evaluated, is shown in Fig. 1. In this experiment, 10 μg of plasmid DNA (pCMVβIL-2[9]) was complexed with different liposome formulations at previously identified optimal ratios[12] and injected directly into small, established human lung tumor xenografts in SCID mice. Tumors were removed surgically 16–20 h after injection and were processed into single-cell suspensions. The cells were then stained for β-galactosidase expression, and the transfection efficiency was determined as described previously. These data demonstrate that DC-cholesterol is superior to other formulations in this model.

Concluding Remarks

In addition to the lipid/DNA ratio, transfection efficiency is influenced by several other variables, which should be considered before therapy studies. One such factor is the tissue tropism displayed by a given liposome formulation. A specific formulation may work well for a certain tumor type but not for another.[12,13] Thus, it is advisable to screen several different formulations for each tumor type to identify the most effective candidate.[12] Another parameter that affects transfection efficacy significantly is the size of the tumor at the time of transfection.[12] Transfection efficiencies are

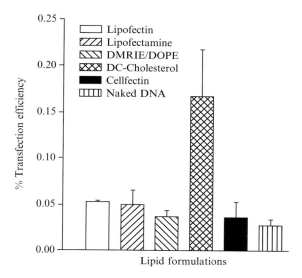

Fig. 1. *In vivo* transfection efficiencies of different cationic lipid formulations. Each liposome formulation was complexed with 10 μg of pCMVβIL-2[12] at the optimal *in vivo* lipid to DNA ratio [4:1 for Lipofectin (Gibco-BRL, Gaithersburg, MD), Cellfectin (Gibco-BRL, Gaithersburg, MD) and DMRIE-DOPE (Vical, Inc., San Diego, CA), 1:1 for DC-Cholesterol (Dr. Leaf Huang, University of Pittsburgh, Pittsburgh, PA) and Lipofectamine(Gibco-BRL, Gaithersburg, MD)] and were delivered to subcutaneous human lung tumor xenografts of similar size (18-30 mg) in SCID mice by direct injection[9]. Naked plasmid DNA (10 μg) was injected into control tumors. Each bar is an average of 3 mice with error bars corresponding to standard deviation. DC-Cholesterol was superior to all other formulations ($p < 0.001$, Student's t-test).

better for smaller tumors than for larger tumors.[12] This may be due either to the presence of necrotic centers in large tumors, which do not transfect well,[16] or simply to the reduced DNA/tumor ratio in larger tumors. Increasing the amount of DNA per milligram of tumor improves transfection efficiency,[13] which supports the notion that the DNA/tumor ratio is important. Finally, the injection technique can be important in achieving high transfection efficiencies. In this case, the injection volume appears to be more important than the injection site.[13]

Acknowledgment

This work was supported by the NIH grant CA54491 to R. B. B.

[16] T. Nomura, S. Nakajima, K. Kawbata, F. Yamashita, Y. Takakura, and M. Hashida, *Cancer Res.* **57**, 2681 (1997).

[34] Cationic Liposome-Mediated Gene Delivery *In Vivo*

By Yong Liu, Sylvia Fong, and Robert J. Debs

Introduction

Cationic liposome–mediated transfer and expression of plasmid-based genes and anti-gene constructs have been used to analyze gene function and gene regulation in a variety of animal model systems. In addition, there are ongoing attempts to use cationic liposomes to develop effective approaches for gene-based therapies directed against a variety of inherited and acquired diseases. Cationic liposome-based *in vivo* gene delivery offers a number of potential advantages over competing viral vector systems. Cationic liposome: DNA complexes (CLDC) are simpler and less expensive to prepare and formulate. CLDC injected into immunocompetent hosts do not appear to induce significant adaptive immune responses. Thus, their repetitive injection into immunocompetent hosts allows full re-expression of delivered genes.[1] Conversely, immune responses can permanently limit re-expression of genes re-injected by many viral vectors.[2] Unlike integrating viral vectors, genes delivered by CLDC show a very low incidence of integration into genomic DNA. This limits the possibility of insertional mutagenesis, which is an increasingly significant issue for integrating viral vectors.[3,4] Cationic liposomes spontaneously complex to and can deliver very large (kb) size DNA vectors in functional form.[5] However, to date, the limited level and duration of expression of genes delivered by CLDC continues to restrict their use for *in vivo* gene delivery.

A number of approaches have increased significantly the efficiency and/or duration of expression of genes delivered *in vivo* by CLDC. These include the development of more efficient cationic lipids,[6,7] the use of

[1] G. Tu, A. L. Kirchmaier, D. Liggitt, Y. Liu, S. Liu, W. H. Yu, T. D. Heath, A. Thor, and R. J. Debs, *J. Biol. Chem.* **275,** 30408 (2000).

[2] D. Chen, B. Murphy, R. Sung, and J. S. Bromberg, *Gene. Ther.* **10,** 991 (2003).

[3] J. Kaiser, *Science* **299,** 495 (2003).

[4] H. Nakai, E. Montini, S. Fuess, T. A. Storm, M. Grompe, and M. A. Kay, *Nat. Genet.* **34,** 297 (2003).

[5] H. F. Willard, *Science* **290,** 1308 (2000).

[6] C. J. Wheeler, P. L. Felgner, Y. J. Tsai, J. Marshall, L. Sukhu, S. G. Doh, J. Hartikka, J. Nietupski, M. Manthorpe, M. Nichols, M. Plewe, X. Liang, J. Norman, A. Smith, and S. H. Cheng, *Proc. Natl. Acad. Sci. USA* **93,** 11454 (1996).

[7] I. Solodin, C. S. Brown, M. S. Bruno, C. Y. Chow, E. H. Jang, R. J. Debs, and T. D. Heath, *Biochemistry* **34,** 13537 (1995).

alternative neutral lipids,[8–10] DNA condensing agents[11] or pure cationic liposomes,[12] the design of more efficiently[13,14] or more durably[1,15] expressing plasmid vectors, and the use of agents that specifically modify the host milieu in order to increase transfection efficiency.[11,16–20] In addition, reducing the innate immune responses elicited by CLDC, either through CpG-depletion of plasmid vectors[21] or co-injecting synthetic anti-sense oligonucleotides targeting inflammatory mediators such as NF-KappaB[20] can also increase significantly the level and duration of *in vivo* expression of CLDC-delivered genes.

CLDC have been used to deliver and express reporter and/or biologically relevant genes locally into the skin,[22] the peritoneal cavity,[23] the airways,[24,25] the fetus *in utero*,[26] or the central nervous system.[27] Intravenous (IV) delivery of CLDC has been shown to transfect a wide variety of tissues and cell types.[8–11,13,17,18,28,29] In each case, the efficiency of

[8] F. Liu, H. Qi, L. Huang, and D. Liu, *Gene. Ther.* **4,** 517 (1997).

[9] N. S. Templeton, D. D. Lasic, P. M. Frederik, H. H. Strey, D. D. Roberts, and G. N. Pavlakis, *Nat. Biotechnol.* **15,** 647 (1997).

[10] Y. Liu, L. C. Mounkes, H. D. Liggitt, C. S. Brown, I. Solodin, T. D. Heath, and R. J. Debs, *Nat. Biotechnol.* **15,** 167 (1997).

[11] S. Li and L. Huang, *Gene. Ther.* **4,** 891 (1997).

[12] T. Ren, Y. K. Song, G. Zhang, and D. Liu, *Gene. Ther.* **7,** 764 (2000).

[13] Y. Liu, D. Liggitt, W. Zhong, G. Tu, K. Gaensler, and R. Debs, *J. Biol. Chem.* **270,** 24864 (1995).

[14] X. Liang, J. Hartikka, L. Sukhu, M. Manthorpe, and P. Hobart, *Gene. Ther.* **3,** 350 (1996).

[15] N. S. Yew, M. Przybylska, R. J. Ziegler, D. Liu, and S. H. Cheng, *Mol. Ther.* **4,** 75 (2001).

[16] C. Tros de Ilarduya, M. A. Arangoa, M. J. Moreno-Aliaga, and N. Düzgüneş, *Biochim. Biophys. Acta* **1561,** 209 (2002).

[17] L. G. Barron, K. B. Meyer, and F. C. Szoka, Jr. *Hum. Gene. Ther.* **9,** 315 (1998).

[18] M. C. Pedroso de Lima, S. Simões, P. Pires, R. Gaspar, V. Slepushkin, and N. Düzgüneş, *Mol. Membr. Biol.* **16,** 103 (1999).

[19] Y. Liu, H. D. Liggitt, S. Dow, C. Handumrongkul, T. D. Heath, and R. J. Debs, *J. Biol. Chem.* **277,** 4966 (2002).

[20] Y. Tan, J. S. Zhang, and L. Huang, *Mol. Ther.* **6,** 804 (2002).

[21] N. S. Yew, H. Zhao, M. Przybylska, I. H. Wu, J. D. Tousignant, R. K. Scheule, and S. H. Cheng, *Mol. Ther.* **5,** 731 (2002).

[22] M. Yokoyama, J. Zhang, and J. L. Whitton, *FEMS. Immunol. Med. Microbiol.* **14,** 221 (1996).

[23] R. Philip, D. Liggitt, M. Philip, P. Dazin, and R. Debs, *J. Biol. Chem.* **268,** 16087 (1993).

[24] R. Stribling, E. Brunette, D. Liggitt, K. Gaensler, and R. Debs, *Proc. Natl. Acad. Sci. USA* **89,** 11277 (1992).

[25] E. W. Alton, P. G. Middleton, N. J. Caplen, S. N. Smith, D. M. Steel, F. M. Munkonge, P. K. Jeffery, D. M. Geddes, S. L. Hart, and R. Williamson, *Nat. Genet.* **5,** 135 (1993).

[26] K. M. Gaensler, G. Tu, S. Bruch, D. Liggitt, G. S. Lipshutz, A. Metkus, M. Harrison, T. D. Heath, and R. J. Debs, *Nat. Biotechnol.* **17,** 1188 (1999).

[27] B. J. Roessler and B. L. Davidson, *Neurosci. Lett.* **167,** 5 (1994).

expression of delivered genes has been shown to depend on the optimization of each component of the CLDC gene delivery system for the specific route of administration. Parameters that can be optimized to improve gene transfer efficiency include the cationic and/or neutral lipids used, the size and the lamellar structure of the cationic liposomes, the diluent in which CLDC are injected, the cationic lipid to DNA ratio injected, the design of the expression plasmid used, and the dose of liposome-DNA complexes injected.[8-11,13,16-19,22-30] In addition, strain-based genetic factors can critically modify both the tissue deposition and the expression of genetic materials delivered by CLDC.[19]

Factors Influencing the Efficiency of Cationic Liposome–Mediated Gene Delivery in Vivo

Animal Host

Strain Differences. A range of animal species can be transfected by CLDC. These include mice, rats, rabbits, goats, and monkeys. Studies in mice have shown that there are strain-based genetic differences in the efficiency of the IV CLDC-based gene transfection and/or expression. Different mouse strains can exhibit significantly different levels of DNA delivery to various tissues, as well as differences in the efficiency at which delivered plasmid DNAs are expressed. For example, ICR/CD1 mice express IV CLDC-delivered genes at significantly higher levels than do either Swiss Webster or FBV mice. These differences can be partially compensated for by pre-treating animals with agents that may alter host pathways important in CLDC-mediated gene transfer and expression.[19] These include dexamethasone and ammonium chloride.[19,31] Therefore, when initiating experiments in vivo, it is important to consider potential strain-based differences in the efficiency and duration of expression of delivered genes.

Routes of Administration

The routes of administration used to deliver genetic materials in vivo via CLDC can be divided into the following three categories:

[28] K. L. Brigham, B. Meyrick, B. Christman, M. Magnuson, G. King, and L. C. Jr., Berry, *Am. J. Med. Sci.* **298,** 278 (1989).
[29] N. Zhu, D. Liggitt, Y. Liu, and R. Debs, *Science* **261,** 209 (1993).
[30] C. Meuli-Simmen, Y. Liu, T. T. Yeo, D. Liggitt, G. Tu, T. Yang, M. Meuli, S. Knauer, T. D. Heath, F. M. Longo, and R. J. Debs, *Hum. Gene. Ther.* **10,** 2689 (1999).
[31] J. W. Wiseman, C. A. Goddard, and W. H. Colledge, *Gene. Ther.* **8,** 1562 (2001).

1. Injection directly into the systemic circulation by IV injection.
2. Injection into a regional circulation, including administering CLDC into the central nervous system (CNS) via intracerebral-ventricular (ICV) injection, or into the respiratory tract via aerosol or intratracheal injection.
3. Injection locally by delivering CLDC into a particular tissue or an isolated compartment of a host, including intramuscular (IM), subcutaneous, or intratumoral injection, or direct injection of CLDC into the fetus.[26]

Systemic Transfection. Generally, the ratio of cationic lipid to plasmid DNA needs to be significantly higher to achieve efficient systemic gene transfer and expression by IV CLDC. Significant levels of expression of IV CLDC-delivered genes can be detected in essentially all organs analyzed. As the first microcapillary bed traversed after injection, the lung routinely exhibits significantly (>10 fold) higher levels of gene expression than any other organ, following IV CLDC.

Gene Delivery Directly into a Regional Circulation or an Organ or Tissue Compartment. Direct gene delivery into a specific tissue offers several potential advantages over systemic gene transfer. It (1) is site-specific, (2) induces fewer systemic side effects, and (3) is often easier to perform technically. For example, delivering CLDC via intraperitoneal (IP) injection transfects significant numbers of T lymphocytes that reside in lymph nodes and spleen.[23] These organs filter particulate material drained from the peritoneal cavity following intraperitoneal administration. Therefore, delivering genes into animals by CLDC-based IP injection can preferentially target delivered genes to organs of the lymphatic system. Injecting CLDC into the cerebral-ventricular cavity produces gene expression along the CNS axis, with minimal gene expression outside the CNS.[30]

Cationic Liposomes

Many different cationic liposome compositions suitable for *in vivo* gene delivery have been described.[10,32,33] These include multilamellar vesicles (MLV), smallunilamellar vesicles (SUV), and extruded liposomes, composed of one of a variety of different cationic lipids with or without a neutral lipid, usually dioleoylphosphatidylethanolamine (DOPE) or cholesterol. IV injection of MLV or extruded liposomes, such as octadecenoyloxyethyl-heptadencenyl-(hydroxyethyl) imidizolium chloride (DOTIM):cholesterol MLV, pure N-[1-(2,3-dioleyloxy) propyl]-n,n,n-trimethylammonium chloride (DOTMA) MLV, pure 1, 2-dioleoyl-3-trimethylammonium-propoane (DOTAP) MLV or DOTAP-cholesterol

[32] Y. K. Song, F. Liu, S. Chu, and D. Liu, *Hum. Gene. Ther.* **8,** 1585 (1997).
[33] S. Li, M. A. Rizzo, S. Bhattacharya, and L. Huang, *Gene. Ther.* **5,** 930 (1998).

TABLE I
LIST OF COMMONLY USED CATIONIC LIPIDS:

Abbreviations	Chemical names
DOTMA	N-[1-(2,3-dioleyloxy)-propyl]-N,N,N-trimethylammoonium chloride
DOTAP	1,2-Dioleyl-3-trimethylammonium-propane
DOTIM	1-[2-9(2)-octadecenoylloxy)ethyl]-2-(8(2)-heptadecenyl)-3-(2-hydroxyethyl)-midizolinium chloride
DDAB	Dimethyldioctadecylammonium Bromide
DOGS	dioctadecylamidoglycylspermine
DMRIE	1,2-dimyristyloxypropyl-3-dimethyl-hydroxyethyl ammonium bromide
DOSPA	2,3-dioleyloxy-N-[2(spermine-carboxamido)ethyl]-N,N-dimethyl-propanaminiumtrifluoroacetate

extruded liposomes display higher transfection efficiencies when compared to IV-injected SUV liposomes of the same lipid composition.[9,10] Multiple different cationic liposome reagents are now commercially available. These include Lipofectin (i.e., DOTMA-DOPE [Bethesda Research Laboratories, Gaithersburg, VA]); DOTAP (Avanti Polar Lipids); Lipofectamine (i.e., DOSPA-DOPE, [GIBGO BRL], and LipofectACE (i.e., DDAB-DOPE [GIBCO-BRL]) (Table I). A study comparing the transfection efficiency of CLDC following IV injection[34] of some of these commercially available cationic liposomes has been reported.

Plasmid DNA Vector

Each of multiple different regulatory elements within expression plasmids has a significant impact on the levels of CLDC-mediated *in vivo* gene expression achieved. The effect of intronic sequences, enhancer and promoter elements, poly A sequences, the bacterial vector backbone, and the antibiotic resistance gene within the expression vector have each been shown to influence the level of expression of genes delivered *in vivo*.[14] To date, among the viral promoters used for CLDC-based, *in vivo* gene delivery, the HCMV promoter-enhancer, or vectors containing mammalian promoters fused with a consensus HCMV enhancer element, generally produce the highest levels of expression of genes delivered by CLDC *in vivo*. Plasmids with an intron 5' to the cDNA coding sequence, or vectors lacking an intron sequence, produce significantly higher levels of gene expression than the analogous vector containing an intron 3' to the cDNA coding sequence.[13]

[34] A. Bragonzi, A. Boletta, A. Biffi, A. Muggia, G. Sersale, S. H. Cheng, C. Bordignon, B. M. Assael, and M. Conese, *Gene. Ther.* **6**, 1995 (1999).

CLDC Complex

Diluents Used in Formulating CLDC. CLDC can be formulated in a wide variety of solutions. Iso-osmotic solutions appropriate for IV injection include 5% glucose or dextrose in water (D5W), 0.9% normal saline, phosphate buffered saline (PBS), lactated Ringer's, and Ringer's-dextran 40 combinations. The presence of EDTA in the diluent can predispose plasmid DNA and cationic liposomes to aggregate, particularly if high lipid: DNA ratios and/or small injection volumes are used. By altering the diluents used to formulate CLDC for *in vivo* injection, some hosts can be selectively converted from low expressors of CLDC-delivered genes towards high expressors of CLDC-delivered genes.[19,32]

Preparation and Mixing of Plasmid DNA and Cationic Liposomes to Form CLDC. The preparation of CLDC involves making cationic liposomes, purifying plasmid DNA, and the subsequent mixing of cationic liposomes and DNA at specific DNA:lipid ratios to form CLDC. The nature of the interactions between negatively charged DNA and positively charged cationic liposomes changes as the DNA:lipid ratio changes. A certain range of DNA:lipid ratios will approach overall electroneutrality, predisposing the resulting CLDC to precipitate. When cationic liposomes and plasmid DNA are mixed properly together at appropriate DNA:lipid ratios, they form a homogeneous-appearing, whitish milky solution. When mixed improperly at non-optimal DNA:lipid ratios, in inadequate volumes of diluent and/or in inappropriate diluents, the resulting plasmid DNA/ cationic liposome mixture aggregates. These aggregated DNA:cationic liposome complexes form large and heterogeneous precipitated particles that are detected readily in solution. The presence of visible precipitates generally abolishes transfection and significantly increases host toxicity.

Precipitation of CLDC on mixing can be avoided or minimized by the following approaches. In addition to the specific DNA:lipid ratio used, factors that can predispose CLDC to precipitate include: (1) high lipid/ DNA concentrations, (2) improper mixing techniques, (3) inadequate volumes of diluent, and/or (4) high salt concentrations in the diluent. To minimize the tendency of plasmid DNA and cationic liposomes to aggregate upon mixing, it is important to: (1) pre-screen a range of DNA:lipid ratios, using small amounts of DNA and liposomes to identify specific DNA:lipid ratios that cause dense precipitation; (2) prepare CLDC in as large a volume as possible, compatible with the site to be injected; (3) add plasmid DNA to liposomes quickly, and in one motion, to avoid unequal mixing of the lipid and DNA components that can cause focal niduses that promote aggregation while mixing; and (4) freshly prepare CLDC just prior to injection. At least partial transfection activity of preformed CLDC can be preserved by lyophilizing CLDC, using various kinds of saccharides as cryoprotectants.[35]

Experimental Procedures

Procedures for Preparing CLDC for IV Injection

Preparation of DOTAP:DOPE or DOTAP: Cholesterol Cationic Liposomes. The cationic lipid DOTAP and DOPE are purchased from Avanti Polar Lipids, and cholesterol from Sigma (St. Louis, MO). Liposomes containing DOTAP (usually prepared and stored as 5 to 20 mM stock solutions in organic solvents) in a 1:1 molar ratio with cholesterol or DOPE are prepared by drying down the lipids to a thin film in a 20 × 160 mm screw-capped glass tube using a rotary evaporator. The dried lipid film is then re-suspended in D5W by incubation in a water bath for 6 hours at 54° to produce MLV. SUV are produced from MLV by placing 1 ml of MLV in a 16 × 160 mm screw-capped glass tube, and sonicating the suspension under argon in a cylindrical sonic bath (Laboratory Supplies, Hicksville, NY) for 15 min. The liposomes are stored under argon (an inert gas) at 4° until use.

DNA preparation. Plasmid DNA can be purified from bacteria using one of a variety of different purification techniques. These include alkaline lysis and ammonium acetate precipitation, ionexchange columns such as Qiagen columns, cesium chloride gradient separation, or PEG precipitation methods.[36] After the last salt precipitation step with each of these methods, the DNA pellet is washed thoroughly with 70% ethanol to remove any residual salt, which may predispose DNA:liposome complexes to precipitate. After the ethanol wash, the plasmid is dissolved in an appropriate diluent to a concentration of approximately 1 to 5 μg/μl. Purified DNA can be dissolved in a variety of diluents, including D5W, PBS, or lactated Ringer's solution. However, using EDTA in the diluents predisposes CLDC to aggregate. The plasmid should be completely dissolved before determining the DNA concentration by OD. The DNA should have an A260/280 ratio between 1.8 to 2 and stored at −20° until thawed for mixing with cationic liposomes to form CLDC. If indicated, kits are available to remove endotoxin from plasmid DNA following its purification (EndoFree Plasmid kits, Qiagen, Valencia, CA). Purifying the same

---

[35] K. Hong, W. Zheng, A. Baker, and D. Papahadjopoulos, *FEBS. Lett.* **400**, 233 (1997).

[36] J. Sambrook, E. F. Fritsch, and T. Maniatis, (eds.) *in* "Molecular Cloning: A Laboratory manual," Cold Spring Harbor Labratory Press, Cold Spring Harbor Labratory Press, Cold Spring Harbor, New York (1989).

plasmid DNA construct using different methods of plasmid purification can alter significantly the DNA:lipid ratios of the resulting CLDC that produce the highest levels of gene expression *in vivo*. Thus, when using published DNA:lipid ratios already determined to produce optimal transfection efficiency for a specific cationic liposome formulation, it is important to use plasmid DNA purified by the specific method cited in that publication or protocol.

Preparation and Mixing of Plasmid DNA:Liposome Complexes. Example: One plans to inject each of five mice intravenously with 20 μg of plasmid DNA complexed to DOTAP:Cholesterol liposomes at a ratio of 1:10 (μg DNA:nmoles total lipid). The concentration of the plasmid DNA stock solution is 1 μg/μl and the concentration of the DOTAP:Cholesterol stock solution is 10 mM. Since 200 μl of CLDC containing 20 μg of plasmid DNA are injected into each mouse, a total volume of CLDC at least 1200 μl is prepared. At least 20% more CLDC than the volume of CLDC that will actually be injected should be prepared, to compensate for syringe dead space, faulty injections, etc. A sample procedure for preparing CLDC is as follows:

1. The specific volumes of plasmid DNA stock solution, liposome stock solution, and diluent needed for the experiment outlined above are calculated. These include:

   a. Total volume of plasmid *DNA* stock solution needed is:
      20 μg DNA $\times$ 5 mice $\times$ 1.2 (120% of total amount injected) = 120 μg DNA needed, 120 μg $\div$ 1 μg/μl DNA = 120 μl DNA stock solution needed.
   b. Total volume of DOTAP:Cholesterol liposome stock solution needed is:
      20 μg DNA $\times$ 10 (DNA:liposome ratio = 1 μg DNA:10 nanomoles total lipid) = 200 nmol of liposomes
      200 nanomoles of liposomes $\times$ 1.2 (120% of total amount injected) = 240 nmol
      240 nmol $\div$ 10 (the liposomal stock solution is 10 mM or 10 nanomoles/μl) = 120 μl liposomal stock solution.
   c. Total volume of diluent (sterile D5W) required to make up a total volume of 200 μl $\times$ 5 mice $\times$ 1.2 (extra) = 1200 μl CLDC.
      1200 μl $-$ 120 μl plasmid DNA solution $-$ 120 μl liposome solution = 960 μl D5W required.
   d. 2 ml sterile polypropylene Eppendorf tube. (Note: Sterile polypropylene or glass tubes are commonly used to mix CLDC. It is recommended that one use a tube with a significantly larger capacity

than the actual volume of solution to be prepared. The use of polystyrene tubes may predispose CLDC to aggregate.[37])

2. The plasmid DNA stock solution (120 μl) is mixed with 480 μl of D5W to bring up the DNA–D5W mixture to 600 μl, or one half of the total volume of CLDC to be prepared, in one 2 ml Eppendorf tube.

3. The DOTAP:Cholesterol solution (120 μl) is mixed with 480 μl D5W to bring up the liposome-D5W mixture to 600 μl, or one half of the total volume of CLDC to be prepared, in another 2 ml Eppendorf tube.

4. The 600 μl of diluted DNA is added to the 2 ml Eppendorf tube containing 600 μl of diluted liposome solution in the following manner: The scale of the pipettor is set to 650 μl (a volume 50 μl greater than the total volume of DNA in D5W). The DNA-containing solution is taken up into the pipette, which is then placed just above the surface of the liposome-containing liquid with the tube kept upright. All of the DNA-containing solution is pipetted quickly into the liposome-containing tube. The mixture is pipetted gently up and down two times. The tube subsequently should not be agitated or vortexed.

5. The resulting solution is examined by holding it against a light source. CLDC should appear as a homogenous milky solution without visible particles or aggregates.

6. Animals are injected approximately 10 min after making the complexes.

7. As mentioned, CLDC tend to aggregate at specific DNA:lipid (μg of plasmid DNA:nanomole total lipid) ratios, particularly those that approach electroneutrality, thereby abolishing charge repulsion between individual lipid:DNA complexes. The specific ratios that may cause CLDC precipitate differ from lipid to lipid, and may differ based on the liposomal structure (MLV, SUV, etc.), the diluent used and/or the method of plasmid DNA purification used. Because of these variables, it is recommend that DNA:lipid ratios already determined to be optimal for the cationic formulation, plasmid DNA preparative technique, and route of injection be used whenever possible.

Delivering Genes into Animals via CLDC

Materials

1. *Animal host*: A range of animal species can be transfected by CLDC. These include mice, rats, rabbits, goats, and monkeys. As mentioned, there are significant, genetically-based variations in transfection

[37] C. E. Holt, N. Garlick, and E. Cornel, *Neuron* **4,** 203 (1990).

efficiency among different mouse strains[19] that may apply to other species as well.

2. *Syringe and needle*: The volume of syringe and the size and length of needle required depends on the kind of host being injected. For example, a 0.5 cc insulin syringe with an attached 28.5 gauge needle is most suitable for injecting mouse tail veins.

3. *Device for restraining animals*: Mouse injection restrainers that can restrain and position a mouse for injection can be used and are commercially available.

4. *Freshly prepared CLDC*: (see methods described previously).

Delivering CLDC into Mice by IV. IV injection of the mouse is a technically challenging procedure that requires careful instruction by an individual skilled in tail vein injections. A significant amount of practice is often required to gain the necessary expertise so that one can perform it efficiently, accurately, and reproducibly.[38] The procedure for IV (tail vein) injection of CLDC into a mouse follows.

1. The animal is secured in an injection chamber with the tail veins exposed. Four laterally-placed blood vessels are seen on the mouse tail, only two of which can be injected. The two injectable lateral veins can be distinguished by their darker color and somewhat larger diameter. To inject, the tail is supported so that the whole tail is roughly parallel to the surface of the bench on which the mouse in the restrainer is placed. The mouse's tail is placed gently between the thumb and index finger so that the distal two-thirds of the tail are placed on the supporting object, with one of lateral veins facing up and roughly parallel to the bench surface.

2. The needle is inserted carefully into the vein at about a 15 degree angle to the vein itself, with the bevel facing up. A small volume of CLDC is injected gently in order to determine whether the needle is in the lumen of the vein. If there is significant resistance to injection, then the needle is likely not in the lumen of the vein, and the material is being injected subcutaneously into the tail, rather than into the systemic circulation. There should be little resistance if the material is being injected into the vein itself. Once it is established that the needle is in the lumen of the vein, then the entire volume of CLDC is injected over 10 to 20 seconds, while exerting gentle, continuous pressure on the plunger of the syringe. CLDC should not be injected more rapidly, as rapid injection may itself cause significant host toxicity or even death.

[38] J. E. Coligan, A. M. Kruisbeek, D. H. Margulies, E. M. Shevach, and W. Strober, (eds.) "Current Protocols In Immunology." John Wiley & Sons, Inc, New York (1991).

3. After removing the needle, the injection site is compressed firmly for about 30 s with a gauze pad or swab to prevent bleeding.

4. The animal is monitored closely after injection for any sign of distress. Most mice recover from the acute effects of CLDC appropriately injected by tail vein in less than 15 minutes, if such effects occur at all.

5. The transfection efficiency is evaluated by assessing the expression of the delivered gene or the inhibition of a target gene if an anti-gene is used, or by a specific phenotype directly linked to expression of the gene product at biologically significant levels. Procedures for analyzing the expression of some commonly used reporter genes are listed below.

Delivering CLDC into Mice by non-IV Routes. DNA:lipid ratios that produce the highest levels of gene expression by non-IV routes of injection are generally significantly higher (contain less net positive charge) than DNA:lipid ratios that produce the highest levels of gene expression after IV injection of CLDC. For most non-IV routes, a DNA:lipid ratio of approximately 1:1 (μg of plasmid DNA:nanomoles total lipid) generally approaches peak levels of gene expression.[23,30,39] The range of CLDC doses and injection volumes for non-intravenous delivery depend on the route and site of injection. Several examples are described below.

ICV INJECTION OF CLDC INTO MICE. All injections are performed using a 0.5 cc insulin syringe with an attached 28.5 gauge needle. Mice are first anesthetized by either methofane inhalation or by injecting a 10:1 mixture of ketamine/xylazine. A dose equivalent to 50–150 mg/kg of ketamine and 5–15 mg/kg of xylazine is injected IM to produce sufficient surgical anesthesia. Adequacy of anesthesia is documented by a lack of response to painful stimuli.[38] The injection site is then shaved and disinfected with 70% ethanol. For ICV injection, the skull skin is incised sagittally, and a hole in the skull 1 mm lateral of the sagittal suture and 2 mm rostral to the coronary suture, with the needle perpendicularly inserted 3 mm deep, is guided by a stereotaxic surgical instrument (David Kopf Instruments, Tijunga, CA). Using this approach, the bevel of the needle should reach the lateral ventricle.[40] Pilot studies should be performed to determine that intraventricular needle placement can be achieved consistently with these coordinates by using a tracking dye, such as Trypan blue. CLDC in ≤ 5 μl of 5% dextrose in water (D5W) are injected into both the left and

[39] M. Meuli, Y. Liu, D. Liggitt, M. Kashani-Sabet, S. Knauer, C. Meuli-Simmen, M. R. Harrison, N. S. Adzick, T. D. Heath, and R. J. Debs, *J. Invest. Dermatol.* **116,** 131 (2001).
[40] K. B. J. Franklin, and G. Paxinos, *in* "The Mouse Brain in Stereotaxic Coordinates" Academic Press, San Diego (1997).

the right lateral ventricles. After successfully placing the needle connected to the syringe containing the CLDC, the needle is held steady and the CLDC solution is injected gently over a few seconds. Special attention needs to be paid to not push the needle any further down during injection. The needle is held in place for 5 seconds after injection to prevent the injected solution from refluxing. The skin incision is closed. The injected mice are placed in a clean warm cage and monitored closely for any signs of adverse reaction until the mice are fully recovered from the anesthesia.

INTRATHECAL (IT) INJECTIONS. Mice are first anesthetized as described previously. A complete laminectomy is then performed between lumbar levels L1–L4 to expose the lumbar spinal column, while leaving its meningeal covering intact. Then, CLDC in ≤ 5 μl of D5W is administered IT.

SUBEPINEURAL INJECTIONS. Mice are first anesthetized as described previously. Subepineural injections are performed using a 0.5cc insulin syringe with an attached 28.5 gauge needle. The injection site is shaved and disinfected with 70% ethanol. Subepineural injection of 25 μg of plasmid DNA within CLDC, diluted in 25 μl of D5W per side, is performed bilaterally on the proximal part of the sciatic nerve after exposure of the nerve via dissection through the posterior proximal thigh.[30]

IM INJECTIONS. Mice are first anesthetized as described previously. The site of injection needs to be consistent and accurate to ensure inter-experimental reproducibility. CLDC are often injected directly into muscle through the skin. When greater accuracy in identifying the injection site is required, a surgical wound can be made to visualize the muscle site to be injected. Surgical wounds are sutured using 7/0 absorbable suture material. Accurate IM injection into mouse quardriceps muscle can also be performed by the following the procedure:

1. The quadriceps muscle is located and immobilized by the thumb and index finger.

2. An insulin syringe is used with a 28.5 gauge needle protected with a collar adjusted to limit the depth of needle insertion. The needle is inserted to a distance of 2 mm into the central part of the rectus muscle at a 60-degree angle relative to the longitudinal axis of the muscle, with the needle pointed toward the hip and the bevel up.

3. The CLDC are injected slowly in a volume of 25–50 μl. The needle is held in place for a few seconds after injecting all of the solution. The needle is pulled out and gentle local pressure is applied to facilitate the dispersal of the injected material.

SKIN TRANSFECTION VIA INTRADERMAL INJECTION.[38] The DNA dose, DNA:liposome ratio, and volume of CLDC injected for intradermal transfection varies from host to host and from lipid to lipid. The dermal layer of small animals such as mice is very thin and therefore very difficult to target specifically for injection. The procedure for intradermal transfection is as follows:

1. Mice are first anesthetized as described above.

2. The injection area is shaved and disinfected with 70% ethanol. The dorsum is a commonly injected site.

3. The injection area is defined with an indelible black marker. For example, one can inject up to 4 sites of skin, each approximately 10 mm in diameter, on the back of one mouse. Using this approach, it is necessary to ensure that each of the four individual injection sites is clearly demarcated and separable from the other three adjacent sites.

4. Using a 0.5 cc insulin syringe with a 28.5 gauge needle, the needle is first inserted at a very shallow angle (5–10 degrees) to the surface of the back. Then the needle is advanced approximately 3–4 mm further.

5. One-fourth of the total volume of CLDC is injected gently. The needle is then retracted slightly without removing it and reinserted 3–4 mm further at a 90-degree angle to the first injected site, in order to inject a second quadrant adjacent to the first injection site. This approach is continued to inject all four quadrants, each injection site at 90 degrees to the original injection site. A bleb or skin bubble should appear at each injection site if the CLDC solution is correctly injected.

6. After the desired period of transfection, the injected skin sample delineated by the black marker is harvested by lifting up the skin from the center of the injection site and cutting the piece of skin along the marked outline. The expression of the delivered gene is then evaluated.

TOPICAL DELIVERY OF CLDC TO THE SKIN. Mice are first anesthetized as described previously. The hair of the dorsal surface of the back is then stripped either by applying Blenderm tape (3 M Health Care, St. Paul, MN) five times and/or brushed using a tooth brush with rounded firm nylon bristles. The number of brushing strokes can be varied from 50–200. Circled areas with a 1 cm diameter are marked around the treated skin site with an indelible stamp. Twenty microliters of CLDC are pipetted onto the circled area and spread evenly using a flame-sealed 1 ml plastic pipette tip. Following stripping and/or brushing, the applied DNA is allowed to completely air dry on the treated skin, which takes approximately 10 min. All of the applied solution is either absorbed or evaporates by this time. Treated mice can be sacrificed from 1 h to 14 days following application when the treated skin is harvested, homogenized, and assayed for the product of the applied gene as described later. Examples of some commonly-used reporter genes are listed later.

Analysis of Activities of Transgene Expression

Analysis of Reporter Gene Activity

Assessment of GFP Protein Expression by Microscopy. Mice are sacrificed by cervical dislocation, and tissues of interest are fixed in 4% buffered paraformaldehyde overnight at 4°, blotted dry of excess paraformaldehyde and kept in 20% sucrose in PBS overnight at 4°. Cryofixation is done by immersion of tissues in ice-cold isopentane for 3 min followed by freezing at −80°. Fixed, frozen tissues are mounted in Tissue-Tek OCT 4583 compound (Sakura Finetek, Torrance, CA) and sectioned. Sections are mounted on slides and analyzed by fluorescence microscopy, using FITC filter and with hematoxylin/eosin staining.

Measurement of Tissue Luciferase Activity. Fresh tissue samples are placed in lysis buffer (Promega Luciferase Assay System): 0.2 ml/mouse spleen, 0.3 ml/liver, and 1.0 ml/lung. Samples are homogenized with a Polytron (Kinematica, Litau-Lucerne, Switzerland) and then spun for 10 min at 8000 rpm at room temperature in a table top Beckman GPR centrifuge. The supernatant (20 μl) are assayed with 0.1 ml luciferase substrate (Promega, Madison, WI) in a Monolight 2010 luminometer (Analytical Luminescence Laboratory, San Diego, CA). Relative light units (RLU) are normalized by micrograms of protein added in the luciferase assay. RLU are then converted to luciferase protein using a reference standard curve derived by measuring purified luciferase protein standards purchased from Analytical Luminescence Laboratory. Luciferase protein levels are corrected for tissue quenching by measuring the luciferase protein standards in tissue extracts from untreated control animals. Tissue protein determinations are performed using a Bradford reagent (BioRad, Hercules, CA).

Thin Layer Chromatography-Based, Radiometric Assay of CAT Activity. Organs are dissected from animals after sacrifice, washed in cold PBS, and homogenized in a hand-held tissue homogenizer containing 250 mM Tris.HCl, pH 7.5, 5 mM EDTA, for lungs and spleen, or 250 mM Tris.HC1, pH 7.5/5 mM EDTA plus the protease inhibitors aprotinin, E-64, and leupeptin (Boehringer Mannheim, San Diego, CA), for liver, heart, and kidneys. These inhibitors prevent degradation of acetylated chloramphenicol species generated during the assay, thereby allowing optimal detection of CAT expression. After homogenization, cells are lysed by three freeze-thaw cycles, and the lysate is heated (65° for 10 min) min) and centrifuged (16,000g, 2 min). The protein concentrations of the extracts are measured with a Coomassie blue G250-based assay (Bio-Rad). Protein concentrations are normalized, and a volume of extract is added to 10 μl of 100 mM acetyl-CoA (Sigma), 0.3 μCi of [^{14}C] chloramphenicol (Amersham) and distilled water to a final volume of 180 μl and allowed to react at 37° for 8–10 h.[21] Following the reaction, the acetylated and unacetylated chloramphenicol species are extracted with cold ethyl acetate, spotted on silica TLC plates, and developed with a chloroform/methanol (95:5, vol/vol) solvent. The TLC plates are exposed to photographic film (Kodax X-Omat) for 1–3 days.

GROSS, HISTOLOGICAL, AND ELECTRON MICROSCOPIC ANALYSES OF β-GALACTOSIDASE ACTIVITY. For gross observation of β-galactosidase expression, lung tissue is infused gently via the trachea with 2% paraformaldehyde and 0.4% glutaraldehyde in PBS at room temperature and fixed undisturbed for approximately 20 min. The tissue is then rinsed twice in PBS and stained with X-gal solution (United States Biochemical, Cleveland, OH), 20 mM potassium ferrocyanide, 20 mM potassium ferricyanide, and 2 mM MgCl in PBS. Tissues are incubated with X-gal at 37° for 1 h and then 14 h at room temperature, rinsed and post-fixed in 10% buffered formalin. For Bluo-gal incubations, tissues are incubated for 12 h at room temperature. For light and electron microscopic analyses, tissues are incubated with X-gal and Bluo-gal (5-bromo-3-indolyl-b-D-galactoside; BRL Life Technologies, Gaithersburg, MD),[31] respectively. Tissues are post-fixed in 4% formaldehyde-1% glutaraldehyde and embedded in Medcast (Ted Pella Inc., Redding, CA). Semi-thin sections are cut for histological analysis (no counterstain). Thin sections are cut and stained with uranyl acetate/lead citrate or left unstained for electron microscopy.

Concluding remarks

Overall, cationic liposomes provide a versatile *in vivo* delivery system for plasmid-based genes and anti-gene constructs. They offer a number of potential advantages over competing viral vector systems, including relative ease and simplicity of formulation, an often reduced incidence of toxic inflammatory and immune-mediated anti-DNA carrier responses, and the ability to deliver very large (Mb) DNA vectors. However, their overall utility as an *in vivo* delivery system remains limited, due to the often low level and duration of expression of CLDC-delivered genes. Although recent advances in both DNA carrier and DNA vector technologies have significantly improved their *in vivo* efficiency, these factors still limit their utility in assessing gene function and gene regulation, as well as their ability to achieve successful gene therapy endpoints. It is likely that ongoing improvements in carrier and vector technologies, as well as improved understanding of the host factors that control CLDC-mediated gene transfer and expression, will produce much more powerful generations of non-viral gene delivery systems *in vivo*.

Author Index

A

Abakumova, O. V., 444, 445(69)
Abe, M., 176, 184(17)
Abell, B., 358
Aberle, A. M., 313, 321
Aboody-Guterman, K. S., 507
Abouhamze, A., 523
Abrams, M. J., 245(29), 246
Adam, M. A., 494
Adams, R. L. P., 443, 444(66)
Adan, J., 402
Adrian, M., 285, 473, 475(21)
Adzick, N. S., 546
Agarwal, S., 344
Agrawal, A. K., 187
Aguerre, O., 362, 365(18)
Aguerre-Charriol, O., 364
Ahkong, Q. F., 209, 244
Ahmad, I., 187
Ahmed, S. A., 378
Ahn, J., 497, 499(38), 503(38), 505(38)
Aikawa, M., 17
Airiau, M., 358, 362, 364, 365(18), 366(11), 368(11)
Akots, G., 250, 258(5)
Alafandy, M., 240
Albani, S., 124(21), 125
Alberson, M., 73
Albritton, N. L., 402
Alford, D. R., 494, 503(11)
Ali, S., 422
Alila, H., 402
Aliño, S. F., 399, 401, 402, 403, 407, 407(1; 2; 32), 409(32), 415(1; 2; 32; 37), 416(1; 2; 37)
Alkan-Önyüksel, H., 204, 205, 206
Allan, C. H., 92
Allen, M., 190

Allen, T. M., 176, 184(14), 187, 199, 213, 510, 520(27)
Allison, A. C., 92, 101
Allison, D. P., 284, 285(9), 286, 286(9)
Allison, J. P., 92, 93, 93(11), 103, 119
Almeida, J. D., 74, 76(1), 79(1)
Almgren, M., 440
Alshoaibi, Z., 435, 438(41), 447(41)
Alt, T., 292(36), 293
Althaus, B., 16, 76, 78(23), 85, 86, 86(23; 38)
Alton, E. W., 537, 538(25)
Alves-Rosa, F., 6
Alving, C. R., 16, 17, 17(1; 2), 18, 18(4), 21, 22, 22(28), 23(34), 24, 24(8), 26(25), 34, 35, 43, 44, 44(1–6; 11; 15), 47(9), 52, 58, 66(22), 136, 137, 139, 140, 140(18; 19), 141(22; 24–28), 142, 144(26; 27), 145(19; 20; 22), 148(22), 149, 149(22), 150(18–20; 24), 151(18; 19)
Alvis, D. L., 215
Aly, O. M., 226, 227(31)
Amagasa, T., 344
Amici, A., 435, 440
Amigorena, S., 18, 102, 103
Amini, A., 508
Amos, N., 240
Amselem, S., 17, 24(8)
Anasetti, C., 93
Anchordoquy, T. J., 340
Anderson, A. R., 210, 211(50), 212(50)
Anderson, K. S., 28
Anderson, M. E., 358
Anderson, W. F., 433(16), 434, 493(4), 494, 497
Andreadis, S. T., 496
Andres, R. Y., 189
Androlewicz, M. J., 28
Anklesaria, P., 497, 498(33), 499(33), 500(33), 505(33), 533

S

Subject Index

A

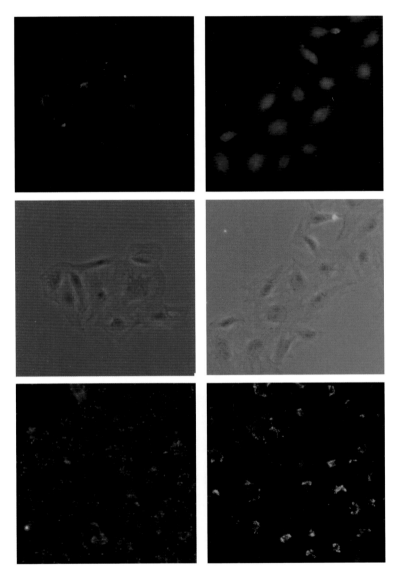

KONO AND TAKAGISHI, CHAPTER 27, FIG. 3. Micrographs of CV1 cells treated with plain EYPC liposomes (left) and SucPG-modified EYPC liposomes (right). (Upper and middle) Fluorescence and phase-contrast micrographs of CV1 cells treated with the calcein-loaded liposomes. (Lower) Fluorescence micrographs of CV1 cells treated with the liposomes containing NBD-PE and Rh-PE.

A

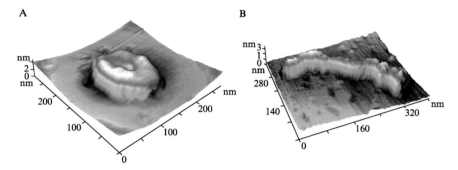

B

ZHDANOV *ET AL.*, CHAPTER 28, FIG. 2. Atomic force microscopy-images of lipoplexes composed of (A) DEGA-3 liposomes and pEGFP-N1, and (B) DEGA-7 liposomes and pCMV-SPORT-β-gal. The lipid/DNA ratio was 7.5:1 (w/w).

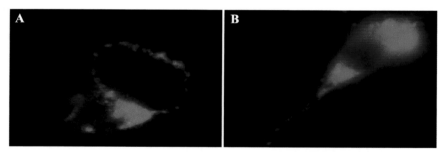

ZHDANOV *ET AL.*, CHAPTER 28, FIG. 3. Interaction of coumarin-labeled DEGA-3 liposomes with MCF-7 cells. The incubation times were (A) 5 min (fluorescence at the cell perimeter) and (B) 15 min (fluorescence at the cell perimeter, with accumulation in cytoplasmic compartments) (1250×).

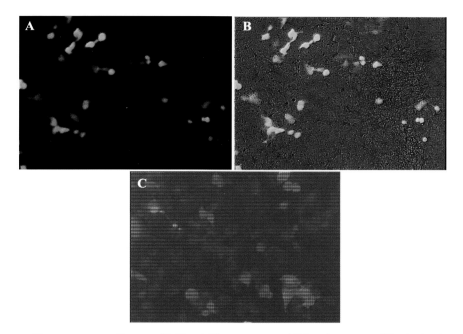

ZHDANOV *ET AL.*, CHAPTER 28, FIG. 4. Fluorescence and light microscopy of lipofection in cell culture with dicationic DEGA-3 and DEGA-7 lipoplexes. (A) 293 cells transfected with DEGA-3/pEGFP-N1 lipoplexes (fluorescence microscopy, 300×); (B) 293 cells transfected with DEGA-3/pCMV-SPORT-β-gal lipoplexes (light microscopy, 300×); (C) 293 cells transfected with Lipofectin^R/pEGFP-N1 lipoplexes (fluorescence microscopy, 300×).

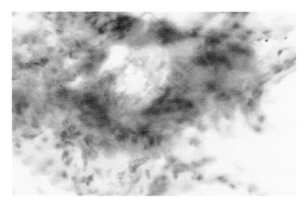

ZHDANOV *ET AL.*, CHAPTER 28, FIG. 5. *In vivo* transfection by DICHOLENIM/p-CMV-SPORT-β-gal lipoplexes following portal vein injection in mice: Expression of the *Lac Z* gene in the lung (light microscopy, 400×).

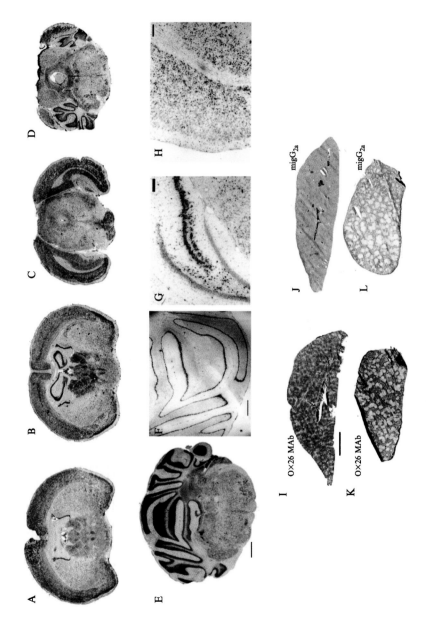

A

B

C

D

E

F

G

H

I   OX26 MAb

J   mIgG$_{2a}$

K   OX26 MAb

L   mIgG$_{2a}$

PARDRIDGE, CHAPTER 32, FIG. 3. β-Galactosidase histochemistry of mouse (A–H) and rat (I–L) tissues 48 h after a single intravenous injection of a β-galactosidase expression plasmid encapsulated in the interior of pegylated immunliposomes (PIL) conjugated with either the 8D3 (A–H) or OX26 (I–L) MAb. (A–D) Coronal sections through the mouse brain at the level of the septostriatum, the rostral diencephalon, the rostral mesencephalon, and the caudal mesencephalon, respectively. Expression of the exogenous gene throughout the brain is shown. (D) A counterstained scanned image of the mouse cerebellum. Light microscopy of selected regions of the mouse brain are shown for the cerebellum (F), the hippocampus (G), and the temporoparietal cortex (H). The β-galactosidase gene is expressed widely throughout the rat liver (I) and spleen (K) following the administration of OX26 PIL. However, if the OX26 MAb is replaced on the pegylated liposome by a mouse IgG_{2a} isotype control, there is no β-galactosidase gene expression in liver (J) or spleen (L). The magnification bars are 600 μm, 250 μm, 70 μm, and 70 μm in E–H, respectively. The magnification bar in I is 2.1 mm. The only section that was counterstained is E.

ISBN: 0-12-182276-1

90000

9 780121 822767